I0057707

Pabulo H. Rampelotto (Ed.)

Extremophiles and Extreme Environments

This book is a reprint of the Special Issue that appeared in the online, open access journal, *Life* (ISSN 2075-1729) from 2012–2013 (available at: http://www.mdpi.com/journal/life/special_issues/life-extremophiles).

Guest Editor
Pabulo H. Rampelotto
Federal University of Rio Grande do Sul
Brazil

Editorial Office
MDPI AG
Klybeckstrasse 64
Basel, Switzerland

Publisher
Shu-Kun Lin

Assistant Managing Editor
Changzhen Fu

1. Edition 2016

MDPI • Basel • Beijing • Wuhan • Barcelona

ISBN 978-3-03842-177-1 (Hbk)
ISBN 978-3-03842-178-8 (PDF)

Table of Contents

List of Contributors

Angeles Aguilera: Astrobiology Center, Spanish Institute for Aerospace Technologies, Carretera de Ajalvir Km 4, Torrejón de Ardoz, 28850 Madrid, Spain.

Ricardo Amils: Centro de Astrobiología. INTA-CSIC. Torrenjón de Ardoz, Madrid 28850, Spain.

Lenin Arias-Rodriguez: División Académica de Ciencias Biológicas, Universidad Juárez Autónoma de Tabasco (UJAT), Villahermosa, Tabasco, CP 86150, Mexico.

Makoto Ashiuchi: Graduate School of Integrated Arts and Sciences, Kochi University, Nankoku, Kochi 783-8502, Japan.

Alessio Ausili: Laboratory for Molecular Sensing, Institute of Protein Biochemistry, CNR, Via Pietro Castellino, 111, Napoli, 80131, Italy.

Dennis A. Bazylinski: University of Nevada at Las Vegas, School of Life Sciences, Las Vegas, Nevada, 89154-4004, USA.

Joan M. Bernhard: Geology and Geophysics Department, Woods Hole Oceanographic Institution, Woods Hole, MA 02543, USA.

Abraham A. M. Bielen: Laboratory of Microbiology, Wageningen University, Dreijenplein 10, 6703 HB Wageningen, The Netherlands.

David Bierbach: Department of Ecology & Evolution, J.W. Goethe University Frankfurt, Max-von-Laue-Straße 13, Frankfurt am Main, D-60438, Germany.

Paul Blum: Beadle Center for Genetics, University of Nebraska, Lincoln, NE 68588-0666, USA.

Sara Borin: Department of Food, Environment and Nutritional Sciences (DeFENS), University of Milan, via Celoria 2, Milan, 20133, Italy.

Brendan P. Burns: Australian Centre for Astrobiology; School of Biotechnology and Biomolecular Sciences, University of New South Wales, Sydney, NSW 2052, Australia.

Carlo Calfapietra: Consiglio NazionaledelleRicercheIstituto di BiologiaAgroambientale e Forestale via Marconi 2-05010 Porano (TR), Italy.

Abigail Calzada: Geology Department, University of Oviedo, Jesús Arias de Velasc, Oviedo 33005, Spain.

Francesco Canganella: Department for Innovation in Biological, Agrofood, and Forest Systems, University of Tuscia, via C. de Lellis, Viterbo 01100, Italy.

Alessandro Capo: Laboratory for Molecular Sensing, Institute of Protein Biochemistry, CNR, Via Pietro Castellino, 111, Napoli, 80131, Italy.

María Esperanza Cerdán: Departamento de Bioloxía Celular e Molecular, Facultade de Ciencias, Universidade da Coruña, 15071 A Coruña, Spain.

James C. Charlesworth: School of Biotechnology and Biomolecular Sciences, University of New South Wales, Sydney, NSW 2052, Australia.

Harald Claus: Institute of Microbiology and Wine Research, Johannes Gutenberg-University, 55099 Mainz, Germany.

Kris Coupland: College of Natural Sciences, Bangor University, Deiniol Road, Bangor, LL57 2UW, UK.

Jonathan Dattelbaum: Department of Chemistry, University of Richmond, Richmond, VA 23173, USA.

Sabato D'Auria: Laboratory for Molecular Sensing, Institute of Protein Biochemistry, CNR, Via Pietro Castellino, 111, Napoli, 80131, Italy.

María de Lourdes Moreno: Department of Microbiology and Parasitology, University of Seville, n°2. 41012, Sevilla, Spain.

Nikki Dellas: Thermal Biology Institute; Department of Plant Sciences and Plant Pathology, Montana State University, Bozeman, MT 59717, USA.

Ewald B. M. Denner: Medical University Vienna, Währingerstrasse 10, 1090 Wien, Austria.

Yan Ding: Department of Biomedical and Molecular Sciences, Queen's University, Kingston Ontario, K7L 3N6, Canada.

Marion Dornmayr-Pfaffenhuemer: Department of Molecular Biology, University of Salzburg, Billrothstr. 11, 5020 Salzburg, Austria.

Virginia P. Edgcomb: Geology and Geophysics Department, Woods Hole Oceanographic Institution, Woods Hole, MA 02543, USA.

Howell G.M. Edwards: Centre for Astrobiology and Extremophiles Research, School of Life Sciences, University of Bradford, Bradford BD7 1DP, UK; Space Research Centre, Department of Physics & Astronomy, University of Leicester, Leicester LE1 7RH, UK.

Josef Elster: Institute of Botany AS CR, Dukelská 135, Třeboň, CZ-379 82, Czech Republic; Faculty of Science, University of South Bohemia, Branišovská 31, České Budějovice, CZ-370 05, Czech Republic.

María Teresa García: Department of Microbiology and Parasitology, University of Seville, n°2. 41012, Sevilla, Spain.

Miguel A. Gonzalez: Institut Laue Langevin, 6, Rue Jules Horowitz, F-38042 Grenoble Cedex 9, France.

María Isabel González-Siso: Departamento de Bioloxía Celular e Molecular, Facultade de Ciencias, Universidade da Coruña, 15071 A Coruña, Spain.

Elena González-Toril: Centro de Astrobiología. INTA-CSIC. Torrenjón de Ardoz, Madrid 28850, Spain.

Claudia Gruber: Department of Molecular Biology, University of Salzburg, Billrothstr. 11, 5020 Salzburg, Austria.

Kevin B. Hallberg: College of Natural Sciences, Bangor University, Deiniol Road, Bangor, LL57 2UW, UK.

Domenica Hamisch: Department of Plant Biology Technical University of Braunschweig, Pockelsstr. 14, Brunschweig, 38092, Germany.

Inga Hänelt: Department of Molecular Microbiology and Bioenergetics, Goethe University Frankfurt am Main, Max-von-Laue-Str. 9, 60438 Frankfurt am Main, Germany.

Robert Hänsch: Department of Plant Biology Technical University of Braunschweig, Pockelsstr. 14, Brunschweig, 38092, Germany.

Nina Herrmann: Institute of Biochemistry & Biology, Unit of Animal Ecology, University of Potsdam, Maulbeerallee 1, Potsdam, 14469, Germany; División Académica de Ciencias Biológicas, Universidad Juárez Autónoma de Tabasco (UJAT), Villahermosa, Tabasco, CP 86150, Mexico.

Jeane Rimber Indy: División Académica de Ciencias Biológicas, Universidad Juárez Autónoma de Tabasco (UJAT), Villahermosa, Tabasco, CP 86150, Mexico.

Ken F. Jarrell: Department of Biomedical and Molecular Sciences, Queen's University, Kingston Ontario, K7L 3N6, Canada.

D. Barrie Johnson: College of Natural Sciences, Bangor University, Deiniol Road, Bangor, LL57 2UW, UK.

Susana E. Jorge-Villar: Area Geodinamica Interna, Facultad de Humanidades y Educacion, Universidad de Burgos, Calle Villadiego, Burgos 9001, Spain; CENIEH, P° Sierra de Atapuerca, s/n, Burgos 09002, Spain.

Tohru Kamei: Graduate School of Integrated Arts and Sciences, Kochi University, Nankoku, Kochi 783-8502, Japan.

Catherine M. Kay: College of Natural Sciences, Bangor University, Deiniol Road, Bangor, LL57 2UW, UK.

Servé W. M. Kengen: Laboratory of Microbiology, Wageningen University, Dreijenplein 10, 6703 HB Wageningen, The Netherlands.

Burkhard Knopf: Frauenhofer-Institut für Molekularbiologie und Angewandte Ökologie, 57392 Schmallenberg, Germany.

Helmut König: Institute of Microbiology and Wine Research, Johannes Gutenberg-University, 55099 Mainz, Germany.

Jana Kviderova: Institute of Botany AS CR, Dukelská 135, Třeboň, CZ-379 82, Czech Republic; Faculty of Science, University of South Bohemia, Branišovská 31, České Budějovice, CZ-370 05, Czech Republic.

C. Martin Lawrence: Thermal Biology Institute; Department of Chemistry and Biochemistry, Montana State University, Bozeman, MT 59717, USA.

Rebecca LeBard: School of Biotechnology and Biomolecular Sciences, University of New South Wales, Sydney, NSW 2052, Australia.

Christopher T. Lefèvre: CEA Cadarache/CNRS/Aix-Marseille Université, UMR7265 Service de Biologie Végétale et de Microbiologie Environnementale, Laboratoire de Bioénergétique Cellulaire, 13108, Saint-Paul-lez-Durance, France.

Andrea Legat: Department of Molecular Biology, University of Salzburg, Billrothstr. 11, 5020 Salzburg, Austria.

Inmaculada Llamas: Department of Microbiology, Faculty of Pharmacy, University of Granada, Campus Universitario de Cartuja, 18071 Granada, Spain; Biotechnology Research Institute, Polígono Universitario de Fuentenueva, University of Granada, 18071 Granada, Spain.

Olalla López-López: Departamento de Bioloxía Celular e Molecular, Facultade de Ciencias, Universidade da Coruña, 15071 A Coruña, Spain.

Yukari Maezato: Beadle Center for Genetics, University of Nebraska, Lincoln, NE 68588-0666, USA.

Salvatore Magazù: Department of Physics, University of Messina, Viale D'Alcontres 31, P.O. Box 55-98166, Messina, Italy.
Sveinn Magnússon: Matis ohf. Food Safety, Environment and Genetics, Vinlandsleid 12, Reykjavik, 113, Iceland.
Francesca Mapelli: Department of Food, Environment and Nutritional Sciences (DeFENS), University of Milan, via Celoria 2, Milan, 20133, Italy.
Ramona Marasco: Department of Food, Environment and Nutritional Sciences (DeFENS), University of Milan, via Celoria 2, Milan, 20133, Italy.
Viggó Marteinsson: Matis ohf. Food Safety, Environment and Genetics, Vinlandsleid 12, Reykjavik, 113, Iceland.
Ikuo Matsui: Biomedical Research Institute, National Institute of Advanced Industrial Science and Technology (AIST), 1-1-1 Higashi, Tsukuba 305-8566, Japan.
Eriko Matsui: Biomedical Research Institute, National Institute of Advanced Industrial Science and Technology (AIST), 1-1-1 Higashi, Tsukuba 305-8566, Japan.
Mauro Medori: Consiglio NazionaledelleRicercheIstituto di BiologiaAgroambientale e Forestale via Marconi 2-05010 Porano (TR), Italy.
Encarnación Mellado: Department of Microbiology and Parasitology, University of Seville, nº2. 41012, Sevilla, Spain.
Federica Migliardo: Department of Physics, University of Messina, Viale D'Alcontres 31, P.O. Box 55-98166, Messina, Italy.
Claudia Mondelli: CNR-IOM-OGG, Institut Laue Langevin, 6, Rue Jules Horowitz, F-38042 Grenoble Cedex 9, France.
Kate Montgomery: School of Biotechnology and Biomolecular Sciences, University of New South Wales, Sydney, NSW 2052, Australia.
Volker Müller: Department of Molecular Microbiology and Bioenergetics, Goethe University Frankfurt am Main, Max-von-Laue-Str. 9, 60438 Frankfurt am Main, Germany.
Divya B. Nair: Department of Biomedical and Molecular Sciences, Queen's University, Kingston Ontario, K7L 3N6, Canada.
Aharon Oren: Department of Plant and Environmental Sciences, The Alexander Silberman Institute of Life Sciences, The Hebrew University of Jerusalem, 91904, Jerusalem, Israel.
Stewart F. Parker: ISIS Facility, Rutherford Appleton Laboratory, Chilton, Oxon, OX11 0QX, UK.
Dolores Pérez: Department of Microbiology and Parasitology, University of Seville, nº2. 41012, Sevilla, Spain.
Peter Pfeiffer: Institute of Microbiology and Wine Research, Johannes Gutenberg-University, 55099 Mainz, Germany.
Martin Plath: Department of Ecology & Evolution, J.W. Goethe University Frankfurt, Max-von-Laue-Straße 13, Frankfurt am Main, D-60438, Germany.
Emilia Quesada: Department of Microbiology, Faculty of Pharmacy, University of Granada, Campus Universitario de Cartuja, 18071 Granada, Spain; Biotechnology Research Institute, Polígono Universitario de Fuentenueva, University of Granada, 18071 Granada, Spain.

Pabulo Henrique Rampelotto: Interdisciplinary Center for Biotechnology Research, Federal University of Pampa, Antônio Trilha Avenue, P.O.Box 1847, 97300-000 São Gabriel–RS, Brazil.

Eyjólfur Reynisson: Matis ohf. Food Safety, Environment and Genetics, Vinlandsleid 12, Reykjavik, 113, Iceland.

Rüdiger Riesch: Department of Animal and Plant Sciences, University of Sheffield, Western Bank, Sheffield S10 2TN, UK.

Laura Rocchetti: Department of Life and Environmental Sciences, Università Politecnica delle Marche, Via Brecce Bianche, 60131 Ancona, Italy.

Owen F. Rowe: Department of Ecology and Environmental Science, Umeå University, SE-901 87 Umeå, Sweden.

Matthias Schulte: Institute of Biochemistry & Biology, Unit of Animal Ecology, University of Potsdam, Maulbeerallee 1, Potsdam, 14469, Germany; División Académica de Ciencias Biológicas, Universidad Juárez Autónoma de Tabasco (UJAT), Villahermosa, Tabasco, CP 86150, Mexico.

Melanie Schwab: Department of Microbiology, Faculty of Pharmacy, University of Granada, Campus Universitario de Cartuja, 18071 Granada, Spain.

Sarah Siu: Department of Biomedical and Molecular Sciences, Queen's University, Kingston Ontario, K7L 3N6, Canada.

Virginia Souza-Egipsy: Centro de Astrobiología. INTA-CSIC. Torrenjón de Ardoz, Madrid 28850, Spain.

Maria Staiano: Laboratory for Molecular Sensing, Institute of Protein Biochemistry, CNR, Via Pietro Castellino, 111, Napoli, 80131, Italy.

Helga Stan-Lotter: Department of Molecular Biology, University of Salzburg, Billrothstr. 11, 5020 Salzburg, Austria.

Ali Tahrioui: Department of Microbiology, Faculty of Pharmacy, University of Granada, Campus Universitario de Cartuja, 18071 Granada, Spain.

Parag Vaishampayan: Biotechnology and Planetary Protection Group, Jet Propulsion Laboratory, California, Institute of Technology, Pasadena, CA 91109, USA.

John van der Oost: Laboratory of Microbiology, Wageningen University, Dreijenplein 10, 6703 HB Wageningen, The Netherlands.

Antonio Varriale: Laboratory for Molecular Sensing, Institute of Protein Biochemistry, CNR, Via Pietro Castellino, 111, Napoli, 80131, Italy.

Marcel R. A. Verhaart: Laboratory of Microbiology, Wageningen University, Dreijenplein 10, 6703 HB Wageningen, The Netherlands.

Beata G. Vertessy: Institute of Enzymology, Research Center for Natural Sciences, Hungarian Academy of Science, Budapest, Hungary; Department of Applied Biotechnology and Food Sciences, University of Technology and Economics, Budapest, Hungary.

Pieter T. Visscher: Center for Integrative Geosciences, University of Connecticut 354 Mansfield Road, Storrs, CT 06269-2045, USA; Australian Centre for Astrobiology, University of New South Wales, Sydney, NSW 2052, Australia.

Gerhard Wanner: LMU Biocenter, Ultrastructural Research, Grosshadernerstrasse 2-4, 82152 Planegg-Martinsried, Germany.

Juergen Wiegel: Department of Microbiology, University of Athens (GA), Athens 10679, USA.

Takashi Yamamoto: Graduate School of Integrated Arts and Sciences, Kochi University, Nankoku, Kochi 783-8502, Japan.

Kazuhiko Yamasaki: Biomedical Research Institute, National Institute of Advanced Industrial Science and Technology (AIST), 1-1-1 Higashi, Tsukuba 305-8566, Japan.

Hideshi Yokoyama: School of Pharmaceutical Sciences, University of Shizuoka, Shizuoka 422-8526, Japan.

Mark J. Young: Department of Microbiology; Thermal Biology Institute; Department of Plant Sciences and Plant Pathology, Montana State University, Bozeman, MT 59717, USA.

Claudia Zimmer: Department of Ecology & Evolution, J.W. Goethe University Frankfurt, Max-von-Laue-Straße 13, Frankfurt am Main, D-60438, Germany.

About the Guest Editor

Pabulo Henrique Rampelotto is a molecular biologist currently developing his research at the Federal University of Rio Grande do Sul (Brazil). Prof. Rampelotto is the founder and Editor-in-Chief of the Springer Book Series **Grand Challenges in Biology and Biotechnology**. In addition, he serves as Editor-in-Chief of **Current Biotechnology** as well as Associate Editor, Guest Editor and member of the editorial board of several scientific journals in the field of Life Sciences and Biotechnology. Prof. Rampelotto is also a member of four scientific advisory boards (Astrobiology/SETI Board, Biotech/Medical Board, Policy Board, and Space Settlement Board) of the Lifeboat Foundation, alongside several Nobel Laureates and other distinguished scientists, philosophers, educators, engineers, and economists. Some of the most distinguished team leaders in the field have published their work, ideas, and findings in his books and special issues.

Preface

Over recent decades, the study of extremophiles has provided groundbreaking discoveries that challenge the paradigms of modern biology and make us rethink intriguing questions such as "what is life?", "what are the limits of life?", and "what are the fundamental features of life?". The mechanisms by which different microorganisms adapt to extreme environments provide a unique perspective on the fundamental characteristics of biological processes present in most species. Extremophiles are also critical for evolutionary studies related to the origins of life, since they form a cluster at the base of the tree of life. Furthermore, the application of extremophiles in industrial processes has opened a new era in biotechnology. The study of extreme environments has become a key area of research for astrobiology. Extremophiles may help us understand what form life takes on other planetary bodies in our own solar system and beyond. These findings and possibilities have made the study of life in extreme environments one of the most exciting areas of research in recent decades. However, despite the latest advances we are just at the beginning of exploring and characterizing the world of extremophiles. This book covers all aspects of life in extreme environments.

Pabulo Henrique Rampelotto
Guest Editor

Extremophiles and Extreme Environments

Pabulo Henrique Rampelotto

Reprinted from *Life*. Cite as: Rampelotto, P.H. Extremophiles and Extreme Environments. *Life* **2013**, *3*, 482-485.

Over the last decades, scientists have been intrigued by the fascinating organisms that inhabit extreme environments. Such organisms, known as extremophiles, thrive in habitats which for other terrestrial life-forms are intolerably hostile or even lethal. They thrive in extreme hot niches, ice, and salt solutions, as well as acid and alkaline conditions; some may grow in toxic waste, organic solvents, heavy metals, or in several other habitats that were previously considered inhospitable for life. Extremophiles have been found depths of 6.7 km inside the Earth's crust, more than 10 km deep inside the ocean—at pressures of up to 110 MPa; from extreme acid (pH 0) to extreme basic conditions (pH 12.8); and from hydrothermal vents at 122 °C to frozen sea water, at −20 °C. For every extreme environmental condition investigated, a variety of organisms have shown that they not only can tolerate these conditions, but that they also often require those conditions for survival.

They are classified according to the conditions in which they grow: As thermophiles and hyperthermophiles (organisms growing at high or very high temperatures, respectively), psychrophiles (organisms that grow best at low temperatures), acidophiles and alkaliphiles (organisms optimally adapted to acidic or basic pH values, respectively), barophiles (organisms that grow best under pressure), and halophiles (organisms that require NaCl for growth). In addition, these organisms are normally polyextremophiles, being adapted to live in habitats where various physicochemical parameters reach extreme values. For example, many hot springs are acid or alkaline at the same time, and usually rich in metal content; the deep ocean is generally cold, oligotrophic (very low nutrient content), and exposed to high pressure; and several hypersaline lakes are very alkaline.

Extremophiles may be divided into two broad categories: extremophilic organisms which require one or more extreme conditions in order to grow, and extremotolerant organisms which can tolerate extreme values of one or more physicochemical parameters though growing optimally at "normal" conditions.

Extremophiles include members of all three domains of life, *i.e.*, bacteria, archaea, and eukarya. Most extremophiles are microorganisms (and a high proportion of these are archaea), but this group also includes eukaryotes such as protists (e.g., algae, fungi and protozoa) and multicellular organisms.

Archaea is the main group to thrive in extreme environments. Although members of this group are generally less versatile than bacteria and eukaryotes, they are generally quite skilled in adapting to different extreme conditions, holding frequently extremophily records. Some archaea are among the most hyperthermophilic, acidophilic, alkaliphilic, and halophilic microorganisms known. For example, the archaeal *Methanopyrus kandleri* strain 116 grows at 122 °C (252 °F, the highest recorded temperature), while the genus *Picrophilus* (e.g., *Picrophilus torridus*) include the most acidophilic organisms currently known, with the ability to grow at a pH of 0.06.

Among bacteria, the best adapted group to various extreme conditions is the cyanobacteria. They often form microbial mats with other bacteria, from Antarctic ice to continental hot springs. Cyanobacteria can also develop in hypersaline and alkaline lakes, support high metal concentrations and tolerate xerophilic conditions (*i.e.*, low availability of water), forming endolithic communities in desertic regions. However, cyanobacteria are rarely found in acidic environments at pH values lower than 5–6.

Among eukaryotes, fungi (alone or in symbiosis with cyanobacteria or algae forming lichens) are the most versatile and ecologically successful phylogenetic lineage. With the exception of hyperthermophily, they adapt well to extreme environments. Fungi live in acidic and metal-enriched waters from mining regions, alkaline conditions, hot and cold deserts, the deep ocean and in hypersaline regions such as the Dead Sea. Nevertheless, in terms of high resistance to extreme conditions, one of the most impressive eukaryotic polyextremophiles is the tardigrade, a microscopic invertebrate. Tardigrades can go into a hibernation mode, called the tun state, whereby it can survive temperatures from −272 °C (1 °C above absolute zero!) to 151 °C, vacuum conditions (imposing extreme dehydration), pressure of 6,000 atm as well as exposure to X-rays and gamma-rays. Furthermore, even active tardigrades show tolerance to some extreme environments such as extreme low temperature and high doses of radiation.

In general, the phylogenetic diversity of extremophiles is high and very complex to study. Some orders or genera contain only extremophiles, whereas other orders or genera contain both extremophiles and non-extremophiles. Interestingly, extremophiles adapted to the same extreme condition may be broadly dispersed in the phylogenetic tree of life. This is the case for different psychrophiles or barophiles, for which members may be found dispersed in the three domains of life. There are also groups of organisms belonging to the same phylogenetic family that have adapted to very diverse extreme or moderately extreme conditions.

Over the last few decades, the fast development of molecular biology techniques has led to significant advances in the field, allowing us to investigate intriguing questions on the nature of extremophiles with unprecedented precision. In particular, new high-throughput DNA sequencing technologies have revolutionized how we explore extreme microbiology, revealing microbial ecosystems with unexpectedly high levels of diversity and complexity. Nevertheless, a thorough knowledge of the physiology of organisms in culture is essential to complement genomic or transcriptomic studies and cannot be replaced by any other approach. Consequently, the combination of improved traditional methods of isolation/cultivation and modern culture-independent techniques may be considered the best approach towards a better understanding of how microorganisms survive and function in such extreme environments.

Based on such technological advances, the study of extremophiles has provided, over the last few years, ground-breaking discoveries that challenge the paradigms of modern biology and make us rethink intriguing questions such as "what is life?", "what are the limits of life?", and "what are the fundamental features of life?". These findings have made the study of life in extreme environments one of the most exciting areas of research, and can tell us much about the fundaments of life.

The mechanisms by which different organisms adapt to extreme environments provide a unique perspective on the fundamental characteristics of biological processes, such as the biochemical limits to macromolecular stability and the genetic instructions for constructing macromolecules that stabilize in one or more extreme conditions. These organisms present a wide and versatile metabolic diversity coupled with extraordinary physiological capacities to colonize extreme environments. In addition to the familiar metabolic pathway of photosynthesis, extremophiles possess metabolisms based upon methane, sulfur, and even iron.

Although the molecular strategies employed for survival in such environments are still not fully clarified, it is known that these organisms have adapted biomolecules and peculiar biochemical pathways which are of great interest for biotechnological purposes. Their stability and activity at extreme conditions make them useful alternatives to labile mesophilic molecules. This is particularly true for their enzymes, which remain catalytically active under extremes of temperature, salinity, pH, and solvent conditions. Interestingly, some of these enzymes display polyextremophilicity (*i.e.,* stability and activity in more than one extreme condition) that make their wide use in industrial biotechnology possible.

From an evolutionary and phylogenetic perspective, an important achievement that has emerged from studies involving extremophiles is that some of these organisms form a cluster on the base of the tree of life. Many extremophiles, in particular the hyperthermophiles, lie close to the "universal ancestor" of all organisms on Earth. For this reason, extremophiles are critical for evolutionary studies related to the origins of life. It is also important to point out that the third domain of life, the archaea, was discovered partly due to the first studies on extremophiles, with profound consequences for evolutionary biology.

Furthermore, the study of extreme environments has become a key area of research for astrobiology. Understanding the biology of extremophiles and their ecosystems permits developing hypotheses regarding the conditions required for the origin and evolution of life elsewhere in the universe. Consequently, extremophiles may be considered as model organisms when exploring the existence of extraterrestrial life in planets and moons of the Solar System and beyond. For example, the microorganisms discovered in ice cores recovered from the depth of the Lake Vostok and other perennially subglacial lakes from Antarctica may serve as models for the search of life in the Jupiter's moon Europa. Microbial ecosystems found in extreme environments like the Atacama Desert, the Antarctic Dry Valleys and the Rio Tinto may be analogous to potential life forms adapted to Martian conditions. Likewise, hyperthermophilic microorganisms present in hot springs, hydrothermal vents and other sites heated by volcanic activity in terrestrial or marine areas may resemble potential life forms existing in other extraterrestrial environments. Recently, the introduction of novel techniques such as Raman spectroscopy into the search of life signs using extremophilic organisms as models has open further perspectives that might be very useful in astrobiology.

With these groundbreaking discoveries and recent advances in the world of exthemophiles, which have profound implications for different branches of life sciences, our knowledge about the biosphere has grown and the putative boundaries of life have expanded. However, despite the latest advances we are just at the beginning of exploring and characterizing the world of extremophiles. This special issue discusses several aspects of these fascinating organisms, exploring their habitats, biodiversity, ecology, evolution, genetics, biochemistry, and biotechnological applications in a collection of exciting reviews and original articles written by leading experts and research groups in the field. I would like to thank the authors and co-authors for submitting such interesting contributions. I also thank the Editorial Office and numerous reviewers for their valuable assistance in reviewing the manuscripts.

Conflict of Interest

The authors declare no conflict of interest.

Chapter 1:
Extremophiles in Extreme Environments

A Laboratory of Extremophiles: Iceland Coordination Action for Research Activities on Life in Extreme Environments (CAREX) Field Campaign

Viggó Marteinsson, Parag Vaishampayan, Jana Kviderova, Francesca Mapelli, Mauro Medori, Carlo Calfapietra, Angeles Aguilera, Domenica Hamisch, Eyjólfur Reynisson, Sveinn Magnússon, Ramona Marasco, Sara Borin, Abigail Calzada, Virginia Souza-Egipsy, Elena González-Toril, Ricardo Amils, Josef Elster and Robert Hänsch

Abstract: Existence of life in extreme environments has been known for a long time, and their habitants have been investigated by different scientific disciplines for decades. However, reports of multidisciplinary research are uncommon. In this paper, we report an interdisciplinary three-day field campaign conducted in the framework of the Coordination Action for Research Activities on Life in Extreme Environments (CAREX) FP7EU program, with participation of experts in the fields of life and earth sciences. *In situ* experiments and sampling were performed in a 20 m long hot springs system of different temperature (57 °C to 100 °C) and pH (2 to 4). Abiotic factors were measured to study their influence on the diversity. The CO_2 and H_2S concentration varied at different sampling locations in the system, but the SO_2 remained the same. Four biofilms, mainly composed by four different algae and phototrophic protists, showed differences in photosynthetic activity. Varying temperature of the sampling location affects chlorophyll fluorescence, not only in the microbial mats, but plants (*Juncus*), indicating selective adaptation to the environmental conditions. Quantitative polymerase chain reaction (PCR), DNA microarray and denaturing gradient gel electrophoresis (DGGE)-based analysis in laboratory showed the presence of a diverse microbial population. Even a short duration (30 h) deployment of a micro colonizer in this hot spring system led to colonization of microorganisms based on ribosomal intergenic spacer (RISA) analysis. Polyphasic analysis of this hot spring system was possible due to the involvement of multidisciplinary approaches.

Reprinted from *Life*. Cite as: Marteinsson, V.; Vaishampayan, P.; Kviderova, J.; Mapelli, F.; Medori, M.; Calfapietra, C.; Aguilera, A.; Hamisch, D.; Reynisson, E.; Magnússon, S.; Marasco, R.; Borin, S.; Calzada, A.; Souza-Egipsy, V.; González-Toril, E.; Amils, R.; Elster, J.; Hänsch, R. A Laboratory of Extremophiles: Iceland Coordination Action for Research Activities on Life in Extreme Environments (CAREX) Field Campaign. *Life* **2013**, *3*, 211-233.

1. Introduction

Research on life in extreme environments (LEXEN) has tremendous potential as a source for new bioactive compounds in biotechnology, but it is also essential to understand how life was established on the early Earth and to speculate about the possibilities for life on other planets. Life and growth of living organisms is governed by numerous physical and chemical factors in their environment. Most life forms thrive on the surface of the Earth, where temperatures are generally moderate, *i.e.*, at temperatures from 4 °C to 40 °C, at pH between pH 5 to 8.5 and where salinity, hydrostatic pressure and ionizing radiation are low. Unlike many organisms that cannot survive outside of temperate conditions, extremophiles thrive optimally when one or several of these parameters are in the extreme range [1,2]. Temperature and pH are probably the most drastic factors for growth. Organisms living in such adverse environmental conditions are assigned to thermophilic, psychrophilic, acidophilic and alkalophilic categories. This classification encompasses several natural biotopes in which extreme environmental conditions are more prevalent than usually found in nature. Evidently, considering the high variety of biotopes on Earth, the physiological responses to the environmental extremes can be observed on a gradual scale from tolerance to absolute requirement.

High-temperature environments are generally associated with volcanic activity, but some are also in man-made industrial complexes. Important biotopes are terrestrial geothermal fields, *i.e.*, alkaline freshwater hot springs, acid solfatara fields and hydrothermal systems in marine coastal, shallow and deep areas. Hot environments often display a wide range of pH, from acid to alkaline, depending on temperature, water availability, gases and ion concentrations [3]. Natural geothermal areas are widely distributed around the globe, but they are primarily associated with tectonically active zones at which the movements of the Earth's crust occur. Due to this localization of geothermal heat sources, hot springs are generally restricted to a few concentrated areas. From the biological perspective, the best known terrestrial sites are Iceland, the Naples area in Italy, Yellowstone National Park in USA, Japan, New Zealand and the Kamchatka Peninsula in Siberia [4,5].

Terrestrial geothermal areas, *i.e.*, in Iceland, can be generally divided into high-temperature and low-temperature fields, according to the nature of the heat source and pH. High temperature vent fields are located within the active volcanic zones, and the heat source is a magma chamber at a depth of 2 to 5 km. In these areas, the water temperature reaches 150 °C to 350 °C at the depths of 500 m to 3,000 m, and steam and volcanic gases are emitted at the surface. Mainly, the gases consist of N_2 and CO_2, but H_2S and H_2 can make up to 10% of the total produced. Traces of CH_4, NH_3 and CO can also be found [6]. On the surface, H_2S is oxidized chemically and biologically first to sulfur and then to sulfuric acid, which acts as the buffering agent in the hot spring environment [1]. As a result, the pH often stabilizes at 2 to 2.5. Because of the high temperature, little liquid water comes to the surface, and the hot springs are usually in the form of fumaroles and steam holes or grey and brown mud pots, resulting from the corrosion of surrounding rocks by the high concentrations of sulfuric acid [6]. Neutral to slightly alkaline sulfide-rich hot springs may also co-exist in high-temperature fields, but are rarer. They appear on the periphery of the active zone and are created if water is abundant at low depths, *i.e.*, by melting of snow or rain or with high levels of the groundwater. The Hveragerdi area in Iceland is a good example of such a field, with a great verity of hot springs with sulfide concentrations as high as 30 mg L^{-1}, and under such conditions, thick microbial mats are formed with precipitated sulfur and make spectacular bright yellow or white colors [6,7].

The low temperature hot spring fields are located outside the active volcanic zones. Extinct or deep lava flows and dead magma chambers serve as heat sources, and the water temperature is usually below 150 °C at depths of 500 m to 3,000 m. Groundwater percolating through these zones warms up and returns to the surface, enriched with high concentrations of dissolved minerals (*i.e.*, silica) and gases (mainly CO_2 and little H_2S). On the surface, CO_2 is blown away, and the silica precipitates, resulting in an increase in pH, often stabilizing at 9 to 10. The hot springs in the low temperature field are characterized by a general stability in temperature, water flow and pH [6].

The CAREX project (Coordination Action for Research Activities on Life in Extreme Environments, EC Grant agreement no.: 211700) started in 2008 and was funded by the European Commission in 2009 [8]. The aim of this program was to improve coordination of research on life in extreme environments (LEXEN) and identify the need for the better coordination of LEXEN research. CAREX objectives were focused on establishing interaction, coordinating activities and promoting a community identity for European research in LEXEN. To reach these very ambitious objectives, there is no better way than a real scientific campaign with scientists from different fields of expertise collaborating in a fieldtrip. With this idea in mind, CAREX designed a task, which was called "Field Procedures Inter-comparisons". One of the main CAREX objectives was to coordinate research interdisciplinary integrated actions as campaigns for studying extreme field sites with multidisciplinary international teams of scientists. Establishing such a community will encourage greater interdisciplinarity and increase knowledge of extreme environments from very different perspectives. This activity was planned to develop fieldtrips; the first of them was organized for a scientific campaign in Río Tinto (South-west Spain), and the second one in Iceland, which is reported in this paper.

The Icelandic field visits were organized in order to promote the interaction of different disciplines in a field setting to demonstrate the use of selected technologies, compare methodologies, exchange research experience and to promote harmonization of techniques and methodologies. The fieldwork aims were focused on developing and evaluating new technologies of common use across LEXEN research, including remote sensing devices and field analysis of ecosystem level processes, focusing on hot spring and glacial techniques.

2. Results and Discussion

2.1. Site Description, "CAREX Hot Spring"

The sampling site was a high-temperature hot spring field, but with some characteristics of low temperature hot spring fields in Iceland. The sampling zone was comprised of acidic, neutral and alkali hot springs in a narrow area. The formation of these various springs in such a small range was due to abundant water supply in low depths in the surroundings. The selected hot spring in this study was designated as "CAREX hot spring" (Figure 1), which was part of larger system. The hot spring system was about 20 m long with many small spring outlets, which were not visible on the surface, and with different ranges of pH and temperature. The whole system formed three surface inter-connected main pools (P1, P2 and P3) and one open hot spring (P0) at the beginning of the system. This high temperature (98 °C) mud pool (P0) was without any surrounding vegetation and had no surface connection to the nearby pools, P1 or P2 and P3, and it was extremely acidic (pH 3.8). The P system was surrounded by vegetation, and temperature ranged between (57 °C to 62 °C) with a low pH (2.9 to 3.2). Pool P1 was 57 °C with pH 2.9, pool P2 was 62 °C and pH 2.9 and pool

P3 was 59 °C and pH 3.2. Two additional hot springs located in the same field, but at approximate 250–300 m away from the CAREX hot spring system, were also measured and used as reference hot springs, HS1 (98 °C, pH 7) and L1 (98 °C, pH 2.0).

2.2. Measurements of Photochemical Activity of Microbial Mats and Higher Plants with FluorCam

2.2.1. Measurement in Microbial Mats

The results of *in situ* measurements of photochemical activities of four different microbial mat communities at different environmental conditions are given in Table 1. The microbial mats were collected along the stream at four cooler sites located on the edge of the stream along the CAREX hot spring (Figure 1 S1, S2, S3 and S4).

The following species composition was detected later in laboratory from samples collected from each site (Figure 1 S1 to S4). (Centre for Phycology, Institute of Botany AS CR, Czech Republic).

- S1 d_1—cf. *Chlamydomonas* sp. d_2—*Klebsormidium* sp., with presence of diatoms, *Euglena* sp.
- S2 d_1—cf. *Zygnematopsis* sp. and d_2—*Klebsormidium* sp.
- S3 d_1—*Klebsormidium* sp., with presence of diatoms
- S4 d_1—*Euglena* sp., no other species observed

Similar species as detected in S1–S4 were also found in an acidic habitat during the CAREX Río Tinto Fieldtrip in Spain, with the exception of *Cyanidium* sp., which dominated in the Río Tinto samples [9]. This study also shows the presence of cf. *Zygnematopsis* sp., which was not found at Río Tinto. Such a difference could reflect the different chemical composition of the water at the study sites, especially heavy metals content [10]. The photochemical performance of the biofilms was evaluated using FluorCam, and the results are summarized in Table 2. Since green algae are dominant in all samples, with the exception of sampling site S4, the parameters could indicate minor stress in the microbial community. Algae in sample S4 was probably photo-inhibited by high irradiance (Table 2). However, for detailed explanations of *in situ* fluorescence parameter measurements, knowledge of the response of individual biofilm species to environmental conditions is crucial, as proposed in Kvíderová [11]. In general, the microbial mats seemed to be well adapted to the given conditions, with the exception of sampling site S4, where the algae were probably subjected to some stress. The stress was probably caused by high irradiance, but low pH effects cannot be excluded. Further laboratory investigations should be performed in defined combinations of temperature, irradiance and pH.

Figure 1. Coordination Action for Research Activities on Life in Extreme Environments (CAREX) hot spring. Picture of the CAREX hot spring system (left photo) and sampling sites, pool 1, 2 and 3, and its specific sampling sites. Different pools of the hot-spring system were sampled: pool 1 (P1), pool 2 (P2), pool 3 (P3) and Pool 0 (P0) (inserted photo in lower left corner). The "Biofilm Catcher" is shown in pool 1 (P1). 16S rRNA polymerase chain reaction (PCR)-denaturing gradient gel electrophoresis (DGGE) profiles of the bacterial communities in the water of the three pools (P1, P2 and P3) are shown on the gel photo in the middle. Circles on the bands indicate the DNA fragments that were excised from the gel and successfully amplified and sequenced. Sampling sites for measurements of photochemical activity of microbial mats and higher plants with FluorCam are marked S1, S2, S3 and S4 in the photo of system and enlarged in four photos on the right side of the figure. An enlarged photo of site S5 is also on the right side of the figure (bottom). Site S6 can't be visualized in the photo of the CAREX hot spring and is therefore shown enlarged on the right corner, at the top of the photo.

Table 1. The environmental conditions at individual sites where microbial mat communities were sampled.

Site	Color of mat	Temperature [°C]	pH	Irradiance [μmol m^{-2} s^{-1}]
S1	Green mat	24.9	3.1	460
S2	Brown mat	22.3	3.1	280
S3	Green biofilm	30.5	3.1	550
S4	Green biofilm	24.8	2.7	1,200

Table 2. The photochemical parameters of individual biofilm samples (mean ± SD, n = 3). F_V/F_M: maximum quantum yield; Φ_{PSII}: actual quantum yield under irradiance of 150 μmol m^{-2} s^{-1}; NPQ: Stern-Volmer non-photochemical quenching; qP: photochemical quenching.

	F_V/F_M	Φ_{PSII}	NPQ	qP
S1	0.54 ± 0.02	0.33 ± 0.02	0.09 ± 0.01	0.62 ± 0.03
S2	0.64 ± 0.17	0.33 ± 0.04	0.65 ± 0.30	0.63 ± 0.13
S3	0.65 ± 0.03	0.26 ± 0.04	0.70 ± 0.04	0.49 ± 0.08
S4	0.44 ± 0.00	0.27 ± 0.02	0.07 ± 0.02	0.64 ± 0.04

2.2.2. Measurement in Juncus Plants

Since the F_V/F_M of plants in optimum conditions is approximately 0.83 [12], the F_V/F_M values indicated that the photochemical activity of plants was not seriously damaged by the environmental conditions, and the influence of temperature on photochemical performance was not observed in *Juncus* plants in the CAREX hot spring (Table 3). Other parameters also confirm only minimum stress on the plants. Moreover, FluorCam and Li-COR provided comparable results (Table 3). The photosynthetic apparatus of *Juncus* does not seem to be stressed by high temperatures, and significant differences were found in photochemical parameters derived from fluorescence measurement (F_V/F_M, Φ_{PSII}, NPQ and qP) and photosynthesis expressed as CO_2 assimilation rate (Table 3 and Figure 2). However, since the measurements had to be performed in water, the samples drifted during the measurement. The precise evaluation will require step-by-step calculations of individual fluorescence signals from the camera record of the fluorescence. Despite this problem, the results from the automatic data processing by FluorCam software indicate that the plants are not seriously damaged by the environmental conditions, and the influence of temperature on photochemical performance was not observed.

2.3. Measurements of Photosynthetic Performance of Thermophilic Biofilms

Results showed differences in the photosynthetic performance of the biofilms analyzed (Figure 1). While *Euglena* sp. cells (Figure 1, S6 and Figure 3) showed photo-inhibition behavior, the biofilms formed by *Klebsormidium* sp. (Figure 1, S5 and Figure 3) are photo-saturated. *Euglena* sp. showed photo-inhibition over the light intensity of 0 to 200 μmol photons m^{-2} $s^{-1,}$ and the *Klebsormidium* sp. sample showed a light-saturated photosynthesis model under irradiations higher that 200 μmol photons m^{-2} s^{-1}. The fitted parameters for the biofilms analyzed are shown in Table 4. All fits showed correlation *r* values higher than 0.85. The highest values of compensation light (Ic) and saturating light (Ik) intensities were shown by *Klebsormidium* sp. biofilm, followed by *Euglena* sp. No significant differences were found in P_{max} values; both biofilms showed *ca.* 15 mgO_2 mg Chl a^{-1} h^{-1} (Chl a: chlorophyll *a*). However, *Euglena* sp. showed higher photosynthetic efficiency values (α = 0.5) than *Klebsormidium* sp. biofilms (α = 0.24). By sampling two areas (Figure 1, S5 and S6) in the same hot spring system, we obtained different results from two dominating species, and the measurements with FluorCam confirmed our previous results. Furthermore, photosynthetic performance shows photoinhibition in the *Euglena* sp. sample, and no serious damage was detected in *Klebsormidium* sp. in microbial mats samples.

Table 3. Measurements on photochemical performance of *Juncus* plants in the CAREX hot spring (mean SD, n = 3).). F_V/F_M: maximum quantum yield; Φ_{PSII}: actual quantum yield under irradiance of 150 μmol m^{-2} s^{-1}; NPQ: Stern-Volmer non-photochemical quenching; qP: photochemical quenching.

	30 °C	40 °C	50 °C	60 °C
F_V/F_M	0.842 ± 0.017	0.828 ± 0.017	0.845 ± 0.010	0.824 ± 0.021
Φ_{PSII}	0.550 ± 0.059	0.584 ± 0.020	0.492 ± 0.132	0.578 ± 0.012
NPQ	0.808 ± 0.229	0.505 ± 0.036	1.288 ± 0.359	0.742 ± 0.311
qP	0.735 ± 0.072	0.768 ± 0.051	0.693 ± 0.166	0.789 ± 0.004

Figure 2. Photosynthesis measured on plants growing near a hot spring at two different temperatures.

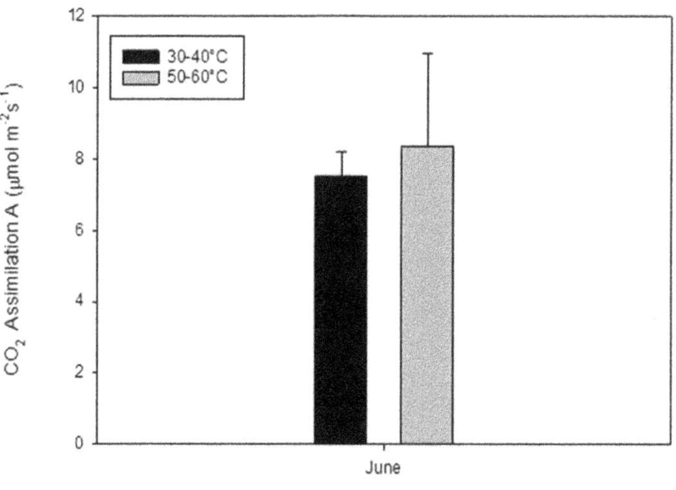

Table 4. Photosynthetic parameters of the different biofilms assayed. Compensation light intensity (Ic) and light saturation (Ik) are expressed on a photon basis (μmol photons m^{-2} s^{-1}). Maximal photosynthesis rate (P_{max}) and photosynthetic efficiency (α) are expressed on a chlorophyll *a* (Chl *a*) basis (mg O_2 mg Chl a^{-1} h^{-1}).

Species	Ic	Ik	Pmax	α
Euglena sp.	22.3 ± 2.4	112.6 ± 12.3	16.1 ± 2.6	0.5 ± 0.01
Klebsormidium sp.	45.8 ± 4.6	197.6 ± 18.6	15.3 ± 3.1	0.24 ± 0.01

Figure 3. Net oxygen production *versus* irradiance curves. Photosynthetic rates were normalized to chlorophyll *a*. The collection sites for the fluorescence measurements are shown in Figure 1, site S5 and site S6).

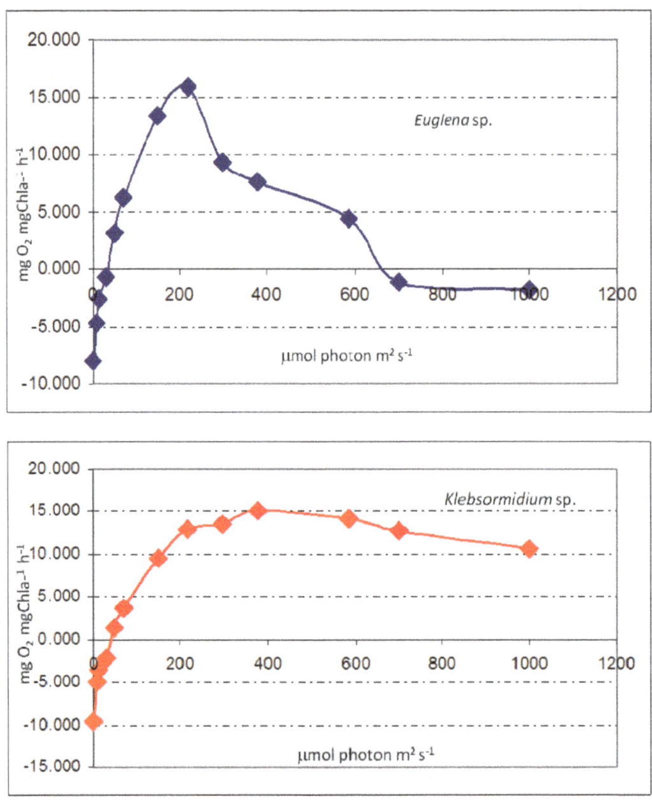

2.4. CO_2 Monitoring at the CAREX Hot Spring

The CO_2 at the top of the stream at site P0 was measured as 399.6 ppm, P1 was 390.2 ppm and P2 388.7 ppm, on average. The measurement was performed to find out if CO_2 was in high concentrations and if it was due to volcanic spring activity. CO_2 is a greenhouse gas naturally present in the atmosphere with a mean concentration of 0.038%. The high value in volcanic areas must either come from geological or biological sources (animals, plants, cells in general). We measured the highest CO_2 concentration in site P0, which was not vegetated and, therefore, suggests that it was of geological origin.

2.5. Volatile Organic Compounds as Carbon Losses in Plants and Their Thermo-Tolerance

The presence of VOCs in the atmosphere influences its composition and contributes to the formation of greenhouse gases and pollutants [13]. The aim was to investigate the VOC emissions from *Juncusalpino articulatus* living in hot-springs (Figure 4), to identify differences in plants living close to the hot spring from those living further and to relate the VOC emission with the physiological status of the plants. The results are presented in Figure 3 to 8 and in more detail by Medori *et al.* [14]. Plants of *Juncus* sp. living in higher water temperature (HGT, 50 °C–60 °C) showed a mean value of CO_2 assimilation of 8.34 μmol m^{-2} s^{-1} (Figure 2). This assimilation was higher than that measured in *Juncus* sp. living in lower water temperature (LGT, 30 °C–40 °C) (Figure 4) with the mean value of assimilations was 7.52 μmol m^{-2} s^{-1} CO2. These high temperatures plants were able to maintain their optimal stomatal conductance (Figure 5). Intercellular CO_2 concentration (Ci) measured on plants growing near a hot spring at two different temperatures shows stress conditions; in this case, the high water temperatures stimulate plants with greater emissions of VOCs (Figure 6). The rate of carbon emitted with α-pinene represents 0.0057% for HGT and 0.0016% for LGT of the carbon assimilated through the photosynthesis (Figure 7). It is also interesting to find that in both detected species, the emission of VOCs was stimulated by the proximity with the hot spring (Figure 8). Clearly, due to the small number of samples, it is difficult to carry out a statistical test, which could produce reliable results. However it is likely that the warmer temperatures could have stimulated the synthesis of these compounds in plants growing nearby the hot springs, regardless of their protective role in plants. It is known indeed that these compounds are highly dependent on temperature.

Figure 4. Plant of *Juncusalpino articulatus* growing in water of a temperature between 50 and 60 °C.

Figure 5. Stomatal conductance measured on plants growing near a hot spring at two different temperatures.

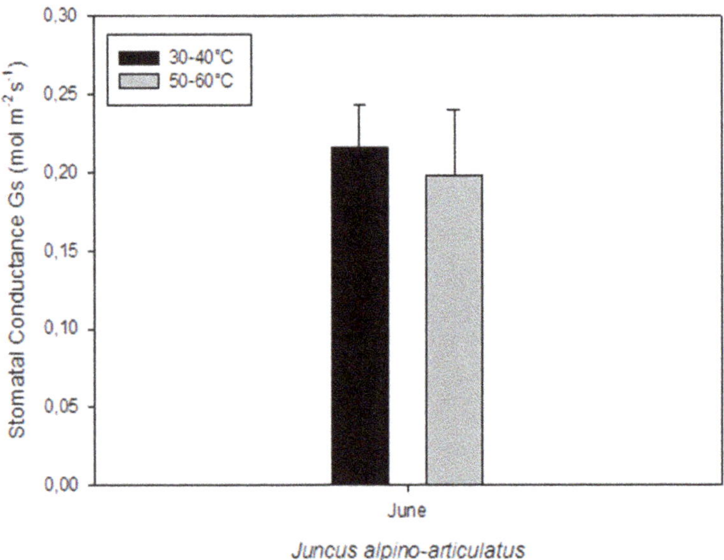

Figure 6. Intercellular CO_2 concentration (Ci) measured on plants growing near a hot spring at two different temperatures.

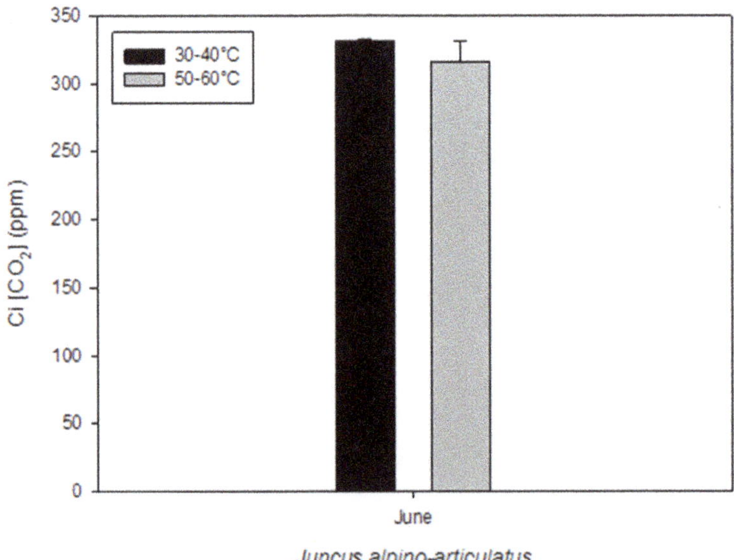

Figure 7. Percentage of Carbon emitted as VOCs in comparison with carbon assimilated through photosynthesis measured on plants growing near a hot spring at two different temperatures.

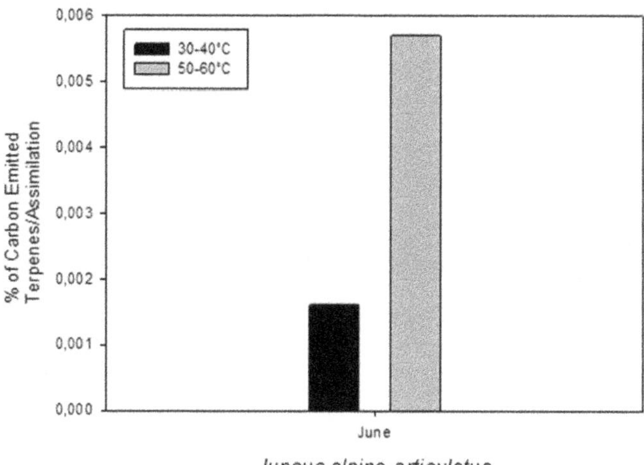

Figure 8. VOCs emitted from plants growing near a hot spring at two different temperatures.

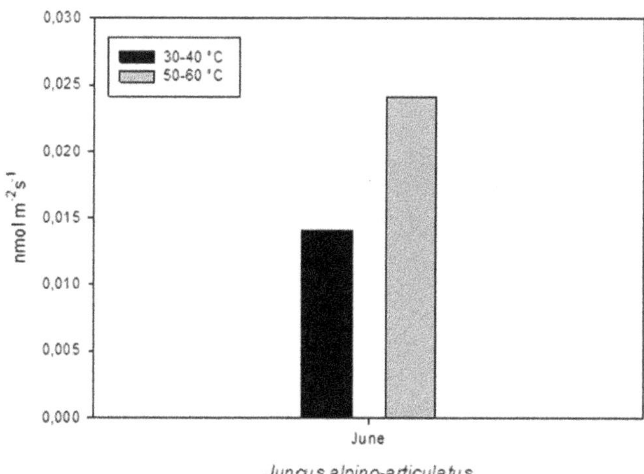

2.6. SO_2 and H_2S Measurements of the CAREX Hot Spring Area

In this study, the control site 50 m away from the study site was measured as SO_2: 0.0037 ppm and H_2S: 0,0085 ppm. Site P0 was SO_2: 0.0111 ppm, H_2S: 1.8 ppm; site P1 SO_2: 0.0034 ppm, H_2S:

0.486 ppm; and site P2 SO_2: 0.0036 ppm, H_2S: 0.0138 ppm. The P0 site showed the highest values in both SO_2 and H_2S (three- and 35-fold higher, respectively), while the other sites, P2 and P3, had similar values in SO_2 compared to the control, but they had much higher values in H_2S, especially P2.

Typical gaseous emissions from geothermal fields and volcanoes are hydrogen sulfide (H_2S) and sulfur dioxide (SO_2), which both have influences on human, animal and environmental health. Plants are able to overcome moderate SO_2 concentrations with sulfite oxidase, a specific enzyme for this purpose: [15]. In volcanic and geothermal areas, most SO_2 is converted into H_2S as a result of the prevalently higher pressure: $SO_2 + 3H_2 \leftrightarrow H_2S + 2H_2O$ [16]. Additionally, it has been reported that SO_2 generation from H_2S is minimal and very slow and *vice versa* [17]. Usually, plants gain their sulfur need out of sulfate available from the soil, but further, they are able to use SO_2 as a sulfur source [18]. Despite the usability of SO_2, excess amounts are of high toxicity and have influences on the whole plant, up to visible injury and death.

2.7. "Biofilm Catcher" in the CAREX Hot Spring, RISA and DGGE Analysis

Water samples collected from the hot spring pools showed higher DNA concentration (ranging from 100 to 200 ng/μL) compared to the different substrates tested by the "Biofilm Catcher", which displayed low DNA concentration (10–15 ng/μL).

Ribosomal intergenic spacer analysis (RISA) produced faint bands on agarose gel from the pool samples and only from a subset of the substrates samples (paper, iron and titanium) deployed through the "Biofilm Catcher" micro-colonizer, but other solid substrates were negative (pyrite, steel, copper and glass). The RISA profiles showed the presence of a few peaks (Figure 1), indicating the occurrence of a microbiome of low bacterial diversity, both on the pool water and on the "Biofilm Catcher" (data not shown). The RISA profiles showed also the presence of partially different bands among pool water compared to the solid substrates (data not shown), suggesting the selection of specific bacteria on the tested solid materials from the total bacterial community that colonize the P1. However, due to the fact that the retrieved bands were very close to the detection limit of the RISA technique, it is not possible to draw any firm conclusion on the "Biofilm Catcher" experiment. The successful utilization of different solid substrates, including glass, stainless steel and polypropylene, to isolate novel bacteria has been recently demonstrated by inoculating freshwater samples in laboratory microcosms [19]. Possibly, longer periods of deployment of the "Biofilm Catcher" in the natural ecosystem could lead to increased adhesion of the biofilm forming prokaryotes on the solid materials and the selection of previously uncultured bacteria.

Taking in consideration the limits showed by the RISA technique, the bacterial community structure of the pool samples was investigated by DGGE fingerprinting. DGGE profiles obtained from the water samples collected at P1, P2 and P3 (Figure 1) showed very similar bacterial communities in the three interconnected pools, as expected, since the environmental condition are basically the same in terms of pH and temperature (Table 5). Partial 16S rRNA gene sequences (500 bp) obtained from the DGGE bands have been deposited in the GeneBank database under the accession numbers HF547636–HF547650.

Table 5. Phylogenetic identification and distribution of bacterial sequences retrieved from 16S rRNA DGGE gel. Identification of the dominant bands in the PCR-DGGE fingerprinting profiles (marked in Figure 1) and their distribution in the three different interconnected pools, P1, P2 and P3, of the hot spring system.

Band	Class (RDP)	Closest Relative (accession number)	%	Environments	Closest Type Strain or Described Cultivable Strain (accession number)	%	Pool		
							P1	P2	P3
12	*Alphaproteobacteria*	Uncultured bacterium (DQ834212)	99	Hot springs, Yellowstone National Park	*Acidicaldus organivorans* (AY140238)	98	**X**	X	X
13	*Alphaproteobacteria*	Uncultured bacterium (DQ834212)	99	Hot springs, Yellowstone National Park	*Acidicaldus organivorans* (AY140238)	99	**X**	X	X
1, 2	*Betaproteobacteria*	*Ralstonia pickettii* (FR873796)	99	Water samples	*Ralstonia pickettii* (AY741342)	99	X	**X**	**X**
15	*Actinobacteria*	*Acidimicrobium* sp. (AY140240)	99	Geothermal sites, Yellowstone National Park	*Acidimicrobium ferrooxidans* (CP001631)	98	X	**X**	X
4, 5	*Bacilli*	*Geobacillus debilis* (AB548612)	99	High temperature compost	*Geobacillus debilis* (AJ564616)	99	X	**X**	**X**
9	*Clostridia*	Uncultured *Bacillus* sp. (EU250948)	89	High temperature compost	*Thermovenabulum ferriorganovorum* (AY033493)	88	X	X	**X**
14	*Clostridia*	Uncultured bacterium (AF523921)	94	Forested wetland	*Sulfobacillus benefaciens* (EF679212)	90	X	X	**X**
6	*Aquificae*	Uncultured *Hydrogenobaculum* sp.(EF156602)	99	Norris Geyser, Yellowstone National Park	*Hydrogenobaculum acidophilum* (D16296)	98	**X**		
10, 11	*Aquificae*	Uncultured *Hydrogenophilus* sp. (EF156602)	98	Norris Geyser, Yellowstone National Park	*Hydrogenobaculum acidophilum* (D16296)	97	X	**X**	**X**
3	Unclassified Bacteria	Uncultured *Rhizobiales* (JF317890)	98	Terrestrial hot spring, 85 °C, pH 5.5	*Ignavibacterium album* (AB478415)	85	**X**		
7, 8	Unclassified Bacteria	Uncultured bacterium (EF464600)	96	Acidic mine tailings (pH 3.5-5)	*Thermosinus carboxydivorans* (AAWL01000046)	84	**X**		**X**

%: percent of identity between the DGGE band sequence and the closest relative sequence in GeneBank. Environment: environment of origin of the closest relative sequence. X: presence of the band in the DGGE profile of each sample; in bold are indicated the bands that were actually sequenced.

The identification of the dominant bands in the PCR-DGGE fingerprinting profiles and their distribution in the three different interconnected pools of the hot spring system (P1, P2, P3) are shown in Figure 1. A widely diversified bacterial community composed by different classes of bacteria colonize the three interconnected pools and are represented by *Alphaproteobacteria* (bands 12, 13,), *Betaproteobacteria* (bands 1, 2), *Actinobacteria* (band 15), *Bacilli* (bands 4, 5), *Clostridia* (bands 9, 14) and *Aquificae* (bands 6, 10, 11), besides three bands (3, 7, 8) described as unclassified bacteria. All the obtained sequences showed a high percentage of identity with 16S rRNA bacterial sequences retrieved from environments similar to the CAREX hot spring, such as hot-springs and geysers of Yellowstone National Park.

2.8. Adenosine Triphosphate (ATP) Based Analysis

Results of total and internal adenosine triphosphate (ATP) results are depicted in Figure 9. The results show the expected biomass in each sites and P0 containing the lowest ATP value of them all. The low ATP value at site P0 could be anticipated, as the temperature was high with low pH, and the surroundings were not vegetated. Moreover, the sample at site P0 was expected to be difficult to measure, as clay in the sample was oily and sticky. Therefore, we estimate that the ATP value is in fact underestimated, which was confirmed with measurement based on 16S rRNA quantitative PCR (qPCR) (Figure 10). It is also possible that the clay in the samples was interfering with the qPCR measurements, and therefore, the bio-burden is higher. Lumitester PD-10N is a hand-held ultra-high sensitive ATP measurement instrument, and this lightweight, rapid assay (10 s) instrument has been extensively used by the food industry to monitor microbial bio-burden. The results in this study demonstrate well that this instrument is an ideal field instrument for rapid estimation of bacterial bio-burden to select biological "hot spots" from a large field area.

2.9. DNA Extraction and qPCR-Based Bacterial Quantification

The qPCR method measuring the 16S rRNA copy number/mL at the sites (P0 to P3) correlated quite well with the ATP results (Figure 9), although some variance was observed, especially for sites P0 and P2 (Figure 10). These variances could have possibly been explained by some difficulty in DNA extractions, especially from the P0 sample. Bacterial cells can lyse differently, depending on degenerating agents and conditions, but environmental chemicals could also interfere with the DNA yield and PCR performance. The 16S rRNA copy number was significantly lower in P0 and HS1 samples, compared to other sites (Figure 10).

2.10. High Density 16S Microarray (PhyloChip) Analysis

The PhyloChip results and other environmental parameters of the sampling sites are shown in Table 6. The Phylochip-based analysis showed the highest diversity of bacteria and archaea at site P1 (313 bacteria and 18 subfamilies of archaea) and P3 (318 bacteria and 18 of subfamilies archaea), but lower diversity at site P2 (127 bacteria and 10 subfamilies of archaea) and very low diversity at site P0 (eight bacteria and zero subfamilies of archaea). More detailed results of the PhyloChip analyses are presented elsewhere by Krebs *et al.* 2013 [20]. The PhyloChip results are in correlation with the ATP and qPCR results, and again, the low diversity at site P2 may be explained by a slightly higher temperature, compared to P1 and P2 samples. The results on archaeal subfamilies detected in the high temperature reference hot springs, HS1 and LS1 (Table 6), show that the PhyloChip technique was sensitive and robust enough for high temperature hot springs by detecting archaea at both neutral pH (7.0) and acidic pH (2.0). Interestingly, the structure of the clay in the acidic reference hot spring, L1, was different from the spring, P0. The clay was brick red in L1, but gray/black in P0, and the clay was more viscous (like oil) in P0 than in L1. The nature of the clay and low water abundance in P0 could possibly explain why low ATP and the 16S rRNA copies were detectable in P0.

Figure 9. Microbial bio-burden of the hot spring pool samples (P0, P1, P2 and P3) based on total and internal adenosine triphosphate (ATP). HS1 and L1 are samples from reference hot springs.

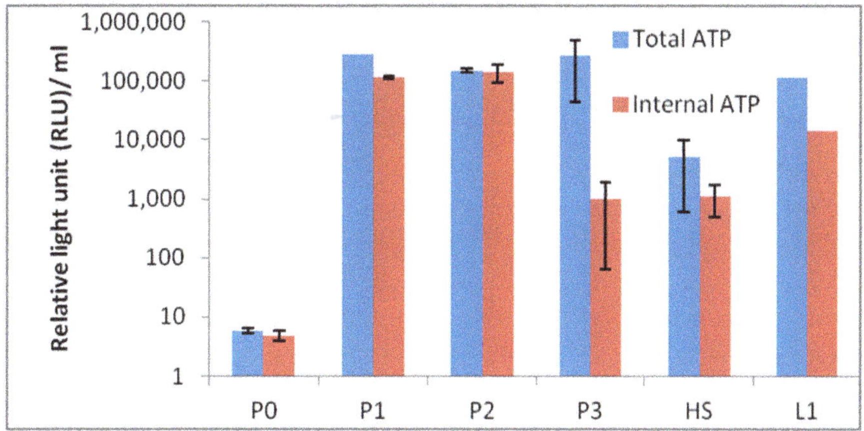

Figure 10. Microbial bio-burden of the hot spring pool samples (P0, P1, P2, P3, HS1 and L1) based on 16S rRNA quantitative polymerase chain reaction (qPCR). Y axis: 16S rRNA copy number/mL.

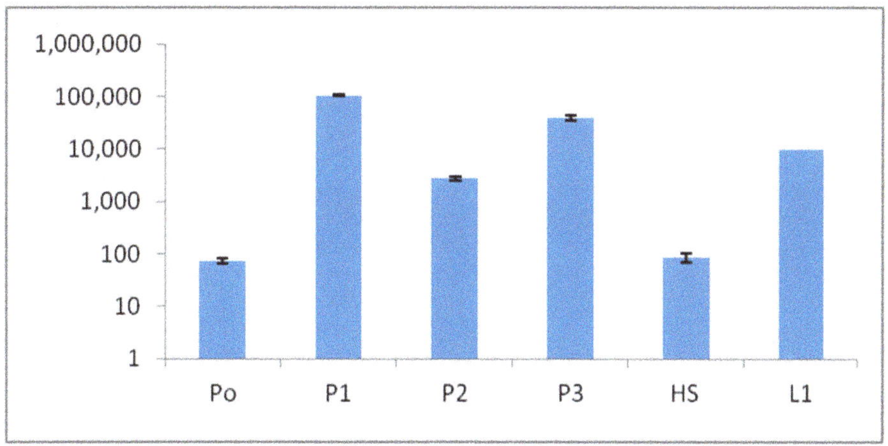

Table 6. Characteristics of the six pool sites in the CAREX hot spring and two additional reference sampling sites in the same area.

Test	P0	P1	P2	P3	L1	HS1
Temperature (°C)	100	59	64	57	98	98
pH	3.8	3.2	2.9	2.9	2	7
Color of the water	gray/black	brick red	brick red	brick red	brick red	colorless
Presence of Vegetation	no	yes	yes	yes	no	yes
Total ATP	5.8	2.8×10^5	1.5×0^5	2.7×10^5	1.1×10^5	5.3×10^3
Internal ATP	4.8	4.0×10^4	1.4×10^5	1.0×10^3	1.4×10^4	3.7×10^2
Bacterial 16S rRNA copies	7.4×10^1	1.0×10^6	2.8×10^4	3.9×10^5	3.4×10^4	8.6×10^2
*Bacterial subfamilies	8	313	127	318	7	8
*Archaeal subfamilies	0	16	10	18	4	2

* Results based on Phylochips [20].

3. Activity Performed on Site and in Laboratory: A Brief Description of Instruments, Materials, Methods and Results.

3.1. Sampling Site

In situ experiments and sampling were performed in the vegetated hot spring system, designated as CAREX hot spring. The hot spring area of Reykir in Hveragerdi was selected to demonstrate the use of the selected technologies to perform a holistic research on eukaryotic and prokaryotic organisms living in an extreme environment. The research area encompasses various types of hot springs that were formed in May 2008 after a series of strong earthquakes in the area. The zone had low open thermal activity before the earthquakes, which is located a few meters from a small botanical research center in Hveragerdi. This makes the spot very well situated for *in situ* experiments, to interact in a field setting and to demonstrate the use of *in situ* field technologies. The whole thermal active area is located just outside a small town, Hveragerdi, which is located about 50 km east from Reykjavik (Figure 1).

3.2. Measurements of Photochemical Activity of Microbial Mats and Higher Plants with FluorCam

The variable chlorophyll *a* fluorescence measurement is a common method in plant physiology (see e.g.,[21–23] for review of a detailed description of the protocols used). The detailed description of photochemical processes related to individual measured fluorescence parameters is summarized in Gomez *et al.* 2011 [9]. Photochemical activities of microbial mat communities at different environmental conditions and leaves of *Juncus* grass growing in a temperature gradient were investigated. Two biofilm samples from each site were collected. One sample was kept in cold and the second one was fixed using final concentration of 4% formaldehyde + 1% (w/v) $CuSO_4.5H_2O$. The species composition was evaluated using an Olympus BX-51 light microscope (Olympus, Japan) in the laboratory. The photochemical performance of the biofilms was evaluated using a FluorCam fluorescence imaging camera (Photon Systems Instruments, Czech Republic). A quenching protocol was applied for the results, as previously described [9].

3.3. *Measurement in* Juncus *Plants*

The *Juncus* plants were collected in the temperature gradient from 30 to 60 °C. The tips of the leaves were cut, and the fluorescence measurements were performed using the same protocol as for biofilms.

3.4. *Measurements of Photosynthetic Performance of Thermophilic Biofilms*

Light-saturated net photosynthesis (P_{max}) was determined as oxygen exchange using an oxygen electrode (Oxytherm, Hansatech Instruments, Norfolk, UK). Two biofilms, mainly composed by filamentous algae, *Klebsormidium* sp., and a phototrophic protist, *Euglena* sp., were analyzed (Figure 1, S5 and S6). Three replicates of each biofilm (*ca.* 1 mg fresh weight) were incubated in the Oxytherm chamber filled with 2 mL of BG11 media at pH 2 at 40 °C and 30 ± 1 °C, respectively, as previously described [24]. Briefly, the biofilms were incubated in darkness for at least 20 min. To estimate P_{max}, the samples were incubated at different light irradiances ranging from 0 to 1,000 μmol m^{-2} s^{-1} of PAR (μmol m^{-2} s^{-1}). Each increase in irradiance was applied, until steady-state oxygen production was observed (5 min). Net photosynthetic rate on the mg Chl *a* basis were determined for each light intensity. Photosynthetic parameters were estimated from the fitting of the equation of Edwards and Walker [25]. The following photosynthetic parameters were estimated by P_{max} (maximum photosynthetic rate under light-saturated condition or photosynthetic capacity), Ic (compensation light intensity), the α value (photosynthetic efficiency) and Ik (light saturation parameter).

3.5. *CO_2 Monitoring at the CAREX Hot Spring*

The device used for *CO_2* measurements was EGM-1 (by PP system, Hitchin, UK), which detects CO_2 (ppm) by infrared analysis and is equipped with a chamber for gas sampling from the soil. Each measurement took 2 minutes, and 4 measurements were taken at each point. A reference point was taken in the grassland, about 20 meters south from the stream. The measurements were taken at 1–2 p.m. and the weather was fairly windy.

3.6. *Contribution of VOCs as Carbon Losses in Plants and Their Thermotolerance*

The VOC emissions were measured in 3 plants of *Juncusalpino articulatus* growing at a higher water temperature (between 50 and 60 °C) and 3 other plants growing at a lower water temperature (between 30 and 40 °C). The sites used for the measurements were located at the end of the CAREX hot spring. Gas exchange was measured using an infrared CO_2 and H_2O analyzer Li-Cor-6400 photosynthesis system. A leaf area of 6 cm^2 was measured inside a chamber with a leaf temperature of 20 °C; a photosynthetic photon flux density (PPFD) of 500 μmol m^{-2} s^{-1}, a CO_2 concentration of 400 ppm and an air flow through the cuvette of 500 ml/min.

3.7. Measurements of the of SO_2 and H_2S in the CAREX Hot Spring Area

Measurements of SO_2 and H_2S concentrations in the air of the CAREX hot spring were determined by using the APSA 370 air pollution monitor (HORIBA, Kyoto, Japan). Air samples were measured in two zones: the area directly associated to the hot springs (assumed spot of SO_2 and H_2S fumigation) and at 20 m distance from the CAREX hot spring, as the control site.

3.8. "Biofilm Catcher" in the CAREX Hot Spring, RISA and DGGE Analysis

Water samples were collected from the three inter-connected pools of the CAREX hot spring. About 100 mL of water were filtered from each pool (P1, P2, P3) through 0.22 μm pore size Sterivex filters. DNA extraction was performed directly from the filters using an established protocol [26]. The "Biofilm Catcher" micro-colonizer was deployed in P1 for about 30 hours. Different substrates were used in the micro colonizer in order to obtain the enrichment of biofilm forming bacteria on different solid surfaces (paper, iron, titanium, pyrite, steel, copper and glass). Total DNA was extracted also from the different materials by using a commercial kit (Power Soil DNA Isolation kit) and following the manufacturer's instruction (MoBio, Carlsbad, CA, USA). DNA concentration was evaluated on agarose gel and by NanoDrop spectrophotometer (Thermo Scientific, Wilmington, MA, USA). Ribosomal intergenic spacer analysis (RISA)-PCR was performed as previously described [27] on the DNA extracted from the different tested solid surfaces and on water samples. Denaturing gradient gel electrophoresis (DGGE) was applied on 16S rRNA to describe the hot-spring water dwelling bacterial community by applying the same procedure as reported before [28].

3.9. Adenosine Triphosphate (ATP) Based Analysis

Lumitester PD-10N, a hand held ultra-high sensitive ATP measurement instrument (Kikkoman Corporation, Japan), was used to determine the bio-burden of the various sampling locations. The bioluminescence assay was used to determine the total ATP and intracellular ATP of all the samples, as described previously [29,30]. Briefly, to determine total ATP (total microbial population), 0.1 mL sample aliquots (4 replicates) were each combined with 0.1 mL of a cell lysing detergent (benzalkonium chloride) and then incubated at room temperature for 1 min prior to the addition of 0.1 mL of luciferin-luciferase reagent. The sample was mixed, and the resulting bioluminescence was measured with a luminometer. To determine intracellular ATP (total viable microbial population), 0.1 mL of an ATP-eliminating reagent (apyrase, adenosine deaminase) was added to a 1 mL portion of the sample, mixed and allowed to incubate for 30 min to remove any extracellular ATP, after which the assay for ATP was carried out, as described above. As previously established, 1 RLU is approximately equal to 1 colony-forming unit (CFU) [30].

3.10. DNA Extraction and qPCR-Based Bacterial Quantification

A matrix of surface water and mud (50 g) was collected in triplicate from each sampling site, and the samples transported from the hot springs site to the laboratory in a cooling box at 4 °C.

Nucleic acid from each sample was extracted in duplicate with a Power Soil DNA extraction kit (MoBio), using the manufacturer's protocol. Real-time quantitative polymerase chain reaction (qPCR) assay was performed in triplicate, targeting the 16S rRNA gene to measure bacterial burden with a BioRad CFX-9600 Q-PCR Instrument. Universal bacterial primers targeting the 16S rRNA gene, 1369F (50-CGG TGA ATACGT TCY CGG-30) and modified 1492R (5'-GGW TAC CTTGTT ACG ACT T-3') were used for this analysis [31]. Each 25 μL reaction consisted of 12.5 μL of BioRad 2X iQ SYBR Green Supermix, 1 μL of each of forward and reverse oligonucleotide primer and 1 μL of template DNA. Reaction conditions were as follows: 95 °C denaturation for 3 min, followed by 35 cycles of denaturation at 95 °C for 15 s and a combined annealing and extension at 55 °C for 35 s [20].

3.11. High Density 16s Microarray (PhyloChip) Analysis

Bacterial and archaeal 16S rRNA genes were amplified from genomic DNA preparations of each sample, as described earlier. Four separate PCR reactions were performed for each sample with the use of a gradient of annealing temperatures (48 °C, 50.1 °C, 54.4 °C and 57.5 °C). A detailed explanation of the processing of the PhyloChip assay has been described elsewhere [32]. Briefly, the pooled PCR product from each sampling event was spiked with known amounts of synthetic 16S rRNA gene fragments and non-16S rRNA gene fragments. Florescent intensities from these controls were used as standards for normalization among samples. Target fragmentation, biotin labeling, PhyloChip hybridization, scanning and staining, as well as background subtraction, noise calculation and detection and quantification criteria, were performed, as previously reported [33]. An OTU was considered present in the sample when 90% or more of its assigned probe pairs for its corresponding probe set were positive (positive fraction >0.90). For each sample, all OTU intensity measurements were normalized by a scaling factor, such that the overall chip intensity was equal among each PhyloChip [20].

4. Conclusions

Very few reports on multidisciplinary field research have been published to date. We report here a successful interdisciplinary research performed in a field campaign with participation of experts in the fields of life and earth sciences. The scientific group successfully selected a hot spring system for ecological studies, such as on environmental factors, chemicals, plants, algae and microbes. The hot spring system was designated as "CAREX hot spring" with a temperature ranging from 30 °C to 98 °C. Measurements of the photochemical activity of microbial mats and higher plants with FluorCam revealed that the microbial mats seemed to be well adapted to the given conditions, but with some exception, as was observed at sampling site S4. Nevertheless, the green algae in S4 were probably stressed by high irradiance, although low pH effects cannot be excluded. The F_V/F_M values indicated that the photochemical activity of the *Juncus* plants was not seriously damaged by the environmental conditions, and the impact of temperature on photochemical performance was not observed. Other parameters obtained with FluorCam and Li-COR also confirm only minimum stress of the plants at these extreme conditions. Moreover, the plants of

Juncus sp. living in the higher water temperature (HGT, 50–60 °C) showed higher CO_2 assimilation than that measured in *Juncus* sp. living in the lower water temperature (LGT, 30–40 °C). It is surprising to notice, despite the high temperatures, how plants are able to maintain an optimal physiological state, both in terms of stomatal conductance and assimilation, with basically no decrease of functionality, as compared to the plants grown at lower temperature conditions. It is also interesting to find that in both experimental species, the emission of VOCs was stimulated by the proximity of the hot spring.

The measurements of photosynthetic performance of biofilms were different between species in different sites. While *Euglena* sp. cells showed photo-inhibition behavior, the biofilms formed by *Klebsormidium* sp. were photo-saturated.

The CO_2 level was significantly higher in the top pool, P0, and less in the other pools. The highest value of CO_2 was at site P0, which was also not vegetated, and therefore, we suggest that the increased CO_2 value was of geological origin. The H_2S concentration was also higher in P0, or 35-fold higher than in other pools, but the SO_2 remained similar in all of them.

A hand-held instrument was successfully used to measure life without visual observation by detecting *in situ* adenosine triphosphate (ATP) in all samples. This shows the advantage of using such instruments in field campaigns. The ATP results correlated well to the results obtained with quantitative polymerase chain reaction (qPCR). Some microbes attached to the different solid surfaces on the "Biofilm Catcher", but it was not possible to draw any firm conclusion on the experiment, and a longer incubation time will be necessary for better assumptions. Nevertheless, it is planned to use the "Biofilm Catcher" for long-term exposure experiments in different aquatic ecosystems in the near future.

DGGE profiles obtained from the water samples collected at pools P1, P2 and P3 showed the presence of very similar bacterial communities in the three interconnected pools, and all the sequences showed a high percentage of identity with 16S rRNA bacterial sequences retrieved from similar environments elsewhere. This was anticipated, since the environmental condition between samples was basically the same in terms of pH and temperature. However, deeper analysis of the DNA from the pools is necessary, *i.e.*, with Phylochip, which will be reported independently. The results obtained with the Phylochip shows a much more detailed distinction of the bacterial and archaea taxa, and it reveals the rare microbiota in the samples.

In this study, we have demonstrated that, using polyphasic analysis on a selected environment, the ecology of extremophiles of diverse origin can be studied simultaneously, providing more extensive understanding on the whole ecosystem, rather than focusing on individual life forms separately. Holistic approaches to study ecosystems in a wider perspective are currently lacking in the field, but are important to include in future studies. This research was only a minor effort in that direction, but much more effort is needed.

Acknowledgments

The authors are grateful to the CAREX Project (Coordination Action for Research Activities on Life in Extreme Environments, FP7-ENV-2007-1 project No. 211700) for funding the fieldtrip to Iceland. A special thanks to N. Walter, ESF and to all other participants in the Icelandic CAREX fieldtrip for fieldwork assistance. The laboratory analyses of FlowCam were supported by a project of the Ministry of Education, Youth and Sports of the Czech Republic, no. LM2010009 CzechPolar—Czech polar stations: Construction and logistic expenses, by project Creating of the Working Team and Pedagogical Conditions for Teaching and Education in the Field of Polar Ecology and Life in Extreme Environment, reg. No. CZ.1.07/2.2.00/28.0190, co-financed by the European Social Fund and the state budget of the Czech Republic and as a long-term research development project no. RVO 67985939. Oxymetric analysis were supported by a Spanish Ministry of Economy and Competitive CGL2011/02254 Grant. F. Mapelli was supported by Università degli Studi di Milano, European Social Found (FSE) and Regione Lombardia (contract "Dote Ricerca").

References

1. Brock, T.D. *Thermophilic Microorganisms and Life at High Temperatures*; Springer-Verlag: New York, Heideldberg, Berlin, USA ,1978;.
2. Madigan, M.T.; Martinko, J.M.; Dunlap, P.V.; Clark, D. *Brock Biology of Microorganisms*; Pearson/Benjamin Cummings: San Francisco, CA, USA, 2009;.
3. Kristjansson, J.K.; Hreggvidsson, G.O. Ecology and habitats of extremophiles. *World J. Microbiol. Biotechnol.* **1995**, *11*, 17–25.
4. Baldantoni, D.; Ligrone, R.; Alfani, A. Macro- and trace-element concentrations in leaves and roots of Phragmitesaustralis in a volcanic lake in Southern Italy. *J. Geochem. Explor.* **2009**, *101*, 166–174.
5. Prieur, D.; Erauso, G.; Jeanthon, C. Hyperthermophiliclifa at deep-sea hydrothermal vents *Planet. Space Sci.* **1995**, *43*, 115–122.
6. Kristjansson, J.K.; Stetter, K. Thermophilic bacteria. In *Thermophilic Bacteria*; CRC Press: Boca Raton, FL, USA, 1992; pp. 1–18.
7. Skirnisdottir, S.; Hreggvidsson, G.O.; Hjorleifsdottir, S.; Marteinsson, V.T.; Petursdottir, S.K.; Holst, O.; Kristjansson, J.K. Influence of sulfide and temperature on species composition and community structure of hot spring microbial mats. *Appl. Environ. Microbiol.* **2000**, *66*, 2835–2841.
8. Ellis-Evans, C.; Walter, N. Coordination action for research activities on life in extreme environments–The CAREX project. *J. Biol. Res. Thessalon.* **2008**, *9*, 11–15.
9. Gomez, F.; Walter, N.; Amils, R.; Rull, F.; Klingelhofer, A.K.; Kviderova, J.; Sarrazin, P.; Foing, B.; Behar, A.; Fleischer, I.; Parro, V.; *et al.* Multidisciplinary integrated field campaign to an acidic Martian Earth analogue with astrobiological interest: Rio Tinto. *Int. J. Astrobiol.* **2011**, *10*, 291–305.
10. Lopez-Archilla, A.I.; Marin, I.; Amils, R. Microbial community composition and ecology of an acidic aquatic environment: The Tinto River, Spain. *Microbial Ecol.* **2001**, *41*, 20–35.
11. Kviderova, J. Photochemical performance of the acidophilic red alga *Cyanidium* sp. in a pH gradient. *Origins Life Evol. B* **2012**, *42*, 223–234.

12. Schreiber, U.; Bilger, W.; Neubauer, C. Chlorophyll fluorescence as a nonintrusive indicator for rapid assesment of *in vivo* photosynthesis. In *Ecophysiology of Photosynthesis*; Schulze, E.D., Caldwell, M.M., Eds.; Springer-Verlag: Berlin, Heildelberg, New York, NY, USA, 1995; pp. 47–70.
13. Monson, R.K.; Holland, E.A. Biospheric trace gas fluxes and their control over tropospheric chemistry. *Annu. Rev. Ecol. Sys.* **2001**, *32*, 547–276.
14. Medori, M.; Michelini, L.; Nogues, I.; Loreto, F.; Calfapietra, C. The Impact of Root Temperature on Photosynthesis and Isoprene Emission in Three Different Plant Species. *Sci. World J.* **2012**, doi: 10.1100/2012/525827.
15. Eilers, T.; Schwarz, G.; Brinkmann, H.; Witt, C.; Richter, T.; Nieder, J.; Koch, B.; Hille, R.; Hansch, R.; Mendel, R.R. Identification and biochemical characterization of Arabidopsis thaliana sulfite oxidase. A new player in plant sulfur metabolism. *J. Biol. Chem.* **2001**, *276*, 46989–46994.
16. Lee, H.-F.; Yang, T.F.; Lan, T.F.; Song, S.-R.; Tsao, S. Fumarolic Gas Composition of the Tatun Volcano Group, Northern Taiwan. *TAO* **2005**, *16*, 843–864.
17. Kristmannsdottir, H.; Sigurgeirsson, M.; Armannsson, H.; Hjartarson, H.; Olafsson, M. Sulfur gas emissions from geothermal power plants in Iceland. *Geothermics* **2000**, *29*, 525–538.
18. Rennenberg, H. The fate of excess sulphur in higher plants. *Annu. Rev. Plant Physiol.* **1984**, *35*, 121–153.
19. Gich, F.; Janys, M.A.; Konig, M.; Overmann, J. Enrichment of previously uncultured bacteria from natural complex communities by adhesion to solid surfaces. *Environ. Microbiol.* **2012**, *14*, 2984–2997.
20. Krebs, J.; Vaishampayan, P.; Probst, A.J.; Tom, L.; Marteinsson, V.; Andersen, G.L.; Venkateswaran, K. Microbial community structures in Icelandic hot springs systems revealed by PhyloChip G3 analysis. *Int. Soc. Microb. Ecol. J.* **2013**, in press.
21. Bohlar-Nordenkampf, H.R.; Long, S.P.; Baker, N.R.; Öquist, G.; Schreiber, U.; Lechner, E.G. Chlorophyll fluorescence as a probe of the photosynthetic competence of leaves in the field: A review of current instrumentation. *Funct. Ecol.* **1989**, *4*, 497–514.
22. Maxwell, K.; Johnson, G.N. Chlorophyll fluorescence—A practical guide. *J. Exp. Bot.* **2000**, *51*, 659–668.
23. Roháček, K. Chlorophyll fluorescence parameters: The definitions, photosynthetic meaning and mutual relationship. *Photosynthetica* **2002**, *40*, 13–29.
24. Souza-Egipsy, V.; Altamirano, M.; Amils, R.; Aguilera, A. Photosynthetic performance of phototrophic biofilms in extreme acidic environments. *Environ. Microbiol.* **2011**, *13*, 2351–2358.
25. Edwards, G.E.; Walker, D.A. *C3, C4 Mechanisms and Cellular and Environmental Regulation of Photosynthesis*; Blackwell Sci. Pub.: Oxford, UK, 1983;.
26. Borin, S.; Brusetti, L.; Mapelli, F.; D'Auria, G.; Brusa, T.; Marzorati, M.; Rizzi, A.; Yakimov, M.; Marty, D.; de Lange, G.J.; *et al.* Sulfur cycling and methanogenesis primarily drive microbial colonization of the highly sulfidicUrania deep hypersaline basin. *Proc. Natl. Acad. Sci.USA* **2009**, *106*, 9151–9156.

27. Daffonchio, D.; Cherif, A.; Brusetti, L.; Rizzi, A.; Mora, D.; Boudabous, A.; Borin, S. Nature of polymorphisms in 16S–23S rRNA gene intergenic transcribed spacer fingerprinting of Bacillus and related genera. *Appl. Environ. Microbiol.* **2003**, *69*, 5128–5137.
28. Marasco, R.; Rolli, E.; Ettoumi, B.; Vigani, G.; Mapelli, F.; Borin, S.; Abou-Hadid, A.F.; El-Behairy, U.A.; Sorlini, C.; Cherif, A.; *et al.* A drought resistance—Promoting microbiome is selected by root system under desert farming. *Plos One* **2012**, *7*, e48479.
29. Venkateswaran, K.; Hattori, N.; La Duc, M.T.; Kern, R. ATP as a biomarker of viable microorganisms in clean-room facilities. *J. Microbiol. Meth.* **2003**, *52*, 367–377.
30. La Duc, M.T.; Osman, S.; Venkateswaran, K. Comparative analysis of methods for the purification of DNA from low biomass samples based on total yield and conserved microbial diveristiy. *J. Rapid Meth. Aut. Microbiol.* **2009**, *17*, 350-368.
31. Suzuki, M.T.; Taylor, L.T.; DeLong, E.F. Quantitative analysis of small-subunit rRNA genes in mixed microbial populations via 5'-nuclease assays. *Appl. Environ. Microbiol.* **2000**, *66*, 4605–4614.
32. Wilson, K.H.; Wilson, W.J.; Radosevich, J.L.; DeSantis, T.Z.; Viswanathan, V.S.; Kuczmarski, T.A.; Andersen, G.L. High-density microarray of small-subunit ribosomal DNA probes. *Appl. Environ. Microbiol.* **2002**, *68*, 2535–2541.
33. Flanagan, J.L.; Brodie, E.L.; Weng, L.; Lynch, S.V.; Garcia, O.; Brown, R.; Hugenholtz, P.; DeSantis, T.Z.; Andersen, G.L.; Wiener-Kronish, J.P.; *et al.* Loss of bacterial diversity during antibiotic treatment of intubated patients colonized with Pseudomonas aeruginosa. *J. Clin. Microbiol.* **2007**, *45*, 1954–1962.

Molecular Mechanisms of Survival Strategies in Extreme Conditions

Salvatore Magazù, Federica Migliardo, Miguel A. Gonzalez, Claudia Mondelli, Stewart F. Parker and Beata G. Vertessy

Abstract: Today, one of the major challenges in biophysics is to disclose the molecular mechanisms underlying biological processes. In such a frame, the understanding of the survival strategies in extreme conditions received a lot of attention both from the scientific and applicative points of view. Since nature provides precious suggestions to be applied for improving the quality of life, extremophiles are considered as useful model-systems. The main goal of this review is to present an overview of some systems, with a particular emphasis on trehalose playing a key role in several extremophile organisms. The attention is focused on the relation among the structural and dynamic properties of biomolecules and bioprotective mechanisms, as investigated by complementary spectroscopic techniques at low- and high-temperature values.

Reprinted from *Life*. Cite as: Magazù, S.; Migliardo, F.; Gonzalez, M.A.; Mondelli, C.; Parker, S.F.; Vertessy, B.G. Molecular Mechanisms of Survival Strategies in Extreme Conditions. *Life* **2012**, *2*, 364-376.

1. Introduction

Extreme environments are widely spread on Earth, encompassing very different regions at every altitude and latitude, such as deserts, volcanoes, seafloors and mountains; analogously, very different forms of life grow and evolve by refining a wide range of survival strategies depending on stress factors, such as temperature, pressure and pH [1]. As a consequence, extremophiles, organisms living in extreme environments, are classified, for example, as hyperthermophiles and thermophiles (very high temperature), psychrophiles (very low temperature) and halophiles (high salt concentrations) [1].

Several extremophiles belonging to different natural kingdoms share analogous strategies to survive under various stress conditions. In this review, the attention will be focused on the study by complementary spectroscopic techniques of some bioprotectant systems, such as a disaccharide, trehalose and an alcohol, glycerol playing a key role under thermal and anhydrobiotic stresses, and on their effects on some proteins, such as lysozyme and dUTPase, found also in extremophiles.

Some strains of *Thermus thermophilus* [2–4] are commonly found in marine hot springs. They grow in media containing 3% to 6% NaCl, and they produce trehalose during salt-induced osmotic stress. In the thermophilic archaeon *Sulfolobus acidocaldarius* [5,6], the pathway for the synthesis of trehalose converts the terminal unit of a glucose polymer to trehalose via maltooligosyltrehalose synthase and maltooligosyltrehalose trehalohydrolase.

In very high saline environments, halophile organisms, such as tardigrades, nematodes and the crustacean *Artemia salina*, can tolerate extreme desiccation by passing into anhydrobiosis, a state characterized by little intracellular water and no metabolic activity. In tardigrades [7,8], a breakdown of lipid and glycogen in the cavity cells and a concomitant increase in intracellular concentrations of trehalose and glycerol occurs in anhydrobiotic conditions. Furthermore, in the nematode *Aphelenchus* [9], trehalose is accumulated during desiccation in 97% relative humidity, while glycerol amounts are found after this phase. The two bioprotectant systems therefore allow the nematode to maintain its metabolic functions even when dehydration occurs.

Furthermore dry cysts of *Artemia salina* [10–13], a crustacean known as the "brine shrimp", are very resistant to extreme temperatures and, in anhydrobiosis, stop trehalose-based energy metabolism. The trehalose utilization and glycogen synthesis that occur during development of fully hydrated cysts are both blocked during desiccation.

Other examples of the accumulation and interplay of trehalose and glycerol have been demonstrated to occur in desiccation and freezing conditions, as in the arctic insect *Megaphorura arctica,* where the natural synthesis of trehalose and glycerol is related to the changes in membrane composition and to the prevention of damage from dehydration [14,15].

The biological relevance of the combined trehalose and glycerol bioprotectant effect on several organisms living in anhydrobiotic and cryobiotic conditions have promoted both experimental and simulation studies [16–22].The cofactors making the combination of trehalose and glycerol so effective in the protein protection under stress conditions has been determined by focusing on the molecular interactions between the two systems. Trehalose and glycerol have been demonstrated to create an environment around proteins that is able to improve their thermal stability and to control their dynamics on the pico- and nano-second timescale. In such a way, the two bioprotectant systems are capable to modulate both the extent of the protein atomic mean square displacements and the onset of the dynamical transition [16,17]. A non-Debye relaxation dynamic, as a result of the combination of the effects of confinement and mixing of the two constituents, has been revealed, as well as an increase of the non-exponential character of the structural relaxation [17,18]. Furthermore, enzymes embedded in mixtures of glycerol and trehalose with various compositions showed longer deactivation times and smaller mean square displacements [19–21]. Finally, the antiplasticizing effect of glycerol on trehalose has been probed by dielectric studies [22].

Several proteins have shown an extraordinary capability to adapt their conformations and motions to exert their biological functions even under stress conditions. Among them, lysozyme is a well known protein that has been extensively studied by theoretical, experimental and simulation methods [23–26] because of its properties that make it a model protein to study more complex biomolecules as those found in extremophiles. It has been also pointed out that in lysozme water solutions, ordered and disordered hydration sites extended over the protein surface, suggesting the presence of a dynamic hydration layer with ionic "flip-flop" occurring between bound waters [24–26]. Small Angle Neutron Scattering (SANS) measurements revealed that the average interparticle distance increases in lysozyme unsaturated solutions, due to the increased interaction between molecules, progressed as the salt concentration decreased, while in supersaturated solutions, crystallization processes are activated [26].

Furthermore, lysozyme is responsible for breaking down the polysaccharide walls of many kinds of bacteria, so providing some protection against infection, and it is also a cold-adapted protein. It has been shown that lysozyme from the insect *Manduca sexta* possesses a higher content of α-helix secondary structure compared to that of hen egg white lysozyme. In addition, the *M. sexta* lysozyme enzymatic activity is higher, in the range of 5 °C−30 °C [27].

One of the few protein factors essential in both the maintenance of stable genetic information and the strict control of the nucleotide pools is dUTPase. dUTPase has been also isolated by hyperthermophilic archaeon *Thermococcus onnurineus* NA1 and in the archaeon *Pyrococcus furiosus*. In *P. furiosus*, a thermostable enzyme has been found to be a multimer of two discrete proteins, P45 and P50, the first one converting dUTP to dUMP and inorganic pyrophosphate. Archaeal dUTPases may play an essential role in preventing dUTP incorporation and inhibition of DNA synthesis by family B DNA polymerases [28].

The present review shows a plethora of spectroscopic data collected on the binary systems trehalose/water mixtures and trehalose/glycerol mixtures, as well as on the ternary systems trehalose/lysozyme/water and trehalose/dUTPase/water in order to elucidate the molecular mechanisms allowing extremophiles to survive under stress conditions.

2. Bioprotection Mechanisms and Extreme Conditions

2.1. Cryobiotic and Cryptobiotic Effects of Trehalose

From the molecular point of view, the manifold aspects of the bioprotective function of trehalose, which can explain its ubiquity, have been investigated in deep detail by using complementary spectroscopic techniques covering very wide space and time ranges [29–38]. The whole body of the collected data pointed out the fundamental role played by the interaction of trehalose with water.

The study of the structural properties of trehalose water mixtures highlighted that water molecules are arranged in the presence of trehalose in a particular configuration, which avoids ice formation, so preserving biomolecules from damage due to freezing and cooling. Neutron diffraction, Raman spectroscopy and Inelastic Neutron Scattering findings [29–32], shown in Figure 1, revealed that the addition of trehalose, with respect to the other disaccharides, completely destroys the tetrahedral intermolecular network of water, which by lowering the temperature would give rise to ice. In the vibrational spectrum of liquid water, one can distinguish the existence of an isosbestic point in the isotropic spectrum of pure water allowing the decomposition of each spectrum into an ''open'' contribution, attributed to the O–H vibration in tetrabonded H_2O molecules that have an ''intact bond", and a ''closed" contribution, corresponding to the O–H vibration of H_2O molecules that have a not fully developed hydrogen bond (distorted bond). One can observe that for the same concentration, the integrated area of the ''open" band is smaller in the trehalose aqueous solution. This allows us to state that a more marked destructuring effect occurs in the presence of trehalose, rather than in the presence of sucrose or maltose [29,30,32]. As a confirmation, neutron diffraction results confirm the changes induced by disaccharides on water tetrahedral structure. In fact, the peak at 4.5 Å in $g_{OO}(r)$ of pure water, which is associated with the "degree of tetrahedrality", in the distribution function of trehalose/H_2O mixture at a concentration value corresponding to 40 H_2O molecules for each trehalose molecule at $T = 300$ K is absent, and the general trend is significantly distorted [20]. Uchida and coworkers [33] detecting freeze-fractured replica images of disaccharide (trehalose, sucrose and maltose) solutions using a field-emission type transmission electron microscope, confirmed that trehalose molecules have a greater inhibitory effect of sucrose on the growth of ice crystals, while Furuki [34] observed that aqueous trehalose has a larger amount of unfrozen water content in comparison with the other disaccharide mixtures and interpreted their different degree of anti-freeze effects in view of the molecular structure of the disaccharides, concluding that the aqueous unfrozen behavior induced by the presence of trehalose depends on the position of the glycosidic linkage between the two constituent units.

On the other hand, the study of the dynamical properties of trehalose water mixtures (Figure 1) has shown that the diffusion of water is strongly affected by trehalose and that trehalose and water form a unique entity, creating a rigid environment where biomolecules can be protected [35–38]. More specifically, Quasi Elastic Neutron Scattering results revealed that the diffusion coefficient of water in the presence of trehalose is similar to that of pure water at lower temperature, so showing that trehalose, besides imposing an order on the tetrahedral hydrogen bond network of water, significantly slows the dynamics of water. The higher slowing down effect of the diffusive dynamics observed for trehalose is evidently linked to its extraordinary capability to "switch off" metabolic functions [36,37]. Furthermore, the elastic intensity and the mean square displacement behaviors of trehalose water mixtures as a function of temperature revealed that a higher onset temperature value for trehalose, as compared to the other disaccharides together with a lower fragility of trehalose water mixtures [35,38]. It is possible to conclude that the trehalose-water system is more "rigid" than maltose-water and sucrose-water systems. From this analysis [29–38], it clearly emerges that trehalose and its water mixtures are characterized with respect to the other disaccharides and their mixtures by a superior structural resistance to thermal stress, which allows them to create a more rigid shell to protect biological structures.

Figure 1. Structural and dynamic properties of trehalose/water mixtures [29–32,35–38].

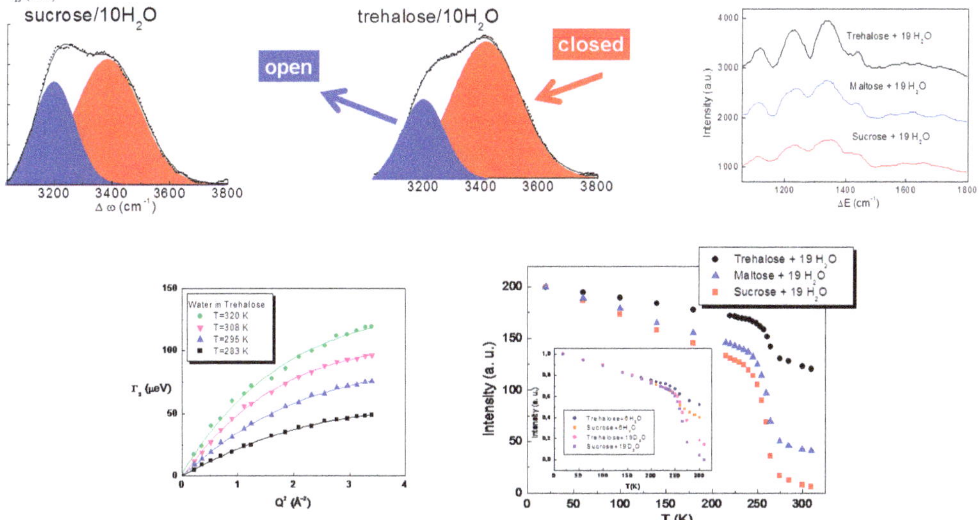

2.2. Increased Bioprotective Effectiveness of Trehalose/Glycerol Mixtures

With the aim to investigate the different bioprotective effectiveness of trehalose and trehalose/glycerol (T/G) mixtures, a systematic study on mixtures at different glycerol concentration values in trehalose (0%, 1.25%, 2.5%, 5%, 7.5% and 10% by weight) has been performed. The goal of these experiments was to investigate the vibrational, relaxational and diffusive dynamics of T/G mixtures by complementary neutron scattering techniques [39–42].

The vibrational spectral region from 0 cm^{-1} to 2500 cm^{-1}, shown in Figure 2, has been investigated by Inelastic Neutron Scattering [39] in order to understand the molecular mechanisms of the trehalose-glycerol interactions at different glycerol content. By the analysis of both the intramolecular and intermolecular motions of pure trehalose and of T/G mixtures, it emerges that at the glycerol content of 2.5%, the hydrogen bonded network strength of trehalose is mostly affected by the presence of glycerol, while a higher amount of glycerol does not have remarkable effects.

Furthermore, the relaxational behavior of T/G mixtures revealed the presence of an excess vibrational contribution at low energy by varying concentration. The R_1 fragility parameter has been evaluated in order to take into account the relative weight between the relaxational to vibrational contribution [40–42], showing that both at low and high temperature values, a minimum at the glycerol content of 2.5%, and then revealing a stronger character of the T/G mixtures at this concentration value. In addition, the decrease of the elastic intensity as a function of Q^2 for the T/G mixture with a glycerol content of 2.5% (see Figure 2) is less marked than pure trehalose and the other mixtures, so confirming a higher rigidity at this glycerol content. Analogously, the derived mean square displacement behavior as a function of concentration for T/G mixtures shows a minimum at the concentration value of 2.5% by weight of glycerol.

Finally, Quasi Elastic Neutron Scattering allowed us to characterize the diffusive dynamics of T/G mixtures by evaluating the translational line width behavior as a function of Q and by extracting the diffusion coefficient values. The results show that the anomaly in the dynamics observed at low frequency is still present, since a minimum in the translational line width behavior occurs at the glycerol concentration value of 2.5% (Figure 2), so revealing a slowing down of the diffusive dynamics at this concentration value. These findings show that both the local and diffusive dynamics, which are linked to the stabilizing action on biomolecules, are suppressed at a very low glycerol concentration value, suggesting that for this glycerol content, the atomic attractive interactions are the strongest among the investigated concentration values.

Figure 2. Vibrational, relaxational and diffusive properties of trehalose/glycerol mixtures [39–42].

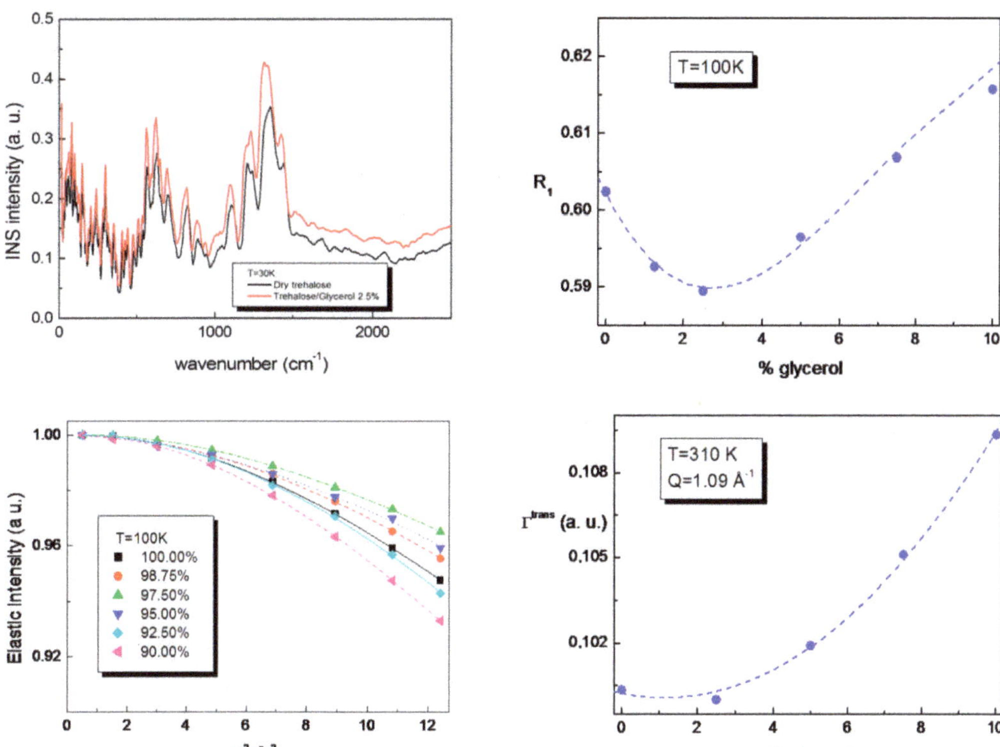

2.3. Interaction Mechanisms between Bioprotectant Systems and Biomolecules

Let us present the results obtained by neutron scattering techniques on lysozyme/trehalose/water mixtures and dUTPase/trehalose/water mixtures [43–46].

The conformational properties of lysozyme in the presence of trehalose as a function of temperature have been investigated by Small Angle Neutron Scattering in order to determine the conformational properties of the protein in presence of the disaccharide. By considering the Guinier relation [45,46], the size of the protein can be evaluated by allowing us to extract the gyration radius of lysozyme/D_2O/trehalose solutions at different temperature values. The R_g values remain almost constant (16.2 Å at T = 310K and R_g = 16.4 Å at T = 333K) even when temperature increases, this thermal change inducing in absence of trehalose a conformational change in lysozyme (Figure 3). This result emphases the stabilizing effect of trehalose on lysozyme and the capability of trehalose to significantly inhibit the swelling of lysozyme induced by thermal stress.

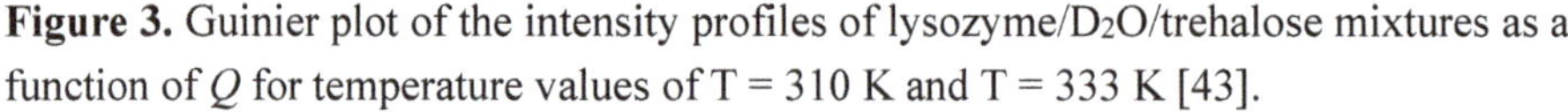

Figure 3. Guinier plot of the intensity profiles of lysozyme/D_2O/trehalose mixtures as a function of Q for temperature values of T = 310 K and T = 333 K [43].

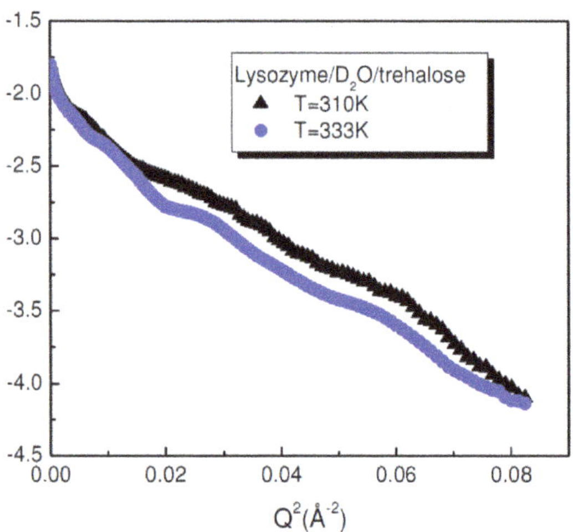

On the other hand, the dynamic properties of dUTPase protein immersed in a trehalose matrix have been investigated in order to study the effect of the host solvent on the protein dynamics. As shown in Figure 4, where the viscosity of the trehalose/H_2O mixtures is plotted as a function of the local mean square displacement of the D_2O-hydrated dUTPase/trehalose system, a linear relationship between the solvent, composed by trehalose and water, and the mean square displacement of hydrated dUTPase/trehalose system is verified. This result is a signature of a strong coupling between protein and the surrounding matrix, showing that a correlation exists between the protein dynamics and the viscosity of the surrounding environment [46].

Figure 4. Linear dependence of viscosity of the trehalose/H_2O mixture on the local mean square displacement of the hydrated dUTPase/trehalose system [44–46].

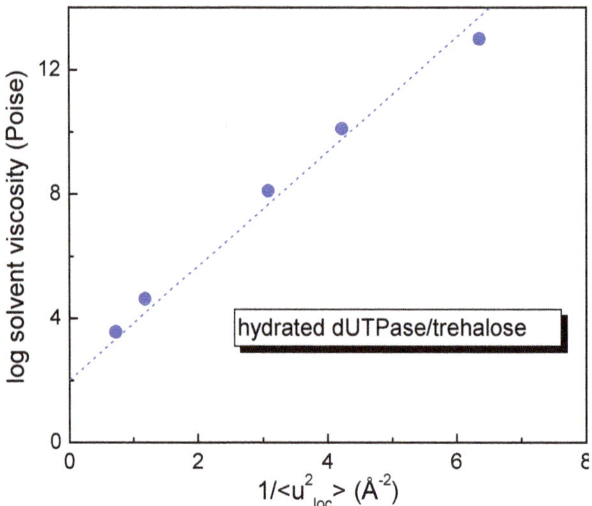

3. Experimental Section

During the last year, several complementary techniques have been used to get the spectroscopic findings shown in the present review article. Here, we present an overview of the experimental details of the performed measurements.

Elastic Neutron Scattering measurements have been carried out across the glass transition temperature values by using the backscattering spectrometer IN13 at the Institute Laue Langevin (ILL, Grenoble, France). The IN13 main characteristics is the relatively high energy of the incident neutrons (16 meV), which makes it possible to span a wide range of momentum transfer Q (≤ 5.5 Å^{-1}) with a very good energy resolution (~8 μeV). Therefore, neutron scattering experiments on IN13 provide information on the motions of the sample hydrogens in a space-time window of 1 Å and 0.1 ns given by its scattering vector modulus, Q, range and energy resolution, and allow us to characterize both flexibility (obtained from the fluctuation amplitudes) and rigidity (obtained from how fluctuations vary with temperature and expressed as a mean environmental force constant).

Inelastic Neutron Scattering measurements have been performed by using the TOSCA indirect geometry time-of-flight spectrometer at the ISIS Pulse Neutron Facility (Rutherford Appleton Laboratory, Oxford, UK). The high energy resolution of TOSCA ($\Delta E/E \approx 1.5\%$–2% for energy transfers up to several hundred meV) coupled with the high intensity of the ISIS source makes TOSCA ideal for studying the dynamics of water and aqueous mixtures below 2000 cm^{-1} (250 meV). TOSCA has revealed itself to be very effective in providing detailed results combined to optical spectroscopic techniques, such as Raman spectroscopy because of its design associating a single momentum transfer with each energy transfer.

Quasi Elastic Neutron Scattering experiments were carried out by using the OSIRIS and IRIS spectrometers at the ISIS Facility (Rutherford Appleton Laboratory, Oxford, UK) and by using the IN4 and IN6 spectrometers at the Institute Laue Langevin (ILL, Grenoble, France). OSIRIS, situated on the N6(B) beam line at ISIS, is an inverted geometry time-of-flight instrument such that neutrons scattered by the sample are energy analyzed by means of Bragg scattering from large-area crystal-analyzer array. It can be used as either a high-resolution, long-wavelength diffractometer or for high-resolution quasi/inelastic neutron scattering spectroscopy. The configuration of OSIRIS used for the INS measurements was: scattering angle range of $11° < 2\theta < 55°$, PG004 graphite with a momentum transfer range of 0.7Å$^{-1} < Q < 3.6$ Å^{-1} and energy resolution of 99 μeV (FWHM). IRIS, which is also an inverted geometry spectrometer, has been used in the high resolution configuration, *i.e.*, graphite 002 and mica 006 analyzer reflections, to measure sets of QENS spectra covering a Q,ω-domain extending from $\hbar\omega = -0.3$ to 0.6 meV and Q = 0.5 to 1.8 Å^{-1}. The used detectors give a mean energy resolution of $\Gamma = 8$ μeV (HWHM) as determined by reference to a standard vanadium plate. The IN4 spectrometer is a time-of-flight spectrometer used for the study of excitations in condensed matter, and it was configured for the measurements with an incident wavelength of 2.96 Å and an energy resolution of 450 μeV. The IN6 spectrometer is a time-of-flight spectrometer designed for quasi-elastic and inelastic scattering for incident wavelengths in the range of 4 to 6 Å. The incident wavelength used for the measurements was 5.12 Å with an energy resolution of 50 μeV.

Small Angle Neutron Scattering experiments have been performed by using the LOQ spectrometer at the ISIS Facility (Rutherford Appleton Laboratory, Oxford, UK) for different contrast values (20 H_2O-80% D_2O, 80 H_2O-20% D_2O and 100% D_2O). The contrast variation technique collecting data at different D_2O/H_2O molar ratio has been employed in order to determine the protein scattering density length. The Q-range covered by the LOQ spectrometer in this experiment is from 0.007 $Å^{-1}$ to 0.287 $Å^{-1}$. Incoming neutrons are monochromatized by a mechanical velocity selector with variable wavelength from 2.2 to 10.0 Å, the wavelength resolution (FWHM) being $8\% < \Delta\lambda/\lambda < 18\%$.

4. Conclusions

All the studies performed on trehalose water mixtures clearly support the hypothesis of a privileged water-disaccharide interaction. Both the results dealing with structural and dynamic properties suggest that on the one hand, trehalose binds more strongly to water molecules, so disrupting their tetrahedral configuration arrangements and slowing down their mobility, and on the other hand, trehalose shows a larger structural resistance to temperature changes and a higher "rigidity" in comparison with its homologues.

The physical picture obtained from the studies performed on trehalose water mixtures shows that the higher bioprotectant effectiveness of trehalose in comparison with the other disaccharides is due to the combined effect of different co-factors. What emerges is that trehalose, besides modifying significantly the structural and dynamical properties of water, forms with H_2O a less fragile entity able to encapsulate biological structures and to protect them in a more rigid environment. Due to the fundamental role of water in living organisms as its major component and as the prerequisite for proteins and cells to exert their biological functions, the elucidation of the bioprotectant-water interactions can explain the bioprotective functions under the harsh conditions encountered in extreme environment.

The whole body of data on trehalose/glycerol mixtures at different glycerol content support the hypothesis that in this small glycerol concentration range, the T/G matrix forms a stronger hydrogen bonding network with respect to that of pure trehalose and to what happens for higher glycerol concentration values. The signature of a strengthening of the hydrogen bonded network created by trehalose and glycerol is recognizable in the trends followed by all the determined physical quantities. More specifically, the increased rigidity revealed by the dynamic features confirms that the hydrogen bonded interactions are rearranged in a stronger network as a consequence of the addition of glycerol. The molecular origin of this anomalous behavior can be linked to the registered minima in the mean square displacement, in the R_l parameter and in the translational line width, which clearly signals the presence of a not-ideal mixing process.

Furthermore, it is to be observed that the inelastic data were collected at very low temperature, therefore, they can have interesting implications about the described combined role of the trehalose/glycerol system as a cryo- and lyo-protectant system. The occurrence of large amounts of trehalose and small amounts of glycerol in several organisms capable of activating a cryoprotective dehydration process can find physical elucidation in the present findings.

It is known that the coupling between the dynamics of the host medium with that of the protein may explain the bioprotectant function. The results on lysozyme and dUTPase in the presence of trehalose emphasize that proteins are complex systems, which are to be considered as dynamic systems that perform motions to execute their functions. These motions actually involve the atoms not just of the biological structure itself, but also of the surrounding medium with which a coupling exists. Therefore, depending on the circumstances, the protein can be considered "slaved" or "sequestered" by the host medium, which may be able to suppress its dynamics, so resulting in a retardation of denaturation processes or a slowing down of biological function, as happens in extreme conditions.

The findings on the binary bioprotectant mixtures and on the ternary bioprotectant/biomolecule systems provide precious information to explain at a molecular level the behavior of biomolecules under stress conditions. Here, since all the shown data have been collected as a function of temperature in a wide range (20 K–400 K), thermal stress plays a key role.

The neutron scattering data at very high temperatures can help in the understanding of the dynamic nature of hyperthermostable proteins due to the unraveling of the mechanism responsible for the balance between rigidity, which is related to heat resistance, and molecular fluctuations at high temperatures, which account for biological function. On the other hand, the findings at very low temperatures can support the hypothesis that cold-temperature adapted proteins from psychrophiles become more rigid, implying that enhancing flexibility can restore function. Oother useful suggestions are furnished by the data on trehalose/glycerol mixtures based on the elucidation of their interaction, this circumstance being crucial for halophile organisms.

Acknowledgments

The authors gratefully acknowledge the ILL facility for the dedicated runs on the IN4, IN6 and IN13 spectrometers and the ISIS facility for the dedicated runs on the TOSCA, LOQ, IRIS and OSIRIS spectrometers.

References

1. Rothschild, L.J.; Mancinelli, R.L. Life in extreme environments. *Nature* **2001**, *409*, 1092–1101.
2. Henne, A.; Bruggemann, H.; Raasch, C.; Wiezer, A.; Hartsch, T.; Liesegang, H.; Johann, A.; Lienard, T.; Gohl, O.; Martinez-Arias, R.; *et al.* The genome sequence of the extreme thermophile *Thermus. thermophilus*. *Nat. Biotechnol.* **2004**, *22*, 547–553.
3. Leuschner, C.; Antranikian, G. Heat-stable enzymes from extremely thermophilic and hyperthermophilic microorganisms. *World J. Microbiol. Biotechnol.* **1995**, *11*, 95–114.
4. Silva, Z.; Alarico, S.; Nobre, A.; Horlacher, R.; Marugg, J.; Boos, W.; Mingote, A.I.; da Costa, M.S. Osmotic adaptation of *Thermus. thermophilus* RQ-1: Lesson from a mutant deficient in synthesis of trehalose. *J. Bacteriol.* **2003**, *185*, 5943–5952.
5. Gueguen, Y.; Rolland, J.L.; Schroeck, S.; Flament, D.; Defretin, S.; Saniez, M.H.; Dietrich, J. Characterization of the maltooligosyl trehalose synthase from the thermophilic archaeon *Sulfolobus. acidocaldarius*. *FEMS Microbiol. Lett.* **2001**, *194*, 201–206.

6. Grogan, D.W. Exchange of genetic markers at extremely high temperatures in the archaeon *Sulfolobus. acidocaldarius*. *J. Bacteriol.* **1996**, *178*, 3207–3211.
7. Hengherr, S.; Brünner, F.; Schill, R.O. Anhydrobiosis in tardigrades and its effects on longevity traits. *J. Zool.* **2008**, *275*, 216–220.
8. Schill, R.O.; Fritz, G.B. Desiccation tolerance in embryonic stages of the tardigrade. *J. Zool.* **2008**, *276*, 103–107.
9. Browne, J.A.; Dolan, K.M.; Tyson, T.; Goyal, K.; Tunnacliffe, A.; Burnell, A.M. Dehydration-specific induction of hydrophilic protein genes in the anhydrobiotic nematode *Aphelenchus. avenae*. *Eukaryot. Cell* **2000**, *3*, 966–975.
10. Gajardo, G.M.; Beardmore, J.A. The Brine Shrimp Artemia: Adapted to Critical Life Conditions. *Front. Physiol.* **2012**, *3*, 185–192.
11. Hebert, P.D.N.; Remigio, E.A.; Colbourne, J.K.; Taylor, D.J.; Wilson, C.C. Accelerated molecular evolution in halophilic crustaceans. *Evolution* **2002**, *56*, 909–926.
12. Gajardo, G.; Beardmore, J.A. Ability to switch reproductive mode in Artemia is related to maternal heterozygosity. *Mar. Ecol. Prog. Ser.* **1989**, *56*, 191–195.
13. Dwivedi, S.N.; Diwan, A.D.; Iftekhar, M.B. Oxygen uptake in the brine shrimp Artemia in relation to salinity. *Ind. J. Fish.* **1987**, *34*, 359–361.
14. He, X.; Fowler, A.; Tonera, M.J. Water activity and mobility in solutions of glycerol and small molecular weight sugars: Implication for cryo- and lyopreservation. *Appl. Phys.* **2006**, *100*, 074702–074712.
15. Michaud, M.R.; Denlinger, D.L. Shifts in the carbohydrate, polyol, and amino acid pools during rapid cold-hardening and diapause-associated cold-hardening in flesh flies (*Sarcophaga. crassipalpis*): a metabolic comparison. *J. Comp. Physiol. B* **2007**, *177*, 753–763.
16. Busselez, R.; Lefort, R.; Guendouz, M.; Frick, B.; Merdrignac-Conanec, O.; Morineau, D. Molecular dynamics of glycerol and glycerol-trehalose bioprotectant solutions nanoconfined in porous silicon. *J. Chem. Phys.* **2009**, *130*, 214502-1–214502-8.
17. Caliskan, G.; Mechtani, D.; Roh, J.H.; Kisliuk, A.; Sokolov, A.P.; Azzam, S.; Cicerone, M.T.; Lin-Gibson, S.; Peral, I. Protein and solvent dynamics: How strongly are they coupled? *J. Chem. Phys.* **2004**, *121*, 1978–1983.
18. Seo, J.A.; Kim, S.; Kwon, H.Y.; Hwang, Y.H. The glass transition temperatures of sugar mixtures *Carbohydr. Res.* **2006**, *341*, 2516–2520.
19. Dirama, T.E.; Carri, G.A.; Sokolov, A.P. Role of hydrogen bonds in the fast dynamics of binary glasses of trehalose and glycerol: A molecular dynamics simulation study. *J. Chem. Phys.* **2005**, *122*, 114505-1–114505-8.
20. Cicerone, M.T.; Soles, C.L. Fast Dynamics and Stabilization of Proteins: Binary Glasses of Trehalose and Glycerol. *Biophys. J.* **2004**, *86*, 3836–3845.
21. Curtis, J.E.; Dirama, T.E.; Carri, G.A.; Tobias, D.J. Inertial suppression of protein dynamics in a binary glycerol-trehalose glass. *J. Phys. Chem. B* **2006**, *110*, 22953–22956.
22. Anopchenko, A.; Psurek, T.; Vanderhart, D.; Douglas, J.F.; Obrzut, J. Dielectric study of the antiplasticization of trehalose by glycerol. *Phys. Rev. E* **2006**, *74*, 031501-1–031501-10.
23. Jolles, P. Animal lysozymes c and g: An overview. *EXS* **1996**, *75*, 3–12.

24. Baker, L.J.; Hansen, A.M.F.; Rao, P.B.; Bryan, W.P. Effects of the presence of water on lysozyme conformation. *Biopolymers* **1983**, *22*, 1637–1640.
25. Svergun, D.I.; Richard, S.; Koch, M.H.J.; Sayers, Z.; Kuprin, S.; Zaccai, G. Protein hydration in solution: experimental observation by X-ray and neutron scattering. *Proc. Natl. Acad. Sci. USA* **1998**, *95*, 2267–2272.
26. Minezaki, Y.; Niimura, N.; Ataka, M.; Katsura, T. Small angle neutron scattering from lysozyme solutions in unsaturated and supersaturated states (SANS from lysozyme solutions). *Biophys. Chem.* **1996**, *58*, 355–363.
27. Sotelo-Mundo, R.R.; Lopez-Zavala, A.A.; Garcia-Orozco, K.D.; Arvizu-Flores, A.A.; Velazquez-Contreras, E.F.; Valenzuela-Soto, E.M.; Rojo-Dominguez, A.; Kanost, M.R. The lysozyme from insect (*Manduca. sexta*) is a cold-adapted enzyme. *Protein Pept. Lett.* **2007**, *14*, 774–778.
28. Hogrefe, H.H.; Hansen, C.J.; Scott, B.R.; Nielson, K.B. Archaeal dUTPase enhances PCR amplifications with archaeal DNA polymerases by preventing dUTP incorporation. *Proc. Natl. Acad. Sci. USA* **2002**, *99*, 596–601.
29. Magazu, S.; Migliardo, F.; Ramirez-Cuesta, A.J. Inelastic neutron scattering study on bioprotectant systems. J. *Royal Soc. Interface* **2005**, *2*, 527–532.
30. Magazu, S.; Migliardo, F.; Telling, M.T.F. Structural and dynamical properties of water in sugar mixtures. *Food Chem.* **2008**, *106*, 1460–1466.
31. Cesaro, A.; Magazu, V.; Migliardo, F.; Sussich, F.; Vadalà, M. Comparative study of structural properties of trehalose water solutions by neutron diffraction, synchrotron radiation and simulation. *Physica B* **2004**, *350*, E367–E370.
32. Magazu, S.; Migliardo, F.; Ramirez-Cuesta, A.J. Changes in vibrational modes of water and bioprotectants in solution. *Biophys. Chem.* **2007**, *125*, 138–142.
33. Uchida, T.; Nagayama, M.; Shibayama, T.; Gohara, K. Morphological investigations of disaccharide molecules for growth inhibition of ice crystals. *J. Cryst. Growth* **2007**, *299*, 125–135.
34. Furuki, T. Effect of molecular structure on thermodynamic properties of carbohydrates. A calorimetric study of aqueous di- and oligosaccharides at subzero temperatures. *Carbohydr. Res.* **2002**, *337*, 441–450.
35. Blazhnov, I.V.; Magazù, S.; Maisano, G.; Malomuzh, N.P.; Migliardo, F. Macro- and microdefinitions of fragility of hydrogen-bonded glass-forming liquids. *Phys. Rev. E* **2006**, *73*, 031201-1–031201-7.
36. Magazu, S.; Migliardo, F.; Telling, M.T.F. Study of the dynamical properties of water in disaccharide solutions. *Eur. Biophys. J.* **2007**, *36*, 163–171.
37. Magazu, S.; Migliardo, F.; Telling, M.T.F. alpha,alpha-Trehalose-water solutions. VIII. Study of the diffusive dynamics of water by high-resolution quasi elastic neutron scattering. *J. Phys. Chem. B* **2006**, *110*, 1020–1025.
38. Magazu, S.; Migliardo, F.; Mondelli, C.; Vadalà, M. Correlation between bioprotective effectiveness and dynamic properties of trehalose-water, maltose-water and sucrose-water mixtures. *Carbohydr. Res.* **2005**, *340*, 2796–2801.

39. Magazù, S.; Migliardo, F.; Parker, S.F. Vibrational Properties of Bioprotectant Mixtures of Trehalose and Glycerol. *J. Phys. Chem. B* **2001**, *115*, 11004–11009.
40. Magazù, S.; Migliardo, F.; Affouard, F.; Descamps, M.; Telling, M.T.F. Study of the Relaxational and Vibrational Dynamics of Bioprotectant Glass-Forming Mixtures by Neutron Scattering and Molecular Dynamics Simulation. *J. Chem. Phys.* **2010**, *132*, 184512-1–184512-9.
41. Magazù, S.; Migliardo, F.; Telling, M.T.F. Dynamics of glass-forming bioprotectant systems. *J. Non-Cryst. Sol.* **2011**, *357*, 691–694.
42. Magazù, S.; Migliardo, F.; Gonzalez, M.A.; Mondelli, C. Inelastic neutron scattering study of dynamical properties of bioprotectant solutions against temperature. *J. Non-Cryst. Sol.* **2012**, *358*, 2635–2640.
43. Magazu, S.; Migliardo, F.; Benedetto, A. Thermal behavior of hydrated lysozyme in the presence of sucrose and trehalose by EINS. *J. Non-Cryst. Sol.* **2011**, *357*, 664–670.
44. Varga, B.; Migliardo, F.; Takacs, E.; Vertessy, B.; Magazù, S.; Telling, M.T.F. Study of solvent-protein coupling effects by neutron scattering. *J. Biol. Phys.* **2010**, *36*, 207–220.
45. Vertessy, B.G.; Magazù, S.; Mangione, A.; Migliardo, F.; Brandt, A. Structure of Escherichia coli dUTPase in Solution: A Small Angle Neutron Scattering Study. *Macromol. Biosci.* **2003**, *3*, 477–481.
46. Varga, B.; Migliardo, F.; Takacs, E.; Vertessy, B.; Magazù, S. Experimental study on dUTPase-inhibitor candidate and dUTPase/disaccharide mixtures by PCS and ENS. *J. Mol. Struct.* **2008**, *886*, 128–135.

Quorum Sensing in Extreme Environments

Kate Montgomery, James C. Charlesworth, Rebecca LeBard, Pieter T. Visscher and Brendan P. Burns

Abstract: Microbial communication, particularly that of quorum sensing, plays an important role in regulating gene expression in a range of organisms. Although this phenomenon has been well studied in relation to, for example, virulence gene regulation, the focus of this article is to review our understanding of the role of microbial communication in extreme environments. Cell signaling regulates many important microbial processes and may play a pivotal role in driving microbial functional diversity and ultimately ecosystem function in extreme environments. Several recent studies have characterized cell signaling in modern analogs to early Earth communities (microbial mats), and characterization of cell signaling systems in these communities may provide unique insights in understanding the microbial interactions involved in function and survival in extreme environments. Cell signaling is a fundamental process that may have co-evolved with communities and environmental conditions on the early Earth. Without cell signaling, evolutionary pressures may have even resulted in the extinction rather than evolution of certain microbial groups. One of the biggest challenges in extremophile biology is understanding how and why some microbial functional groups are located where logically they would not be expected to survive, and tightly regulated communication may be key. Finally, quorum sensing has been recently identified for the first time in archaea, and thus communication at multiple levels (potentially even inter-domain) may be fundamental in extreme environments.

Reprinted from *Life*. Cite as: Montgomery, K.; Charlesworth, J.C.; LeBard, R.; Visscher, P.T.; Burns, B.P. Quorum Sensing in Extreme Environments. *Life* **2013**, *3*, 131-148.

1. Introduction

Quorum sensing is a type of microbial communication that regulates gene expression in high cell densities [1]. It relies on the production of signaling molecules that are released from the cell into the surrounding environment. Each cell produces these molecules constitutively, and it is when they reach a critical concentration that gene transcription is initiated. Quorum sensing is considered to be a process by which the microbial population as a whole can monitor and regulate gene expression and hence physiology (including metabolism), as the characteristics controlled by quorum sensing are unproductive when undertaken by a single cell alone [1–3].

Quorum sensing is known to be responsible for the regulation of bioluminescence, cell competency and horizontal gene transfer, virulence, motility, the formation of biofilms and the production of antibiotics and other secondary metabolites [4,5]. A number of quorum sensing systems have now been well characterized and extensively documented, such as those that regulate the production of bioluminescence in the marine bacteria *Vibrio harveyi* and *Allivibrio fischeri* [1,6]. There still remains a wide range of organisms and environments in which quorum sensing has yet to be identified or characterized, with extremophiles being one of these groups of organisms.

The extremophiles represent a variety of organisms known for their ability to survive in and adapt to "extreme" environmental conditions [7]. Though this is an anthropogenic definition, organisms in this group demonstrate the breadth of environments that life can survive in. This includes high and low pH levels, extremes of hot and cold, high-pressure levels, salinity (high and low), nutrient limitations, or combinations of the above. Short-term fluctuations in environmental parameters can also be considered as an extreme condition, as microorganisms need to be able to rapidly adapt to survive in a given niche. This is particularly relevant in an environment of interest, microbial mats, to be covered later in this review. Extremophiles have been studied intensively in recent years for their insight into environmental adaptation. Of particular interest for researchers is the potential for application of extremophilic metabolites and extracellular enzymes in industry and biotechnology as demands in these areas increase [8]. The role of quorum sensing in extreme environments is one not currently investigated in detail. The importance of quorum sensing in the adaptation of microorganisms in general, and particularly "extremophiles" to their environment has been studied in a limited number of individual organisms, however the role of quorum sensing in the extended microbial biosphere is still relatively unknown. The advent of bioinformatic technologies and extensive databases has allowed for a relative wealth of information regarding the extremophiles. As they live in such harsh conditions culturing in the laboratory is often difficult, though not impossible. Consequently, it has been largely the recent ability to perform genomic and proteomic studies on environmental samples that has allowed considerable insight into the capabilities of these microbes.

This review aims to assess our current understanding of quorum sensing in extreme environments and present the evidence for its potential role and function in these ecosystems.

2. Quorum Sensing Systems

Quorum sensing was first identified in the organisms *V. harveyi* and *A. fischeri* that were noted to produce a luminescent quality when cells reached a particular level of density. These and other Gram-negative bacteria were found to have a common genetic system that is known as the LuxI-LuxR, or AI-1, quorum sensing system. From these initial observations a second more universal sensing system was identified, though whether it is a true signaling system is a source of debate. This is the autoinducer-2 (AI-2) system [1,3,6].

The LuxI-LuxR system as observed in the *Vibrio* spp. makes use of acylated homoserine lactones (AHLs) as autoinducers. It is these molecules that are released and received in this quorum sensing system. AHL-based signaling has now been identified in more than 70 microbial species [2,3]. Interestingly while this system was first observed in the proteobacterial phylum, it has been recently identified in cyanobacteria [9] and most recently within archaea [10]. AHLs are produced by the LuxI synthase or its homologue. AHLs all contain a central ring structure that remains constant, but they differ considerably in their side-chains. These vary in length and may possess oxo or hydroxyl groups, allowing the signaling molecules to be species specific [2]. The molecules exit and re-enter the cells by either active transport or passive diffusion depending on the size of the molecule and environmental conditions.

The chemical properties of the AHL side chain (e.g., the number of carbon atoms in the alkyl side chain), has a profound impact on signaling efficiency. The stability of long chain alkyl groups (e.g., C10–C14) at elevated pH (> 8.2) is several orders of magnitude greater than that of short chain ones. This suggests that changing ratios of short and long chain sides (caused by the pH of the environment) can be used by cells to determine this physicochemical value. When extracellular concentrations of the signals reach a critical level, the AHLs bind to the *lux* box, a promoter element, resulting in transcription of the associated genes [11]. Interestingly AHLs have been shown to readily degrade under high temperatures and alkaline pH conditions, with the lactone ring coming under nucleophilic attack. This lactone ring has the capability of reforming if the pH is lowered substantially to a pH of 2.0 [12]. Longer chain AHLs appear to be more resistant to chemical degradation and as such may be utilized by microbes that live in harsher conditions [12,13]. Furthermore, the stability of long chain alkyl groups (e.g., C10–C14) at elevated pH (> 8.2) is several orders of magnitude greater than that of short chain alkyl groups. By detecting a change in the ratio between long and short chain AHLs, microorganisms may be able to determine the actual pH of their environment. This mechanism has been proposed for microbial mats that are discussed in a later section.

Unlike the signaling molecules found in the LuxI-LuxR system, the autoinducer molecules utilized in AI-2 quorum sensing system are all identical [14]. This has led to the suggestion that the AI-2 quorum sensing system is a universal system, allowing both inter- and intra-species communication. The AI-2 system is moderated by the *luxS* gene that encodes for S-ribosyl homocysteine lyase, a Fe^{2+} dependent enzyme that cleaves bonds in the S-ribosylhomocysteine (SRH) to produce the precursors to AI-2 signals. The function of the AI-2 signal varies and depends on the associated genes [5]. The autoinducer 2 or AI-2 system that utilizes furanosyl borate diesters as a messenger molecule was initially described as a bacterial "Esperanto" or universal language, due to ubiquitous nature of *luxS*, the protein that synthesizes the diesters [15]. However, this idea however has been criticized as it is uncertain whether the diesters are indeed acting as signaling molecules in all instances. The *luxS* gene is a part of a biochemical pathway that recycles S-adenosyl-L-methionine, and as such, it is possible the AI-2 molecule is merely a byproduct rather than a true signal [16].

As a generalization, Gram-negative bacteria use AHLs as autoinducers while Gram-positive bacteria use peptide-based signaling systems. The latter consists of processed peptides usually fewer than 40 amino acids long that are assembled within the cell, and then transported to the extracellular space by active transport. External sections of membrane-bound sensor proteins interact with the signal molecules, eliciting an intra-cellular response [1]. The production of these molecules has been observed to be cell density dependent and so this has become known to be a form of quorum sensing. Peptide based signaling offers an extreme advantage in that the molecules display high thermostability [17]. AHLs are subject to thermal degradation and thus a peptide-based signaling system may be advantageous in a hyperthermal environment, for example.

Some bacteria such as *Pseudomonas aeruginosa* are able to produce and respond to multiple quorum sensing signals, including species specific systems utilizing the quinolone molecule [16]. Others have been observed to use more than one quorum sensing system. *V. harveyi* for example uses a highly integrated network of three different quorum sensing systems to control bioluminescence and biofilm formation [2]. These multi-component systems appear to be limited to the *Vibrio* spp. [11] although this may be more widespread.

Although studies initially focused on those microbes capable of sending quorum sensing signals, it has been observed that a number of bacteria are unable to send signals, but are still able to receive and respond to them. *Salmonella* sp. for instance has no LuxI-LuxR homologue but do have a LuxR-like receptor, SdiA, which allows a response to cues produced by other microbes [18]. This has become known as the concept of "eavesdropping" and adds further questions to the consideration of quorum sensing as a community event, particularly in mixed species culture [2].

3. Detection of Quorum Sensing—Biosensors

In order to search for the presence of quorum sensing molecules such as AHLs or furanosyl borate diesters, biosensors are often employed. Biosensors are strains of organisms engineered to produce a measureable phenotype, e.g. luminescence or pigment production, in response to stimulus from a quorum sensing molecule [19]. Importantly biosensor strains do not produce quorum sensing molecules of their own, rather rely on exogenous sources to activate. While being useful tools for the study of quorum sensing behavior, biosensors do have drawbacks that need to be considered, particularly when examining extreme environments where little is known. There exists a wide range of biosensors for AHLs [19] and furanosyl borate diester systems [20]. These biosensors can have varying ranges of sensitivity, for instance some biosensors are better suited to detect short chain AHLs as opposed to longer chain varieties [19]. Biosensors can also be activated and inhibited by molecules that are not related to quorum sensing and these molecules could be searched for using analytical chemical techniques to confirm any putative results.

Another factor to consider before selecting a biosensor is to consider which environment the organism that is being examined is sourced from. For example, the AHL biosensor *Chromobacterium violaceum* CV026 produces a purple pigment in response to AHL stimulus [21]. *C. violaceum* CV026 has been known to be quite sensitive to salt conditions and therefore modifications to biosensor protocols might be considered when examining halophilic organisms [22]. Extreme environmental conditions can also complicate extraction procedures for quorum sensing molecules; for example *Natronococcus occultus* thrives in alkaline saline conditions, and this alkalinity would also contribute to short lifespans of AHL molecules. In order to extract AHLs from alkaline conditions, acidification steps can be pursued to re-form the molecules [12,23].

4. Quorum Sensing in Specific Extremophile Groups

4.1. Halophiles

Halophiles are organisms that thrive in environments with high salt concentrations. In addition to salinity, due to the dynamics of alkaliphilic (high pH) environments, the two conditions of salinity and alkalinity are often seen in tandem [24]. The halophiles are represented by microbes from all three domains of life: bacteria, archaea and eukarya. While the non-halophiles are able to grow in the absence of salt and in minimal concentrations, the halophiles prefer environments containing approximately 2.5 M salt concentrations and require these salts for growth [25]. Some halophiles are also capable of surviving in high temperatures and the alkaliphiles are considered to be those which require a pH level of >9 for survival [7,26].

The production of AHLs was investigated in *Halomonas* isolates from various locations and it was found that all four species examined were able to produce these molecules (Table 1) [22]. Although the study has proven successful in identifying the production of AHLs in culture, nothing is yet known about the purpose of the signaling molecules in these microbial species. The authors suggested a role in the formation of biofilms and exopolysaccharide (EPS) production, and in fact, EPS is known to protect cells from desiccation and enhances communication through formation of specific channels [27,28]. Of particular interest in these results in the observation that the bacteria could each produce more than one type of AHL and that, with the exception of *Halomonas ventosae*, they all produced the same AHLs.

The moderately halophilic *Halobacillus halophilus* is a Gram-positive bacterium isolated from a salt marsh on the coast of Germany. It has become a model organism for studying salt adaptation because of its strict Cl^- dependence. Growth and cell division of *H. halophilus* is entirely dependent upon the presence of chloride ions with optimal growth occurring at 0.8–1.0 M Cl^-. Flagella production, motility and a number of other physiological processes were also shown to be dependent upon the anion concentration [29]. The *luxS* operon in *H. halophilus* codes for a number of molecules involved in the production of putative AI-2 signals. The expression of the operon is growth-phase dependent and highly reliant upon the presence of Cl^- ions [29]. Maximum expression was observed during mid-exponential phase in 2.0 M NaCl. This is the first recorded demonstration of LuxS as a chloride dependent system. A potential link has been suggested between the LuxS signaling system and cell motility [30], though further work is needed to confirm.

Finally, eukaryotic algae have also been shown to engage in the quorum sensing process in saline environments. The micro-algae *Dunaliella salina* is a eukaryote found in hypersaline salterns [31], and it has been shown to produce quorum quenching molecules that inhibit the function of quorum sensing signals [32].

4.2. Acidophiles & Heavy Metal Resistant Microbes

The extreme acidophiles are a group largely investigated for their ability to withstand high concentrations of heavy metals as this has considerable industrial applications. One member of this group, *Ferroplasma acidarmanus* Fer1 is an acidophilic archaeon isolated from the Iron Mountain mine in California. It is typically found in mixed-species biofilm formations, though it will often dominate these by up to 85% of cellular mass. It is considered that the biofilm mode of life confers a competitive advantage in these environments, allowing the microbes to remain sheltered from the acidic conditions [37]. The genome of *F. acidarmanus* was analyzed for potential quorum sensing genes and was found to contain many genes related to biofilm formation and motility, though a direct functional link still needs to be made, and no LuxR or LuxS homologues were identified.

Distinct morphological changes in biofilm formations suggest a type of cellular response system. These changes were observed in single-species culture, strongly suggesting a role for intra- but not inter-species cell signaling [37].

Table 1. Summary of findings of AHL production in *Halomonas* species.

Bacterial Species	Optimum Salt Concentration	pH Growth Range	Quorum Sensing
Halomonas eurihalina	7.5%	7.2	Production of three different AHLs observed on Thin Layer Chromatography (TLC) [33]
Halomonas maura	1%–15% Salt required for growth Optimum growth at 7.5%–10%	6–9 Optimum growth at 7.2	Activation of indicator strain suggesting AHL production. Production of three different AHLs similar to those of *H. eurihalina* observed on TLC [34]
Halomonas ventosae	3%–15% Salt required for growth	6–10	Very low levels of AHLs detected by TLC [35]
Halomonas anticariensis	0.5%–15% Optimum growth at 7.5%	6–9	Activation of indicator strain on culture media. AHL production of the same 3 AHLs as *H. eurihalina* and *H. maura* in significantly larger amounts and is clearly growth phase dependent [36]

The genome of the extremely acidophilic bacterium *Acidithiobacillus ferrooxidans* contains the divergently orientated genes *afeI* and *afeR* that are linked and predicted to produce proteins similar to the LuxI-LuxR proteins. This microbe prefers environments with a pH range of 1–2 and is often associated with bioleaching operations. Theses genes were initially identified by bioinformatics due to their high similarity to the Lux proteins [4] and have since been studied in detail. In addition to tolerating acidic environments, *A. ferrooxidans* is highly tolerant of heavy metals. Cu^{2+} is a trace element essential to life however it can be toxic in high concentrations [38]. Studies have investigated the effect of both high concentrations of Cu^{2+} and synthetic furanones on the expression of these genes and their products. Furanones are molecules known to interrupt quorum sensing [39]. Results demonstrated significant reduction in the tolerance of Cu^{2+} ions when furanone compounds were present, strongly suggesting that quorum sensing plays a vital role in heavy-metal resistance [38].

To support these observations the production of AHLs by *A. ferrooxidans* was also assessed in this study. It had been shown in previous studies that *A. ferrooxidans* is capable of producing a diverse range of AHLs and in this instance a range of long-chain AHLs were detected that are known to be stable under acidic conditions [13]. The presence of furanones significantly reduced the amount of these AHLs produced by the bacterium [38]. *A. ferrooxidans* has now been shown to possess a second putative quorum sensing system by genomic analysis. An orthologue of *hdtS* that encodes AHL synthase in *Pseudomonas fluorescens* was identified and termed *act*. Its similarity to known genes suggests that it plays a role in membrane synthesis and fluidity. It is suggested that the two individual quorum sensing systems regulate the ability of the microbes to utilize different energy sources [40].

Acidithiobacillus thiooxidans and *Leptospirillum ferrooxidans* are both acidophilc microbes used in biomining. Studies have attempted to identify quorum sensing systems in these organisms due to their close phylogenetic and functional relationship to *A. ferrooxidans*. It was found that *A. thiooxidans* produces AHLs while *L. ferrooxidans* does not. On genomic analysis however, a LuxI-LuxR homologue was found in *L. ferrooxidans*, composed of two divergent genes, *lttI* and *lttR*. The putative proteins produced by these genes have a high level of similarity to known quorum sensing molecules in the bacterium *Geobacter uraniireducens* [41]. Similarities have also been drawn to *Eschericia coli* genes involved in cell growth, biofilm formation and motility including chemotaxis and flagellum production [42].

The human body, with a range of acidic, oxic and anoxic conditions, could certainly be considered an "extreme" environment, and provides many challenges for microbes. Vibrio cholerae is the bacterium that causes the disease cholera, which is endemic in many regions, particularly in the developing world. It is highly virulent and the nature of the disease is a result of the toxins it produces. It is the highly developed quorum sensing system of V. cholerae that gives it such virulence and allows its survival in the human host. The genetic characteristics of V. cholerae that allow it to successfully survive the human host have been well documented, in which quorum sensing plays a significant role. A large number of these genes affect biofilm formation and the production of Vibrio polysaccharide (VPS). VPS is an extracellular compound vital to bacterial attachment to a surface and biofilm formation. Significant differences have been noted in *V. cholerae* biofilms depending on the environment and many different genes have been associated with these variances [43]. It is this ability of V. cholerae to change its biofilm structure in response to environmental changes that allows it to successfully colonize and infect the human body. One of the hurdles, which must be overcome by bacteria attempting to enter the human body *via* the alimentary canal, is the highly acidic environment of the stomach ($pH < 1$). Upon entering the stomach V. cholerae form thick, glutinous biofilms by production of excess VPS. This is achieved by a lack of Hap, a quorum sensing regulator that inhibits expression of the VPS operon, with CqsA acting as an autoinducer synthase. When the cells have passed out of the stomach and the protection of the biofilm is no longer required, production of HapR resumes, causing conformational change of the biofilm [43,44].

Another bacterium known to inhabit the human stomach is *Helicobacter pylori*, an opportunistic pathogen. This bacterium has become highly adapted to this niche environment and it is only in relatively recent years that they could be cultured within a laboratory due to the difficulties in replicating these conditions. *H. pylori* displays a *luxS* homologue and has demonstrated AI-2 production responsible for regulation of flagella gene transcription leading to immotility [45].

4.3. Thermophiles

Quorum sensing was initially thought impossible in the even moderately thermophilic environment due to the heat-labile nature of the AHLs lactone ring [46]. It has since been suggested that quorum sensing plays a vital role in these environments despite this initial observation.

The thermophilic bacterium *Thermus* sp. GH5 has demonstrated a role for AHL signaling in response to cold shock. It is during the early phase of the cold shock response that quorum sensing signals have been detected, though this is a condition when AHLs would be most stable. The AHL synthesis cycle was induced and the chemical precursors to AHLs were overexpressed. The production of these AHL precursors during cold shock was linked to biofilm formation. A gene coding for a short chain amino acid was located on the genome of *Thermotoga maritima* that was expressed at a considerably higher rate in higher cell densities, and was also considered a potential quorum sensing molecule [17].

The hyperthermophilic archaeon *Pyrococcus furiosus* has been studied for its potential quorum sensing abilities as it has been seen to form symbiotic relationships with sessile microbes, suggesting some form of cellular communication. It was suggested that quorum sensing was involved in this process, but the idea was quickly disregarded as the genome does not code for the LuxI/R or LuxS-type proteins, meaning that traditional models of quorum sensing are unlikely to be found [47]. Despite these initial thoughts, it was found that when *P. furiosus* was cultured with *T. maritima*, the two together could produce an AI-2 type signal through a series of biotic and abiotic steps. *T. maritima* also lacks a LuxS-encoding homologue in its genome. Although this signal was detected, no observable phenotypic change occurred in response to this molecule [47].

Further examinations of *T. maritima* have identified a pathway for EPS production that suggests the existence of peptide-based quorum sensing. *T. maritima* was grown in co-culture with *Methanocaldococcus jannaschii*, both of which have no *luxS* gene and no AHLs were found in the culture media. However, it was observed that production of EPS was considerably raised in higher cell densities and it was considered that quorum sensing might play a role despite this lack of known genetic requirements [17]. It has been reported that *T. maritima* displays a transcriptome-based stress response typical of that observed in AI-2 signaling and therefore proposed that these putative quorum sensing signals play a role in the heat shock response [47].

4.4. Psychrophiles

The psychrophiles are a group of organisms particularly lacking in information regarding their potential for quorum sensing. The ecological importance of cold-adapted microbes is something that is being studied closely, although very few discussions touch on the role of quorum sensing in these environments. This is an area in which the evolution of bioinformatics has provided insight into the capabilities of microorganisms, but the functionality and interaction of these organisms with their environment remains to be fully understood.

The psychrophile *Pseudoalteromonas haloplanktis* was observed to contain the *mtnN* gene that is implicated in the production of putative AI-2 signals, yet no LuxS homologue was identified [48]. The genome does encode a number of different genes that can be indicative of an alternative, lesser-known quorum sensing system. This includes the gene PSHAa0159 that codes for a multidomain putative aconitate hydratase, which can act as a signal in the stationary growth phase. It also contains genes known to be involved in the production of diffusible signaling factors in the Gammaproteobacteria [48].

Analysis of the genome of the psychrophile *Psychromonas ingrahamii* has allowed for identification of many different regulatory mechanisms. An orthologue of *LuxR* is reported though further details are unknown. The authors hint at an important role for biofilms in the resilience of this organism in the sea-ice in which it is found, however this is not something that has been investigated to date. It is suggested that the production of EPS allows the microbes to lower the freezing point in the surrounding environment, increasing availability of water for growth [49].

4.5. Piezophiles

Previously known as barophiles, the piezophiles are a group of microorganisms characterized by their preference for high-pressure conditions. A number of microbes have now been successfully isolated from high-pressure environments, primarily deep in the ocean.

Photobacterium profundum SS9, a bacterium isolated from 3, 600 m depth of the sea, has become a model organism for the study of piezophiles. Being of the family Vibrionaceae, *P. profundum* is closely related to those organisms in which quorum sensing was first identified, *V. harveyi* and *A. fischeri*. Comparative genomic studies have attempted to identify AI-2 signaling systems in *P. profundum*, finding that, although a LuxS homologue is present, it appears to have a metabolic function only [50]. *P. profundum* is also stated to contain a new quorum sensing system that has yet to be fully identified. This putative quorum sensing system shows approximately 35% sequence identity with the LuxMN and AinSR systems in *V. harveyi* and *A. fischeri*, giving strength to the idea that quorum sensing may play a vital role in high pressure environments [51], and demonstrating that other genomes need to be reevaluated when new quorum sensing systems are discovered.

Shewanella benthica and *Shewanella violacea* are piezophilic microorganisms also known to inhabit the deep-sea environment. The *Shewanella* spp. are all documented as containing the *luxS* gene [52] though its function in these piezophiles has not yet been investigated. Many *Shewanella* spp. have demonstrated the ability to degrade AHLs and interrupt quorum sensing in other species. Some of these microorganisms were able to significantly affect cross-domain signaling by inhibiting the settlement of zoospores [53]. Studies examining the potential for LuxS-type signaling *via* genome analyses of the *Shewanella* spp., concluded that this is more closely dependent upon phylogenetic affiliation than it is on a microbe's environment. As such there is a strong possibility for both quorum quenching and quorum sensing functions in the piezophilic *Shewanella* spp. [52].

Many gaps remain in our knowledge and understanding of this unique group of microorganisms. Cultivating microbes under high-pressure conditions is not without its difficulties, and thus our understanding of this unique environment is limited. While modern molecular genetic techniques have allowed some insight into their physiology, much remains unknown about their interactions with the environment.

4.6. Radiation Resistant Organisms

The bacterium *Deinococcus radiodurans* is one of many known to be able to endure high levels of radiation due to unique DNA repair mechanisms. *D. radiodurans* is known to survive inside nuclear reactors and its genes have been isolated and examined for use in industry. While the complete genome of *D. radiodurans* has been annotated and reported, little is yet known about its multi-cellular behavior(s). It is known to contain a *luxS* homologue with a role in the recycling of S-adenosyl-homocysteine (SAH), which produces AI-2 signals. It has been found to contain a two-step Pfs/LuxS pathway for production of AI-2 signals, though the function of this system and its products has not been investigated [54].

Another radioresistant bacterium, *Deinococcus gobiensis*, has shown a number of molecular responses immediately following exposure to UV irradiation [55]. One of these that is yet to be investigated is the *gidA* gene that encodes the glucose-inhibited cell division protein A that, in *Pseudomonas aeruginosa*, controls the post-transcriptional regulation of quorum sensing genes. In *P. aeruginosa* this occurs *via* the RhlR-dependent and RhlR-independent pathways similar to those observed in many quorum sensing soil symbionts. The exact role of this gene in *D. gobiensis* is yet to be investigated, though its close homology to these known quorum sensing systems suggests a role for quorum sensing in the UV resistant bacteria [55].

4.7. Archaea

While the archaea are often considered "extremophiles" and can be found in most of the environments described above, until recently there was no evidence for quorum sensing in the archaeal domain. Initial studies attempted to identify traditional LuxI-LuxR and AI-2 quorum sensing systems in archaea with no success [11,13]. However, several recent studies have yielded interesting results. Sulfur-reducing archaea from the crenarcheaota phylum have been shown to directly interact with AHL-based signaling systems, such as those currently found in bacteria, *via* enzymatic (lactonase) degradation of this signal [56]. The environmental role of this lactonase and the impact upon microbial community around it is currently unknown. Recent findings have also indicated the potential for AHLs to serve as messenger molecules for archaeal quorum sensing systems. The earliest indication of a potential AHL based quorum sensing system in archaea came from a haloalkaliphile *Natronococcus occultus*.

N. occultus is a haloalkaliphilic archaeon that has been found to produce an extracellular protease in the late exponential and stationary growth phases as well as during starvation [23]. It was hypothesized that production of this protease was quorum sensing dependent as it was observed in the stages of growth allowing for optimal cell density [23]. Furthermore, extracts from *N. occultus* at these particular time points were able to activate an AHL biosensor. This activation of an AHL biosensor, while not conclusive, suggests the role of AHLs in mediating the extracellular enzyme production, a phenomenon known to be often quorum regulated in bacteria.

The haloalkaliphile *Natrialba magadii* is an archaeon isolated from Lake Magadi, a saline lake in Kenya. It is known to produce Nep, a halolysin-like protease that is stable in high salt concentrations [26]. Nep is produced in the stationary phase of growth and this production is considered to be in response to low availability of nutritional requirements. AHLs and quorum sensing have been implicated in the up-regulation of Nep synthesis, however this has yet to be confirmed. A potential quorum sensing molecule was identified but it could not be purified [26], and bioinformatic analysis of the *N. magadii* genome elicited no genes similar to traditional quorum sensing genes.

Most recently, a study found that the methanogenic archaeon *Methanothrix (Methanosaeta) harundinacea* produces carboxylated AHLs and filament formation is induced by these signals [10]. Specifically, three compounds were detected: N-carboxyl-decanoyl-homoserine lactone (carboxylated C10 HSL), N-carboxyl-dodecanoyl-homoserine lactone (carboxylated C12 HSL) and N-carboxyl-tetradecanoyl-homoserine lactone [10]. This is the first direct evidence of quorum sensing in archaea. These carboxyl AHLs were detected by analytical chemistry techniques, and this particular class of AHL has not yet been observed in bacteria. The carboxyl group appears to be attached to the amino group of the homoserine lactone (HSL) ring [10]. It remains unclear what effect carboxylation has on chemical stability and hence the potential signal in the environment. Carboxyl AHLs from this archaeon were shown to activate bacterial biosensors, however bacterial AHLs were not able to induce filament production in *M. harundinacea* [10], indicating the potential for one-way cross talk. Thus the potential for interspecies and even interdomain signaling is quite significant and an area worthy of further investigation.

Both *N. occultus* and *N. magadii* are sourced from extremely alkaphillic environments, and it implies that modifications to the traditional quorum sensing molecules would be necessary for them to function under the extreme environmental conditions, as short-chain AHLs are unstable in alkaline conditions [13]. One such modification could be an increase in chain length, which as mentioned previously, can enable AHLs to survive longer under alkaline conditions [12,13]. Other chemical modifications could too play a role such as the carboxylation described above in *M. harundinacea* [10].

Finally, it is possible that differing molecules such as diketopiperazines are functioning as quorum sensing molecules, as they have been previously indicated to activate biosensors [57] and are present in haloarchaea [58].

5. Quorum Sensing in an Extreme Environment—Microbial Mats

Although this review has focused primarily on quorum sensing in individual organisms, this discussion will now illustrate this phenomenon in a particular environmental setting, that of microbial mats. Microbial mats can be considered a good example of an extreme environment, with resident microbes subjected to a range of fluctuating parameters [59,60]. These organosedimentary systems contain copious amounts of EPS [59], which may play an important role in modulating cell signaling [28]. As they are composed of a diverse community of microbial functional groups they experience significant changes in O_2, H_2S and pH in response to diel cycles. The processes of photosynthesis and aerobic respiration dominate the metabolism of the microbial mat during the light while fermentation and anaerobic respiration (sulfate reduction) do so during the dark, which result in large fluctuations in geochemical and physicochemical conditions [24]. This creates for example, a large shift in the pH values within the mat, which can range from >11 during peak photosynthesis and < 5.5 during the night [13]. As a result of this quorum, sensing must be efficient, effective and highly adapted to these fluctuating conditions.

AHLs have been shown in early studies to be sensitive to elevated pH so it is likely that the diel fluctuations typical for microbial mats causes degradation of AHLs especially during the afternoon, resulting in disruption of cell signaling. It was shown that the shorter-chain (<7) AHLs were present at significantly lower concentrations during the day when compared to the nighttime scenario [13]. The efficacy of the short chain AHLs is therefore limited during the night, which could explain the presence of usually long acyl side chains (*i.e.*, C12–C14) in samples from natural mats and organisms isolated form mats. Many different AHLs have been detected in microbial mats and a pattern has been suggested in which longer-chain AHLs are produced during the day and shorter chains produced at night. This would allow the bacteria within the biofilm to tailor their gene expression to the times when it is most favorable or appropriate [13].

Confocal microscopy has revealed dense clusters of microorganisms, particularly within the upper layers of the microbial mats [60]. The complexity of the microbial communication system is clear when considering the need for alteration and diffusion of these signals within the microbial mat. Larger molecules cannot diffuse as far as smaller ones and the need to communicate with cells of the same species at a distance, despite immediate high cell density, causes some unique challenges for quorum sensing systems within microbial mats [60]. It has been suggested that in mixed species environments such as the microbial mats that have a high diversity of AHLs, quorum sensing may be used to monitor the diversity (and metabolism) of species in the environment rather than individuals of the same species. The possibility remains that under these conditions the LuxR-LuxI type proteins may mediate inter-species communication [6].

It has also been shown in several studies that oxygen-sensitive sulfate reduction peaks during the maximum of oxygenic photosynthesis, which necessitates physiological adaptations [61]. It was hypothesized that inter-species communication between sulfate-reducing bacteria (SRB) and sulfide-oxidizing bacteria (SOB) would enable both physiological groups to coexist in an environment with supersaturated O_2 concentrations: preliminary experiments have indeed shown that a mixture of long-chain AHLs, like the ones produced by SRB stimulate the sulfide oxidation by SOB in mats. The latter metabolism would remove the toxic sulfide produced by SRB and O_2 resulting from oxygenic photosynthesis. Exopolymeric substances are an important part of the microbial mat, providing the matrix for a 3D architecture and allowing dense cell clusters to exist. Certain EPS properties, such as nanochannels may prove critical in communication on these systems [60]. Furthermore, protection of quorum sensing compounds against desiccation, excessive UV radiation, H_2O_2, OH radicals and singlet O_2 by the EPS matrix could be instrumental in the functioning of the mat.

6. Biotechnological Applications

The biotechnological applications for extremophiles and their products are just as vast as the range of organisms themselves. In many cases, such as seen with the acidophiles, it appears as though study of the microbes has begun with a perceived application within industry. Ranging from industrial processes to aquaculture, bioremediation, medical interventions and uses within the food industry, the extremophiles provide an extensive range of possibilities.

The application of quorum sensing systems often entails the promotion of beneficial microbial associations by changing environmental conditions. This concept has been applied in the agricultural industry to promote beneficial growth in the soil and has similarly been used in aquaculture processes. Halophilic organisms such as the micro-alga *Dunaliella salina* may have the potential to supplement aquaculture due to their production of quorum sensing inhibitors [32].

As previously mentioned, *Acidithiobacillus ferrooxidans* is used often in bioleaching processes [4] as well as for the recovery of metals such as copper, gold and uranium from metal ores in the process of biohydrometallurgy [38]. Similarly, *Ferroplasma acidarmanus* is used in biomining in which the metal sulfide oxidation undertaken by the microbes is exploited to release metals from sulfide minerals [37]. There is the possibility of using the quorum sensing systems of both these organisms to optimise these processes for which they have already been highly selected, such as seen in the agricultural industry.

Enzymes and other biomolecules from extremophiles are already heavily used in industry. These range from amylases and ligases as well as plasmids and maltose-binding proteins, which are all isolated from the acidophiles alone [62]. Table 2 summarizes some of the biological products from extremophiles that are being used or show potential for industrial applications. There is great potential for the production of these compounds to be optimized through the exploitation of quorum sensing. Some extracellular enzymes taken from these extreme environments show potential value to the biotechnological sector, and it is possible that these extracellular enzymes may be quorum controlled [23].

There is a strong link between quorum sensing and antibiotic production, and as such, one of the unique opportunities for the application of quorum sensing is the medical sector [26]. AHLs and their derivatives are being studied intensively for their use in the medical field as antimicrobial platforms. They offer an attractive alternative to traditional antibiotics as they interrupt bacterial colonization without killing all native microbes and incurring antibiotic resistance factors [2]. The use of extremophiles in this context would allow for the use of acid-stable molecules as would be required in the digestive system, for example.

Table 2. Extremophiles and their biological products with industrial applications.

Organism	Biological Products
P. furiosus [63]	DNA polymerase and hydrogenase
T. aquaticus [62]	Taq polymerase, α-glucosidase
N. magadii [11,64]	Biocatalysts, Nep (solvent tolerant enzyme)
N. amylolyticus [65]	α-amylase

Food microbiology is an extensive area of research and practice in which quorum sensing has been investigated. Quorum sensing signals have been identified in many food products and, as it is often microbial contamination of food that causes spoilage, it may be possible to use quorum sensing inhibitors to prevent microbial growth. For example, the bacterium *Serratia proteamaculans* has been implicated in the spoilage of milk. When a mutant strain that did not produce AHLs was added to milk cultures spoilage of the milk did not occur within the same time frame [66]. High pressure is an alternative method of sterilization in food processing considered beneficial as it preserves the color and flavor of the food. This process involves the short-term elevation of pressure that kills or inactivates microbes. While this is in general beneficial, it also kills those microbes used within the food industry for their ability to add to taste amongst other desired properties. While there are no current applications for piezophiles in the food industry, it has been suggested that the use of their metabolites in these high-pressure procedures may be of considerable benefit [63], though it remains to be seen whether they are quorum-regulated.

An application for quorum sensing has also been implicated in the production of consumables. Wine for example relies heavily upon the activity of yeast and bacteria to produce particular flavors. The balance of yeast-yeast interactions against the yeast-bacterial interactions can have a significant impact on the end product. It has been suggested that assessing quorum sensing as a means of maintaining a balance in this relationship may be a new way of achieving desired results of certain product qualities [67].

7. Conclusions

The extremophiles are a highly diverse and well-studied group, yet our knowledge of their ability to produce and receive quorum sensing signals is extremely limited. The development of genomic sequencing and bioinformatic databases has allowed for the identification of quorum sensing genes in microbes and conjecture as to their ability to use these to interact with their environment. Many studies show traditional models of quorum sensing to be insufficient, as though organisms demonstrate the ability to produce AHLs and other quorum sensing molecules, in many cases we still do not have a clear understanding of the genetic basis of their production/regulation. It is clear that there is a large gap in our understanding of the quorum sensing abilities of the extremophiles.

Quorum sensing systems continue to be identified in novel environments and appear to play key roles in regulating an array of phenotypes that assist survival in these environments. Quorum sensing has yet to be fully investigated in extremophiles mainly due to the difficulty in culturing these organisms. There are major ecological implications of quorum sensing through evolutionary time, particularly in respect to inter-species or even inter-domain signaling. It is likely that novel molecules and receptors exist and waiting to be discovered. We need to better understand the role of the environment in modifying quorum sensing signals and also the extent to which such modified molecules can still result in a phenotype change. Through the use of bioinformatics, genomic sequencing and new techniques such as plasmid based biosensors, we are beginning to reveal the secrets of quorum sensing in extreme environments.

Acknowledgments

This work was funded by the Australian Research Council. No competing financial interests exist.

References

1. Bassler, B. Small Talk: Cell-to-Cell Communication in Bacteria. *Cell* **2002**, *109*, 421–424.
2. Dobretsov, S.; Teplitski, M.; Paul, V. Mini-review: Quorum sensing in the marine environment and its relationship to biofouling. *Biofouling* **2012**, *25*, 413–427.
3. Fuqua, C.; Parsek, M.; Greenberg, P. Regulation of gene expression by cell-to-cell communication: acyl-homoserine lactone quorum sensing. *Ann. Rev. Gen.* **2001**, *35*, 439–468.
4. Rivas, M.; Seeger, M.; Holmes, D.; Jedlicki, E. A Lux-like quorum sensing system in the extreme acidophile *Acidithiobacillus. ferrooxidans*. *Biol. Res.* **2005**, *38*, 283–297.
5. Chaphalkar, A.; Salunkhe, N. Phylogenetic analysis of nitrogen-fixing and quorum sensing bacteria. *Int. J. Bioinf. Res.* **2010**, *2*, 17–32.
6. Fuqua, C.; Winans, S.; Greenberg, E. Census and consensus in bacterial ecosystems—the LuxR-LuxI family of quorum-sensing transcriptional regulators. *Ann. Rev. Microbiol.* **1996**, *50*, 727–751.
7. Pituka, E.V.; Hoover, R.B. Microbial extremophiles at the limits of life. *Crit. Rev. Microbiol.* **2007**, *33*, 183–209.
8. Van den Burg, B. Extremophiles as a source for novel enzymes. *Curr. Opin. Microbiol.* **2003**, *6*, 213–218.
9. Sharif, D.I.; Gallon, J.; Smith, C.J.; Dudley, E. Quorum sensing in cyanobacteria: N-octanoyl-homoserine lactone release and response, by the epilithic colonial cyanobacterium *Gloeothece* PCC6909. *ISME J.* **2008**, *2*, 1171–1182.
10. Zhang, G.; Zhang, F.; Ding, G.; Li, J.; Guo, X.; Zhu, J.; Zhou, L.; Cai, S.; Liu X.; Luo, Y.; *et al.* Acyl homoserine lactone-based quorum sensing in a methanogenic archaeon. *ISME J.* **2012**, *6*, 1–9.
11. Miller, M.; Bassler, B. Quorum Sensing in Bacteria. *Ann. Rev. Microbiol.* **2001**, *55*, 165–199.
12. Yates, E.A.; Philipp, B.; Buckley, C.; Atkinson, S.; Chhabra, S.; Sockett, R.E.; Goldner, M.; Dessaux, Y.; Camara, M.; Smith, H.; Williams, P.; *et al.* N-acylhomoserine lactones undergo lactonolysis in a pH-, temperature-, and acyl chain length-dependent manner during growth of *Yersinia pseudotuberculosis* and *Pseudomonas aeruginosa*. *Infect. Immun.* **2002**, *70*, 5635–5646.
13. Decho, A.W.; Visscher, P.T.; Ferry, J.; Kawaguchi, T.; He, L.; Przekop, K.M.; Norman, R.S.; Reid, P. Autoinducers extracted from microbial mats reveal a surprising diversity of N-acylhomoserine lactones (AHLs) and abundance changes that may relate to diel pH. *Environ. Microbiol.* **2009,** *11*, 409–20.
14. Xavier, K.; Bassler, B. Interference with AI-2-mediated bacterial cell–cell communication. *Nature* **2005**, *437*, 750–753.
15. Chen, X.; Schauder, S.; Potier, N.; Dorsselaer, A.V.; Pelczer, I.; Bassler, B.; Hughson, F. Structural identification of a bacterial quorum-sensing signal containing boron. *Nature* **2002**, *5*, 545–549.

16. Diggle, S.P.; Cornelis, P.; Williams, P.; Cámara, M. 4-quinolone signalling in *Pseudomonas aeruginosa*: Old molecules, new perspectives. *IJMM* **2006**, *296*, 83–91.
17. Johnson, M.; Montero, C.; Connors, S.; Shockley, K.; Bridger, S.; Kelly, R. Population density-dependent regulation of exopolysaccharide formation in the hyperthermophilic bacterium *Thermotoga. maritima. Mol. Microbiol.* **2005**, *55*, 664–674.
18. Smith, M.B.; Smith, J.N.; Swift, S.; Heffron, F.; Ahmer, B.M. SdiA of *Salmonella enterica* is a LuxR homolog that detects mixed microbial communities. *J. Bacteriol.* **2001**, *183*, 5733–5742.
19. Steindler, L.; Venturi, V. Detection of quorum-sensing N-acyl homoserine lactone signal molecules by bacterial biosensors. *FEMS Microbiol. Lett.* **2007**, *266*, 1–9.
20. Rajamani, S.; Zhu, J.; Pei, D.; Sayre, R. A LuxP-FRET-based reporter for the detection and quantification of AI-2 bacterial quorum-sensing signal compounds. *Biochemistry* **2007**, *46*, 3990–3997.
21. McClean, K.; Winson, M.K.; Fish, L.; Taylor, A.; Chhabra, S.R.; Camara, M.; Daykin, M.; Lamb, J.H.; Swift, S.; Bycroft, B.W.; Stewart, G.S.; Williams, P. Quorum sensing and *Chromobacterium violaceum*: Exploitation of violacein production and inhibition for the detection of N-acylhomoserine lactones. *Mol. Microbiol.* **1997**, *143*, 3703–3711.
22. Llamas, I.; Quesada, E.; Martínez-Cánovas, M.J.; Gronquist, M.; Eberhard, A.; González, J.E. Quorum sensing in halophilic bacteria: detection of N-acyl-homoserine lactones in the exopolysaccharide-producing species of *Halomonas*. *Extremophiles* **2005**, *9*, 333–341.
23. Paggi, R.; Martone, C.; Fuqua, C.; de Castro, R. Detection of quorum sensing signals in the haloalkaliphilic archaeon *Natronococcus. occultus*. *FEMS Microbiol. Lett.* **2003**, *221*, 49–52.
24. Visscher, P.T.; Dupraz, C.; Braissant, O.; Gallagher, K.L.; Glunk, C.; Casillas, L.; Reed, R.E. Biogeochemistry of carbon cycling in hypersaline mats: Linking the present to the past through biosignatures. In *Cellular Origin, Life in Extreme Habitats and Astrobiology :Microbial Mats*; Seckbach, J., Oren, A., Eds.; Springer Verlag: Berlin, Germany, 2010; Volume 14, pp 443–468.
25. Margesin, R.; Schinner, F. Potential of halotolerant and halophilic microorganisms for biotechnology. *Extremophiles.* **2001**, *5*, 73–83.
26. Penesyan, R.; Madrid, E.; D'Alessandro, C.; Cerletti, M.; de Castro, R. Growth phase-dependent biosynthesis of Nep, a halolysin-like protease secreted by the alkaliphilic haloarchaean *Natrialba. magadii*. *Lett. Appl. Microbiol.* **2010**, *51*, 36–41.
27. Decho, A.W. Microbial exopolymer secretions in ocean environments: their role(s) in food webs and marine processes. *Mar. Biol. Ann. Rev.* **1990**, *28*, 73–153.
28. Decho, A.W. Microbial biofilms in intertidal systems: An overview. *Continental Shelf Res.* **2000**, *20*, 1257–1273.
29. Averhoff, B.; Muller, V. Exploring research frontiers in microbiology- recent advances in halophilic and thermophilic extremophiles. *Res. Microbiol.* **2010**, *161*, 506–514.
30. Sewald, X.; Saum, S.; Palm, P.; Pfeiffer, F.; Oesterhelt, D.; Muller, V. Autoinducer-2-Producing Protein LuxS, a Novel Salt- and Chloride-Induced Protein in the moderately halophilic *Bacterium Halobacillus. halophilus*. *Appl. Environ. Microbiol.* **2007**, *73*, 371–379.

31. DasSarma, S.; DasSarma, P. Halophiles. In *Encyclopedia of Life Sciences*; Wiley: London, UK, 2006.
32. Natrah, F.; Kenmegne, M.; Wiyoto, W.; Sorgeloos, P.; Bossier, P.; Defoirdt, T. Effects of micro-algae commonly used in aquaculture on acyl-homoserine lactone quorum sensing. *Aquaculture* **2011**, *317*, 53–57.
33. Calvo, C.; Martinez-Checa, F.; Mota, A.; Bejar, V.; Quesada, E. Effects of cations, pH and sulfate on the viscosity and emulsifying activity of the *Halomonas. eurihalina* exopolysachharide. *J. Ind. Microbiol. Biotechnol.* **1998**, *20*, 205–209.
34. Bouchotroch, S.; Quesada, E.; del Moral, A.; Llamas, I.; Bejar, V. *Halomonas. maura* sp. nov., a novel moderately halophilic, exopolysachharide-producing bacterium. *Int. J. Syst. Evol. Microbiol.* **2001**, *51*, 1625–1632.
35. Matinez-Canovas, M.; Quesada, E.; Llamas, I.; Bejar, V. *Halomonas. ventosae* sp. nov., a moderately halophilic, denitrifying, exopolysachharide-producing bacterium. *Int. J. Syst. Evol. Microbiol.* **2004**, *54*, 733–737.
36. Martinez-Canovas, M.; Bejar, V.; Martinez-Checa, F.; Quesada, E. *Halomonas. anticariensis* sp. nov., from Fuente de Piedra, a saline-wetland wildfowl reserve in Malaga, southern Spain. *Int. J. Syst. Evol. Microbiol.* **2004**, *54*, 1329–1332.
37. Baker-Austin, C.; Potrykus, J.; Wexler, M.; Bond, P.L.; Dopson, M. Biofilm development in the extremely acidophilic archaeon *Ferroplasma. acidarmanus* Fer1. *Extremophiles* **2010**, *14*, 485–491.
38. Wenbin, N.; Dejuan, Z.; Feifan, L.; Lei, Y.; Peng, C.; Xiaoxuan, Y.; Hongyu, L. Quorum-sensing system in *Acidithiobacillus. ferrooxidans* involved in its resistance to Cu^{2+}. *Lett. Appl. Microbiol.* **2011**, *53*, 84–91.
39. Penesyan, A.; Kjelleberg, S.; Egan, S. Development of novel drugs from marine surface associated microorganisms. *Mar. Drugs* **2010**, *8*, 438–459.
40. Rivas, M.; Seeger, M.; Jedlicki, E.; Holmes, D. Second acyl homoserine lactone production system in the extreme acidophile *Acidithiobacillus. ferrooxidans*. *Appl. Environ. Microbiol.* **2007**, *73*, 3225–3231.
41. Ruiz, L.; Valenzuela, S.; Castro, M.; Gonzalez, A.; Frezza, M.; Soulere, L.; Rohwerder, T.; Queneau, Y.; Doutheau, A.; Sand, W.; Jerez, C.; Guiliani, N. AHL communication is a widespread phenomenon in biomining bacteria and seems to be involved in mineral-adhesion efficiency. *Hydrometallurgy* **2008**, *94*, 133–137.
42. Moreno-Paz, M.; Gomez, M.; Arcas, A.; Parro, V. Environmental transcriptome analysis reveals physiological differences between biofilm and planktonic modes of life of the iron oxidising bacteria *Leptospirillum.* spp. in their natural microbial community. *BMC Genomics* **2010**, *11*, 404–418.
43. Hammer, B.; Bassler, B. Quorum sensing controls biofilm formation in *Vibrio. cholera*. *Mol. Microbiol.* **2003**, *50*, 101–114.
44. March, J.; Bentley, W. Quorum sensing and bacterial cross-talk in biotechnology. *Curr. Opin. Biotechnol.* **2004**, *15*, 495–502.

45. Rader, B.; Campagna, S.; Semmelhack, M.F.; Bassler, B.; Guillemin, K. The quorum-sensing molecule autoinducer 2 regulates motility and flagellar morphogenesis in *Helicobacter pylori*. *J. Bacteriol.* **2007**, *189*, 6109–6117.
46. Schopf, S.; Wanner, G.; Rachel, R.; Wirth, R. An archaeal bi-species biofilm formed by *Pyrococcus. furiosus* and *Methanopyrus. kandleri*. *Arch. Microbiol.* **2008**, *190*, 371–377.
47. Nichols, J.; Johnson, M.; Chou, C.; Kelly, R. Temperature, not LuxS, mediates AI-2 formation in hydrothermal habitats. *FEMS Microbiol. Ecol.* **2009**, *68*, 173–181.
48. Medigue, C.; Krin, E.; Pascal, G.; Barbe, V.; Bernsel, A.; Bertin, P.; Cheung, F.; Cruveiller, S.; D'Amico, S.; Duillo, A.; *et al.* Coping with cold: The genome of the versatile marine Antarctica bacterium *Pseudoalteromonas. haloplanktis* TAC125. *Genome Res.* **2005**, *15*, 1325–1335.
49. Riley, M.; Staley, J.; Danchin, A.; Wang, T.Z.; Brettin, T.S.; Hauser, L.J.; Land, M.L.; Thompson, L.S. Genomics of an extreme psychrophile, *Psychromonas. ingrahamii*. *BMC Genomics* **2008**, *9*, 1–19.
50. Rezzonico, F.; Duffy, B. Lack of genomic evidence of AI-2 receptors suggests a non-quorum sensing role for luxS in most bacteria. *BMC Microbiol.* **2008**, *8*, 1–19.
51. Reen, F.; Almagro-Moreno, S.; Ussery, D.; Boyd, E. The genomic code: inferring *Vibrionaceae.* niche specialization. *Nat. Rev. Microbiol.* **2006**, *4*, 1–8.
52. Bodor, A.; Elxnat, B.; Thiel, V.; Schulz, S.; Wagner-Dobler, I. Potential for luxS related signalling in marine bacteria of autoinducer-2 in the genus *Shewanella*. *BMC Microbiol.* **2008**, *8*, 1–9.
53. Tait, K.; Williamson, H.; Atkinson, S.; Williams, P.; Camara, M.; Joint, I. Turnover of quorum sensing signal molecules modulates cross-kingdom signaling. *Environ. Microbiol.* **2009**, *11*, 1792–1802.
54. Sun, J.; Daniel, R.; Wagner-Dobler, I.; Zeng, A. Is autoinducer-2 a universal signal for interspecies communication- a comparative genomic and phylogenetic analysis of the synthesis and signal transduction pathways. *BMC Evol. Biol.* **2004**, *4*, 1–11.
55. Yuan, M.; Chen, M.; Zhang, W.; Lu, W.; Wang, J.; Yang, M.; Zhao, P.; Tang, R.; Li, X.; Hao, Y.; *et al.* Genome sequence and transcriptome analysis of the radioresistant bacterium *Deinococcus. gobiensis:* Insights into the extreme environmental adaptations. *PLos One.* **2012**, *7*, 34458–34551.
56. Ng, F.S.W.; Wright, D.M.; Seah, S.Y.K. Characterization of a phosphotriesterase-like lactonase from *Sulfolobus. solfataricus* and its immobilization for disruption of quorum sensing. *Appl. Environ. Microbiol.* **2011**, *77*, 1181–1186.
57. Holden, M.T.; Chhabra, S.R.; Nys, R.; Stead, P.; Bainton, N.; Hill, P.; Manefield, M.; Kumar, N.; Labatte M.; England, D.; *et al.* Quorum-sensing cross talk: isolation and chemical characterization of cyclic dipeptides from *Pseudomonas aeruginosa* and other gram-negative bacteria. *Mol. Microbiol.* **1999**, *33*, 1254–66.
58. Tommonaro, G.; Abbamondi, G.R.; Iodice, C.; Tait, K.; de Rosa, S. Diketopiperazines produced by the halophilic archaeon, *Haloterrigena hispanica*, activate AHL bioreporters. *Microb. Ecol.* **2012**, *63*, 490–495.

59. Braissant, O.; Decho, A.W.; Przekop, K.M.; Gallagher, K.L.; Glunk, C.; Dupraz, C.; Visscher, P.T. Characteristics and turnover of exopolymeric substances (EPS) in a hypersaline microbial mat. *FEMS Microbiol. Ecol.* **2009**, *67*, 293–307.
60. Decho, A.; Norman, S.; Visscher, P.T. Quorum sensing in natural environments: Emerging views from microbial mats. *Trends Microbiol.* **2010**, *18*, 73–80.
61. Visscher, P.T.; Prins, R.A.; van Gemerden, H. Rates of sulfate reduction and thiosulfate consumption in a marine microbial mat. *FEMS Microbiol. Ecol.* **1992**, *86*, 283–294.
62. Sharma, A, Kawarabayasi, Y.; Satyanarayana, T. Acidophilic bacteria and archaea: acid stable biocatalysts and their potential applications. *Extremophiles.* **2012**, *16*, 1–19.
63. Abe, F.; Horikoshi, K. The biotechnological potential of piezophiles. *Trends Biotechnol.* **2001**, *19*, 102–108.
64. D'Alessandro, C.; de Castra, R.; Gimenez, M.; Paggi, R. Effect of nutritional conditions on extracellular protease production by the haloalkaliphilic archaeon *Natrialba. magadii*. *Lett. Appl. Microbiol.* **2007**, *44*, 637–642.
65. Kanai, H.; Kobayashi, T.; Aono, R.; Kudo, T. *Natronococcus. amylolyticus* sp. nov., a haloalkaliphilic archaeon. *Int. J. Syst. Bacteriol.* **1995**, *45*, 762–766.
66. Bai, A.; Rai, V. Bacterial Quorum Sensing and Food Industry. *Comprehensive Rev. Food Sci. Food Saf.* **2011**, *10*, 184–194.
67. Fleet, G. Yeast interactions and wine flavor. *Int. J. Food Microbiol.* **2003**, *86*, 11–22.

Microorganism Response to Stressed Terrestrial Environments: A Raman Spectroscopic Perspective of Extremophilic Life Strategies

Susana E. Jorge-Villar and Howell G.M. Edwards

Abstract: Raman spectroscopy is a valuable analytical technique for the identification of biomolecules and minerals in natural samples, which involves little or minimal sample manipulation. In this paper, we evaluate the advantages and disadvantages of this technique applied to the study of extremophiles. Furthermore, we provide a review of the results published, up to the present point in time, of the bio- and geo-strategies adopted by different types of extremophile colonies of microorganisms. We also show the characteristic Raman signatures for the identification of pigments and minerals, which appear in those complex samples.

Reprinted from *Life*. Cite as: Jorge-Villar, S.E.; Edwards, H.G.M.. Microorganism Response to Stressed Terrestrial Environments: A Raman Spectroscopic Perspective of Extremophilic Life Strategies. *Life* **2013**, *3*, 276-294.

1. Introduction

Since the first reports in the literature of the Raman spectroscopic analyses of the biological colonization of toxic mineral pigments in Renaissance frescoes [1] in 1991 and that of an endolithic colonization of Beacon sandstone from the Antarctic peninsula [2,3] in 1997, the analytical interrogation of extremophilic systems with a variety of geological substrates has resulted in novel information being provided about the nature of the synthetic protective chemicals and the strategies being employed for survival of the colonies in terrestrial stressed environments using this technique. At first, the identification of the chemical complexes formed as a result of the reaction of lichen metabolic waste products upon calcareous substrates in the form of oxalates was a primary objective [4–8] and it was not until much later that the characterization of the organic by-products was accomplished through the adoption of longer wavelength laser excitation and comparison with extracted materials [9–14]. The successful survival of extremophiles in stressed environments is dependent on their adaptation to the prevailing conditions of high or low temperatures, extreme desiccation, high energy ultraviolet insolation, high or low barometric pressures, extremes of pH in the range of <1 to >12 and on the presence of toxic chemicals and ions such as mercury (II), antimony (III), lead (II), barium (II), arsenic (III) and copper (II), which are found in many minerals. It was also apparent that the production of protective chemicals by extremophiles as a response to environmental stresses was supported by their ability to adapt their geological matrices and substrates, often resulting in biogeological signatures, which have remained in the geological record long after the extinction of the biological colonies.

It was appreciated at the outset that a major advantage of Raman spectroscopy as an analytical technique for the study of extremophilic colonization and survival was its ability to interrogate a system microscopically across a horizontal or vertical transect without any physical separation or chemical and mechanical pretreatment such as coating, grinding or polishing [15–18]. Hence, information about the interaction between the biological and geological components is accessible naturally without extraction or modification of the biogeological system. This was appreciated in the analysis of works of art, which had been subjected to lichen deterioration and required conservation [19–22].

The recognition of key definitive Raman spectral signatures in the same spectrum from the organic and inorganic components is a desirable outcome for the projected deployment of a miniaturized spectrometer unit as part of the life detection instrumentation suite on the planetary rover vehicle on the ESA ExoMars mission scheduled for 2018. In this mission, the Raman spectrometer will be a first-pass probe of selected samples removed from the Martian surface and subsurface (down to a depth of two meters) for the detection of extremophilic spectral signatures from extinct or extant colonies in niche environments. The assimilation of Raman spectral data and signatures into a database relating to extremophiles from terrestrial Mars analogue sites is therefore of vital importance, and this has been undertaken in recent years and is still ongoing [23–30]. Similarly, Raman spectroscopic analyses of other extreme terrestrial sites, which have relevance to the remote exploration of other planets and their satellites in our Solar System have been reported, such as glaciers, deep-sea smokers, hot geysers and snowfields, which have an ambience with planetary icy moons such as Europa, Titan, Io and Enceladus that are also noteworthy in this respect.

The purpose of the present paper is to survey and to critically examine comprehensively the existing wealth of Raman spectral data and their interpretation for terrestrial extremophiles; from this analysis, which has never been undertaken comprehensively hitherto, the following deductions can then be made:

- the commonality of Raman spectral signatures between various sites and extreme environments;
- the factors which affect the recording of Raman spectral signatures from the extremophilic colonization of terrestrial sites;
- the definition of Raman data for evidence of extinct or extant life in the geological record.

2. Results and Discussion

2.1. Technical Considerations of the Use of Raman Spectroscopy for the Study of Extremophiles

Raman spectroscopy is an analytical technique, which provides molecular information for either organic or inorganic compounds [31]. The Raman effect allows one to probe the internal structure and bonds between atoms in a molecule.

When monochromatic laser radiation falls upon a given compound, molecules can jump to a higher energy level (Figure 1); most of these molecules fall back to the same energy level, with the emission of a photon that has the same frequency as the incident light leading to Rayleigh scattering. In contrast, some molecules, which are already in higher or lower energy levels, can either emit or absorb a photon upon irradiation and then make a transition to lower or higher energy levels giving rise to Stokes and anti-Stokes Raman radiation. The result is a spectrum which is characteristic of each molecule and which exhibits a definitive number of key spectral signatures at characteristic wavenumbers, so providing a molecular fingerprint that is analytically useful for molecular species characterization.

Figure 1. Molecular light scattering; Raman Stokes, the molecule absorbs energy; Rayleigh: there is neither absorption nor loss of energy; Raman anti-Stokes: the molecule loses energy.

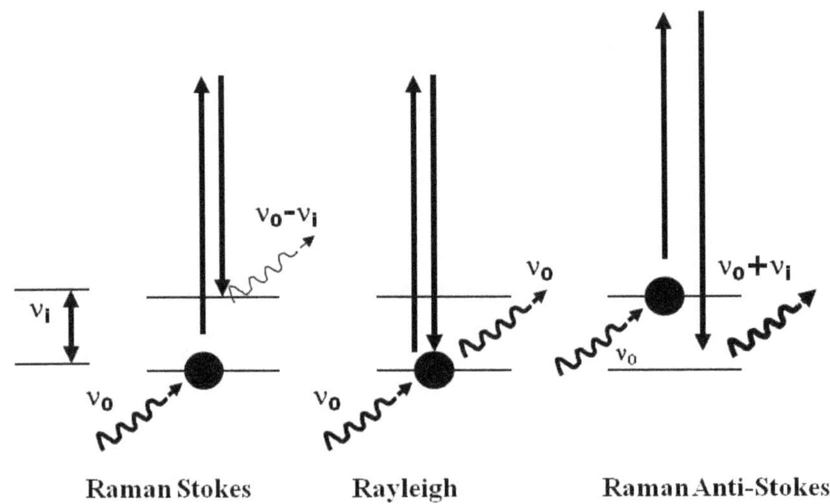

The use of Raman spectroscopy for the study of extremophiles has some advantages with regard to other techniques. No sample preparation, neither chemical nor physical, is required, and further no dilution, concentration, staining, grinding, or desiccation is needed, so the technique is non destructive. Of course, it is necessary to expose the surface for analysis, and, nowadays, with portable instruments, samples can be analyzed in the field nondestructively, a major advantage for further analytical procedures to be undertaken with micro- and macro- analyses made possible by using different sample illumination or collection lenses. Furthermore, sample size is not a problem, as a microscope can be used for the interrogation of small sample regions and an optical fiber head can be attached and used to study very large specimens. Moreover, the recent development of portable instruments allows the in-field use of Raman spectrometers under environmental conditions, which obviates changes that may occur to the sample during the acquisition of samples from the field and their transportation and subsequent storage prior to spectroscopic analysis.

However, although Raman spectroscopy is a very useful technique, it has some disadvantages. Some compounds give a strong fluorescence emission, which can be several orders of magnitude greater than the Raman signal and masks the inherently weaker Raman bands. As this technique is based upon the use of a laser, usually operating in the infrared or visible region of the electromagnetic spectrum, absorption of the radiation can induce sample heating, which can result in molecular or mineralogical changes or, in the worst-case scenario, even sample degradation or burning. It is clear then that the laser power used for spectral acquisition should be used with the lowest power possible consistent with the acquisition of the spectrum even though this may require correspondingly greater spectral accumulation times. Furthermore, the laser wavelength used for exciting the sample can influence the fluorescence emission and also the relative intensity of the Raman bands: Hence, a strong signature observed using a particular laser wavelength can become only of medium or weak intensity when an alternative laser wavelength is used. This effect can cause problems in spectrum interpretation.

A particular difficulty lies in the nature of the extremophiles themselves. These are organisms usually living in or on various supports, ranging from ice to stone, soil, rocks, wood, and synthetic substrates such as plastics, glass, metals and ceramics. There is a complex mixture of signatures from organic and inorganic compounds arising from the extremophiles themselves and from their substrates that can be observed in the Raman spectrum. Each compound will show a variable number of Raman signatures in admixture with those from other compounds, and the observed intensity of the significant Raman bands for any given compound in the spectrum is related directly to the proportional amount of the compound in the sample. A further complication is that the range of molecular scattering factors for different species in the Raman effect confers a different response of the molecular material to its interaction with the incident laser radiation, and this can additionally dependent upon the laser wavelength; an example of this is the intensity enhancement of carotenoid signatures in extremophile spectra excited using a green or blue laser ascribed to resonance Raman scattering processes.

The critical functionality region for organic molecules lies in the range between 1,800 to 1,000 cm^{-1} in the Raman spectrum, whereas for minerals, this is most characteristic in the 1,100 to 100 cm^{-1} region. Minerals do not usually show bands at higher wavenumbers than this, with the exception of water and hydroxyl bands; spectral bands in the 3,000–3,500 cm^{-1} region are characteristic of coordinated water and OH groups; in the lower spectral wavenumber region below 1,000 cm^{-1}, the Raman spectral signatures of both organic and inorganic compounds occur together.

When the Raman spectrum of an extremophilic organism shows the spectrum of a mixture of possible compounds, either organic and/or mineral, the superposition and overlapping of Raman signatures is realizable and, then, individual bands become broader and a peak band envelope can be observed to move to lower or higher wavenumbers. In this case, comparison with the database for molecular characterization based on pure materials is fraught with difficulty and care must be taken in the interpretation of the spectral data.

Another important consideration for the spectral interpretation lies in the fact that the Raman database of biomolecules from natural samples is rather limited and is naturally based upon detailed spectra from extracted and purified materials; hence, it is not always possible to assign all bands to a particular compound in a complex mixture unless characteristic features of each species are known and several features necessarily remain tentative or unassigned. More supporting studies linked to the identification of components extracted from extremophiles need to be carried out in the future for enlarging the Raman database and facilitating the unambiguous identification of biomolecular signatures in complex systems.

Extremophiles are described as those organisms able to live and survive under extreme conditions such as desiccation, extremes of temperature, pressure, pH, high salinity, radiation insolation and restricted nutrient and oxygen availability [32–35]. Most of these organisms are unicellular, mainly prokaryotics, although bacteria are also widespread [32]. There is a general idea that organisms living under similar hazardous environmental conditions develop similar adaptive survival strategies. Those strategies necessarily combine biological and geological mechanisms, such as the production of pigments, changes in cellular membranes (bio-strategies) or migration into the rock substratum and mineralogical alterations (geo-strategies) in their adaptation strategies for successful survival under hostile environments [33,36–39].

The analyses of extreme organisms using Raman spectroscopy started first in the 1990s. Up to the present time, most of the specimens studied have been related to cold and hot deserts, where the availability of water, together with extreme temperatures are the main restrictive environmental parameters (Table 1)—see Table 1 for references. No work, to our knowledge, has been undertaken on the characterization of extremophilic survival strategies without extraction using Raman spectroscopy for lack of oxygen or high pH habitats. For high pressure in the deep ocean environments, papers focus on the identification of minerals and sulfur, sometimes related to biological activity [40–42]. For high levels of cosmic radiation [43], only one paper has been published with the aim of detecting the degradation of biosignatures by Raman spectroscopy. In acid environments [44], only one paper, to our knowledge, has addressed the residual organic signatures in rocks.

The interest in the study of extremophiles rests mainly on the potential use of these organisms in industrial biotechnology and medical fields and in extraterrestrial life signature characterization. The search for life beyond the Earth has encouraged the survey of terrestrial extremophilic strategies, and most of the studies on extremophiles using Raman spectroscopy have been carried out with this purpose. The capabilities of Raman spectroscopy as an analytical technique have led it to be adopted as a part of the instrumental suite to be sent to Mars in the ExoMars-C 2018 mission led by the European Space Agency (ESA) [45].

2.2. Analysis of the Data

Although a large number of relevant papers have been surveyed for this review, many of them are not specifically linked to the stated theme because they analyze extremophiles grown under laboratory conditions [46,47]. Other studies do not identify compounds but specific signatures related to generically important but nonspecific organic biomaterials, such as those characteristic of some proteins, aminoacids, fatty acids, *etc* [47,48]; in other articles related to the study of extremophiles, the use of Raman spectroscopy was directed towards mineral identification [49,50], or only carbon was identified as organic compound and other analytical techniques were utilized for the organic molecule characterization [51]. Studies of bacterial specimens have also been carried out using surface enhanced Raman spectroscopy (SERS) [52–54], a Raman application in which the signal is enhanced by using a specially designed colloidal silver or gold substrate over which the sample is fixed by adsorption; in this case, as sample preparation is required, the analytical method is not strictly non-destructive. Although several of these analyses have studied extremophiles, it should be noted that extraction of the relevant organic compounds or sample manipulation was required for the analyses [55–57].

The analyses of the published papers reported here are related to conventional Raman spectroscopy of naturally occurring extremophilic microorganisms, in which the molecular characterization of the spectral signatures was accomplished without sample manipulation; these are summarized in Tables 1 and 2. References to the research works are given in Table 1 [58–76].

Table 1 shows the specimens related to origin/climate, environmental conditions and their substratum. Most of the samples came from the cold deserts of Antarctica and the Arctic, specifically from Svalbard (samples SP1A, SP1B, SP2 and SP10) or the north of Canada (SP7).

Extremophilcs may be characterized as epiliths, chasmoliths or endoliths, depending on whether the organisms are surface dwelling, have colonized cracks in rocks or, finally, have created subsurface colonies (geostrategies) within the pores of rocks. Usually, endoliths create a colored band a few millimeters below the surface of porous rocks. Although we do not present here a statistical study giving the range of endoliths analyzed, we should note that the most frequent endolithic rock in this review is a sandstone. This is believed to reflect the characteristics of this rock, since it has very well connected pores, which enable air and water to move easily across the rock transect. Another interesting characteristic is that it is composed of quartz, a translucent mineral that allows sunlight to reach the subsurface colony and initiate the chlorophyll function.

Light penetration is a limiting factor for this chlorophyll function in the case of organisms living inside a rock (either in pores, cracks or vacuoles). Organisms have solved this problem either by living as close to the surface as possible or through the production of complementary light-harvesting pigments, such as c-phycocyanin, which are accessory pigments to chlorophyll for the harnessing of photoactive radiation (bio-strategies). In some cases, mineralogical changes as well as mineral mobilization (geo-strategies), directly related with the colonized layer, have been noticed. It is possible to observe these physical changes in the system from the Raman spectroscopic data and relate them to to a biogeological strategy.

When a rock is not porous, such as applite or basalt in Table 1, extremophiles must colonize the surface, or inhabit a crack, become located beneath translucent light colored minerals or inside a vacuole connected with the exterior by a pore (geo-strategies). In this last case, gas and water interchange can occur through the pore, and light can also reach the organisms through it. However, very few chasmoliths have been studied by Raman spectroscopy; hence, only seven specimens from the forty-five described in this review are chasmolithic in origin. Organisms living on the surface of a rock (epilith) are more exposed to the hazardous environmental conditions and they therefore require the operation of supplementary bio-strategies, since they are denied the useful protection offered by rock substrates. In these cases, some microorganisms produce one or more additional protective pigments.

Table 1. Extremophile specimens and related substratum (End = endolith; Chas = chasmolith; Epi = epilith; Sand = sandstone; MR = magmatic rock; Mag = Magnesite; Dol = dolomite; Mar = marble; Gyp = gypsum; Apl = aplite; Fum = fumarole; LB = lava basalt; bact mats = bacteria mats). The numbers in superscript in each specimen indicate the reference paper.

Specimen	Origin/Climate	Environmental Conditions	End	Chas	Epi	Sand	MR	Mag	Dol	Mar	Gyp	Salt	Apl	Fum	LB
ANT1 [58]	Antarctic		X			X									
ANT2 [58]	Antarctic		X			X									
ANT5 [58]	Antarctic		X			X									
EN1 [58]	Antarctic		X			X									
EN2 [58]	Antarctic		X			X									
RW [58]	Antarctic		X			X									
SP1A [59]	Arctic				X				X	X					
SP1B [59]	Arctic			X					X	X					
SP1C [59]	Tropical	Primary colonization			X									X	
SP2A [60]	Arctic			X		X									
SP2B [60]	Arctic			X		X									
SP2C [60]	Arctic		X			X									
SP2D [60]	Arctic		X			X									
SP2E [60]	Arctic				X	X									
SP3A [61]	Antarctic			X									X		
SP3B [61]	Antarctic			X									X		
SP4A [62]	Hot desert	Primary colonization			X	X									
SP4B [62]	Hot desert	Primary colonization			X		X								
SP5A [63]	Antarctic		X			X									
SP5B [63]	Antarctic	Exposed mats													
SP6A [64]			X					X							
SP7A [65]	Arctic	Halotrophic	X								X				
SP7B [65]	Arctic	Halotrophic	X								X				

Table 1. *Cont.*

Specimen	Origin/Climate	Environmental Conditions	End	Chas	Epi	Sand	MR	Mag	Dol	Mar	Gyp	Salt	Apl	Fum	LB
SP8A [66]	Antarctic	Sediments													
SP9A [67]	Arctic	High temperature/ Biofilm													
SP10A [68]	Arctic			X											X
SP10B [68]	Arctic		X												X
SP11A [69]	Hot desert	Halotrophic/ Natron	X									X			
SP12A [70]	Hot desert	Halotrophic/ Halite			X							X			
SP12B [70]	Hot desert	Halotrophic/ Halite	X									X			
SP12C [70]	Hot desert	Halotrophic/ Halite	X									X			
SP12D [70]	Hot desert	Halotrophic/ Halite	X									X			
SP13A [71]	Antarctic	Snow algae													
SP13B [71]	Antarctic	Snow algae													
SP14A [72]	Antarctic		X			X									
SP14B [72]	Antarctic		X			X									
SP14C [72]	Antarctic		X			X									
SP15A [73]	Hot desert	Halotrophic	X									X			
SP15B [73]	Hot desert	Halotrophic	X								X	X			
SP15C [73]	Hot desert	Halotrophic/bact mats													
SP16A [74]	Antarctic			X					X	X					
SP17A [75]		Deep see vents/mats													
SP18A [76]	Hot desert	Halotrophic	X									X			
SP18B [76]	Hot desert	Halotrophic	X									X			
SP18C [76]	Hot desert	Halotrophic	X									X			

Table 2. Extremophile specimens and their strategies detected by Raman spectroscopy.

	ANT1	ANT2	ANT5	EN1	EN2	RW	SP1A	SP1B	SP1C	SP2A	SP2B	SP2C	SP2D	SP2E
Chlorophyll	X	X	X	X	X			X		X	X		X	
One carotenoid	X	X	X			X			X		X		X	
Two or more carotenoids								X		X		X		
Scytonemin						X	X						X	
C-Phycocyanin												X		
Compound 1		X	X	X										
Compound 2									X					
Calcium oxalate monohydrate Whewellite	X			X	X									
Calcium oxalate dihydrate Weddellite	X													
Hematite	X		X	X		X				X				X
Limonite/goethite										X				
Calcite						X								X
Hydrocerussite						X								
Gypsum						X								
Pyrophyllite			X											
Rutile													X?	X?
	SP3A	**SP3B**	**SP4A**	**SP4B**	**SP5A**	**SP5B**	**SP6A**	**SP7A**	**SP7B**	**SP8A**	**SP9A**	**SP10A**	**SP10B**	**SP11A**
Chlorophyll		X	X	X			X	X	X	X		X	X	X
One carotenoid	X	X	X		X		X	X	X		X			
Two or more carotenoids												X	X	X
Scytonemin			X			X	X	X						
C-Phycocyanin					X								X	
Atranorin or parietin				X										
Cholesterol					X									
Parietien									X					
Compound 3							X							
Calcium oxalate monohydrate Whewellite				X										
Hematite					X									
Calcite					X									
Aragonite	X						X							
Quartz									X					
Realgar											X			

Table 2. *Cont.*

	SP12A	SP12B	SP12C	SP12D	SP13A	SP13B	SP14A	SP14B	SP14C	SP15A	SP15B	SP15C	SP16A	SP17A
Chlorophyll	X	X	X		X	X	X			X		X	X	
One carotenoid	X	X			X	X	X	X			X	X		X
Two or more carotenoids			X										X	
Scytonemin	X	X	X		X							X	X	
C-Phycocyanin								X					X	
Atranorin						X								
Phycobiliprotein			X											
Proteins							X							
Cellulose								X						
Compound 4				X										
Compound 5								X						
Compound 6								X						
Compound 7											X			
Compound 8													X	
Calcium oxalate monohydrate Whewellite							X	X	X					
Calcium oxalate dihydrate Weddellite							X	X						
Calcite								X						

	SP18A	SP18B	SP18C
Chlorophyll		X	
Scytonemin			X
C-Phycocyanin			
Compound 9	X		

Extremophilic microorganisms living as biofilms, bacterial mats (microorganim sheets formed by organic compounds, bacteria, archaea and minerals mainly), in or on ice or snow, limestone and other sedimentary rocks have been studied using Raman espectroscopy, Table 1, but rather more analyses have been carried out on salt crystals, particularly halite and gypsum.

Most of the specimens analyzed by Raman spectroscopy originated from cold deserts, Table 2, from the Arctic and Antarctic regions, followed by those which came from hot deserts, which are, in most of the cases, related to halotrophic environments. Primary colonizations are also considered extremophilic because of the nutrient shortage. Despite this, specimens SERS2B, SERS2C and SERS2D were collected under cold climate conditions, with the limiting parameter being the high temperature, since they are all related with a hot spring environment.

From Table 1, it can be inferred that deserts, either cold or hot, have been the main environments studied. The current search for life in extra-terrestrial environments is currently focused on the planet Mars and on the planetary satellites Europa and Titan [77–80]. As water is undoubtedly the determining factor for life, the search for extraterrestrial life is focused on regions where liquid water could have existed. Nowadays, it is generally accepted that liquid water existed in the past on Mars, but this water disappeared in its ancient geological history into the subsurface or evaporated into space—although there is clear evidence recently for subsurface fluvial activity and water of crystallization tied up in mineral occlusions. Now, Mars is a cold and dry planet (at least on the surface) but, if life ever existed on Mars, it should have adapted to the ever more extreme environmental conditions assuming that it had the time to do so before the prevailing climate became too inhospitable. It is reasonable to assume that extremophilic microorganisms and their analogues are the most probable source if there is any chance of finding life signals on Mars, either extinct or extant.

Sandstone, carbonated and saline rocks are the most investigated terrestrial substrata with biological colonization, Table 1. These sedimentary rock groups are widely spread on the Earth's surface and have also been described on Mars. Although it is an advantage that water does not interfere with the Raman signals from extremophilic colonies and optimal results can be obtained from the study of microorganisms living in water, ice or snow, surprisingly few studies have been carried out on these systems.

Biomolecules and related minerals produced by the specimens described here have been summarized in Table 2a,b. In several extremophilic specimens, some kind of mineralogical change or mineral mobilization is observed, such as hematite accumulation on the rock surface, a deficiency of hematite noted around the organic colony, iron oxide transformation closely related to the microorganisms and changes between calcium carbonate phases (calcite to aragonite) where the organisms are detected; these alterations, after identification by Raman spectroscopy, can then be ascribed to a geo-strategic change. Hence, in these cases, minerals have also been included in the Table 2, but it is important to note that the aim of most of the investigations carried out and published has been focused only on microorganisms and their biomolecules, and not on the study of minerals, so naturally, the mineral occurrence has not been generally described.

Minerals related with extremophiles have also in some cases been deposited from surrounding areas by wind and rain and thereby become associated with the biological colony (SP7B). Other specimens show a clear relationship between aragonite and the microorganism area, such as in (SP3A), where calcite appears as a white dust in the crack surface and only in the immediate area of the colony does aragonite occur. A similar phenomenon happens in SP6A, where in a hydromagnesite rock, aragonite is present in the vicinity of the microorganisms. SP14B and SP14C show evidence of a calcium carbonate phase related with some of the microorganism layers, in spite of the fact that the rock substratum is, in both cases, a sandstone. An interesting case is the specimen RW, from Antarctica, where microorganisms colonize pores a few millimeters below the surface, but a layer of calcite, hydrocerussite and hematite appears, not in the organic layer itself, as in the previous examples, but just surrounding them; since hydrocerussite is a lead carbonate and lead is a toxic element, it was suggested that this mineral could play a role in the prevention of biological attacks upon the colony by predators. All these are examples of geo-strategic biological colonization.

Rutile appears in SP2D and E specimens, both in a sandstone substratum in the vicinity of the colonization zone; despite its nonoccurrence in the bulk of the rock, it is not possible to conclude that there is a biological transformation between anatase (a rutile polymorph) and rutile, nor that there is a direct relationship between its presence and the biological activity. Hematite appears in some of the sandstones described, giving a reddish-orange color to the rock; it is remarkable that microorganisms seem to have mobilized this mineral, creating a lighter colored area around them and concentrating the mineral on the surface of the rock, such as in ANT1, ANT5, EN1, SP2A or SP5A. It is interesting that, to our knowledge, there has never been a description of the presence of hematite in the colonized area, and rather when an iron oxide appears directly related with the microorganisms it is in the form of a goethite phase, such as in SP2A.

Chlorophyll and carotenoids are the most prevalent biomolecules found in the Raman spectra of extremophiles. However, in many papers, the precise Raman identification of the type of carotenoid has not been assigned specifically—a possible explanation for this is that the unambiguous identification of a carotenoid can be very difficult in natural samples because of several factors which operate in complex mixtures and which cause quite large discrepancies to occur between the observed carotenoid bands and their pure isolated counterparts [81]. What can be assessed from the Raman spectrum is the presence of one or more carotenoids in a system and this is well documented. Carotenoids play very important roles as protective pigments: As DNA repair agents, free-radical quenching molecules and UV-radiation screens (bio-strategies). Production of c-phycocyanin (bio-strategy), a light harvesting pigment, was mainly found in those extremophiles that have colonized dark rocks, such as in SP10B, or deep areas within the rock (SP2C, SP5A, SP14B, SP16A); other organisms with c-phycocyanin but no data on the rock color or depth are provided for the specimens SERS2C, SERS2D. Another frequently found pigment is scytonemin, which has been demonstrated to play an important role as a UV screening extracellular compound.

The characteristic Raman signatures for each biomolecule and mineral identified for each extremophile described in this review are shown in Table 3.

There are some Raman bands observed which as yet are not assigned, but these have been cited in Tables 2 and 3 as Compounds 1 to 9. In most cases, no indication of a possible source was supplied, except for Compound 1 (a chlorophyll-like compound) and Compound 7 (an aromatic polyphenolic compound) but for Compounds 5 (aliphatic) and 6 (aromatic) only a rather vague idea is given.

Table 3. Raman characteristic signatures of the biomolecules and minerals related with extremophile microorganisms.

COMPOUNDS	RAMAN SIGNATURES
Chlorophyll	1326, 1285, 987, 916, 755, 744, 516
Carotenoids	1550–1500, 1145–1160, 990–1015
Scytonemin	1591, 1553, 1432, 1381, 1321, 1170, 984, 752, 675, 574
C-Phycocyanin	1638, 1582, 1463, 1369, 1272, 815
Atranorin or parietin (antraquinone)	1630, 963
Cholesterol	1445, 1437, 875, 539
Atranorin	1666, 1658, 1303, 1294, 1266, 588
Parietien	1671, 1613, 1553, 1387, 1370, 1277, 1255, 926, 458
Phycobiliprotein	1638, 1586, 1468, 1369, 1281, 1236, 1049, 665
Proteins	1659 (amide I)1240-1290 (amideIII)
Cellulose	1380,1292
Compound 1 chlorophyll-like	1637, 1568, 1480, 1407, 1343, 1320, 789, 688, 554, 499
Compound 2	1449, 1342, 952, 834, 748, 680, 595, 438, 257, 173
Compound 3	1468, 1439, 948, 884
Compound 4	1531, 1452, 1342, 1143, 748, 681
Compound 5 (aliphatic)	940
Compound 6 (aromatic)	1005
Compound 7 aromatic polyphenolic compound	1442, 1629, 1757
Compound 8	1544, 1500, 1436, 1356, 1316, 1306, 1154
Compound 9	1690, 1645, 1520, 1441, 1250
Calcium oxalate monohydrate (whewellite)	1490, 1463, 896, 504
Calcium oxalate dihydrate (weddellite)	1475, 910, 506
Hematite	610, 405, 292, 223
Limonite/goethite	555, 395, 299, 203
Calcite	1086, 713, 282, 156
Aragonite	1086, 708, 203, 156
Hydrocerussite	1051, 681, 412
Gypsum	1132, 1008, 679, 618, 492, 413
Pyrophyllite	813, 705, 353, 259, 214, 195, 171
Rutile	609, 442
Quartz	463, 205, 128
Realgar	342, 220, 192, 182

3. Conclusions

Raman spectroscopy has been demonstrated to be an appropriate technique for the study of extremophilic microorganisms owing to its capabilities, particularly because it is a non-destructive technique, involving no chemical or textural structural changes in the sample and being able to identify organic and mineral compounds in admixture without any physical or chemical sample preparation.

From the specimens studied using this technique, it is possible to conclude that particular biomolecules are generally not related to any specific protective strategy apart from phycocyanins involved in photochemical light enhancement collection; chlorophyll and carotene are the most extensively occurring pigments and can be relatively easily detected using Raman spectroscopy. However, only extremophiles from cold and hot deserts have been most widely studied followed by halophiles and hot spring organisms; with such few studies having been undertaken overall, we can conclude that there are not enough data as yet to make unequivocal conclusions about the existence of common extremophilic survival strategies related to climate or substratum.

Raman spectroscopy is now vitally important as an analytical tool for the search for life on Mars and for related space missions, since it can primarily detect organic and inorganic compounds in admixture without their separation being effected; at the same time, it is important to create a large database of terrestrial extremophiles from as diverse a range of possible hostile environments as possible and to indentify the survival bio- and geo-strategies to provide an understanding of the key biomolecular protection chemicals operative prior to the exploration of planetary surfaces; then it should be possible to devise schemes to understand their strategic methodologies and mechanisms for survival; in this, Raman spectroscopic analysis will occupy a crucial role.

References and Notes

1. Edwards, H.G.M.; Farwell, D.W.; Seaward, M.R.D.; Giacobini, C. Preliminary Raman Microscopic Analyses of a Lichen Encrustation Involved in the Biodeterioration of Renaissance Frescoes in Central Italy. *Int. Biodeterior.* **1991**, *27*, 1–9.
2. Edwards, H.G.M.; Holder, J.M.; Russell, N.C.; Wynn-Williams, D.D. *Spectroscopy of Biological Molecules: Modern Trends*; Carmona, P., Navarro, R., Hernanz, A., Eds.; Kluwer Academic: Dordrecht, The Netherlands, 1997; pp. 509–510.
3. Edwards, H.G.M.; Russell, N.C.; Wynn-Williams, D.D. FT-Raman Spectroscopic and Scanning Electron Microscopic Study of Cryptoendolithic Lichens from Antarctica. *J. Raman Spectrosc.* **1997**, *28*, 685–690.
4. Edwards, H.G.M.; Farwell, D.W.; Seaward, M.R.D. Raman Spectra of Oxalates in Lichen Encrustations on Renaissance Frescoes. *Spectrochim. Acta* **1991**, *47*, 1531–1539.
5. Edwards, H.G.M.; Farwell, D.W.; Jenkins, R.; Seaward, M.R.D. Vibrational Raman Spectroscopic Studies of Calcium Oxalate Monohydrate and Dihydrate in Lichen Encrustations on Renaissance Frescoes. *J. Raman Spectrosc.* **1992**, *23*, 185–189.
6. Edwards, H.G.M.; Seaward, M.R.D. Raman Spectroscopy and Lichen Biodeterioration *Spectrosc. Eur.* **1993**, *5*, 16–20.
7. Edwards, H.G.M.; Edwards, K.A.E.; Farwell, D.W.; Lewis, I.R.; Seaward, M.R.D. An Approach to Stone and Fresco Lichen Biodeterioration through FT-Raman Microscopic Investigation of Thallus-Substratum Encrustations. *J. Raman Spectrosc.* **1994**, *25*, 99–103.

8. Edwards, H.G.M.; Seaward, M.R.D. *Biodeterioration and Biodegradation 9*; Bousher, A., Chandra, M., Edyvean, R., Eds.; Institute of Chemical Engineers Publication, Hobbs Printers: Totton, Hampshire, UK, 1995; pp. 199–203.
9. Edwards, H.G.M.; Holder, J.M.; Wynn-Williams, D.D. Comparative FT-Raman Spectroscopy of *Xanthoria.* Lichen-Substratum Systems from Temperate and Antarctic Habitats. *Soil Biol. Biochem.* **1998**, *30*, 1947–1953.
10. Wynn-Williams, D.D.; Edwards, H.G.M.; Garcia-Pichel, F. Functional Biomolecules of Antarctic Stromatolitic and Endolithic Cyanobacterial Communities. *Eur. J. Phycol.* **1999**, *34*, 381–391.
11. Edwards, H.G.M.; Garcia-Pichel, F.; Newton, E.M.; Wynn-Williams, D.D. Vibrational Raman Spectroscopic Study of Scytonemin, the UV-Protective Cyanobacterial Pigment. *Spectrochim. Acta Part A* **2000**, *56*, 193–200.
12. Holder, J.M.; Wynn-Williams, D.D.; Rull Perez, F.; Edwards, H.G.M. Raman Spectroscopy of Pigments and Oxalates *In Situ* within Epilithic Lichens: *Acarospora* from the Antarctic and Mediterranean. *New Phytol.* **2000**, *145*, 271–280.
13. Edwards, H.G.M.; Wynn-Williams, D.D.; Newton, E.M.; Coombes, S.J. Molecular Structural studies of Lichen Substances I: Parietin and Emodin. *J. Mol. Struct.* **2003**, *648*, 49–59.
14. Edwards, H.G.M.; Newton, E.M.; Wynn-Williams, D.D. Molecular Structural Studies of Lichen Substances II: Atranorin, Gyrophoric Acid, Fumarprotocetraric Acid, Rhizocarpic Acid, Calycin, Pulvinic Dilactone and Usnic Acid. *J. Mol. Struct.* **2003**, *651*, 27–37.
15. Marshall, C.P.; Edwards, H.G.M.; Jehlicka, J. Understanding the application of Raman spectroscopy to the detection of traces of life. *Astrobiology* **2010**, *10*, 229–243.
16. Vandenabeele, P.; Jehlicka, J.; Vitec, P.; Edwards, H.G.M. On the definition of Raman spectroscopic detection limits for the analyses of biomarkers in solid matrices. *Planet. Space Sci.* **2012**, *62*, 48–54.
17. Edwards, H.G.M.; Wynn-Williams, D.D.; Little, S.J.; de Oliveira, L.F.C.; Cockell, C.S.; Ellis-Evans, J.C. Stratified Response to Environmental Stress in a Polar Lichen Characterised with FT-Raman Microscopic Analysis. *Spectrochim. Acta Part. A* **2004**, *60*, 2029–2033.
18. Edwards, H.G.M.; Newton, E.M.; Wynn-Williams, D.D.; Lewis-Smith, R.I. Nondestructive Analysis of Pigments and other Organic Compounds in Lichens Using Fourier-Transform Raman Spectroscopy: A Study of Antarctic Epilithic Lichens. *Spectrochim. Acta Part A* **2003**, *59*, 2301–2309.
19. Russ, J.; Palma, R.L.; Loyd, D.H.; Farwell, D.W.; Edwards, H.G.M. Analysis of the Rock Accretions in the Lower Pecos Region of Southwest Texas. *Geoarchaeology* **1995**, *10*, 43–63.
20. Seaward, M.R.D.; Edwards, H.G.M. Lichen-Substratum Interface Studies with Particular Reference to Raman Microscopic Analysis. I. The Deterioration of Works of Art by *Dirina Massiliensis* forma *Sorediata. Cryptogam. Bot.* **1995**, *5*, 282–287.
21. Edwards, H.G.M.; Rull Perez, F. Lichen Biodeterioration of the Convento de la Peregrina, Sahagun, Spain. *Biospectroscopy* **1999**, *5*, 47–52.
22. Villar S.E.J.; Edwards, H.G.M.; Seaward, M.R.D. Lichen Biodeterioration of Ecclesiastical Monuments in Northern Spain. *Spectrochim. Acta Part A* **2004**, *60*, 1229–1237.

23. Wynn-Williams, D.D.; Edwards, H.G.M. Proximal Analysis of Regolith Habitats and Protective Biomolecules *In situ* by Laser Raman Spectroscopy: Overview of Terrestrial Antarctic Habitats and Mars Analogs. *Icarus* **2000**, *144*, 486–503.
24. Wynn-Williams, D.D.; Edwards, H.G.M. Antarctic Eco-Systems as Models for Extra-Terrestrial Surface Habitats. *Planet. Space Sci.* **2000**, *48*, 1065–1075.
25. Wynn-Williams, D.D.; Edwards, H.G.M. *Astrobiology: The Quest for the Conditions of Life*; Horneck, G., Baumstark-Khan, C., Eds.; Springer-Verlag: Berlin, Germany, 2002; pp. 245–260.
26. Edwards, H.G.M.; Newton, E.M.; Wynn-Williams, D.D.; Dickensheets, D.; Schoen, C.; Crowder, C. Laser Wavelength Selection for Raman Spectroscopy of Microbial Pigments *in situ* in Antarctic Desert Ecosystem Analogues of Former Habitats on Mars. *Int. J. Astrobiol.* **2003**, *1*, 333–348.
27. Ellery, A.; Kolb, C.; Lammer, H.; Parnell, J.; Edwards, H.; Richter, L.; Patel, M.; Romstedt, J.; Dickensheets, D.; Steele, A. Astrobiological Instrumentation for Mars—The Only Way is Down! *Int. J. Astrobiol.* **2003**, *1*, 365–380.
28. Edwards, H.G.M.; Newton, E.M.; Dickensheets, D.L.; Wynn-Williams, D.D. Raman Spectroscopic Detection of Biomolecular Markers from Antarctic Materials: Evaluation for Putative Martian Habitats. *Spectrochim. Acta Part A* **2003**, *59*, 2277–2290.
29. Edwards, H.G.M. Raman Spectroscopic Protocol for the Molecular Recognition of Key Biomarkers in Astrobiological Exploration. *Orig. Life Evol. Biospheres* **2004**, *34*, 3–11.
30. Ellery, A.; Wynn-Williams, D.D.; Parnell, J.; Edwards, H.G.M.; Dickensheets, D. The Role of Raman Spectroscopy as an Astrobiological Tool in the Exploration of Mars. *J. Raman Spectrosc.* **2004**, *35*, 441–457.
31. Ferraro, J.R.; Nakamoto, K.; Brown, C.W. *Introductory Raman Spectroscopy*, 2nd ed.; Academy Press: London, UK, 2003.
32. Canganella, F.; Wiegel, J. Extremophiles: From abyssal to terrestrial ecosystems and possibly beyond. *Naturwissenschaften* **2011**, *98*, 253–279.
33. Dong, H.; Yu1, B. Geomicrobiological processes in extreme environments: A review. *Episodes* **2007**, *30*, 202–216.
34. Fujiwara, S. Extremophiles: Developments of their special functions and potential resources. *J. Biosci. Bioeng.* **2002**, *94*, 518–525.
35. Rothschild, L.J.; Mancinelli, R.L. Life in extreme environments. *Nature* **2001**, *409*, 1092–1100.
36. Friedmann, E.I. Endolithic microorganisms in the Antarctic cold desert. *Science* **1982**, *4536*, 1045–1053.
37. Dong, H. Mineral-microbe interactions: A review. *Front. Earth. Sci. China* **2010**, *4*, 127–147.
38. Bennett, P.C.; Rogers, J.R.; Choi, W.J.; Hiebert, F.K. Silicates, silicate weathering and microbial ecology. *Geomicrobiol. J.* **2001**, *18*, 3–19.
39. Rogers, J.R.; Bennett, P.C. Mineral stimulation of subsurface microorganisms: Release of limiting nutrients from silicates. *Chem. Geol.* **2004**, *203*, 91–108.
40. Breier, J.A.; White, S.N.; German, C.R. Mineral-microbe interactions in deep-sea hydrothermal systems: A challenge for Raman spectroscopy. *Phil. Trans. R. Soc. A* **2010**, *368*, 3067–3086.

41. Himmel, D.; Maurin, L.C.; Mansont, J.L. Raman microspectrometry sulfur detection and characterization in the marine ectosymbiotic nematode *Eubostrichus* dianae (Desmodoridae, Stilbonematidae). *Biol. Cell.* **2009**, *101*, 43–54.
42. Oger, P.M.; Daniel, I.; Picard, A. *In situ* Raman and X-ray spectroscopies to monitor microbial activities under high hydrostatic pressure. *Ann. NY Acad. Sci.* **2010**, *1189*, 113–120.
43. Dartnell, L.R.; Page, K.; Jorge-Villar, S.E.; Wright, G.; Munshi, T.; Scowen, I.J.; Ward, J.M.; Edwards, H.G.M. Destruction of Raman biosignatures by ionising radiation and the implications for life detection on Mars. *Anal. Bioanal. Chem.* **2012**, *403*, 131–144.
44. Edwards, H.G.M.; Vandenabeele, P.; Jorge-Villar, S.E.; Carter, E.A.; Rull Perez, F.; Hargreaves, M. The Rio Tinto Mars analogue site: An extremophilic Raman spectroscopic study. *Spectrochim. Acta Part A* **2007**, *68*, 1133–1137.
45. ESA Robotic Exploration of Mars. Available online: http://exploration.esa.int/science-e/www/object/index.cfm?fobjectid=46048/ (accessed on 27 November 2012).
46. Goodwin, J.R.; Hafner, L.M.; Fredericks, P.M. Raman spectroscopic study of the heterogeneity of microcolonies of a pigmented bacterium. *J. Raman Spectrosc.* **2006**, *37*, 932–936.
47. Jehlicka, J.; Oren, A.; Vitek, P. Use of Raman spectroscopy for the identification of compatible solutes in halophilic bacteria. *Extremophiles* **2012**, *16*, 507–514.
48. Fisk, M.R.; Storrie-Lombardi, M.C.; Douglas, S.; Popa, R.; McDonald, G.; di Meo-Savoie, C. Evidence of biological activity in Hawaiian subsurface basalts. *Geochem. Geophys. Geosyst.* **2003**, *4*, 1–24.
49. Cockell, C.S.; Clasteren, P.V.; Mosselmans, J.F.W.; Franchi, I.A.; Gilmour, I.; Kelly, L.; Olsson-Francis, K.; Johnson, D. Microbial endolithic colonization and the geological environment in young seafloor basalts. *Chem. Geol.* **2010**, *279*, 17–30.
50. Gleeson, D.F.; Pappalardo, R.T.; Anderson, M.S.; Grasby, S.E.; Mielke, R.E.; Wrigth, K.E.; Templeton, A.S. Biosignature detection at an Arctic analog to Europa. *Astrobiology* **2012**, *12*, 135–150.
51. Cockell, C.S.; Osinki, G.R.; Banerjee, N.R.; Howard, K.T.; Gilmour, I.; Watson, J.S. The microbe-mineral environment and gypsum neogenesis in a weathered polar evaporite. *Geobiology* **2010**, *8*, 293–308.
52. Efrima, S.; Zeiri, L. Understanding SERS of Bacteria. *J. Raman Spectrosc.* **2008**, *40*, 277–288.
53. Zeiri, L.; Efrima, S. Surface-enhanced Raman spectroscopy of bacteria: The effect of excitation wavelength and chemical modification of the colloidal milieu. *J. Raman Spectrosc.* **2005**, *36*, 667–675.
54. Xie, W.; Su, L.; Shen, A.; Maternyc, A.; Hua, J. Application of surface-enhanced Raman scattering in cell analysis. *J. Raman Spectrosc.* **2011**, *42*, 1248–1254.
55. Bowden, S.A.; Wilson, R.; Cooper, J.M.; Parnell, J. The use of surface-enhanced Raman scattering for detecting molecular evidence of life in rocks, sediments and sedimentary deposits. *Astrobiology* **2010**, *10*, 629–641.
56. Laucks, M.L.; Sengupta, A.; Jungle, K.; Davis, E.J.; Swanson, B.D. Comparison of psychro-active Arctic marine bacteria and common mesophillic bacteria using surface-enhanced Raman spectroscopy. *Appl. Spectrosc.* **2005**, *59*, 1222–1228.

57. Wilson, R.; Monaghan, P.; Bowden, S.A.; Parnell, J.; Cooper, J.M. Surface-enhanced Raman signatures of pigmentation of cyanobacteria from within geological samples in a spectroscopic-microfluidic flow cell. *Anal. Chem.* **2007**, *79*, 7036–7041.
58. Jorge-Villar, S.E.; Edwards, H.G.M.; Cockell, C.S. Raman spectroscopy of endoliths from Antarctic cold desert environments. *Analyst* **2005**, *130*, 156–162.
59. Jorge-Villar, S.E.; Edwards, H.G.M. Raman spectroscopy in Astrobiology. *Anal. Bioanal. Chem.* **2006**, *384*, 100–113.
60. Jorge-Villar, S.E.; Edwards, H.G.M.; Benning, L.G. AMASE 2004 team. Raman spectroscopic analysis of Arctic nodules: Relevance to the astrobiological exploration of Mars. *Anal. Bioanal. Chem.* **2011**, *401*, 2927–2933.
61. Jorge-Villar, S.E.; Edwards, H.G.M.; Wynn-Williams, D.D.; Worland M.R. FT-Raman spectroscopic analysis of an Antarctic endolith. *Int. J. Astrobiol.* **2003**, *1*, 349–355.
62. Edwards, H.G.M.; Moody, C.D.; Jorge-Villar, S.E.; Mancinelli, R. Raman spectroscopy of desert varnishes and their rock substrata. *J. Raman Spectrosci.* **2004**, *35*, 475–479.
63. Edwards, H.G.M.; Moody, C.D.; Jorge-Villar, S.E.; Wynn-Williams, D.D. Raman spectroscopic detection of key biomarkers of cyanobacterial and lichens symbiosis in extreme Antarctic habitats: Evaluation for Mars lander missions. *Icarus* **2005**, *174*, 560–571.
64. Edwards, H.G.M.; Moody, C.D.; Newton, E.M.; Jorge-Villar, S.E.; Russell, M.J. Raman spectroscopic analysis of cyanobacterial colonization of hydromagnesite, a putative martian extremophile. *Icarus* **2005**, *175*, 372–381.
65. Edwards, H.G.M.; Jorge-Villar, S.E.; Parnell, J.; Cockell, C.S.; Lee, P. Raman spectroscopic analysis of cyanobacterial gypsum halotrophs and relevance for sulfate deposits on Mars. *Analyst* **2005**, *130*, 917–923.
66. Moody, C.D.; Jorge-Villar, S.E.; Edwards, H.G.M.; Hodgson, D.A.; Doran, P.T.; Bishop, J.L. Biogeological Raman spectroscopic studies of Antarctic lacustrine sediments. *Spectrochim. Acta Part A* **2005**, *61*, 2413–2417.
67. Jorge-Villar, S.E.; Edwards, H.G.M.; Worland, M.R. Comparative evaluation of Raman spectroscopy at different wavelengths for extremophile exemplars. *Orig. Life Evol. Biospheres* **2005**, *35*, 489–506.
68. Jorge-Villar, S.E.; Edwards, H.G.M.; Benning, L.G. Raman spectroscopic and scanning electron microscopic analysis of a novel biological colonisation of volcanic rocks. *Icarus* **2006**, *184*, 158–169.
69. Edwards, H.G.M.; Currie, K.J.; Ali, H.R.H.; Jorge-Villar, S.E.; David, A.R.; Denton, J. Raman spectroscopy of natron: Shedding light on ancient Egyptian mummification. *Anal. Bioanal. Chem.* **2007**, *388*, 683–689.
70. Vitek, P.; Edwards, H.G.M.; Jehlicka, J.; Ascaso, C.; de los Rios, A.; Vallea, S.; Jorge-Villar, S.E.; Davila, A.F.; Wierzchos, J. Microbial colonization of halite from the hyper-arid Atacama desert studied by Raman spectroscopy. *Phil. Trans. R. Soc.* **2010**, *368*, 3205–3221.

71. Edwards, H.G.M.; Oliveira, L.F.C.; Cockell, C.S.; Ellis-Evans, J.C.; Wynn-Williams, D.D. Raman spectroscopy of senescing snow algae: Pigmentation changes in an Antarctic cold desert extremophile. *Int. J. Astrobiol.* **2004**, *3*, 125–129.
72. Russell, N.C.; Edwards, H.G.M.; Wynn-Williams, D.D. FT-Raman spectroscopic analysis of endolithic communities from Beacon sandstone in Victoria Land, Antarctica. *Antarct. Sci.* **1998**, *10*, 63–74.
73. Edwards, H.G.M.; Mohsin, M.A.; Sadooni, F.N.; Hassan, N.F.; Munshi, T. Life in the Sabkha: Raman spectroscopy of halotrophic extremophiles of relevance to planetary exploration. *Anal. Bioanal. Chem.* **2006**, *385*, 46–56.
74. Edwards, H.G.M.; Jorge-Villar, S.E.; Pullan, D.; Hargreaves, M.D.; Hofmann, B.A.; Westall, F. Morphological biosignatures from relict fossilised sedimentary geological specimens: A Raman spectroscopic study. *J. Raman Spectrosc.* **2007**, *38*, 1325–1361.
75. White, S.N.; Dunk, R.M.; Peltzer, E.T.; Freeman, J.J.; Brewer, P.G. *In situ* Raman analyses of deep-sea hydrothermal and cold seep systems (Gorda Ridge and Hydrate Ridge). *Geochem. Geophys. Geosyst.* **2006**, *5*, 1–12.
76. Edwards, H.G.M.; Sadooni, F.; Vitek, P.; Jehlicka, J. Raman spectroscopy of the Dukhan sabkha: Identification of geological and biogeological molecules in an extreme environment. *Phil. Trans. R. Soc. A* **2012**, *368*, 3099–3107.
77. Smith, J.A.; Onstott, T.C. Follow the Water: Steve Squyres and the Mars Exploration Rovers. *J. Franklin Inst.* **2011**, *348*, 446–452.
78. Baker, V.R. Water and the Martian landscape. *Nature* **2001**, *412*, 228–236.
79. Squyres, S.W.; Kasting, J.F. Early Mars: How warm and how wet? *Science* **1994**, *265*, 744–749.
80. Niles, P.B.; Catling, D.C.; Berger, G.; Chassefière, E.; Ehlmann, B.L.; Michalski, J.R.; Morris, R.; Ruff, S.W.; Sutter, B. Geochemistry of carbonates on Mars: Implications for climate history and nature of aqueous environments. *Space Sci. Rev.* **2013**, *174*, 301–328.
81. De Oliveira, V.E.; Castro, H.V.; Edwards, H.G.; de Oliveira, L.F.C. Carotenes and carotenoids in natural biological samples: A Raman spectroscopic analysis. *J. Raman Spectrosc.* **2010**, *41*, 642–650.

Survival of the Fittest: Overcoming Oxidative Stress at the Extremes of Acid, Heat and Metal

Yukari Maezato and Paul Blum

Abstract: The habitat of metal respiring acidothermophilic lithoautotrophs is perhaps the most oxidizing environment yet identified. Geothermal heat, sulfuric acid and transition metals contribute both individually and synergistically under aerobic conditions to create this niche. Sulfuric acid and metals originating from sulfidic ores catalyze oxidative reactions attacking microbial cell surfaces including lipids, proteins and glycosyl groups. Sulfuric acid also promotes hydrocarbon dehydration contributing to the formation of black "burnt" carbon. Oxidative reactions leading to abstraction of electrons is further impacted by heat through an increase in the proportion of reactant molecules with sufficient energy to react. Collectively these factors and particularly those related to metals must be overcome by thermoacidophilic lithoautotrophs in order for them to survive and proliferate. The necessary mechanisms to achieve this goal are largely unknown however mechanistics insights have been gained through genomic studies. This review focuses on the specific role of metals in this extreme environment with an emphasis on resistance mechanisms in Archaea.

Reprinted from *Life*. Cite as: Maezato, Y.; Blum, P. Survival of the Fittest: Overcoming Oxidative Stress at the Extremes of Acid, Heat and Metal. *Life* **2012**, *2*, 229-242.

1. Introduction

The presence of heavy metals in extreme microbial habitats is common. This juxtaposition offers an important opportunity to investigate resistance and toxicity of diverse heavy metals towards natural communities and individual taxa. Heavy metal resistance in bacteria has been widely reviewed [1–4]. For bacteria, the ecology, chemistry and biologic mechanisms of resistance associated with arsenic, selenium and copper have also been described [3,5]. Heavy metals promote highly oxidative environments which lead to a differential level of toxicity of the metal [6,7]. The response to toxic elements involves diverse strategies including: redox-based metal trans formation [8]; trafficking within the cytoplasm mediated by metal chaperones [9,10]; protein sequestration [11,12]; metal efflux [8]; and by metal-phosphate symport [13,14]. Horizontal gene transfer also plays a crucial part in the evolution of heavy metal resistance within bacterial communities [15]. Archaea also occupy diverse extreme habitats including those that are metal rich. Examples include acidic and sulfidic geothermal pools and soils, marine hydrothermal vents emanating from metal rich ores, and hypersaline pools and soil crusts saturated with metals and other elements. In some of these habitats particularly ore heaps, thermophilic archaea proliferate because of the intrinsic heat generated by chemical and biological oxidation that also promotes solubilization of metal sulfide complexes. However, though archaea flourish in such environments, much less is understood about their mechanisms of survival. Analysis of extant archaeal genomes identifies many genes involved with defenses againts toxic metals while studies using model archaeal taxa have begun to yield mechanistic details underlying metal resistance. Therefore, this review examines the literature related to metal biology of archaea.

2. Ecological Considerations of Archaea in Metal Rich Environments

Many toxic metals including copper (Cu) cadmium (Cd), zinc (Zn), uranium (U) and arsenic (As) are commonly called heavy metals but are also assigned to the category of "soft metals" because of their physical properties and their large ratio of ionic charge to ionic radius [16]. In contrast, alkali metals and alkali earth metals such as potassium (K) and calcium (Ca) are placed in the category of "hard metals". The stronger interaction of soft metals with proteins compared to the weaker ionic interaction with hard metals, is one factor underlying their toxicity [16] and is a reflection of the tendency of soft acids and bases to interact [17]. Despite this effect, soft metals are also important at low concentrations in diverse biological processes [18] such as those required between enzymes and their co-factors. To balance the critical need for trace metals against their potential toxicity, archaea have evolved multiple regulatory mechanisms to control metal exposure in their environments.

Archaea, representing one of the three domains of life, have been found in diverse environments [19]. The discovery and isolation of numerous archaeal species from environments with high concentrations of heavy metals which also contributes to the highly oxidative environment such as mining sites, salterns, and metal contaminated soils, has accelerated interest in studying metal resistance in these organisms. Importantly, archaea as a fraction of total cells are abundant in many habitats from benign to extreme (Table 1). The apparent abundance of archaeal cells implicates them as agents capable of mediating metal transformations. The best characterized archaeal biotypes provide an important context for much of the information that is available linking heavy metals with these organisms. While these biotypes are generalizations that have been expanded through culture-independent studies on archaeal biodiversity, they continue as physiologic paradigms providing a basis for mechanistic information.

2.1. Haloarchaea

Halophilic archaea, including the obligate and extremely halophilic taxa, belong to the phylum *Euryarchaeota*. They occur in areas with high concentrations of salt (>2 M), such as the Dead Sea, salt lakes, inland seas, and evaporating ponds of seawater. These hypersaline habitats are also rich in heavy metals [31,32], and many extreme halophiles have developed strategies to overcome metal toxicity [33]. For example, *Halobacterium* strain NRC-1 has multiple mechanisms for responding to arsenic. In the presence of arsenite (As (III)), these organisms oxidize the metal and then export it via metal specific translocators. Alternatively they alkylate (methylate) this metal as an alternative means of detoxification [32]. High level arsenic resistance in *Halobacterium* strain NRC-1 is mediated by the plasmid encoded *ars* operons (*arsADRC*, and *M*), rather than the chromosome encoded *arsB* gene [32].

Table 1. Relative abundance of archaea in the environment.

Habitat	Abundance (Maximum)	Dominant archaeal type	Method of determination	Citation
Aquatic				
Marine	24%–34% of total prokaryotic rRNA	Crenarchaeota	Measuring of amplification of group specific ribosomal RNA	[20]
Shallow water hydrothermal vent	12% of total prokaryotic rRNA sampled	Euryarchaeota (*i.e.*, Thermococcus, Pyrococcus, Pyrobaculum, methanococcus)	Ribosomal RNA hybridization, Fluorescent in situ Hybridization (FISH)	[21]
Deep-sea Sulfide Chimney	33%–60%	Marine group I Crenarchaeota and Euryarchaeota (*i.e.*, Thermococcales)	Fluorescent in situ Hybridization (FISH), 16SrRNA analysis, and RFLP anlaysis	[22]
Holo-oligomictic Lake	47%		Catalyzed Reporter Deposition- Fluorescence In Situ Hybridization (CARD-FISH) analysis	[23]
Terrestrial				
Geothermal, solfatara	70%	Crenarchaeota (*i.e.*, Sulfolobales)	Fluorescent in situ Hybridization (FISH) analysis	[24]
Solar (Saltern) pond	80%–86% of total prokaryotic rRNA	Euryarchaeaota (*i.e.*, Haloarchaea)	DAPI counts and Fluorescent in situ Hybridization (FISH) analysis	[25]
Soil	~2% of total sampled 16S rRNA	Thaumarchaeota	16S rRNA analysis and barcoded pyrosequencing analysis	[26,27]
Metal Rich environment				
Acid mine drainage (AMD)/Mining sites	Up to 80% of sampled 16S rRNA	Euryarchaea (*i.e.*, Ferroplasma)	16S rRNA library sequencing and Fluorescent in situ Hybridization (FISH) analysis	[28–30]

2.2. Methanoarchaea

Methanogenic archaea produce methane and also are classified as members of the *Euryarchaeota*. Methanoarchaea are found in diverse environments such as deep sea sediments [34], polar caps [20], agricultural soils, and sewage sludge [35]. However, the molecular mechanisms of heavy metal resistance in methanogens are not well characterized. Bioinformatic surveys of various genomes reveal the presence of gene homologs involved with heavy metal resistance. These include arsenic resistance genes (*arsA*, *C*, *D*, *R* and *M*) and copper resistance genes (*copA*, *R*). One molecular genetic study of copper resistance in *Methanobacterium bryantii* suggests involvement of the copper inducible protein Crx (copper response extracellular protein) [36]. A cadmium and copper resistant novel species, *Methanocalculus pumilus*, has also been described [37] though the responsible mechanisms are not yet characterized. Methanogens are important facilitators of geochemical cycling of various heavy metals through their ability to form methylated forms of heavy metals in the environment [38]. Methylation of heavy metals is probably accomplished enzymatically through the action of putative methytransferases such as ArsM [39].

2.3. Hyperthermophilic Archaea

In general, hyperthemophilic archaea and bacteria grow at or above 80 °C. Hyperthermophilic archaea inhabit high temperature habitats such as deep sea hydrothermal vents, geothermal springs and solfataras (sulfur rich volcanic pond) as well as various metal mining sites [40–43]. Many of the thermophilic and hyperthermophilic archaea are also extremely acidophilic and tolerate pH values less than one [44]. The occurrence of heavy metals (e.g., Cu, Hg, Cd, and As) in hydrothermal and geothermal habitats is well known [45–48], and the microbes residing in such niches are faced with the constant challenge of coping with toxic metals. For example *Metallosphaera prunae* an extremely thermoacidophilic species belonging to the phylum *Crenarchaeota*, was isolated from a uranium mining site [42]. Uranium (U) occurs primarily in two redox states, U^{+4} is insoluble while U^{+6} is soluble and may have implications for metal resistance mechanisms. The mesophilic acidophile, *Ferroplasma acidarmanus*, exhibits bacterial-like arsenite (As(III)) resistance mechanisms involving *arsA* (ATPase), *arsB* (efflux pump for As (III)), *arsD* (chaperone) and *arsR* (transcriptional regulator). However, resistance to arsenate (As(V)) is likely to involve a novel mechanism because the *F. acidarmanus* genome lacks an apparent arsenate reductase (*arsC*) necessary to overcome high intracellular levels of oxidized metal [49] perhaps an example of a divergent but related activity.

Other acidophilic archaea notably the hyperthermoacidophile, *Sulfolobus solfataricus*, occur in metal rich habitats such as Coso Hot Springs (CSH) an abandoned mercury mining area at the edge of the Mojave desert in southeastern California, USA. This natural geothermal environment contains high concentrations of Hg ranging from 2.0 mg/li of pool water to nearly 1 g/kg sediment derived from cinnabar (mercuric sulfide) the primary ore of mercury [24]. Mercury is a toxic metal for all three domains of life with minimal inhibitory concentrations (MIC) ranging from micromolar to millimolar concentrations depending on the domain. The mechanism of toxicity of Hg(II)) towards *S. solfataricus* arises from the inactivation of TFIIB one of the core archaeal generalized transcription factors required for gene transcription [50]. This mechanism is identical to that observed in eukaryotes and unlike that observed in bacteria. Unlike all eukaryotes however, *S. solfataricus* also encodes genes that detoxify this metal. A mercury resistance operon including *merR* (regulator), *merH* (TRASH domain, metal chaperone), and *merA* (metal reductase) was characterized using directed recombination to generate gene knockouts [51]. These studies demonstrated that *merA* was required for metal resistance and that the transcription of these genes was responsive to Hg(II) exposure [51,52].

3. Levels of Resistance Towards Heavy Metals

Archaeal taxa especially acidophiles, exhibit relatively high tolerance to many heavy metals. For example, the highly arsenic resistant *Ferroplasma acidarmanus* can tolerate ~130 mM As in its environment (Table 2). Hyperthermophilic archaea also exhibit higher resistance towards various metals such as Zn, Cu and Cd. Interestingly, the MIC of As in haloarchaea strain NRC1 is relatively high. This may arise from the presence of multiple As resistance mechanisms including the plasmid and chromosome encoded As efflux system and the As methylation detoxification system (ArsM). Although an ArsM like protein is also evident in hyperthermophilic archaeal genomes, ArsC homologs are lacking suggesting there are either divergent orthologs or an alternative resistance mechanism.

Table 2. Minimum inhibitory concentration of metals.

Microorganism	MIC (mM)							Citation
	Zn (II)	Cu (II)	Ag (I)	Hg (II)	As (III)	U (VI)	Cd (II)	
Haloarchaea								
Halobacterium sp.	0.5	1–2.5	0.5	0.01	~20	NA[a]	≤2.5	[53]
Halocula sp.	0.05	2.5	0.05	0.01	10	NA	0.05	[53]
Hyperthermophilic archaea								
Sulfolobus solfataricus	50	1~5	0.008	0.002	NA	NA	10	[54,55]
Sulfolobus metallicus	300	16	0.09	0.05	1.3	0.4	0.9	[56]
Metallosphaera sedula	150	100	0.09	0.05	1	0.4	1	[43]
Methanoarchaea								
Methanocalculus pumilus	NA	1	NA	NA	NA	NA	1	[37]
Methanobacterium bryantii	NA	~1.8	NA	NA	NA	NA	NA	[57]
Acidophilic archaea								
Ferroplasma acidarmanus Fer1	NA	312	NA	NA	133	NA	1	[49,54,58,59]
Bacteria								
Acidithiobacillus ferrooxidans	750	~800	0.9	0.5	1.3	0.4	0.09	[43,49]
E. coli ASU7	10	1~3	~>1	0.013	1	2	5	[7,60–62]

Extremely thermoacidophilic members of *Metallosphaera* exhibit among the higher levels of resistance towards copper of all prokaryotes (Table 2). This distinguishing trait is of interest because of its importance to the mining industry (see below). In the case of copper resistance, *copA* (efflux pump) knockout mutants in *M. sedula* show reduced levels of resistance to cupric ion (Cu (II)), however such mutants retain considerable levels of metal resistance [63]. As is the case for As resistance in *Haloarchaea*, perhaps there are multiple mechanisms orchestrating copper resistance in *M. sedula*.

In contrast to copper resistance in *M. sedula* where the trait correlates with environmental metal abundance, this is not the case with Hg and the related *S. solfataricus*. Since Hg levels are elevated in the Coso Hot Springs habitat, it was likely that *S. solfataricus* could elaborate high levels of mercury resistance. Unexpectedly, endogenous isolates of this organism were not unusually metal resistant (0.3–2 μM) and instead were comparable to other mercury resistant bacterial species [24,64,65]. These findings indicate that other mechanisms perhaps dependent on solute transport or internal redox homeostasis must be operative that spare endogenous archaea from metal toxicity.

4. Strategies of Heavy Metal Resistance of the Archaea

Prolonged ecologic and biotechnologic interests in the archaea promote ongoing studies of metal resistance in these organisms. Development of bioremediation strategies have also motivated detailed studies of archaeal metal resistance and the biodiversity of archaeal taxa resident in heavy metal mining sites and metal contaminated habitats [28,42,43,66,67]. The occurrence of thermophilic taxa in some of these habitats particularly ore heaps arises from the intrinsic heat generated by chemical and biological oxidation. This creates a variable environment which develops into one containing high concentrations of solubilized metals dissociated from metal sulfide complexes.

Whether in ore heaps or saline ponds, changes in heavy metal concentration elicit differential gene regulation in archaea. Comparisons of global gene expression patterns arising from heavy metal challenges in *Halobacterium sp.* strain NRC-1 have been demonstrated [33]. In some case, identical genes are up-regulated by different metals such as *yvgX* in the presence of Cu and Zn, and *pstC1* in the presence of cobalt (Co) and nickle (Ni). On the other hand, some genes are uniquely up-regulated in the presence of only certain metals such as *potA1* and *hemeV2* in the presence of manganese (Mn) [33]. This suggests that the cellular perception of metal involves complex signal transduction mechanisms. Multiple mechanisms of Cu resistance also have been identified based on gene expression analysis in the hyperthermophilic acidophile *S. solfataricus*. These include a copper efflux system by *copA* and *B* (P-type ATPase), *copR* (regulator), and *copT* (chaperone) [68–70], and an inorganic polyphosphate transport system in the related *S. metallicus* [13]. The distribution of copper resistance genes among archaea is broad. Many archaeal genomes encode *copA* (P-type ATPase) homologs. However, while many bacterial taxa have metal resistance genes encoded on both plasmids and the genome, there are no CopA homologs annotated as being encoded on archaeal plasmids with the exception of *Haloarcula marismortui* ATCC 43049 mega plasmid pNG600. In this partcular case, five CopA homologs are apparent on this plasmid.

Hexavalant chromium (Cr(VI)) is another example of a heavy metal with environmental relevance. Like mercury, chromium is not an essential trace metal. Various studies have reported that many taxa have the ability to reduce the soluble toxic form of chromium (Cr(VI)) to the less toxic and insoluble form (Cr (III)) [71–73]. This transformation is mediated by a reductase, ChrR, or when Cr(VI) is used as an electron acceptor [74,75]. Once Cr (VI) is reduced to Cr (III), it can be exported by the transmembrane efflux pump ChrA [76]. ChrR is a member of the NAD(P)H dependent FMN-reductase family that has a wide substrate specificity [74]. ChrR appears to be widely distributed among archaea (Figure 1). Interestingly, only a few archaeal species contain *chrA* (efflux transporter), suggesting that archaea may use different strategies to detoxify this metal, such as by activation of oxidative stress mechanisms, or via novel efflux pumps. Studies on the determinants controlling the activity of ChrR show reduction of uranium species could be accomplished by this protein [77]. The potential benefit of a single pathway that can reduce the action of multiple toxic metals could facilitate bioremediation efforts of heavy metal contaminated soils.

Uranium is a radioactive metal of growing interest as a carbon net-negative energy source, for defense-related weapons, and for general ecological considerations. However, in archaeal genomes there are as yet no uranium specific reductases or efflux transporters that can be identified. A recent study describes the formation of insoluble uranium precpitates using cultures of the hyperthermophilic crenarchaeote, *Pyrobaculum islandicus* [78] perhaps reminiscent of the well characterized reduction of this metal by bacterial taxa belonging to *Geobacter* and *Shewanella spp.* [79–83].

Figure 1. Protein phylogenetic tree of archaeal ChrR. The tree was constructed using MEGA 4.0. The bootstrap values were based on 1000 replicates, and *E. coli* ChrA was used as an outgroup. Bootstrap value greater than 50% are shown.

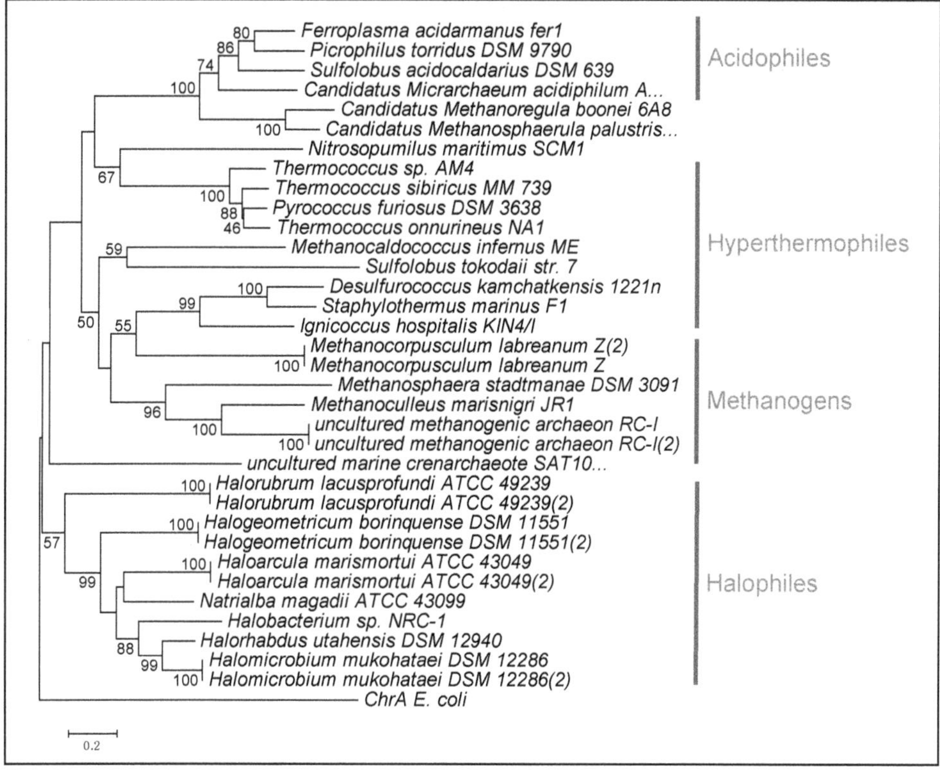

5. Environmental Applications Using Metal Resistant Archaea

There are increasing efforts underway to mitigate acid mine drainage (AMD) from active and abandoned mines. Similarly, there are ever increasing demands for metal commodities to meet the needs of growing societies. For this reason the relationship between metals and archaeal taxa is of great interest. A number of archaeal taxa have been isolated from such sites and their analysis has fostered a better understanding of the microbial community dynamics that occur in such environments using culture independent methods [84].

The role of thermoacidophilic archaea in biomining (bioleaching) of sulfidic metals is another area of growing interest [85]. Hyperthermophilic archaea have the capacity to immobilize metals, such as uranium [78] suggesting uses for bioremedation of contaminated sites. In addition to these benefits, genetic engineering tools for some archaea are established [86] and may lead to biological systems with metal-leaching specificity and with increased rates of metal solubilization.

6. Concluding Remarks

Archaea are globally distributed microorganisms inhabiting extreme environments as well as environments rich in heavy metals. These features foster interest in their genomics. The abundance of archaeal taxa in established environments rich in heavy metals has important ecologic and environmental implications. Elevated temperatures arising from mining in the deep subsurface promise to enhance availability of new thermophilic and hyperthermophilic species with novel metal metabolisms. There are several outstanding questions that remain to be answered in this field. What is the overlap between the unique cell biology of archaea and metal resistance? What is the ancient and modern day role of archaea in biogeochemical process? What unexplored biodiversity remains to be discovered in these extreme environments. Future studies integrating the use of genetic systems with model organisms will be critical to establish cause and effect relationships about metal biology and the archaea.

Acknowledgements

Funding for this study was provided by the U.S. Department of Energy (DEPS0208ERO812; DEFG3608GO88055), Department of Defense (HDTRA1-90030) and National Institutes of Health (1R01GM090209).

References

1. Bruins, M.R.; Kapil, S.; Oehme, F.W. Microbial resistance to metals in the environment. Ecotoxicol. *Environ. Saf.* **2000**, *45*, 198–207.
2. Haferburg, G.; Kothe, E. Microbes and metals: Interactions in the environment. *J. Basic Microbiol.* **2007**, *47*, 453–467.
3. Rensing, C.; Grass, G. Escherichia coli mechanisms of copper homeostasis in a changing environment. *FEMS Microbiol. Rev.* **2003**, *27*, 197–213.

4. Silver, S.; Phung, L.T. Bacterial heavy metal resistance: New surprises. *Annu. Rev. Microbiol.* **1996**, *50*, 753–789.
5. Stolz, J.F.; Basu, P.; Santini, J.M.; Oremland, R.S. Arsenic and selenium in microbial metabolism. *Annu. Rev. Microbiol.* **2006**, *60*, 107–130.
6. Hallas, L.E.; Thayer, J.S.; Cooney, J.J. Factors affecting the toxic effect of tin on estuarine microorganisms. *Appl. Environ. Microbiol.* **1982**, *44*, 193–197.
7. Spain, A. Implications of microbial heavy metal tolerance in the environment. *Rev. Undergrad. Res.* **2003**, *2*, 1–6.
8. Nies, D.H. Efflux-mediated heavy metal resistance in prokaryotes. *FEMS Microbiol. Rev.* **2003**, *27*, 313–339.
9. Morin, I.; Cuillel, M.; Lowe, J.; Crouzy, S.; Guillain, F.; Mintz, E. Cd2+- or Hg2+-binding proteins can replace the Cu+-chaperone Atx1 in delivering Cu+ to the secretory pathway in yeast. *FEBS Lett.* **2005**, *579*, 1117–1123.
10. Tottey, S.; Harvie, D.R.; Robinson, N.J. Understanding how cells allocate metals using metal sensors and metallochaperones. *Acc. Chem. Res.* **2005**, *38*, 775–783.
11. Tanaka, Y.; Tsumoto, K.; Nakanishi, T.; Yasutake, Y.; Sakai, N.; Yao, M.; Tanaka, I.; Kumagai, I. Structural implications for heavy metal-induced reversible assembly and aggregation of a protein: The case of Pyrococcus horikoshii CutA. *FEBS Lett.* **2004**, *556*, 167–174.
12. Yang, J.; Li, Q.; Yang, H.; Yan, L.; Yang, L.; Yu, L. Overexpression of human CUTA isoform2 enhances the cytotoxicity of copper to HeLa cells. *Acta Biochim. Pol.* **2008**, *55*, 411–415.
13. Remonsellez, F.; Orell, A.; Jerez, C.A. Copper tolerance of the thermoacidophilic archaeon Sulfolobus metallicus: Possible role of polyphosphate metabolism. *Microbiology* **2006**, *152*, 59–66.
14. Seufferheld, M.J.; Alvarez, H.M.; Farias, M.E. Role of polyphosphates in microbial adaptation to extreme environments. *Appl. Environ. Microbiol.* **2008**, *74*, 5867–5874.
15. Nemergut, D.R.; Martin, A.P.; Schmidt, S.K. Integron diversity in heavy-metal-contaminated mine tailings and inferences about integron evolution. *Appl. Environ. Microbiol.* **2004**, *70*, 1160–1168.
16. Ghosh, M.; Rosen, B. Microbial Resistance Mechanisms for Heavy Metals and Metalloids. In Heavy Metals in the Environment; Sakar, B., Ed.; Marcel Dekker, Inc: New York, NY, USA, **2002**; pp. 531–548.
17. Pearson, R.G. Hard and soft acids and bases. *J. Amer. Chem. Soc.* **1963**, *85*, 3533–3539.
18. Wackett, L.P.; Dodge, A.G.; Ellis, L.B. Microbial genomics and the periodic table. *Appl. Environ. Microbiol.* **2004**, *70*, 647–655.
19. Pikuta, E.V.; Hoover, R.B.; Tang, J. Microbial extremophiles at the limits of life. *Crit. Rev. Microbiol.* **2007**, *33*, 183–209.
20. DeLong, E.F.; Wu, K.Y.; Prezelin, B.B.; Jovine, R.V. High abundance of Archaea in Antarctic marine picoplankton. *Nature* **1994**, *371*, 695–697.

21. Sievert, S.M.; Ziebis, W.; Kuever, J.; Sahm, K. Relative abundance of Archaea and Bacteria along a thermal gradient of a shallow-water hydrothermal vent quantified by rRNA slot-blot hybridization. *Microbiology* **2000**, *146*, 1287–1293.
22. Schrenk, M.O.; Kelley, D.S.; Delaney, J.R.; Baross, J.A. Incidence and diversity of microorganisms within the walls of an active deep-sea sulfide chimney. *Appl. Environ. Microbiol.* **2003**, *69*, 3580–3592.
23. Callieri, C.; Corno, G.; Caravati, E.; Rasconi, S.; Contesini, M.; Bertoni, R. Bacteria, archaea, and crenarchaeota in the epilimnion and hypolimnion of a deep holo-oligomictic lake. *Appl. Environ. Microbiol.* **2009**, *75*, 7298–7300.
24. Simbahan, J.; Kurth, E.; Schelert, J.; Dillman, A.; Moriyama, E.; Jovanovich, S.; Blum, P. Community analysis of a mercury hot spring supports occurrence of domain-specific forms of mercuric reductase. *Appl. Environ. Microbiol.* **2005**, *71*, 8836–8845.
25. Eilmus, S.; Rosch, C.; Bothe, H. Prokaryotic life in a potash-polluted marsh with emphasis on N-metabolizing microorganisms. *Environ. Pollut.* **2007**, *146*, 478–491.
26. Borneman, J.; Triplett, E.W. Molecular microbial diversity in soils from eastern Amazonia: Evidence for unusual microorganisms and microbial population shifts associated with deforestation. *Appl. Environ. Microbiol.* **1997**, *63*, 2647–2653.
27. Buckley, D.H.; Graber, J.R.; Schmidt, T.M. Phylogenetic analysis of nonthermophilic members of the kingdom crenarchaeota and their diversity and abundance in soils. *Appl. Environ. Microbiol.* **1998**, *64*, 4333–4339.
28. Bruneel, O.; Pascault, N.; Egal, M.; Bancon-Montigny, C.; Goni-Urriza, M.S.; Elbaz-Poulichet, F.; Personne, J.C.; Duran, R. Archaeal diversity in a Fe-As rich acid mine drainage at Carnoules (France). *Extremophiles* **2008**, *12*, 563–571.
29. Edwards, K.J.; Bond, P.L.; Gihring, T.M.; Banfield, J.F. An archaeal iron-oxidizing extreme acidophile important in acid mine drainage. *Science* **2000**, *287*, 1796–1799.
30. Sandaa, R.A.; Enger, O.; Torsvik, V. Abundance and diversity of Archaea in heavy-metal-contaminated soils. *Appl. Environ. Microbiol.* **1999**, *65*, 3293–3297.
31. Stiller, M.; Sigg, L. Heavy Metals in the Dead Sea and thier coprecipitation with halite. *Hydrobiologia* **1990**, *197*, 23–33.
32. Wang, G.; Kennedy, S.P.; Fasiludeen, S.; Rensing, C.; DasSarma, S. Arsenic resistance in Halobacterium sp. strain NRC-1 examined by using an improved gene knockout system. *J. Bacteriol.* **2004**, *186*, 3187–3194.
33. Kaur, A.; Pan, M.; Meislin, M.; Facciotti, M.T.; El-Gewely, R.; Baliga, N.S. A systems view of haloarchaeal strategies to withstand stress from transition metals. *Genome Res.* **2006**, *16*, 841–854.
34. Biddle, J.F.; Fitz-Gibbon, S.; Schuster, S.C.; Brenchley, J.E.; House, C.H. Metagenomic signatures of the Peru Margin subseafloor biosphere show a genetically distinct environment. *Proc. Natl. Acad. Sci. USA* **2008**, *105*, 10583–10588.
35. Michalke, K.; Wickenheiser, E.B.; Mehring, M.; Hirner, A.V.; Hensel, R. Production of volatile derivatives of metal(loid)s by microflora involved in anaerobic digestion of sewage sludge. *Appl. Environ. Microbiol.* **2000**, *66*, 2791–2796.

36. Kim, B.K.; Pihl, T.D.; Reeve, J.N.; Daniels, L. Purification of the copper response extracellular proteins secreted by the copper-resistant methanogen Methanobacterium bryantii BKYH and cloning, sequencing, and transcription of the gene encoding these proteins. *J. Bacteriol.* **1995**, *177*, 7178–7185.
37. Mori, K.; Hatsu, M.; Kimura, R.; Takamizawa, K. Effect of heavy metals on the growth of a methanogen in pure culture and coculture with a sulfate-reducing bacterium. *J. Biosci. Bioeng.* **2000**, *90*, 260–265.
38. Meyer, J.; Michalke, K.; Kouril, T.; Hensel, R. Volatilisation of metals and metalloids: An inherent feature of methanoarchaea? *Syst. Appl. Microbiol.* **2008**, *31*, 81–87.
39. Qin, J.; Rosen, B.P.; Zhang, Y.; Wang, G.; Franke, S.; Rensing, C. Arsenic detoxification and evolution of trimethylarsine gas by a microbial arsenite S-adenosylmethionine methyltransferase. *Proc. Natl. Acad. Sci. USA* **2006**, *103*, 2075–2080.
40. Amo, T.; Paje, M.L.; Inagaki, A.; Ezaki, S.; Atomi, H.; Imanaka, T. Pyrobaculum calidifontis sp. nov., a novel hyperthermophilic archaeon that grows in atmospheric air. *Archaea* **2002**, *1*, 113–421.
41. Brock, T.D.; Brock, K.M.; Belly, R.T.; Weiss, R.L. Sulfolobus: A new genus of sulfur-oxidizing bacteria living at low pH and high temperature. *Arch. Mikrobiol.* **1972**, *84*, 54–68.
42. Fuchs, T.; Huber, H.; Teiner, K.; Burggraf, S.; Stetter, K.O. Metallosphaera prunae, sp. nov., a novel metal-mobilizing, thermoacidophilic archaeum, isolated from a uranium mine in Germany. *Syst. Appl. Microbiol.* **1995**, *18*, 560–566.
43. Huber, G.; Spinnler, C.; Gambacorta, A.; Stetter, K.O. Metallosphaera sedula gen. and sp. nov. epresents a new genus of aerobic, metal-mobilizing thermoacidophilic archaebacteria. *Syst. Appl. Microbiol.* **1989**, *12*, 38–47.
44. Schleper, C.; Puehler, G.; Holz, I.; Gambacorta, A.; Janekovic, D.; Santarius, U.; Klenk, H.P.; Zillig, W. Picrophilus gen. nov., fam. nov.: a novel aerobic, heterotrophic, thermoacidophilic genus and family comprising archaea capable of growth around pH 0. *J. Bacteriol.* **1995**, *177*, 7050–7059.
45. Arnorsson, A. The distribution of some trace elements in thermal waters in Iceland. *Geothermics* **1970**, *2*, 542–546.
46. Edgcomb, V.P.; Molyneaux, S.J.; Saito, M.A.; Lloyd, K.; Boer, S.; Wirsen, C.O.; Atkins, M.S.; Teske, A. Sulfide ameliorates metal toxicity for deep-sea hydrothermal vent archaea. *Appl. Environ. Microbiol.* **2004**, *70*, 2551–2555.
47. Spear, J.R.; Walker, J.J.; McCollom, T.M.; Pace, N.R. Hydrogen and bioenergetics in the Yellowstone geothermal ecosystem. *Proc. Natl. Acad. Sci. USA* **2005**, *102*, 2555–2560.
48. Weissberg, W.G. Gold-silver ore-grade precipitates from New Zealand thermal waters. *Econ. Geology* **1969**, *64*, 95.
49. Baker-Austin, C.; Dopson, M.; Wexler, M.; Sawers, R.G.; Bond, P.L. Molecular insight into extreme copper resistance in the extremophilic archaeon 'Ferroplasma acidarmanus' Fer1. *Microbiology* **2005**, *151*, 2637–2646.

50. Dixit, V.; Bini, E.; Drozda, M.; Blum, P. Mercury inactivates transcription and the generalized transcription factor TFB in the archaeon Sulfolobus solfataricus. *Antimicrob. Agents Chemother.* **2004**, *48*, 1993–1999.
51. Schelert, J.; Dixit, V.; Hoang, V.; Simbahan, J.; Drozda, M.; Blum, P. Occurrence and characterization of mercury resistance in the hyperthermophilic archaeon Sulfolobus solfataricus by use of gene disruption. *J. Bacteriol.* **2004**, *186*, 427–437.
52. Schelert, J.; Drozda, M.; Dixit, V.; Dillman, A.; Blum, P. Regulation of mercury resistance in the crenarchaeote Sulfolobus solfataricus. *J. Bacteriol.* **2006**, *188*, 7141–7150.
53. Nieto, J.J.; Ventosa, A.; Ruiz-Berraquero, F. Susceptibility of halobacteria to heavy metals. *Appl. Environ. Microbiol.* **1987**, *53*, 1199–1202.
54. Dopson, M.; Baker-Austin, C.; Koppineedi, P.R.; Bond, P.L. Growth in sulfidic mineral environments: Metal resistance mechanisms in acidophilic micro-organisms. *Microbiology* **2003**, *149*, 1959–1970.
55. Grogan, D.W. Phenotypic characterization of the archaebacterial genus Sulfolobus: Comparison of five wild-type strains. *J. Bacteriol.* **1989**, *171*, 6710–6719.
56. Huber, G.; Stetter, K.O. Sulfolobus metallicus, sp. nov., a novel strictly chemolithoautotrophic thermophilic archaeal species of metal-mobilizer. *Syst. Appl. Microbiol.* **1991**, *14*, 372–378.
57. Kim, B.K.; de Macario, E.C.; Nolling, J.; Daniels, L. Isolation and characterization of a copper-resistant methanogen from a copper-mining soil sample. *Appl. Environ. Microbiol.* **1996**, *62*, 2629–2635.
58. Baker-Austin, C.; Dopson, M.; Wexler, M.; Sawers, R.G.; Stemmler, A.; Rosen, B.P.; Bond, P.L. Extreme arsenic resistance by the acidophilic archaeon 'Ferroplasma acidarmanus' Fer1. *Extremophiles* **2007**, *11*, 425–434.
59. Dopson, M.; Baker-Austin, C.; Hind, A.; Bowman, J.P.; Bond, P.L. Characterization of Ferroplasma isolates and Ferroplasma acidarmanus sp. nov., extreme acidophiles from acid mine drainage and industrial bioleaching environments. *Appl. Environ. Microbiol.* **2004**, *70*, 2079–2088.
60. Abskharon, R.N.; Hassan, S.H.; Gad El-Rab, S.M.; Shoreit, A.A. Heavy metal resistant of E. coli isolated from wastewater sites in Assiut City, Egypt. *Bull. Environ. Contam. Toxicol.* **2008**, *81*, 309–315.
61. Kaur, S.; Kamli, M.R.; Ali, A. Diversity of arsenate reductase genes (arsC Genes) from arsenic-resistant environmental isolates of E. coli. *Curr. Microbiol.* **2009**, *59*, 288–294.
62. Li, X.Z.; Nikaido, H.; Williams, K.E. Silver-resistant mutants of Escherichia coli display active efflux of Ag+ and are deficient in porins. *J. Bacteriol.* **1997**, *179*, 6127–6132.
63. Maezato, Y.; Blum, P. University of Nebraska, Lincoln, NE, USA. Unpublished work, 2012.
64. Chatziefthimiou, A.D.; Crespo-Medina, M.; Wang, Y.; Vetriani, C.; Barkay, T. The isolation and initial characterization of mercury resistant chemolithotrophic thermophilic bacteria from mercury rich geothermal springs. *Extremophiles* **2007**, *11*, 469–479.
65. Vetriani, C.; Chew, Y.S.; Miller, S.M.; Yagi, J.; Coombs, J.; Lutz, R.A.; Barkay, T. Mercury adaptation among bacteria from a deep-sea hydrothermal vent. *Appl. Environ. Microbiol.* **2005**, *71*, 220–226.

66. Almeida, W.I.; Vieira, R.P.; Cardoso, A.M.; Silveira, C.B.; Costa, R.G.; Gonzalez, A.M.; Paranhos, R.; Medeiros, J.A.; Freitas, F.A.; Albano, R.M.; Martins, O.B. Archaeal and bacterial communities of heavy metal contaminated acidic waters from zinc mine residues in Sepetiba Bay. *Extremophiles* **2009**, *13*, 263–271.
67. Xie, X.; Xiao, S.; He, Z.; Liu, J.; Qiu, G. Microbial populations in acid mineral bioleaching systems of Tong Shankou Copper Mine, China. *J. Appl. Microbiol.* **2007**, *103*, 1227–1238.
68. Deigweiher, K.; Drell, T.L.T.; Prutsch, A.; Scheidig, A.J.; Lubben, M. Expression, isolation, and crystallization of the catalytic domain of CopB, a putative copper transporting ATPase from the thermoacidophilic archaeon Sulfolobus solfataricus. *J. Bioenerg. Biomembr.* **2004**, *36*, 151–159.
69. Ettema, T.J.; Brinkman, A.B.; Lamers, P.P.; Kornet, N.G.; de Vos, W.M.; van der Oost, J. Molecular characterization of a conserved archaeal copper resistance (cop) gene cluster and its copper-responsive regulator in Sulfolobus solfataricus P2. *Microbiology* **2006**, *152*, 1969–1979.
70. Villafane, A.A.; Voskoboynik, Y.; Cuebas, M.; Ruhl, I.; Bini, E. Response to excess copper in the hyperthermophile Sulfolobus solfataricus strain 98/2. *Biochem. Biophys. Res. Commun.* **2009**, *385*, 67–71.
71. Daulton, T.L.; Little, B.J.; Lowe, K.; Jones-Meehan, J. In Situ Environmental Cell-Transmission Electron Microscopy Study of Microbial Reduction of Chromium(VI) Using Electron Energy Loss Spectroscopy. *Microsc. Microanal.* **2001**, *7*, 470–485.
72. Kamaludeen, S.P.; Arunkumar, K.R.; Avudainayagam, S.; Ramasamy, K. Bioremediation of chromium contaminated environments. *Indian J. Exp. Biol.* **2003**, *41*, 972–985.
73. Opperman, D.J.; van Heerden, E. Aerobic Cr(VI) reduction by Thermus scotoductus strain SA-01. *J. Appl. Microbiol.* **2007**, *103*, 1907–1913.
74. Ramirez-Diaz, M.I.; Diaz-Perez, C.; Vargas, E.; Riveros-Rosas, H.; Campos-Garcia, J.; Cervantes, C. Mechanisms of bacterial resistance to chromium compounds. *Biometals* **2008**, *21*, 321–332.
75. Tandukar, M.; Huber, S.J.; Onodera, T.; Pavlostathis, S.G. Biological chromium(VI) reduction in the cathode of a microbial fuel cell. *Environ. Sci. Technol.* **2009**, *43*, 8159–8165.
76. Cervantes, C.; Campos-Garcia, J.; Devars, S.; Gutierrez-Corona, F.; Loza-Tavera, H.; Torres-Guzman, J.C.; Moreno-Sanchez, R. Interactions of chromium with microorganisms and plants. *FEMS Microbiol. Rev.* **2001**, *25*, 335–347.
77. Barak, Y.; Ackerley, D.F.; Dodge, C.J.; Banwari, L.; Alex, C.; Francis, A.J.; Matin, A. Analysis of novel soluble chromate and uranyl reductases and generation of an improved enzyme by directed evolution. *Appl. Environ. Microbiol.* **2006**, *72*, 7074–7082.
78. Kashefi, K.; Moskowitz, B.M.; Lovley, D.R. Characterization of extracellular minerals produced during dissimilatory Fe(III) and U(VI) reduction at 100 degrees C by Pyrobaculum islandicum. *Geobiology* **2008**, *6*, 147–154.
79. Finneran, K.T.; Housewright, M.E.; Lovley, D.R. Multiple influences of nitrate on uranium solubility during bioremediation of uranium-contaminated subsurface sediments. *Environ. Microbiol.* **2002**, *4*, 510–516.

80. Lovley, D.R.; Phillips, E.J.P.; Gorby, Y.A.; Landa, E.R. Microbial reduction of uranium. *Nature* **1991**, *350*, 413–415.
81. Lovley, D.R.; Holmes, D.E.; Nevin, K.P. Dissimilatory Fe(III) and Mn(IV) reduction. *Adv. Microb. Physiol.* **2004**, *49*, 219–286.
82. Wall, J.D.; Krumholz, L.R. Uranium reduction. *Annu. Rev. Microbiol.* **2006**, *60*, 149–166.
83. Wilkins, M.J.; Verberkmoes, N.C.; Williams, K.H.; Callister, S.J.; Mouser, P.J.; Elifantz, H.; N'Guessan, A,L.; Thomas, B.C.; Nicora, C.D.; Shah, M.B.; et al. Proteogenomic monitoring of Geobacter physiology during stimulated uranium bioremediation. *Appl. Environ. Microbiol.* **2009**, *75*, 6591–6599.
84. Baker, B.J.; Tyson, G.W.; Webb, R.I.; Flanagan, J.; Hugenholtz, P.; Allen, E.E.; Banfield, J.F. Lineages of acidophilic archaea revealed by community genomic analysis. *Science* **2006**, *314*, 1933–1935.
85. Rawlings, D.E. Heavy metal mining using microbes. *Annu. Rev. Microbiol.* **2002**, *56*, 65–91.
86. Sowers, K.R.; Blum, P.H.; DasSarma, S. Gene Transfer in Archaea. In Methods for General and Molecular Microbiology, 3rd ed.; Reddy, C.A., Beveridge, T.J., Breznak, J.A., Marzluf, G.A., Schmidt, T.M., Eds.; American Society for Microbiology: Honolulu, HI, USA, 2007; pp. 800–824.

Magnetotactic Bacteria from Extreme Environments

Dennis A. Bazylinski and Christopher T. Lefèvre

Abstract: Magnetotactic bacteria (MTB) represent a diverse collection of motile prokaryotes that biomineralize intracellular, membrane-bounded, tens-of-nanometer-sized crystals of a magnetic mineral called magnetosomes. Magnetosome minerals consist of either magnetite (Fe_3O_4) or greigite (Fe_3S_4) and cause cells to align along the Earth's geomagnetic field lines as they swim, a trait called magnetotaxis. MTB are known to mainly inhabit the oxic–anoxic interface (OAI) in water columns or sediments of aquatic habitats and it is currently thought that magnetosomes function as a means of making chemotaxis more efficient in locating and maintaining an optimal position for growth and survival at the OAI. Known cultured and uncultured MTB are phylogenetically associated with the *Alpha-*, *Gamma-* and *Deltaproteobacteria* classes of the phylum *Proteobacteria*, the *Nitrospirae* phylum and the candidate division OP3, part of the *Planctomycetes-Verrucomicrobia-Chlamydiae* (PVC) bacterial superphylum. MTB are generally thought to be ubiquitous in aquatic environments as they are cosmopolitan in distribution and have been found in every continent although for years MTB were thought to be restricted to habitats with pH values near neutral and at ambient temperature. Recently, however, moderate thermophilic and alkaliphilic MTB have been described including: an uncultured, moderately thermophilic magnetotactic bacterium present in hot springs in northern Nevada with a probable upper growth limit of about 63 °C; and several strains of obligately alkaliphilic MTB isolated in pure culture from different aquatic habitats in California, including the hypersaline, extremely alkaline Mono Lake, with an optimal growth pH of >9.0.

Reprinted from *Life*. Cite as: Bazylinski, D.A.; Lefèvre, C.T. Magnetotactic Bacteria from Extreme Environments. *Life* **2013**, *3*, 295-307.

1. Introduction

A number of eukaryotic organisms are known to use the Earth's magnetic field for orientation and navigation (e.g., pigeon, salmon, bats, *etc.*), a behavior referred to as magnetoreception [1]. Although the mechanisms involved in magnetoreception are not well understood, the presence of magnetic mineral crystals has been established in some of these organisms [1]. The most well understood example of magnetoreception is found within the prokaryotes and involves a group of motile, aquatic bacteria known as the magnetotactic bacteria (MTB). In MTB, the magnetoreceptive behavior is called magnetotaxis and is a result of the cells' ability to biomineralize intracellular, membrane-bounded, tens-of-nanometer-sized crystals of a magnetic mineral consisting of either magnetite (Fe_3O_4) or greigite (Fe_3S_4) [2]. These structures, termed magnetosomes, cause cells to align along the Earth's geomagnetic field lines as they swim: the definition of magnetotaxis [3]. MTB are known to mainly inhabit the oxic–anoxic interface (OAI) of aquatic habitats [2] and it is currently thought that the magnetosomes function as a means of making chemotaxis more efficient in locating and maintaining an optimal position for growth and survival at the OAI [3].

Known cultured and uncultured MTB are phylogenetically associated with the *Alpha*-, *Gamma*- and *Deltaproteobacteria* classes of the phylum *Proteobacteria*, the *Nitrospirae* phylum and the candidate division OP3 of the *Planctomycetes-Verrucomicrobia-Chlamydiae* (PVC) superphylum [4–6]. MTB are ubiquitous in almost all types of aquatic environments [2] and are cosmopolitan in distribution as they have been found on every continent [7]. However, despite their broad phylogenetic diversity and wide geographic distribution, no MTB have been classified as extremophilic until recently because: (1) most known cultured MTB are mesophilic with regard to growth temperature and do not grow much above 30 °C (e.g., *Magnetospirillum* species, *Desulfovibrio magneticus*, *Magnetococcus marinus*, *Magnetospira thiophila* and *Magnetovibrio blakemorei*; [8–12]); (2) virtually all studies on uncultured MTB involved sampling sites that were at 30 °C and below; and (3) cultured MTB grow only at about neutral pH while uncultured MTB had never been found in strongly alkaline or acidic habitats. The purpose of this chapter is to present what is known regarding newly discovered MTB that can be considered extremophilic, moderately extremophilic or tolerant to stressors and also to discuss about the potential existence of MTB in other extreme environments and their potential presence on extraterrestrial bodies such as other planets as was once suggested for Mars [13,14]. It is important to note that the work described in this review is focused upon MTB recovered or isolated directly from environmental samples, rather than those inferred indirectly via the detection of magnetotaxis-specific genes in metagenomes.

2. Thermophilic Magnetotactic Bacteria

In a recent environmental study, Lefèvre *et al.* [15] discovered an uncultured, moderately thermophilic magnetotactic bacterium in hot springs located in northern Nevada. This strain, designated HSMV-1 (*Candidatus* Thermomagnetovibrio paiutensis), was found in mud and water samples collected from the Great Boiling Springs (GBS) geothermal field in Gerlach, Nevada [15] (Figure 1A). GBS is a series of hot springs that range from ambient temperature to ~96 °C [16,17]. The geology, chemistry and microbial ecology of the springs have been described in some detail [16,17].

Microscopic examination of the samples collected at GBS showed the presence of a single morphotype of MTB in samples taken from nine springs where the temperatures ranged from 32 °C to 63 °C. Cells were small (1.8 ± 0.4 by 0.4 ± 0.1 μm), Gram-negative, vibrioid-to-helicoid in morphology and possessed a single polar flagellum (Figure 2). MTB were not observed in water and mud collected from springs that were 67 °C and higher suggesting the maximum survival and perhaps growth temperature for strain HSMV-1 is about 63 °C. When the water temperature in these pools was <30 °C, cells of strain HSMV-1 were not observed although several other types of MTB were present including greigite-producing, rod-shaped bacteria [15]. Cells of HSMV-1 biomineralize a single chain of bullet-shaped magnetite magnetosomes that traverse the cells along their long axis (Figure 2).

Figure 1. Picture showing the sampling of sediment and water (A) at the Great Boiling Springs (GBS) geothermal field in Gerlach, Nevada, in September 2009. (B) Sampling at Mono Lake, a hypersaline, hyperalkaline endorheic lake situated in California, in February 2010. Tufas, large carbonaceous concretions coming out of the lake can be seen in the background.

Figure 2. Transmission electron microscope (TEM) images of cells and magnetosomes of the thermophilic magnetotactic bacterium strain HSMV-1 (*Candidatus* Thermomagnetovibrio paiutensis. (A) TEM image of a cell of HSMV-1 showing a single polar flagellum (black arrow) and a single chain of bullet-shaped magnetosomes (white arrow). (B) High magnification TEM image of a magnetosome chain of strain HSMV-1.

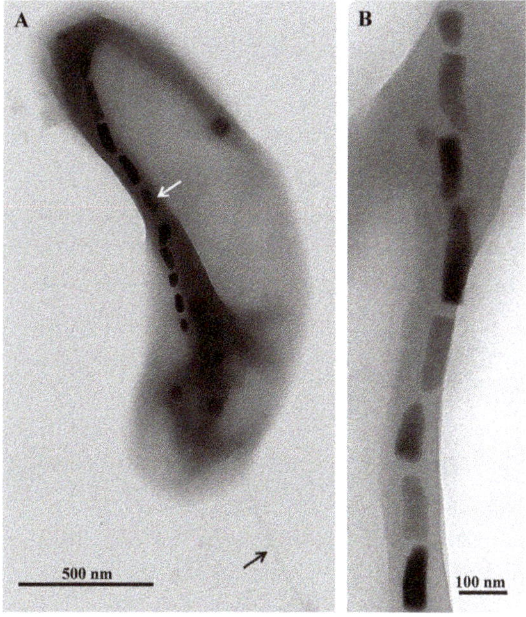

The 16S rRNA gene sequence of strain HSMV-1 places the organism phylogenetically in the phylum *Nitrospirae* with its closest relative in culture being *Thermodesulfovibrio hydrogeniphilus* [18], a non-magnetotactic, thermophilic sulfate-reducing bacterium isolated from a terrestrial Tunisian hot spring with an optimal growth temperature of 65 °C.

Nash [19] also reported the presence of thermophilic MTB in microbial mats at about 45–55 °C adjacent to the main flow in Little Hot Creek and in other springs up to 58 °C all on the east side of the Sierras in California. Cells biomineralized bullet-shaped crystals of magnetite and were phylogenetically affiliated with the phylum *Nitrospirae*. Few additional details were provided [19]. It thus seems likely that these moderately thermophilic MTB are not confined to the GBS and are present in hot springs around the world depending on temperature. Moreover, these studies clearly show that some MTB can be considered at least moderately thermophilic and extend the upper temperature limit for environments where MTB exist and grow and where magnetosome magnetite is deposited.

3. Alkaliphilic Magnetotactic Bacteria

While MTB have never been associated with either strongly alkaline or acidic habitats, three strains of obligately alkaliphilic, anaerobic, sulfate-reducing, MTB belonging to the *Deltaproteobacteria* class of the *Proteobacteria*, were recently described [20]. These new magnetotactic strains, designated ML-1, ZZ-1 and AV-1, were isolated in pure culture from three different highly alkaline environments in California, USA. Each has an optimal growth pH of 9.0–9.5.

Strain ML-1 was isolated from the hypersaline, hyperalkaline Mono Lake which is a well-characterized lake located on the arid eastern side of the Sierra Nevada Mountains in California (Figure 1B). The waters of this endorheic, monomictic basin display high alkalinity (pH 9.2 to 10) and salinity (75 g/L to 90 g/L) [21]. Weathering of the surrounding volcanic rocks and hydrothermal inflow results in high sodium and carbonate concentrations [21,22]. Microbial sulfate reduction accounts for 41% of the mineralization of annual primary production in Mono Lake [23]. The pH and salinity of the sample taken from Mono Lake from which strain ML-1 was isolated was 9.8 and 68–70 ppt (parts per thousand), respectively. Strain ZZ-1 was isolated from Soda Spring, a small alkaline spring situated at the Desert Studies Center, a field station of the California State University system, located at the end of Zzyzx Road south of Interstate 15 in California. The salinity at this location was ~27 ppt and the pH 9.5. The third site, where strain AV-1 was isolated from, is an unnamed small pond in Armagosa Valley situated at Death Valley Junction near the border of Nevada and California. This seasonal alkaline pond probably results from underground water flowing through the alkaline desert uplands of Ash Meadows National Wildlife Refuge [24,25] which is in close proximity. This pond is brackish with a salinity of 3 ppt and had a pH of 9.5 at the time of sampling.

After sampling, the mud and water collected from these three highly alkaline sites contained a significant population (>10^3 cells/mL) of MTB of a single, morphological type based on light microscopic observations. Cells were helical, possessed a single polar flagellum and contained one or two parallel chains of bullet-shaped, magnetite-containing magnetosomes (Figure 3). The 16S rRNA genes of magnetically-purified, uncultured cells from Mono Lake and the brackish pool at

Death Valley Junction were amplified using the polymerase chain reaction and sequenced, and showed that these organisms were closely related (16S rRNA gene sequence identities ≥98.9%) to the non-magnetotactic, alkaliphilic, sulfate-reducing deltaproteobacterium, *Desulfonatronum thiodismutans*, originally isolated from Mono Lake [26]. Magnetically-purified cells were used as inocula in a growth medium for the enrichment of anaerobic, alkaliphilic, sulfate-reducing bacteria modified from Pikuta *et al.* [26] to reflect the salinities of the sampling sites. Cells with the same morphology as those found in the mud and water samples from all sites grew in this growth media. All strains reduced sulfate and used formate and hydrogen as electron donors and were capable of chemolithoautotrophic growth with hydrogen as the electron donor with bicarbonate as the sole carbon source.

The presence of MTB, based on microscopic observations of cells displaying magnetotaxis, has been previously reported from highly alkaline habitats, for example, in Lonar Lake in Maharashtra, India [27,28]. This crater lake [29] is thought to be formed as a result of a meteoritic impact about 50,000 years ago and is a closed basin lake characterized by high alkalinity and salinity. However, no evidence of the presence of magnetosomes and no phylogenetic data were presented for the MTB in Lonar Lake. Further studies are clearly warranted to determine the geographic distribution of alkaliphilic MTB.

Figure 3. Scanning-transmission electron microscope (STEM) and TEM images of alkaliphilic magnetotactic bacteria. STEM images of cells from (A) the hypersaline Mono Lake, California and (B) a brackish pool at Death Valley Junction (Micrographs courtesy of Tanya Prozorov Ames Laboratory, U.S. Department of Energy, Ames, USA). (C) TEM image of a bullet-shaped magnetosome chain of strain ZZ-1 (Reprinted, with permission, from [20]).

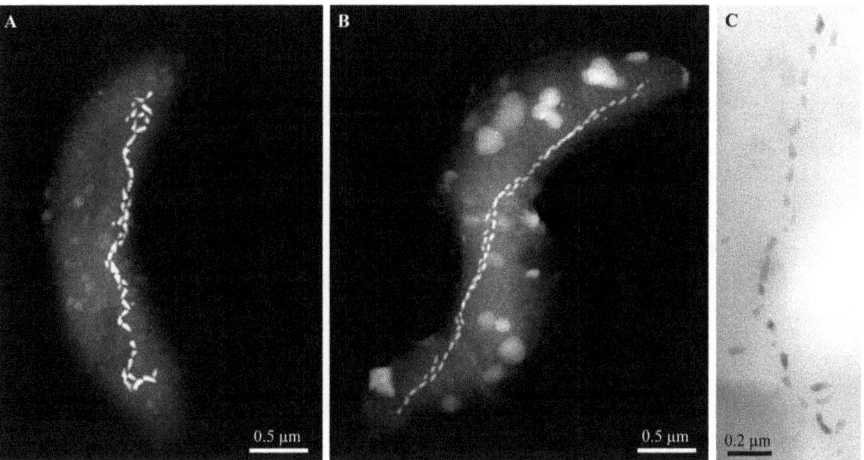

Alkaliphilic bacteria must cope with a number of potentially physiological problems living at high pH which includes maintaining their intracellular pH close to neutrality living in an environment where the external pH is >9.0. These organisms have developed a number of interesting metabolic processes to overcome these problems (e.g., higher cytoplasmic buffering capacities) [30]. In the specific case of alkaliphilic MTB, how these prokaryotes synthesize large numbers of magnetosomes in natural highly alkaline environments is particularly interesting in that iron is extremely insoluble at high pH [31]. Therefore, MTB must possess highly efficient mechanisms of iron uptake under these conditions. The fact that strains ML-1, ZZ-1 and AV-1 exist in the reducing anoxic zone in these environments where iron is more likely to be in the more soluble ferrous form, rather than the oxic zone of their habitat probably obviates part of this problem.

4. Potential for the Presence of MTB in Other Extreme Environments

In general, MTB are gradient-loving (e.g., oxygen and/or sulfide concentration gradients), microaerophilic or anaerobic microorganisms that are primarily located at or just below the OAI in water columns and sediments of chemically-stratified aquatic environments [2]. In theory, all chemically-stratified, aquatic environments having gradients with the appropriate physical-chemical conditions such as a suitable redox potential and the presence of enough soluble iron, could contain MTB. Thus there is no known reason why extremophilic including acidophilic, piezophilic, halophilic or psychrophilic bacteria could not have acquired the ability to biomineralize magnetosomes during their evolution. In fact, it is possible that these organisms have even been isolated in the past without the realization that the isolated strains were magnetotactic. For example, in an earlier section, we described three strains of alkaliphilic, sulfate-reducing MTB as strains of *Desulfonatronum thiodismutans* based on 16S rRNA gene sequences and phenotypic characteristics. This finding raises the question whether the type species of *D. thiodismuans* was magnetotactic when first isolated from the environment but lost the magnetotactic trait during continued cultivation [20]. This is certainly possible as many cultivated MTB are known to lose this trait relatively easily in culture [32–34], sometimes from the loss of a magnetosome genomic island in which the genes for magnetite biomineralization are located [32,34]. In turn, a larger, perhaps more interesting and important question is whether and how many magnetotactic prokaryotic organisms have been characterized from the environment or isolated and deposited in culture collections but have never been recognized as magnetotactic for various reasons. This may be most applicable to the sulfate-reducing bacteria as all cultured dissimilatory sulfate-reducing MTB including *Desulfovibrio magneticus* and the alkaliphic MTB described above, display only a weak magnetotactic response and ability to biomineralize magnetite magnetosomes when grown on sulfate as a terminal electron acceptor [35]. This is most likely due to scavenging of the available iron in the growth medium by sulfide produced during sulfate-reduction.

4.1. Piezophilic and Psychrophilic MTB

The deep-sea piezosphere comprises the volume of the oceans at a depth of 1,000 m and greater, with hydrostatic pressures of greater than 100 atmospheres or 10 megapascal (1 atm = 1.013 bar = 0.1 MPa), and accounts for about 75% of the total volume of the world's oceans [36]. Living MTB have been found in shallow hemipelagic sediments, collected at the depth of ~600 m, from the Santa Barbara Basin in the eastern Pacific with a temperature of 8 °C [37]. No phylogenetic data is available from this study however, only the morphology of the cells and the shape of their magnetosomes were reported. Cells were vibrioid and rod-shaped but the most common observed morphology was coccoid. Magnetosome crystals in these bacteria were composed of magnetite and were cuboidal, elongated prismatic (rectangular in projection) or irregular in shape. The presence of biogenic, ultra-fine-grained, single-domain magnetite (presumably from MTB) in the surface sediments of Santa Barbara Basin was also reported. These magnetite crystals were typical of MTB in their morphologies which included cuboidal, elongated prismatic and teardrop- and bullet-shaped. This study was the first to describe MTB in deep sediments and it was suggested that MTB are the source of the primary remanence carrier in marine sediment.

Petermann and Bleil [38] later reported the presence of MTB in pelagic and hemipelagic sediments of the eastern South Atlantic Ocean. In this study, MTB of different morphologies (cocci, spirilla, vibrioid and rod-shaped) were found at water depths to about 3,000 m in the African continental margin (off Namibia, between the equator and 30 °S) and on the Walvis Ridge in a pelagic environment (about 1,400 km off the coast) situated on a seamount at a water depth of 1,007 m. They showed that the number of MTB in sediment core samples stored at ~25 °C for 12 h significantly declined and those that survived showed less swimming behavior. In samples stored at ~2 °C, MTB could still be detected after several months. It was concluded that at least some marine MTB might be at least facultatively psychrophilic. To test for piezophilic potential in marine MTB, samples of intertidal sediments of the North Sea containing MTB were brought to a water depth of 3,100 m for 24 h [38]. No decrease in numbers or swimming activity of individual cells was detected after recovery indicating that these bacteria are somewhat piezotolerant.

Living MTB were recently discovered in lakes in the Antarctic [39] clearly demonstrating the existence of at least facultatively psychrophilic MTB. In sum, these studies suggest that the possibility of the existence of obligate piezophilic and psychrophilic bacteria cannot be dismissed.

4.2. Acidophilic MTB

Acidophiles are defined as those bacteria having an optimal growth pH of ≤ 3 [40]. To our knowledge, there are no reports of MTB in highly acidic environments such as acid mine drainage or bogs. The formation of magnetite under these conditions might pose a problem for prokaryotes as magnetite produced through chemical means or through biologically induced biomineralization [41] by non-magnetotactic bacteria (e.g., dissimilatory iron-reducing bacteria) does not appear to be thermodynamically favored at very low pH [42]. However, despite living at very low pH, acidophilic bacteria must maintain an intracellular pH of near neutrality [40] and thus it seems like it may be possible for certain acidophilic species to be capable of biomineralizing intracellular magnetosomes. The fact that iron is more soluble and thus likely more bioavailable under acidic conditions suggests that soluble iron in such environments should not be a limiting factor in magnetosome biomineralization by MTB.

4.3. Halophilic Magnetotactic Bacteria

There are no reports specific to halophilic MTB; however there are several studies describing the existence of MTB in hypersaline environments (having salinity levels greater than that of seawater ~35 ppt). Multicellular magnetotactic prokaryotes (MMP) such as *Candidatus* Magnetoglobus multicellularis thrive in the hypersaline Araruama lagoon which has a maximum salinity of 60 ppt [43]. MMPs and other types of MTB are also present in the hypersaline Salton Sea (salinity ~50 ppt) [6,44]. The site from where the sample containing MTB at the Salton Sea was collected was a reddish pool containing large numbers of halophilic prokaryotes as determined by microscopic observation. One species of MTB, designated strain SS-5, from this pool was isolated in pure culture but this strain does not appear to require high concentration of salt to grow. The site from which the alkaliphilic magnetotactic strain ML-1 was isolated in Mono Lake had a salinity of 68 ppt and thus increased the upper limit of salinity where MTB have been found [20]. However like strain SS-5, high salinity is not required for the growth of strain ML-1. Thus some MTB can be considered moderately halophilic and it is possible that halophilic MTB exist as some halophilic strains are known to have a similar metabolism to known MTB (e.g., dissimilatory sulfate reduction) [45] and to be phylogenetically related to MTB (e.g., strain SS-5 related to *Thiohalocapsa marina* [6]).

4.4. Magnetotactic Bacteria as Putative Biosignatures for Extraterrestrial Life

MTB have had a major impact on the field of astrobiology. Magnetite crystals morphologically similar to those present in some magnetosomes of MTB living in the present have been found in the Martian meteorite ALH84001 [14,46–49]. These crystals, putative remains of MTB, have been referred to as "magnetofossils" and have been used as evidence for the past presence of MTB in the meteorite ALH84001 as well as in ancient sediments on Earth dated to about 2 billion years ago [50]. The presence and interpretation of these crystals in Martian meteorite ALH84001 have evoked great controversy and debate particularly because magnetite crystals similar to the ones found in this meteorite can be synthesized in the laboratory without the presence of microorganisms [51,52]. If the magnetite crystals in this meteorite are indeed biogenic, the implication is that bacterial life had existed on ancient Mars [13,14,46–49,51–53]. In turn, this debate has led to a number of criteria to be used to distinguish biogenic magnetite from inorganically-produced magnetite [31,46,54–57].The discovery and isolation of obligately alkaliphilic MTB is discussed in a previous section and clearly demonstrates that some magnetotactic species can be considered extremophilic. Because MTB had never been considered to inhabit extreme environments, highly alkaline habitats have apparently not been searched for magnetofossils. Chemical analyses of soil samples of Mars indicate a period of highly alkaline conditions on the planet in the past [58–60]. Moreover equilibrium modeling based on measured Ca^{2+} and Mg^{2+} concentrations were consistent with carbonate equilibrium for a saturated solution [59,60] and thus carbonate buffering appears to be significant in some Martian soils as it is in Mono Lake. Mono Lake has been used by researchers at a number of institutions, including the National Aeronautics and Space Administration (NASA), as a model for extreme environments that might be comparable to those on the planet Mars [61]. It would be interesting to determine whether bullet-shaped magnetite crystals like those in strains ML-1 are incorporated and preserved as magnetofossils in carbonate minerals, such as the unusual carbonate structures known as tufas, abundant in Mono Lake, as they appear to do in sedimentary carbonates in marine environments [62,63] and in carbonates in the Martian meteorite ALH84001 [47]. Between 2018 and 2023, a mission is planned by NASA to collect rock and dust samples from Mars and to return them to Earth for analysis. It will be interesting and intriguing if magnetofossils are found in these samples.

5. Conclusions

Almost 50 years after the initial discovery of MTB by Salvatore Bellini in 1963 [64,65] and 40 years after the rediscovery and formal publication of these organisms by Richard P. Blakemore [66], it is only recently that extremophilic MTB have been observed and described. These findings clearly raise the possibility that magnetotactic microorganisms might exist in other extreme environments that have never been sampled and examined for their presence. Do MTB exist in environments characterized by very high pressure, by extreme cold, highly acidic or highly saline? To address this question, more sampling missions in extreme environments will need to be done by researchers using specific culture- and non-culture-based techniques. However, based on the recent results described in this chapter, the potential of finding MTB in other extreme environments is high. Moreover, considering the relatively small number of groups that study these intriguing organisms, it seems that the known ecological diversity of MTB is seriously underestimated.

Acknowledgements

Much of the work described here was supported by U.S. National Science Foundation grant EAR-0920718 awarded to D.A.B. C.T.L. was the recipient of an award from the Foundation pour la Recherche Médicale FRM: SPF20101220993.

References

1. Kirschvink, J.L.; Walker, M.M.; Diebel, C.E. Magnetite-based magnetoreception. *Curr. Opin. Neurobiol.* **2001**, *11*, 462–467.
2. Bazylinski, D.A.; Frankel, R.B. Magnetosome formation in prokaryotes. *Nat. Rev. Microbiol.* **2004**, *2*, 217–230.
3. Frankel, R.B.; Bazylinski, D.A.; Johnson, M.S.; Taylor, B.L. Magneto-aerotaxis in marine coccoid bacteria. *Biophys. J.* **1997**, *73*, 994–1000.
4. Amann, R.; Peplies, J.; Schüler, D. Diversity and taxonomy of magnetotactic bacteria. In *Magnetoreception and Magnetosomes, in Bacteria*; Schüler, D., Ed.; Microbiology Monographs; Springer: Berlin/Heidelberg, Germany, 2007; Volume 3, pp. 25–36.
5. Kolinko, S.; Jogler, C.; Katzmann, E.; Wanner, G.; Peplies, J.; Schüler, D. Single-cell analysis reveals a novel uncultivated magnetotactic bacterium within the candidate division OP3. *Environ. Microbiol.* **2012**, *14*, 1709–1721.
6. Lefèvre, C.T.; Viloria, N.; Schmidt, M.L.; Pósfai, M.; Frankel, R.B.; Bazylinski, D.A. Novel magnetite-producing magnetotactic bacteria belonging to the *Gammaproteobacteria. ISME J.* **2012**, *6*, 440–450.
7. Bazylinski, D.A.; Schübbe, S. Controlled biomineralization by and applications of magnetotactic bacteria. *Adv. Appl. Microbiol.* **2007**, *62*, 21–62.

8. Schleifer, K.H.; Schüler, D.; Spring, S.; Weizenegger, M.; Amann, R.; Ludwig, W.; Kohler, M. The genus *Magnetospirillum* gen. nov. description of *Magnetospirillum gryphiswaldense* sp. nov. and transfer of *Aquaspirillum magnetotacticum* to *Magnetospirillum magnetotacticum* comb. nov. *Syst. Appl. Microbiol.* **1991**, *14*, 379–385.
9. Sakaguchi, T.; Arakaki, A.; Matsunaga, T. *Desulfovibrio magneticus* sp nov., a novel sulfate-reducing bacterium that produces intracellular single-domain-sized magnetite particles. *Int. J. Syst. Evol. Microbiol.* **2002**, *52*, 215–221.
10. Bazylinski, D.A.; Williams, T.J.; Lefèvre, C.T.; Berg, R.J.; Zhang, C.L.; Bowser, S.S.; Dean, A.J.; Beveridge, T.J. *Magnetococcus marinus* gen. nov., sp. nov., a marine, magnetotactic bacterium that represents a novel lineage (*Magnetococcaceae* fam. nov.; *Magnetococcales* ord. nov.) at the base of the *Alphaproteobacteria*. *Int. J. Syst. Evol. Microbiol.* **2013**, *63*, 801–808.
11. Williams, T.J.; Lefèvre, C.T.; Zhao, W.; Beveridge, T.J.; Bazylinski, D.A. *Magnetospira thiophila* gen. nov., sp. nov., a marine magnetotactic bacterium that represents a novel lineage within the *Rhodospirillaceae* (*Alphaproteobacteria*). *Int. J. Syst. Evol. Microbiol.* **2012**, *62*, 2443–2450.
12. Bazylinski, D.A.; Williams, T.J.; Lefèvre, C.T.; Trubitsyn, D.; Fang, J.; Beveridge, T.J.; Moskowitz, B.M.; Ward, B.; Schübbe, S.; Dubbels, B.L.; Simpson, B. *Magnetovibrio blakemorei*, gen. nov. sp. nov., a new magnetotactic bacterium (*Alphaproteobacteria*: *Rhodospirillaceae*) isolated from a salt marsh. *Int. J. Syst. Evol. Microbiol.* **2013**, doi: 10.1099/ijs.0.037697-0.
13. McKay, D.S.; Gibson, E.K.; Thomas-Keprta, K.L.; Vali, H.; Romanek, C.S.; Clemett, S.J.; Chillier, X.D.F.; Maechling, C.R.; Zare, R.N. Search for past life on Mars: Possible relic biogenic activity in Martian meteorite ALH84001. *Science* **1996**, *273*, 924–930.
14. Thomas-Keprta, K.L.; Clemett, S.J.; Bazylinski, D.A.; Kirschvink, J.L.; McKay, D.S.; Wentworth, S.J.; Vali, H.; Gibson, E.K.; McKay, M.F.; Romanek, C.S. Truncated hexa-octahedral magnetite crystals in ALH84001: Presumptive biosignatures. *P. Natl. Acad. Sci. USA* **2001**, *98*, 2164–2169.
15. Lefèvre, C.T.; Abreu, F.; Schmidt, M.L.; Lins, U.; Frankel, R.B.; Hedlund, B.P.; Bazylinski, D.A. Moderately thermophilic magnetotactic bacteria from hot springs in Nevada. *Appl. Environ. Microbiol.* **2010**, *76*, 3740–3743.
16. Anderson, J.P. A geochemical study of the southwest part of the Black Rock Desert and its geothermal areas; Washoe, Pershing, and Humboldt Counties, Nevada. *Colo. School Mines Q.* **1978**, *73*, 15–22.
17. Costa, K.C.; Navarro, J.B.; Shock, E.L.; Zhang, C.L.; Soukup, D.; Hedlund, B.P. Microbiology and geochemistry of great boiling and mud hot springs in the United States Great Basin. *Extremophiles* **2009**, *13*, 447–459.
18. Haouari, O.; Fardeau, M.-L.; Cayol, J.-L.; Fauque, G.; Casiot, C.; Elbaz-Poulichet, F.; Hamdi, M.; Ollivier, B. *Thermodesulfovibrio hydrogeniphilus* sp. nov., a new thermophilic sulphate-reducing bacterium isolated from a Tunisian hot spring. *Syst. Appl. Microbiol.* **2008**, *31*, 38–42.

19. Nash, C. Mechanisms and evolution of magnetotactic bacteria. Ph.D. thesis, California Institute of Technology, Pasadena, CA, USA, 2008.
20. Lefèvre, C.T.; Frankel, R.B.; Pósfai, M.; Prozorov, T.; Bazylinski, D.A. Isolation of obligately alkaliphilic magnetotactic bacteria from extremely alkaline environments. *Environ. Microbiol.* **2011**, *13*, 2342–2350.
21. Oremland, R.S.; Dowdle, P.R.; Hoeft, S.; Sharp, J.O.; Schaefer, J.K.; Miller, L.G.; Blum, J.S.; Smith, R.L.; Bloom, N.S.; Wallschlaeger, D. Bacterial dissimilatory reduction of arsenate and sulfate in meromictic Mono Lake, California. *Geochim. Cosmochim. Acta.* **2000**, *64*, 3073–3084.
22. Kulp, T.R.; Han, S.; Saltikov, C.W.; Lanoil, B.D.; Zargar, K.; Oremland, R.S. Effects of imposed salinity gradients on dissimilatory arsenate reduction, sulfate reduction, and other microbial processes in sediments from two California soda lakes. *Appl. Environ. Microbiol.* **2007**, *73*, 5130–5137.
23. Hoeft, S.E.; Kulp, T.R.; Stolz, J.F.; Hollibaugh, J.T.; Oremland, R.S. Dissimilatory arsenate reduction with sulfide as electron donor: experiments with mono lake water and isolation of strain MLMS-1, a chemoautotrophic arsenate respirer. *Appl. Environ. Microbiol.* **2004**, *70*, 2741–2747.
24. Wiemeyer, S. Metals and trace elements in water, sediment, and vegetation at Ash Meadows National Wildlife Refuge-1993. Technical Report for the United States Fish and Wildlife Service, Reno, NV, USA, July 2005.
25. Al-Qudah, O.; Woocay, A.; Walton, J. Identification of probable groundwater paths in the Amargosa Desert vicinity. *Appl. Geochem.* **2011**, *26*, 565–574.
26. Pikuta, E.V.; Hoover, R.B.; Bej, A.K.; Marsic, D.; Whitman, W.B.; Cleland, D.; Krader, P. *Desulfonatronum thiodismutans* sp. nov., a novel alkaliphilic, sulfate-reducing bacterium capable of lithoautotrophic growth. *Int. J. Syst. Evol. Microbiol.* **2003**, *53*, 1327–1332.
27. Chavadar, M.S.; Bajekal, S.S. Magnetotactic bacteria from Lonar lake. *Curr. Sci.* **2009**, *96*, 957–959.
28. Chavadar, M.S.; Bajekal, S.S. Microaerophilic magnetotactic bacteria from Lonar Lake, India. *J. Pure Appl. Microbiol.* **2010**, *4*, 681–685.
29. Rajasekhar, R.P.; Mishra, D.C. Analysis of gravity and magnetic anomalies over Lonar Lake, India: An impact crater in a basalt province. *Curr. Sci.* **2005**, *88*, 1836–1840.
30. Krulwich, T.A. Alkaliphilic prokaryotes. In *The Prokaryotes*; Dworkin, M., Falkow, S., Rosenberg, E., Schleifer, K.-H., Stackebrandt, E., Eds.; Springer: New York, NY, USA, 2006; pp. 283–308.
31. Jimenez-Lopez, C.; Romanek, C.S.; Bazylinski, D.A. Magnetite as a prokaryotic biomarker: A review. *J. Geophys. Res. Biogeosci.* **2010**, *115*, G00G03.
32. Schübbe, S.; Kube, M.; Scheffel, A.; Wawer, C.; Heyen, U.; Meyerdierks, A.; Madkour, M.H.; Mayer, F.; Reinhardt, R.; Schüler, D. Characterization of a spontaneous nonmagnetic mutant of *Magnetospirillum gryphiswaldense* reveals a large deletion comprising a putative magnetosome island. *J. Bacteriol.* **2003**, *185*, 5779–5790.

33. Dubbels, B.L.; DiSpirito, A.A.; Morton, J.D.; Semrau, J.D.; Neto, J.N.E.; Bazylinski, D.A. Evidence for a copper-dependent iron transport system in the marine, magnetotactic bacterium strain MV-1. *Microbiology* **2004**, *150*, 2931–2945.
34. Schüler, D. Genetics and cell biology of magnetosome formation in magnetotactic bacteria. *FEMS Microbiol. Rev.* **2008**, *32*, 654–672.
35. Pósfai, M.; Moskowitz, B.M.; Arató, B.; Schüler, D.; Flies, C.; Bazylinski, D.A.; Frankel, R.B. Properties of intracellular magnetite crystals produced by *Desulfovibrio magneticus* strain RS-1. *Earth Planet Sci. Lett.* **2006**, *249*, 444–455.
36. Fang, J.; Zhang, L.; Bazylinski, D.A. Deep-sea piezosphere and piezophiles: geomicrobiology and biogeochemistry. *Trends Microbiol.* **2010**, *18*, 413–422.
37. Stolz, J.; Chang, S.B.R.; Kirschvink, J.L. Magnetotactic bacteria and single-domain magnetite in hemipelagic sediments. *Nature* **1986**, *321*, 849–851.
38. Petermann, H.; Bleil, U. Detection of live magnetotactic bacteria in South Atlantic deep-sea sediments. *Earth Planet. Sci. Lett.* **1993**, *117*, 223–228.
39. Abreu, F.; Lins, U. Universidade Federal do Rio de Janeiro, Rio de Janeiro, Brazil. Personal communication, 2013
40. Baker-Austin, C.; Dopson, M. Life in acid: pH homeostasis in acidophiles. *Trends Microbiol.* **2007**, *15*, 165–171.
41. Frankel, R.B.; Bazylinski, D.A. Biologically induced mineralization by bacteria. *Rev. Mineral. Geochem.* **2003**, *54*, 95–114.
42. Bell, P.E.; Mills, A.L.; Herman, J.S. Biogeochemical conditions favoring magnetite formation during anaerobic iron reduction. *Appl. Environ. Microbiol.* **1987**, *53*, 2610–2616.
43. Martins, J.L.; Silveira, T.S.; Silva, K.T.; Lins, U. Salinity dependence of the distribution of multicellular magnetotactic prokaryotes in a hypersaline lagoon. *Int. Microbiol.* **2009**, *12*, 193–201.
44. Lefèvre, C.T.; Abreu, F.; Lins, U.; Bazylinski, D.A. Nonmagnetotactic multicellular prokaryotes from low-saline, nonmarine aquatic environments and their unusual negative phototactic behavior. *Appl. Environ. Microbiol.* **2010**, *76*, 3220–3227.
45. Ollivier, B.; Caumette, P.; Garcia, J.L.; Mah, R.A. Anaerobic bacteria from hypersaline environments. *Microbiol. Rev.* **1994**, *58*, 27–38.
46. Thomas-Keprta, K.L.; Bazylinski, D.A.; Kirschvink, J.L.; Clemett, S.J.; McKay, D.S.; Wentworth, S.J.; Vali, H.; Gibson, E.K.; Romanek, C.S. Elongated prismatic magnetite crystals in ALH84001 carbonate globules: Potential Martian magnetofossils. *Geochim. Cosmochim. Acta* **2000**, *64*, 4049–4081.
47. Thomas-Keprta, K.L.; Clemett, S.J.; Bazylinski, D.A.; Kirschvink, J.L.; McKay, D.S.; Wentworth, S.J.; Vali, H.; Gibson, E.K.; Romanek, C.S. Magnetofossils from ancient Mars: a robust biosignature in the Martian meteorite ALH84001. *Appl. Environ. Microbiol.* **2002**, *68*, 3663–3672.
48. Buseck, P.R.; Dunin-Borkowski, R.E.; Devouard, B.; Frankel, R.B.; McCartney, M.R.; Midgley, P.A.; Posfai, M.; Weyland, M. Magnetite morphology and life on Mars. *P. Natl. Acad. Sci. USA* **2001**, *98*, 13490–13495.

49. Clemett, S.J.; Thomas-Keprta, K.L.; Shimmin, J.; Morphew, M.; McIntosh, J.R.; Bazylinski, D.A.; Kirschvink, J.L.; Wentworth, S.J.; McKay, D.S.; Vali, H.; *et al.* Crystal morphology of MV-1 magnetite. *Am. Mineral.* **2002**, *87*, 1727–1730.
50. Chang, S.B.R.; Kirschvink, J.L. Magnetofossils, the magnetization of sediments, and the evolution of magnetite biomineralization. *Annu. Rev. Earth Planet Sci.* **1989**, *17*, 169–195.
51. Golden, D.C.; Ming, D.W.; Morris, R.V.; Brearley, A.J.; Lauer, H.V., Jr.; Treiman, A.H.; Zolensky, M.E.; Schwandt, C.S.; Lofgren, G.E.; McKay, G.A. Evidence for exclusively inorganic formation of magnetite in Martian meteorite ALH84001. *Am. Mineral.* **2004**, *89*, 681–695.
52. Martel, J.; Young, D.; Peng, H.-H.; Wu, C.-Y.; Young, J.D. Biomimetic properties of minerals and the search for life in the Martian meteorite ALH84001. *Annu. Rev. Earth Planet. Sci.* **2012**, *40*, 167–193.
53. Weiss, B.P.; Kim, S.S.; Kirschvink, J.L.; Kopp, R.E.; Sankaran, M.; Kobayashi, A.; Komeili, A. Magnetic tests for magnetosome chains in Martian meteorite ALH84001. *P. Natl. Acad. Sci. USA* **2004**, *101*, 8281–8284.
54. Arato, B.; Szanyi, Z.; Flies, C.; Schüler, D.; Frankel, R.B.; Buseck, P.R.; Pósfai, M. Crystal-size and shape distributions of magnetite from uncultured magnetotactic bacteria as a potential biomarker. *Am. Mineral.* **2005**, *90*, 1233–1240.
55. Kopp, R.E.; Kirschvink, J.L. The identification and biogeochemical interpretation of fossil magnetotactic bacteria. *Earth Sci. Rev.* **2008**, *86*, 42–61.
56. Gehring, A.U.; Kind, J.; Charilaou, M.; Garcia-Rubio, I. The detection of magnetotactic bacteria and magnetofossils by means of magnetic anisotropy. *Earth Planet. Sci. Lett.* **2011**, *309*, 113–117.
57. Kind, J.; Gehring, A.U.; Winklhofer, M.; Hirt, A.M. Combined use of magnetometry and spectroscopy for identifying magnetofossils in sediments. *Geochem. Geophys. Geosyst.* **2011**, *12*, Q08008.
58. Kempe, S.; Degens, E. An early soda ocean? *Chem. Geol.* **1985**, *53*, 95–108.
59. Hecht, M.H.; Kounaves, S.P.; Quinn, R.C.; West, S.J.; Young, S.M.M.; Ming, D.W.; Catling, D.C.; Clark, B.C.; Boynton, W.V.; Hoffman, J.; *et al.* Detection of perchlorate and the soluble chemistry of martian soil at the Phoenix lander site. *Science* **2009**, *325*, 64–67.
60. Kounaves, S.P.; Hecht, M.H.; Kapit, J.; Gospodinova, K.; DeFlores, L.; Quinn, R.C.; Boynton, W.V.; Clark, B.C.; Catling, D.C.; Hredzak, P.; *et al.* Wet chemistry experiments on the 2007 Phoenix Mars Scout Lander mission: Data analysis and results. *J. Geophys. Res. Planets* **2010**, *115*.
61. Kempe, S.; Kazmierczak, J. A terrestrial model for an alkaline martian hydrosphere. *Planet Space Sci.* **1997**, *45*, 1493–1499.
62. McNeill, D.F.; Ginsburg, R.N.; Chang, S.B.R.; Kirschvink, J.L. Magnetostratigraphic dating of shallow-water carbonates from San-Salvador, Bahamas. *Geology* **1988**, *16*, 8–12.
63. Sakai, S.; Jige, M. Characterization of magnetic particles and magnetostratigraphic dating of shallow-water carbonates in the Ryukyu Islands, northwestern Pacific. *Isl. Arc.* **2006**, *15*, 468–475.

64. Bellini, S. On a unique behavior of freshwater bacteria. *Chin. J. Oceanol. Limn.* **2009**, *27*, 3–5.
65. Bellini, S. Further studies on "magnetosensitive bacteria". *Chin. J. Oceanol. Limn.* **2009**, *27*, 6–12.
66. Blakemore, R.P. Magnetotactic bacteria. *Science* **1975**, *190*, 377–379.

Predator Avoidance in Extremophile Fish

David Bierbach, Matthias Schulte, Nina Herrmann, Claudia Zimmer, Lenin Arias-Rodriguez, Jeane Rimber Indy, Rüdiger Riesch and Martin Plath

Abstract: Extreme habitats are often characterized by reduced predation pressures, thus representing refuges for the inhabiting species. The present study was designed to investigate predator avoidance of extremophile populations of *Poecilia mexicana* and *P. sulphuraria* that either live in hydrogen sulfide-rich (sulfidic) springs or cave habitats, both of which are known to have impoverished piscine predator regimes. Focal fishes that inhabited sulfidic springs showed slightly weaker avoidance reactions when presented with several naturally occurring predatory cichlids, but strongest differences to populations from non-sulfidic habitats were found in a decreased shoaling tendency with non-predatory swordtail (*Xiphophorus hellerii*) females. When comparing avoidance reactions between *P. mexicana* from a sulfidic cave (Cueva del Azufre) and the adjacent sulfidic surface creek (El Azufre), we found only slight differences in predator avoidance, but surface fish reacted much more strongly to the non-predatory cichlid *Vieja bifasciata*. Our third experiment was designed to disentangle learned from innate effects of predator recognition. We compared laboratory-reared (*i.e.*, predator-naïve) and wild-caught (*i.e.*, predator-experienced) individuals of *P. mexicana* from a non-sulfidic river and found no differences in their reaction towards the presented predators. Overall, our results indicate (1) that predator avoidance is still functional in extremophile *Poecilia* spp. and (2) that predator recognition and avoidance reactions have a strong genetic basis.

Reprinted from *Life*. Cite as: Bierbach, D.; Schulte, M.; Herrmann, N.; Zimmer, C.; Arias-Rodriguez, L.; Indy, J.R.; Riesch, R.; Plath, M. Predator Avoidance in Extremophile Fish. *Life* **2013**, *3*, 161-180.

1. Introduction

Falling victim to predation excludes an individual from future reproductive opportunities, thus underpinning the importance of appropriate anti-predator behavior to prevent predator-related mortalities [1]. Before a prey species can react to a predator, however, initial recognition is required either through visual [2–4], olfactory [5], tactile [6,7], or auditory cues [8]. Upon this initial detection, the prey then has to assess the likelihood of an attack, which is crucial for triggering an appropriate avoidance response [2]. For the purpose of the present study, we define "predator avoidance" as both detection and identification of predatory stimuli (here: visual cues from piscivorous cichlids) that elicit an avoidance response in prey (*Poecilia mexicana*, a small, neotropical livebearing fish).

Anti-predator behavior is typically associated with some kind of cost [9], and populations are predicted to rapidly loose these expensive behaviors when colonizing low-predation or predator-free environments, in which the costs outweigh potential benefits [10,11]. Economic considerations dictate that this reduction process should be even faster in habitats with low resource availability. Nonetheless, some anti-predator behaviors are known to persist throughout generations following initial isolation from predators [12–14]. On top of that, several studies exemplified that anti-predator responses in natural populations represent a combination of both innate and experiential (*i.e.*, learned) components [15–18]. Learned responses are lost after one generation, and thus need to be reacquired in every generation [17], while genetically based behaviors may persist for many thousand generations, but loss may be permanent [15].

A unique opportunity to study the evolution of anti-predator response mechanisms is provided near the Southern Mexican cities of Teapa and Tapijulapa, which are located in the state of Tabasco. Here, *P. mexicana* inhabits various streams and rivers with diverse fish faunas including several piscivorous species [19,20]. Additionally, locally adapted populations in at least three different tributaries of the Río Grijalva drainage also inhabit springs characterized by the lack of piscivorous fishes in waters containing high amounts of hydrogen sulfide (H_2S) [19,20]. H_2S is acutely toxic to most metazoans even in micromolar amounts as it inhibits aerobic respiration, but also leads to extreme hypoxia in the water [21,22]. To cope with this toxicity and the H_2S-induced hypoxia, locally adapted *P. mexicana* in sulfidic habitats resort to aquatic surface respiration (ASR), thus exploiting the more oxygen-rich air-water interface using their gills (e.g., [23,24]). However, ASR also exposes fish in sulfidic habitats to avian predation, which is up to twenty times higher than in non-sulfidic habitats [25]. Another feature of those extreme habitats is low energy availability [23]. While the predominant food source of non-sulfidic surface-dwelling *P. mexicana* is detritus and green algae, diets of conspecifics in the sulfidic surface and cave streams are dominated by chemoautotrophic (sulfur) bacteria and aquatic invertebrates [26]. Fish in the sulfidic habitats spend up to 85% of their time performing ASR, which strongly reduces the time afforded for feeding [23]. In combination with the energy demanding detoxification of H_2S [22], this leads to lower general body conditions (weight-length ratios [19], abdominal distention [27], and body fat content [28,29]) observed in H_2S-adapted *P. mexicana*. As a result, energy limitation along with reduced piscine predation rates should favor the rapid reduction of costly anti-predator behaviors in all extremophile *P. mexicana* populations.

In the present study, we investigated the time spent near or away from several predatory (and non-predatory) fishes as an estimate of anti-predator behavior in *Poecilia* spp. from three independent sulfur systems. In two of them, the Río Tacotalpa and Río Puyacatengo drainages, *P. mexicana* inhabits several sulfidic surface springs, but in the Río Tacotalpa drainage this species has also successfully colonized a sulfidic cave [30,31] (Figure 1). The sulfidic springs of the Baños del Azufre in the Río Pichucalco drainage, on the other hand, are inhabited by the sulfur molly (*Poecilia sulphuraria* [32]; Figure 1), a highly H_2S-adapted sister species of *P. mexicana* [31].

Although the sulfur springs in these systems can be considered to be essentially free of predatory fish species, several small non-sulfidic streams drain into them, and predatory cichlids (e.g., *Cichlasoma salvini*), a predatory characid (*Astyanax aeneus*), as well as other poeciliids (e.g., *Heterandria bimaculata* and *Xiphophorus hellerii*) can be observed in the less toxic mixing zones [18–20]. The Cueva del Azufre, by contrast, is completely devoid of fish species other than *P. mexicana*, the sole exception being occasional sightings of the synbranchid eel *Ophisternon aenigmaticum* [20]. Nevertheless, *P. mexicana* inhabiting this cave face high predation pressure through a giant water bug, *Belostoma* sp. [33,34], several pisaurid, ctenid and theraphosid spiders [35], one species of trichodactylid freshwater crab, *Avotrichodactylus bidens* [36], possibly also an as yet undetermined pseudothelphusid freshwater crab (M.P., unpublished data), and most likely terrestrial mammals that regularly venture into the cave (based on mammal scat regularly found in different parts of the cave; all authors, personal observation).

Figure 1. Overview of the study area and detailed view of the collection sites. (**1**) Baños del Azufre (sulfidic); (**2**) Río El Azufre (non-sulfidic); (**3**) Río Ixtapangjoya (non-sulfidic); (**4**) Río Puyacatengo (non-sulfidic); (**5**) La Lluvia (sulfidic); (**6**) Cueva del Azufre (cave, sulfidic); (7) El Azufre II (sulfidic).

We tested for differences in anti-predator behavior of *Poecilia* spp. from several extreme and benign habitats and thus analyzed avoidance reactions elicited by predatory fish species that naturally co-occur with *P. mexicana* in non-sulfidic sites [*C. salvini* and *Petenia splendida* (Cichlidae)]. We further included non-piscivorous species [green swordtails, *X. hellerii* (Poeciliidae) and *Vieja bifasciata* (Cichlidae)] that are also common in South Mexican sulfide-free habitats [37] in our experiments as stimuli for control treatments. We predicted that at least *X. hellerii* should not elicit an avoidance reaction [18] if focal fish were able to distinguish threatening and non-threatening stimuli.

Our first question was whether predator identification and subsequent avoidance were still functional in *Poecilia* spp. inhabiting sulfidic springs. We, therefore, compared predator avoidance reactions of populations from non-sulfidic habitats to those of populations inhabiting sulfidic springs in two of the three drainages. We predicted that fish from non-sulfidic waters would show strong avoidance reactions when confronted with the omnivorous *C. salvini* [37,38] and especially when confronted with the piscivorous *P. splendida* [37], while reactions of H_2S-adapted *Poecilia* spp. would be less predictable.

In a second experiment, we looked for differences in the avoidance reactions towards the aforementioned predators between *P. mexicana* from the sulfidic Cueva del Azufre and the adjacent sulfidic surface creek, the El Azufre (Figure 1). Cave mollies are known to have acquired several adaptations related to their life in perpetual darkness: reduced shoaling [39] and reduced aggressive behavior [40–42]. However, while cave mollies have reduced eye size [43–45], their eyes do not differ in spectral sensitivity from those of their surface-dwelling counterparts, and light-reared fish can perceive and react to visual cues [46]. Both the sulfidic surface and the cave habitats are largely devoid of piscivorous fishes, but *P. mexicana* from the sulfidic El Azufre experience immense avian predation [25], and could at least occasionally come into contact with predatory fishes by entering the less sulfidic mixing zones that connect the El Azufre to the nearest non-sulfidic creeks [18–20]. *Poecilia mexicana* from the inside of the Cueva del Azufre, on the other hand, are highly unlikely to ever encounter piscivorous fishes or birds (see above). Thus, we predicted that the cave population—the sole example of a population that most likely evolved without any contact to predatory fishes—would show a reduction in predator avoidance reactions.

Any differences in predator avoidance found in our two previous experiments could probably either be the result of heritable differences between populations, namely, reductions of predator recognition and avoidance or, if predator recognition and avoidance are intact, of a lack of predator experience, leading to an altered evaluation of threat levels [2,18]. Our third experiment was, therefore, designed to disentangle learned from innate effects on predator avoidance reactions [2]. We used laboratory-reared (*i.e.*, predator-naïve) descendants of *P. mexicana* originated from the non-sulfidic Río Ixtapangajoya (Río Puyacatengo drainage; Figure 1) and compared their behavior to that of wild-caught (*i.e.*, predator-experienced) individuals from the same origin. Kelley and Magurran [4] found predator avoidance in the Trinidadian guppy (*Poecilia reticulata*) to be influenced by experience when comparing wild-caught and laboratory-reared (thus predator-naïve) fish. Ferrari *et al.* [47] could show that fathead minnows (*Pimephales promelas*) were able to learn to recognize predators when exposed to predator odors and even generalize their anti-predator response to new, closely-related predators. However, a recent study further revealed that predator-naïve *P. mexicana* females responded with a stronger change in mating preferences when exposed to a set of predators (and non-predators) than wild-caught, predator-experienced females [18]. The authors suggest innate predator recognition mechanisms that are fine-tuned by experience. Predator-experienced *P. mexicana* may be more capable to estimate the predator's motivation to prey—which was assumed to be low since live predators might have been stressed by the experimental handling [18]. Accordingly, we predicted predator-experienced individuals to show weaker predator avoidance responses to visual stimuli from live predatory cichlids than naïve ones.

2. Experimental Section

2.1. Study Organisms, Sampling Sites and Maintenance of the Test Animals

Poeciliid fishes are livebearers, and males use their transformed anal fin, the gonopodium, to transfer sperm [48]. Females store sperm to fertilize several consecutive, monthly broods, and sperm competition is intense [49]. The Atlantic molly (*Poecilia mexicana*) is widespread in various streams, lakes and lagoons along the Central American Atlantic coast [37]. While *P. mexicana* females have a cryptic body coloration, large males show conspicuous black vertical bars, and dominant males may even become completely black with yellowish to orange margins on the dorsal and anal fins [40]. Smaller males are typically less conspicuous in coloration. *Poecilia mexicana* males do not court females [50]; their pre-mating behavior consists of so-called genital nipping where males try to gather chemical information on female receptivity by making oral contact with the female's genital opening [40,50]. Nipping typically, but not always, precedes copulation and is the most frequent sexual behavior in *P. mexicana* [40].

Atlantic molly males usually establish dominance hierarchies, and dominant (typically the largest) males monopolize small shoals of females, which they aggressively defend against rivals [40]. For this study, *Poecilia* spp. were collected in sulfidic springs and adjacent non-sulfidic habitats in the Mexican states of Tabasco and Chiapas, particularly in the region around the city of Teapa. Here, the mountains of the Sierra Madre de Chiapas meet the wide floodplains of northern Tabasco. The six spring complexes known to be inhabited by *Poecilia* are located in the foothills of the Sierra Madre and distributed across three major tributaries of the Río Grijalva. In the upper reaches where the sulfidic springs are located, these tributaries (Ríos Tacotalpa, Puyacatengo, and Pichucalco) are separated by mountains, while they all eventually join the Río Grijalva and are widely interconnected in the lowlands at least during the wet season. In the Río Pichucalco drainage, the sulfidic ecotype has been described as a distinct species, *P. sulphuraria* (Alvarez 1948), which is endemic to sulfide spring complexes at the Baños del Azufre and Rancho La Gloria and represents a more ancient lineage of sulfide spring fish [31]. In the Tacotalpa and Puyacatengo drainages, H_2S-adapted ecotypes cluster phylogenetically within Southern Mexican *P. mexicana*. Nonetheless, all three lineages share a series of traits characteristic of sulfidic spring fish [31]. In the Río Tacotalpa drainage, *P. mexicana* also colonize a sulfidic cave, the Cueva del Azufre [30,51]. The Cueva del Azufre is divided into 13 different chambers, with Chamber XIII being the innermost chamber (after [30]). Several springs in the cave (mainly in Chamber X) release sulfidic water, and the creek that flows through the cave eventually leaves the cave and turns into the sulfidic El Azufre.

For Experiments 1 and 2, we used wild-caught fish from different sulfidic and non-sulfidic surface habitats as well as fish from Chamber II of the Cueva del Azufre [19,30,31] (Table 1). Light enters the front parts of that chamber through several holes in the ceiling [45], so fish could be tested for a visual response under light conditions (even though the skylights are not sufficient to illuminate the chamber fully). Upon capture, fish were transferred into closed and aerated 38 L (43 × 31 × 32 cm) black Sterilite® containers and brought immediately to the Tropical Aquaculture Laboratory at the División Académica de Ciencias Biológicas from the Universidad Juárez Autónoma de Tabasco (UJAT) in Villahermosa, Tabasco, Mexico (DACBIOL-UJAT). Here, they were kept separated by sex in well aerated 70 L tanks (filled with aged tap water) for 24 hours at 30.0 ± 1.0 °C, with approximately 12:12 hours light:dark cycle and could acclimate to the water conditions in the laboratory.

For Experiment 3, we used descendants of wild-caught fish of the second to fourth laboratory generation originating from the Río Ixtapangajoya (Table 1). Fish were reared in 1000-L tanks at DACBIOL-UJAT under semi-natural conditions, in absence of predators. In the laboratory all fish were fed once a day *ad libitum* with commercially available pellet and flake food. Experiments were conducted between 15 April and 15 July 2011.

Table 1. Standard length (SL ± S.E.M.), sample size (*N*, number of individuals) and GPS data for the eight populations studied.

Population	Species	Habitat / Treatment	SL [mm]		Sampling point	
			males	females	Latitude	Longitude
Río Ixtapangajoya (Río Puyacatengo drainage)	*Poecilia mexicana*	wc*, n, sf	28.4 ± 0.4 (*N* = 28)	28.9 ± 0.9 (*N* = 24)	17.49450	−92.99763
Río Ixtapangajoya (Río Puyacatengo drainage)	*P. mexicana*	lab, n, sf	39.5 ± 1.2 (*N* = 28)	43.67 ± 2.0 (*N* = 24)	17.49450	−92.99763
Río El Azufre (Río Pichucalco dranage)	*P. mexicana*	wc, n, sf	40.5 ± 2.2 (*N* = 28)	43.8 ± 1.6 (*N* = 24)	17.55634	−93.00762
Baños del Azufre (Río Pichucalco drainage)	*Poecilia sulphuraria*	wc, s, sf	24.6 ± 0.6 (*N* = 28)	26.04 ± 1.1 (*N* = 24)	17.55225	−92.99859
RíoPuyacatengo	*P. mexicana*	wc, n, sf	39.2 ± 1.3 (*N* = 28)	40.4 ± 2.3 (*N* = 25)	17.47000	−92.89573
La Lluvia (Río Puyacatengo drainage)	*P. mexicana*	wc, s, sf	28.2 ± 0.5 (*N* = 28)	35.8 ± 0.9 (*N* = 25)	17.46387	−92.89541
Cueva del Azufre (Río Tacotalpa drainage)	*P. mexicana*	wc, s, sf	31.1 ± 0.6 (*N* = 28)	39.3 ± 1.2 (*N* = 24)	17.44225	−92.77447
El Azufre II (Río Tacotalpa drainage)	*P. mexicana*	wc, s, ca	33.3 ± 0.9 (*N* = 28)	35.8 ± 1.1 (*N* = 24)	17.43843	−92.77476

*Habitat/treatment variables are defined as follows: wc, wild-caught; lab, laboratory-reared; n, non-sulfidic; s, sulfidic; sf, surface; ca, cave.

2.2. Experimental Design

Tests were conducted in two identical test tanks (42.6 × 30.0 × 16.5 cm) made of UV-transparent plexiglas. Each tank was visually divided into three equally-sized zones by black marks on the outside. The central zone was designated the neutral zone, the two lateral zones as preference zones. Predators were presented in one of two small auxiliary tanks (19.5 × 30.0 × 14.5 cm) on either side of the test tank. Hence, the focal individual could spend time in the zone near a predator or in the zone furthest away from a predator. In order to reduce disturbance from the outside, the experimental setups were placed in large oval tubs that were filled with aged tap water to the level inside the test tanks. The entire set-up was placed on a shelf of about 1 m height, and the observer was standing approximately 1.5 to 2 m away from the experimental setup and observed the fish diagonally from above.

To initiate a trial, we introduced the focal fish into the central tank and let it habituate to the test apparatus for 5 min. We then placed a predatory fish in either the right or the left auxiliary tank. The test tanks used in this study were relatively small, so focal fish were able to see the predator throughout the course of the experiment. Test fish would typically freeze on the bottom of the test tank for a few seconds (to some minutes) after the stimulus was introduced, so a trial began only after the focal fish had started to swim freely in the water column. We measured the time the focal fish spent in each preference zone during a 5-min observation period (*i.e.*, in the zone closest to or the zone furthest away from the predator). To detect side biases, the predator was switched between sides immediately after the first 5-min observation period and measurement was repeated.

In Treatment 1, focal fish were presented with a green swordtail (*Xiphophorus hellerii*) female (48.78 ± 1.25 mm standard length), which served as a control, since *X. hellerii* is a related, non-predatory species of similar body size, appearance and ecology to *P. mexicana*. In the second treatment, we presented focal individuals with *Cichlasoma salvini* (93.59 ± 3.16 mm) which is a native omnivorous cichlid in Southern Mexico and also includes mollies in its diet [37,52]. For the third treatment, we chose the algi- and detrivorous cichlid *Vieja bifasciata* (119.75 ± 1.97 mm) as a stimulus, and in Treatment 4, we presented a purely piscivorous predator, *Petenia splendida* (145.31 ± 3.19 mm) to focal individuals. All stimulus fish are common in natural *P. mexicana* habitats [37,38].

2.3. Statistical Analyses

We calculated an "avoidance score" as the dependent variable for the statistical analyses as: (time spent in the preference zone near the predator-time spent in the opposite preference zone). Scores were checked for assumptions of normality, homogeneity of variance and sphericity prior to conducting the statistical tests, which were performed using SPSS 13. All data are presented as mean ± S.E.M. (standard error of the mean); all graphical illustrations show estimated marginal means derived from the respective models outlined below.

2.3.1. Experiment 1: Predator Recognition in Populations from Sulfide Springs

In our first experiment, we tested wild-caught *Poecilia* spp. from non-sulfidic and sulfidic habitats located in the Río Puyacatengo and Río Pichucalco drainages (see Table 1). While *P. mexicana* inhabits both non-sulfidic and sulfidic habitats in the Río Puyacatengo and non-sulfidic habitats in the Río Pichucalco, sulfidic springs along the Río Pichucalco are inhabited by the sulfur-endemic *P. sulphuraria*, a highly sulfide-adapted sister taxon to *P. mexicana* [31]. We compared patterns of predator avoidance using the avoidance score as dependent variable in a univariate General Linear Model (GLM) with drainage (Río Puyacatengo and Río Pichucalco), H_2S (absent or present), predator type (four presented predators), sex (male or female), as well as all their interactions as independent variables. We included the predators' body size (standard length, SL) and the focal individuals' body size (SL) as covariates. Interactions higher than the second order were removed from our final model because they were not statistically significant ($F \leq 1.35$, $P \geq 0.37$). We also removed the covariate "focal individuals" body size' as it had no significant effect on avoidance scores ($F_{1,\,188} = 2.33$, $P = 0.13$).

2.3.2. Experiment 2: Predator Recognition in Cave-Dwelling *P. mexicana*

In our second experiment we compared predator avoidance reactions between individuals from a surface- and a cave-dwelling population in the Río Tacotalpa drainage. As representatives of a surface-dwelling population, we used fish from the sulfidic El Azufre, collected downstream the outflow of the sulfidic Cueva del Azufre. As representatives of a cave-adapted population, fish from cave Chamber II of the Cueva del Azufre were chosen for the experiment (Table 1; Figure 1). Predator avoidance scores were compared in a univariate GLM with predator type, sex, light regime (surface or cave), and their interactions as independent variables. Predators' body size and focal individuals' body size were included as covariates. We removed the non-significant third order interaction term ($F_{3,86} = 0.22$, $P = 0.88$) as well as focal individual body size ($F_{1,89} = 1.01$, $P = 0.32$) from the final model.

2.3.3. Experiment 3: Influence of Predator Experience

In our third experiment we analyzed predator avoidance between predator-experienced (wild-caught) and naïve (laboratory-reared) individuals from the non-sulfidic Río Ixtapangajoya (Río Puyacatengo drainage). Avoidance scores were analyzed in a univariate GLM with predator type, sex, predator experience (laboratory-reared or wild-caught), and their interactions as independent variables. We also included predators' body size and focal individuals' body size as covariates but removed focal individuals' body size from the final model as it had no significant effect ($F_{1,96} = 1.35$, $P = 0.25$). Furthermore, we excluded all interactions from our final model, as none were statistically significant ($F \leq 2.00$; $P \geq 0.12$).

3. Results

3.1. Reduced Predator Avoidance in H_2S-Adapted Poecilia?

In our first experiment we compared patterns of predator avoidance between populations living either in non-sulfidic or sulfidic water from two different drainages. In our GLM, "predator type" was highly significant (Table 2A), indicating that test fish showed different responses towards the four stimulus species; more specifically, test fish showed avoidance behavior when confronted with the cichlids while they associated with swordtail females (Figure 2A). "Drainage" and "sex" were also significant, suggesting overall differences in the responsiveness between the Río Puyacatengo and Río Pichucalco drainage as well as between males and females (Table 2A).

Table 2. Results from General Linear Models (GLMs) with avoidance scores as the dependent variable. (**A**) Experiment 1; comparison between *Poecilia* spp. from sulfidic and non-sulfidic waters. (**B**) Experiment 2; comparison between surface and cave mollies. (**C**) Experiment 3; comparison between predator-experienced and naïve surface-dwelling *P. mexicana*. Significant effects are in boldface.

	df	*F*	*P*	partial eta^2
(A) Experiment 1				
predator type	**3**	**21.72**	**<0.001**	**0.256**
drainage	**1**	**9.89**	**0.002**	**0.050**
H_2S	1	0.36	0.55	0.002
sex	**1**	**19.28**	**<0.001**	**0.093**
predator body size (SL)	**1**	**16.00**	**<0.001**	**0.078**
predator type× drainage	3	1.92	0.13	0.030
predator type × H_2S	**3**	**4.25**	**0.006**	**0.063**
predator type× sex	3	1.30	0.28	0.020
drainage× H_2S	1	0.61	0.44	0.003
drainage× sex	1	2.13	0.15	0.011
H_2S × sex	**1**	**5.95**	**0.016**	**0.030**
error	189			
(B) Experiment 2				
predator type	**3**	**14.43**	**<0.001**	**0.325**
sex	1	0.61	0.44	0.007
light regime	1	0.84	0.36	0.009
predator body size (SL)	**1**	**9.01**	**0.003**	**0.091**
predator type × sex	3	0.90	0.44	0.029
predator type × light regime	**3**	**3.20**	**0.027**	**0.096**

Table 2. *Cont.*

	df	*F*	*P*	partial eta^2
(B) Experiment 2				
sex × light regime	1	3.33	0.071	0.036
error	90			
(C) Experiment 3				
predator type	**3**	**16.01**	**<0.001**	**0.331**
sex	1	0.05	0.82	0.001
experience	1	0.01	0.91	<0.001
predator body size (SL)	**1**	**12.93**	**0.001**	**0.118**
error	97			

Most importantly, the interaction term "predator type × H_2S" was significant (Table 2A), which suggests that fish from non-sulfidic and sulfidic habitats reacted differently towards the stimulus fishes (Figure 2A). Populations from sulfidic habitats showed a weaker predator avoidance reaction, but the strongest difference compared to populations from non-sulfidic habitats was found in a weaker association tendency with the swordtail females in the control treatment (Figure 2A). Also the interaction term of "sex × H_2S" was found to have a significant effect (Table 2A) due to a stronger difference between the overall reaction of males and females from sulfidic habitats compared to a less pronounced difference between males and females from non-sulfidic waters (Figure 2B). The covariate "predator body size" had a significant effect (Table 2A) and we found a significant negative correlation between this covariate and standardized residuals obtained from our final model (Pearson correlation; $r_P = -0.159$, $P = 0.022$). We, thus, present all avoidance scores as estimated marginal means derived from our final model, which are corrected for predators' body size.

Overall, "predator type" had the strongest effect in the GLM (partial eta$^2 = 0.26$), while the other two factors ("sex" and "drainage") and all significant interactions were only of minor importance (partial eta$^2 < 0.09$; Table 2A).

3.2. Reduced Predator Avoidance in Cave Mollies?

Our GLM analyzing predator avoidance scores of surface and cave-dwelling mollies from the Río Tacotalpa drainage revealed a significant effect of the factor "predator type" (Table 2B) indicating a general difference in the reaction towards the four predators presented in our experiment. The factors "sex" and "light regime" had no overall effect (Table 2B), and neither did the interaction terms "predator type × sex" and "sex × light regime" (Table 2B). Nevertheless, the interaction term "predator type × light regime" was significant (Table 2B), which was mainly due to the strong difference between the reaction of surface and cave-dwelling fish towards *V. bifasciata* (Figure 3). The covariate "predator body size" had a significant effect (Table 2B), and we found a significant negative correlation between this covariate and standardized residuals obtained from our final model ($r_P = -0.15$, $P = 0.013$).

3.3. Influence of Predator Experience on Predator Avoidance Responses in P. mexicana

The GLM comparing avoidance scores of predator-naïve and predator-experienced surface mollies detected a significant effect of the main factor "predator type" (Table 2C; Figure 4) suggesting a general difference in individuals' responses to the different predators used in our study. Both other main factors as well as their interactions had no significant effect (Table 2C) indicating that neither naïve and experienced fish nor males and females differed in their general responsiveness to the predator treatments. The covariate "predator body size" had a significant effect (Table 2C) and we found a significant negative correlation between this covariate and standardized residuals obtained from our final model ($r_P = -0.207$, $P = 0.03$).

Figure 2. (A) Avoidance reactions of *Poecilia* spp. from sulfidic and non-sulfidic waters towards four different stimuli fish species. **(B)** Avoidance reactions of males and females adapted to either sulfidic or non-sulfidic waters. Positive values indicate that focal fish spent more time in the proximity of a stimulus while negative values indicate that focal fish avoided the proximity of a stimulus. Depicted are estimated marginal means (± S.E.M) of the avoidance score.

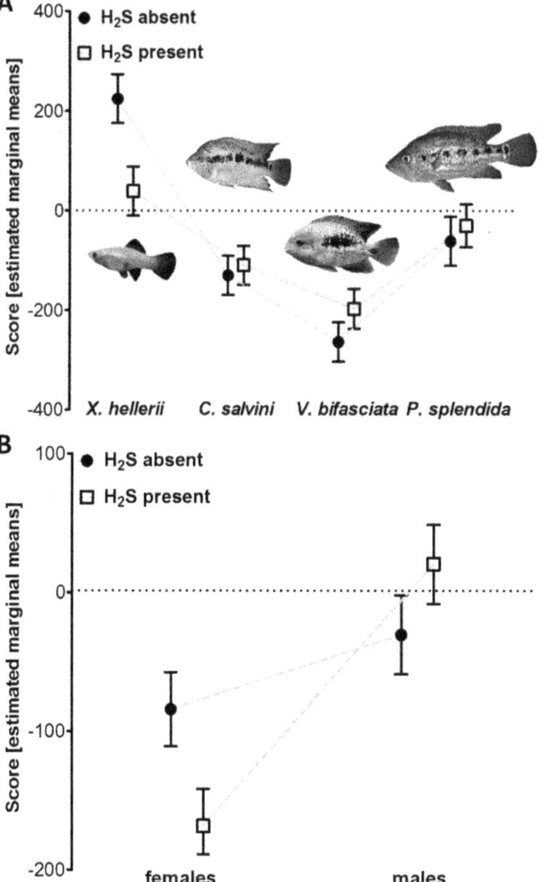

Figure 3. Avoidance reactions of surface- and cave-dwelling *P. mexicana* towards four different stimulus fish species. Positive values indicate that focal fish spent more time in the proximity of a stimulus while negative values indicate that focal fish avoided the proximity of a stimulus. Depicted are estimated marginal means (± S.E.M) of the avoidance score.

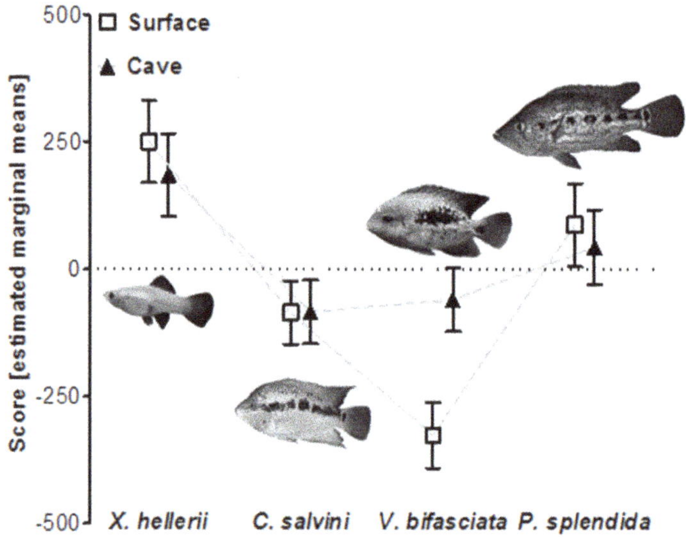

Figure 4. Avoidance reactions of predator-naïve (laboratory-reared) and predator-experienced (wild-caught) *P. mexicana* towards four different stimulus fish species. Positive values indicate that focal fish spent more time in the proximity of a stimulus while negative values indicate that focal fish avoided the proximity of a stimulus. Depicted are estimated marginal means (± S.E.M) of the avoidance score.

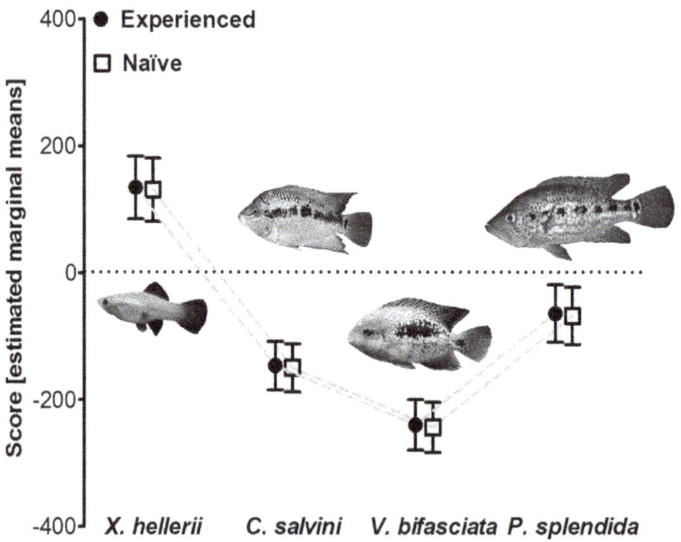

4. Discussion

Our study investigated predator avoidance reactions in extremophile *Poecilia* spp. that either inhabit sulfidic springs or a sulfidic cave in southern México. In addition, we also investigated experiential effects on predator avoidance by comparing laboratory-reared and wild-caught fish. We found that *Poecilia* spp., regardless of origin or level of experience, avoided the proximity of cichlid fishes but associated with the poeciliid *X. hellerii*. The avoidance reaction towards cichlid stimuli correlated negatively with the body size of the stimulus fish, *i.e.*, larger cichlids elicited stronger responses. The results confirm our prediction that mollies from non-sulfidic waters avoid the proximity of the predatory cichlids *C. salvini* and *P. splendida*. Surprisingly, reactions were comparably strong or even stronger towards the non-predatory cichlid *V. bifasciata*. However, reactions towards non-predatory swordtail (*X. hellerii*) females were less consistent: Fish from sulfidic habitats in our first experiment showed a reduced tendency to associate with swordtails, while fish from a sulfidic spring in another drainage (Río Tacotalpa) did not show such a pattern in our second experiment. Furthermore, our results revealed that sex differences in the avoidance reactions are more pronounced in *Poecilia* spp. from sulfidic habitats, with females showing a stronger avoidance response than males. The comparison of predator-naïve (laboratory-reared) and predator-experienced (wild-caught) *P. mexicana* in our third experiment did not reveal any difference in predator avoidance reactions.

We initially proposed that energy limitation in concert with reduced piscine predation rates might favor the rapid reduction of costly anti-predator behaviors in extremophile *Poecilia* spp. populations. However, our results showed that avoidance reactions of fish from sulfidic and non-sulfidic habitats were only slightly reduced. This could be a sign of "ghost of predators past" mechanisms [12,53], assuming that not enough time has elapsed since the colonization of predator-devoid habitats to loose recognition mechanisms. Molecular phylogenetic evidence suggests that the sulfur creeks in southern Mexico were independently invaded by ancestral forms of *P. mexicana/P. sulphuraria*-like fish not adapted to life in sulfidic waters [31]. Nonetheless, as outlined above, fish in all of these extreme habitats are probably energy-limited, which in turn, should impose strong selection on the reduction of costly and dispensable behaviors, like predator avoidance. Under these circumstances, the "ghost of the predators past" hypothesis seems unlikely. While we cannot exclude that anti-predator behaviors are pleiotropically linked to other behavioral traits that need to be maintained regardless of the presence or absence of predators [12,13], it is also possible that mollies adapted to sulfidic habitats at least occasional encounter some cichlid predators, as shoals of the predatory cichlid *C. salvini* can be observed in the mixing zones of non-sulfidic and sulfidic waters [18,19]. Still, most parts of the sulfidic springs are essentially devoid of predatory fishes, so we are inclined to argue in favor of another hypothesis—the "multi-predator hypothesis" assumes the presence of any type of predator to be sufficient to maintain anti-predator behaviors—even for missing predators [9]. Riesch *et al.* [25] found *P. mexicana* from sulfidic habitats to experience 20-fold increased avian predation rates. This increased predation pressure by birds could promote the persistence of avoidance reactions towards predatory fishes.

The strongest difference between mollies from sulfidic habitats and those from non-sulfidic habitats in our first experiment was a reduced tendency of mollies from sulfidic habitats to associate with swordtail females. In all sulfidic habitats, we regularly observe enormous shoals of mollies (often more than 1000 individuals) performing ASR [23,24]. This raises the question why mollies from those habitats showed a reduced tendency to associate with a non-predatory, similar-sized fish, which would be interpretable as a reduced shoaling tendency. In the Trinidadian guppy (*Poecilia reticulata*) it was found that an association preference for swordtails could be induced when juveniles were raised together, and imprinting was suggested as a possible mechanism [54]. No preference was found when guppies had not had the ability for social learning, *i.e.*, when both species were not raised together [54]. In our case, this explanation seems to be rather unlikely though, as swordtails are absent from the sulfidic springs and only occasionally found in the less toxic mixing zones [19,20]. Still, cave mollies and surface mollies from the sulfidic creek (El Azufre) in Experiment 2, as well as laboratory-reared surface mollies in Experiment 3—all of which had no prior experience with swordtails—did show the same high tendency to associate with *X. hellerii* females. However, *Poecilia* spp. in Experiment 1 did associate with the swordtail females at least to some degree, which is in stark contrast to the avoidance of any cichlid stimuli, demonstrating that focal fish were able to evaluate the swordtail as not threatening. We do not have a compelling explanation for this pattern at hand but suggest that shoaling behavior (*i.e.*, the tendency to associate with other fish evaluated as not representing a threat) might be reduced in sulfide-dwelling fish inhabiting the Ríos Puyacatengo and Pichucalco. Why this is so, and why fish from the sulfidic El Azufre do not show such a reduction in shoaling requires further experimentation in the future.

We also found more pronounced sex differences in predator avoidance in mollies from sulfidic habitats, and we suggest population- and sex-specific differences in boldness to account for this pattern: Riesch *et al.* [20] found females from sulfidic habitats to be more cautious than their counterparts from non-sulfidic habitats, while males from both non-sulfidic and sulfidic waters—overall bolder than females—did not differ in their mean boldness. In the Trinidadian guppy, boldness traits differ in relation to the ambient predation regime, with females from habitats with low-predation risk (due to the lack of cichlid predators) being more cautious compared to females from high predation habitats, while males indifferently are bolder [55]. In line with these findings, our current results hint towards piscine predation as the driving force in the evolution of population- and sex-specific variation of boldness in our system.

In our second experiment we asked whether surface and cave forms of *P. mexicana* differed in their level of predator avoidance. Apart from a major difference in the reaction to *V. bifasciata*, surface- and cave-dwelling fish showed very similar responses to the different stimuli, comparable to those observed in fish from non-sulfidic surface habitats in Experiments 1 and 3 (*i.e.*, predator avoidance in response to the other cichlids and shoaling in response to swordtail females). This is contrary to our prediction that cave fish, which live in an energy-limited environment devoid of piscivorous fishes, should rapidly lose their predator avoidance responses through regressive evolution or a lack of predator-experience. However, the Cueva del Azufre, though lacking piscine predators, is not entirely predator-free [33,34], and as outlined for our first experiment, we cannot properly exclude any of the three hypotheses, but argue that the "multi-predator hypothesis" [9] most likely explains the persistence of anti-predator responses in cave fish.

Interestingly, both surface and cave fish showed no reduced tendency to associate with swordtail females, which was predicted for cave mollies based on the results of previous studies testing for species discrimination of *P. mexicana* females when given the choice between a conspecific and a swordtail stimulus female [56] and generally reduced shoaling behavior in cave mollies [39]. Indeed, shoaling in the absence of piscine predators (as in the Cueva del Azufre) loses its benefits but still comprises costs related to intraspecific competition [57]. As the size of the stimulus shoal in the previous study (four conspecific fish [40]) implies higher intraspecific costs than a single swordtail female in our current study, we tentatively suggest that our experimental design may not be appropriate to detect differences in shoaling behavior.

Several studies on teleost fish investigated how different anti-predator behaviors are shaped by experience (*P. reticulata* [2]; *P. mexicana* [18]; *Gobiusculus flavescens* [3]). However, results are not congruent, and the relative importance of experience in shaping anti-predator behavior remains unclear. In our third experiment, we, therefore, asked if there are differences in predator avoidance between laboratory-reared and wild-caught individuals, *i.e.*, between predator-naïve and predator-experienced fish. We could not detect significant differences between groups, which leads us to assume that mere predator avoidance (moving away from a predator), as measured in our experimental design, could possibly be the simplest form of anti-predator behavior not shaped through experience. For example, predator-naïve two-spotted gobies (*G. flavescens*) appear to react strongly to the presentation of any kind of big fish, but are incapable of recognizing predator-specific odors, unless they experienced them in conjunction with a threat cue [3]. Botham *et al.* [58] found avoidance reactions in guppies to be shaped by experience only in populations that evolved in predator-rich environments. Our experiment uncovered no effect of learning on the predator avoidance behavior of fish from a predator-rich (non-sulfidic) site, but the question of whether the learning component of predator avoidance differs between ecotypes of *Poecilia* spp. remains to be studied in detail.

Surprisingly, focal fish in all three experiments (except cave mollies) showed the strongest avoidance when presented with the non-predatory cichlid *V. bifasciata*. We qualitatively observed that individuals of *V. bifasciata* were very active throughout the trials while the two other cichlid species remained rather calm when transferred into the test tank. We, thus, hypothesize that focal fish might have perceived *V. bifasciata* as more threatening, and the observed stronger responses could be an artifact of our test design. Furthermore, a study on coral reef fishes [59] showed that predatory fish share common morphological features like broad heads and big mouths that are used by prey fishes as cues to recognize predators. We hypothesize that the typical "cichlid-shape" of *V. bifasciata* leads the test fish to anticipate a predator and, in combination with the higher activity levels, elicits a strong response. We propose the use of predator models (*i.e.*, video animations of calmly swimming predators) in future studies to standardize the testing procedure (e.g., to exclude the effects of activity differences) and to analyze common features (in body shape or coloration) of predators that might elicit avoidance responses.

Our present study concentrated on visual cues in predator recognition, while future studies will have to elucidate a potential role of alarm pheromone-based predator recognition in extremophile *Poecilia* spp., as chemical cues have been shown to play an important role for predator recognition in poeciliid fishes (see [60,61]). However, it remains unclear what effect the presence of H_2S would have on such chemically mediated cues. For example, studies in the poeciliid genus *Xiphophorus* suggest that disruption of pheromone-based female choice may be caused by (anthropogenic) water pollution [62]. Testing for the responses to "alarm cues" may provide a fruitful opportunity to test an alternative hypothesis explaining—at least in part—our present results: The lack of a difference in the response to (predatory) cichlids between fish from benign and extreme habitats could be explained by our focus on visual cues, as visual presentation of a potential predator represents an extreme risk to the prey fish. In this context one can hypothesize that the response towards chemical cues (representing lower predation risk) could indeed be reduced in extremophile fish.

5. Conclusions

In summary, *Poecilia* spp. living in extreme habitats with impoverished piscine predator regimes did not show reduced predator avoidance responses when presented with predatory cichlids that are found in regular non-sulfidic habitats. Nevertheless, extreme habitats are not predator-free environments, as either birds and/or invertebrates prey on extremophile mollies. Thus, we hypothesize that predator avoidance is still functional in line with the "multi-predator hypothesis" which assumes the presence of *any* kind of predator to be sufficient to maintain anti-predator behavior. A general avoidance of cichlids can be ascribed to cichlid-specific morphological features shared by non-predatory and predatory cichlid species. It remains to be determined what causes the observed variance in the tendency to associate with the non-predatory poeciliid *X. hellerii*. Differences in the response to different predator species cannot be explained through learnt predator recognition, as predator-naïve and predator-experienced fish did not differ in their responses. Our current study, therefore, highlights the importance of inherited avoidance mechanisms towards piscine predators, which are not reduced even when populations invade energy-limited environments largely devoid of predatory fishes.

Acknowledgments

The present study was financially supported by the research funding program "LOEWE – Landes-Offensive zur Entwicklung Wissenschaftlich-ökonomischer Exzellenz" of Hesse's Ministry of Higher Education, Research, and the Arts and by DFG (Deutsche Forschungsgemeinschaft, PL 470/3-1). Experiments were run under the federal permits from Mexican agencies SAGARPA/CONAPESCA (DGOPA.09004.041111.3088) and SEMARNAT/Direction General de Vida Silvestre (SGPA/DGVS/04315/1).

References and Notes

1. Magnhagen, C. Predation risk as a cost of reproduction. *Trends Ecol. Evol.* **1991**, *6*, 183–186.
2. Kelley, J.L.; Magurran, A.E. Learned predator recognition and antipredator responses in fishes. *Fish. Fish.* **2003**, *4*, 216–226.
3. Utne-Palm, A. Response of naive two-spotted gobies *Gobiusculus flavescens* to visual and chemical stimuli of their natural predator, cod *Gadus morhua. Mar. Ecol. Prog. Ser.* **2001**, *218*, 267–274.
4. Kelley, J.L.; Magurran, A.E. Effects of relaxed predation pressure on visual predator recognition in the guppy. *Behav. Ecol. Sociobiol.* **2003**, *54*, 225–232.
5. Chivers, D.P.; Smith, R.J.F. Chemical alarm signalling in aquatic predator-prey systems: A review and prospectus. *Écoscience* **1998**, *5*, 338–352.
6. Crowl, T.A.; Covich, A.P. Responses of a freshwater shrimp to chemical and tactile stimuli from a large decapod predator. *J. N. Am. Benthol. Soc.* **1994**, *13*, 291–298.
7. Mahon, A.R.; Amsler, C.D.; McClintock, J.B.; Baker, B.J. Chemo-tactile predator avoidance responses of the common Antarctic limpet *Nacella concinna. Polar Biol.* **2002**, *25*, 469–473.
8. Wisenden, B.; Pogatshnik, J.; Gibson, D.; Bonacci, L.; Schumacher, A.; Willett, A. Sound the alarm: learned association of predation risk with novel auditory stimuli by fathead minnows (*Pimephales promelas*) and glowlight tetras (*Hemigrammus erythrozonus*) after single simultaneous pairings with conspecific chemical alarm cues. *Environ. Biol. Fish.* **2008**, *81*, 141–147.
9. Blumstein, D.T. The multipredator hypothesis and the evolutionary persistence of antipredator behavior. *Ethology* **2006**, *112*, 209–217.
10. Endler, J.A. Natural selection on color patterns in *Poecilia reticulata. Evolution* **1980**, *34*, 76–91.
11. Endler, J.A.; Houde, A.E. Geographic variation in female preferences for males traits in *Poecilia reticulata. Evolution* **1995**, *49*, 456–468.
12. Byers, J.A. *American pronghorn: Social adaptations and the ghosts of predators past*; University of Chicago Press: Chicago, IL, USA, 1997.
13. Coss, R.G. Effects of relaxed natural selection on the evolution of behavior. In *Geographic Variation in Behavior: Perspectives on Evolutionary Mechanisms*; Foster, S.A.; Endler, J.A., Eds.; Oxford Univ. Press: Oxford, UK, 1999; p. 180—208.
14. Blumstein, D.T.; Daniel, J.C. Isolation from mammalian predators differentially affects two congeners. *Behav. Ecol.* **2002**, *13*, 657–663.
15. Griffin, A.S.; Blumstein, D.T.; Evans, C. Training captive-bred or translocated animals to avoid predators. *Conserv. Biol.* **2000**, *14*, 1317–1326.
16. Magurran, A.E. The causes and consequences of geographic variation in antipredator behavior: perspectives from fish populations. In *Geographic Variation in Behavior: Perspectives on Evolutionary Mechanisms* Foster, S.A.; Endler, J.A., Eds.; Oxford Univ. Press: New York, USA, 1999; pp. 139–163.

17. Brown, G.E.; Chivers, D.P.; Smith, R.J.F. Differential learning rates of chemical *versus* visual cues of a northern pike by fathead minnows in a natural habitat. *Environ. Biol. Fish.* **1997**, *49*, 89–96.
18. Bierbach, D.; Schulte, M.; Herrmann, N.; Tobler, M.; Stadler, S.; Jung, C.T.; Kunkel, B.; Riesch, R.; Klaus, S.; Ziege, M.; Indy, J.R.; Arias-Rodriguez, L.; Plath, M. Predator-induced changes of female mating preferences: innate and experiential effects. *BMC Evol. Biol.* **2011**, *11*, 190.
19. Tobler, M.; Schlupp, I.; Heubel, K.U.; Riesch, R.; Garcia de León, F.J.; Giere, O.; Plath, M. Life on the edge: hydrogen sulfide and the fish communities of a Mexican cave and surrounding waters. *Extremophiles* **2006**, *10*, 577–585.
20. Riesch, R.; Duwe, V.; Herrmann, N.; Padur, L.; Ramm, A.; Scharnweber, K.; Schulte, M.; Schulz-Mirbach, T.; Ziege, M.; Plath, M. Variation along the shy–bold continuum in extremophile fishes (*Poecilia mexicana*, *Poecilia sulphuraria*). *Behav. Ecol. Sociobiol.* **2009**, *63*, 1515–1526.
21. Bagarinao, T. Sulfide as an environmental factor and toxicant: tolerance and adaptations in aquatic organisms. *Aquat. Toxicol.* **1992**, *24*, 21.
22. Grieshaber, M.K.; Völkel, S. Animal adaptations for tolerance and exploitation of poisonous sulfide. *Annu. Rev. Physiol.* **1998**, *60*, 33–53.
23. Plath, M.; Tobler, M.; Riesch, R.; Garcia de León, F.J.; Giere, O.; Schlupp, I. Survival in an extreme habitat: the roles of behaviour and energy limitation. *Naturwissenschaften* **2007**, *94*, 991–996.
24. Tobler, M.; Riesch, R.; Tobler, C.M.; Plath, M. Compensatory behavior in response to sulphide-induced hypoxia affects time budgets, feeding efficiency, and predation risk. *Evol. Ecol. Res.* **2009**, *11*, 935–948.
25. Riesch, R.; Oranth, A.; Dzienko, J.; Karau, N.; Schießl, A.; Stadler, S.; Wigh, A.; Zimmer, C.; Arias-Rodriguez, L.; Schlupp, I.; Plath, M. Extreme habitats are not refuges: Poeciliids suffer from increased aerial predation risk in sulphidic southern Mexican habitats. *Biol. J. Linn. Soc.* **2010**, *101*, 417–426.
26. Roach, K.A.; Tobler, M.; Winemiller, K.O. Hydrogen sulfide, bacteria, and fish: a unique, subterranean food chain. *Ecology* **2011**, *92*, 2056–2062.
27. Plath, M.; Heubel, K.U.; Garcia de León, F.J.; Schlupp, I. Cave molly females (*Poecilia mexicana*, Poeciliidae, Teleostei) like well-fed males. *Behav. Ecol. Sociobiol.* **2005**, *58*, 144–151.
28. Riesch, R.; Plath, M.; Schlupp, I. Toxic hydrogen sulphide and dark caves: pronounced male life-history divergence among locally adapted *Poecilia mexicana* (Poeciliidae). *J. Evol. Biol.* **2011**, *24*, 596–606.
29. Riesch, R.; Plath, M.; Schlupp, I. Toxic hydrogen sulfide and dark caves: life-history adaptations in a livebearing fish (*Poecilia mexicana*, Poeciliidae). *Ecology* **2010**, *91*, 1494–1505.
30. Gordon, M.S.; Rosen, D.E. A cavernicolous form of the poeciliid dish *Poecilia sphenops* from Tabasco, México. *Copeia* **1962**, *1962*, 360–368.

31. Tobler, M.; Palacios, M.; Chapman, L.J.; Mitrofanov, I.; Bierbach, D.; Plath, M.; Arias-Rodriguez, L.; Garcia de León, F.J.; Mateos, M. Evolution in extreme environments: replicated phenotypic differentiation in livebearing fish inhabiting sulfidic springs. *Evolution* **2011**, *65*, 2213–2228.
32. Tobler, M.; Riesch, R.; Garcia de León, F.J.; Schlupp, I.; Plath, M. Two endemic and endangered fishes, *Poecilia sulphuraria* (Alvarez, 1948) and *Gambusia eurystoma* Miller, 1975 (Poeciliidae, Teleostei) as only survivors in a small sulphidic habitat. *J. Fish. Biol.* **2008**, *72*, 523–533.
33. Tobler, M.; Franssen, C.M.; Plath, M. Male-biased predation of a cave fish by a giant water bug. *Naturwissenschaften* **2008**, *95*, 775–779.
34. Tobler, M.; Schlupp, I.; Plath, M. Predation of a cave fish (*Poecilia mexicana*, Poeciliidae) by a giant water-bug (*Belostoma*, Belostomatidae) in a Mexican sulphur cave. *Ecol. Entomol.* **2007**, *32*, 492–495.
35. Horstkotte, J.; Riesch, R.; Plath, M.; Jäger, P. Predation on a cavefish (*Poecilia mexicana*) by three species of spiders in a Mexican sulfur cave. *Bull. Br. Arachnol. Soc.* **2010**, *15*, 55–58.
36. Klaus, S.; Plath, M. Predation on a cave fish by the freshwater crab *Avotrichodactylus bidens* (Bott, 1969) (Brachyura, Trichodactylidae) in a Mexican sulfur cave. *Crustaceana* **2011**, *84*, 411–418.
37. Miller, R.R. *Freshwater Fishes of Mexico*; University of Chicago Press: Chicago, IL, USA, 2006.
38. Conkel, D. *Cichlids of North. and Central America*; T.F.H. Publications: Neptune City, NJ, USA, 1993.
39. Plath, M.; Schlupp, I. Parallel evolution leads to reduced shoaling behavior in two cave dwelling populations of Atlantic mollies (*Poecilia mexicana*, Poeciliidae, Teleostei). *Environ. Biol. Fish.* **2008**, *82*, 289–297.
40. Parzefall, J. Zur vergleichenden Ethologie verschiedener *Mollienesia*-Arten einschließlich einer Höhlenform von *Mollienesia sphenops*. *Behaviour* **1969**, *33*, 1–38.
41. Parzefall, J. Rückbildung aggressiver Verhaltensweisen bei einer Höhlenform von *Poecilia sphenops* (Pisces, Poeciliidae). *Z. Tierpsychol.* **1974**, *35*, 66–84.
42. Bierbach, D.; Klein, M.; Sassmannshausen, V.; Schlupp, I.; Riesch, R.; Parzefall, J.; Plath, M. Divergent evolution of male aggressive behaviour: another reproductive isolation mechanism in extremophile poeciliid fishes. *Int. J. Evol. Biol.* **2012**, *2012*.
43. Peters, N.; Peters, G.; Parzefall, J.; Wilkens, H. Über degenerative und konstruktive Merkmale bei einer phylogenetisch jungen Höhlenform von *Poecilia sphenops* (Pisces, Poeciliidae). *Int. Rev. Gesamten Hydrobiol. Hydrogr.* **1973**, *58*, 417–436.
44. Plath, M.; Hauswaldt, J.S.; Moll, K.; Tobler, M.; Garcia de León, F.J.; Schlupp, I.; Tiedemann, R. Local adaptation and pronounced genetic differentiation in an extremophile fish, *Poecilia mexicana*, inhabiting a Mexican cave with toxic hydrogen sulphide. *Mol. Ecol.* **2007**, *16*, 967–976.
45. Fontanier, M.; Tobler, M. A morphological gradient revisited: cave mollies vary not only in eye size. *Environ. Biol. Fish.* **2009**, *86*, 285–292.

46. Körner, K.E.; Schlupp, I.; Plath, M.; Loew, E.R. Spectral sensitivity of mollies: comparing surface- and cave-dwelling Atlantic mollies, *Poecilia mexicana. J. Fish. Biol.* **2006**, *69*, 54–65.
47. Ferrari, M.C.O.; Gonzalo, A.; Messier, F.; Chivers, D.P. Generalization of learned predator recognition: an experimental test and framework for future studies. *Proc. Roy. Soc. B* **2007**, *274*, 1853–1859.
48. Greven, H. Gonads, genitals, and reproductive biology. In *Ecology and Evolution of Poeciliid Fishes*; Evans, J., Pilastro, A., Schlupp, I., Eds.; Chicago University Press: Chicago, IL, USA, 2011.
49. Evans, J.; Pilastro, A. Postcopulatory sexual selection. In *Ecology and Evolution of Poeciliid Fishes*; Evans, J.; Pilastro, A.; Schlupp, I., Eds.; Chicago University Press: Chicago, IL, USA, 2011.
50. Plath, M.; Makowicz, A.M.; Schlupp, I.; Tobler, M. Sexual harassment in live-bearing fishes (Poeciliidae): Comparing courting and noncourting species. *Behav. Ecol.* **2007**, *18*, 680–688.
51. Parzefall, J. A review of morphological and behavioural changes in the cave molly, *Poecilia mexicana*, from Tabasco, Mexico. *Environ. Biol. Fish.* **2001**, *62*, 263–275.
52. Kullander, S.O. *Cichlidae (Cichlids)*; EDIPUCRS: Porto Alegre, Brasil, 2003.
53. Peckarsky, B.L.; Penton, M.A. Why do *Ephemerella* nymphs scorpion posture - A ghost of predation past. *Oikos* **1988**, *53*, 185–193.
54. Warburton, K.; Lees, N. Species discrimination in guppies: learned responses to visual cues. *Anim. Behav.* **1996**, *52*, 371–378.
55. Harris, S.; Ramnarine, I.W.; Smith, H.G.; Pettersson, L.B. Picking personalities apart: estimating the influence of predation, sex and body size on boldness in the guppy *Poecilia reticulata. Oikos* **2010**, *119*, 1711–1718.
56. Riesch, R.; Schlupp, I.; Tobler, M.; Plath, M. Reduction of the association preference for conspecifics in cave-dwelling Atlantic mollies, *Poecilia mexicana. Behav. Ecol. Sociobiol.* **2006**, *60*, 794–802.
57. Krause, J.; Ruxton, G.D. *Living in groups*; Oxford University Press: Oxford, 2002.
58. Botham, M.S.; Hayward, R.K.; Morrell, L.J.; Croft, D.P.; Ward, J.R.; Ramnarine, I.; Krause, J. Risk-sensitive antipredator behavior in the Trinidadian guppy, *Poecilia reticulata. Ecology* **2008**, *89*, 3174–3185.
59. Karplus, I.; Algom, D. Visual cues for predator face recognition by reef fishes. *Z. Tierpsychol.* **1981**, *55*, 343–364.
60. Brown, G.E.; Chivers, D.P. Learning about danger: chemical alarm cues and local risk assessment in prey fishes. In *Fish Cognition and Behaviour*; Brown, C.; Laland, K.N.; Krause, J., Eds.; Blackwell: London, UK, 2006; pp. 49–69.
61. Kelley, J.L.; Brown, C. Predation risk and decision-making in poeciliid prey. In *Ecology and Evolution of Poeciliid Fishes* Evans, J.P.; Pilastro, A.; Schlupp, I., Eds.; University of Chicago Press: Chicago, IL, USA, 2011; pp. 174–184.
62. Fisher, H.S.; Wong, B.B.M.; Rosenthal, G.G. Alteration of the chemical environment disrupts communication in a freshwater fish. *Proc. Roy. Soc. B* **2006**, *273*, 1187–1193.

Chapter 2:
Extremophiles in Specific Environments

Anaerobic Thermophiles

Francesco Canganella and Juergen Wiegel

Abstract: The term "extremophile" was introduced to describe any organism capable of living and growing under extreme conditions. With the further development of studies on microbial ecology and taxonomy, a variety of "extreme" environments have been found and an increasing number of extremophiles are being described. Extremophiles have also been investigated as far as regarding the search for life on other planets and even evaluating the hypothesis that life on Earth originally came from space. The first extreme environments to be largely investigated were those characterized by elevated temperatures. The naturally "hot environments" on Earth range from solar heated surface soils and water with temperatures up to 65 °C, subterranean sites such as oil reserves and terrestrial geothermal with temperatures ranging from slightly above ambient to above 100 °C, to submarine hydrothermal systems with temperatures exceeding 300 °C. There are also human-made environments with elevated temperatures such as compost piles, slag heaps, industrial processes and water heaters. Thermophilic anaerobic microorganisms have been known for a long time, but scientists have often resisted the belief that some organisms do not only survive at high temperatures, but actually thrive under those hot conditions. They are perhaps one of the most interesting varieties of extremophilic organisms. These microorganisms can thrive at temperatures over 50 °C and, based on their optimal temperature, anaerobic thermophiles can be subdivided into three main groups: thermophiles with an optimal temperature between 50 °C and 64 °C and a maximum at 70 °C, extreme thermophiles with an optimal temperature between 65 °C and 80 °C, and finally hyperthermophiles with an optimal temperature above 80 °C and a maximum above 90 °C. The finding of novel extremely thermophilic and hyperthermophilic anaerobic bacteria in recent years, and the fact that a large fraction of them belong to the *Archaea* has definitely made this area of investigation more exciting. Particularly fascinating are their structural and physiological features allowing them to withstand extremely selective environmental conditions. These properties are often due to specific biomolecules (DNA, lipids, enzymes, osmolites, *etc.*) that have been studied for years as novel sources for biotechnological applications. In some cases (DNA-polymerase, thermostable enzymes), the search and applications successful exceeded preliminary expectations, but certainly further exploitations are still needed.

Reprinted from *Life*. Cite as: Canganella, F.; Wiegel, J. Anaerobic Thermophiles. *Life* **2014**, *4*, 77-104.

1. Introduction

Among anaerobic and thermophilic microorganisms, anaerobic thermophilic Archaea are certainly the most "extreme" in terms of inhabited ecosystems. They represent the deepest, least evolved branches of the universal phylogenetic tree (Figure 1). They often use substrates, which are thought to have been dominant in the primordial terrestrial makeup, indicating that they could have been the first living forms on this planet [1–6]. Studies into how they manage thermostability at the protein and membrane structural level have elucidated many traits of protein, membrane and nucleic acid structure; however, there is not yet a full understanding of the principles of thermostability [7–11]. The development of better genetic tools for the use of these organisms is the key for more practical applications in the future [12–14].

Figure 1. Phylogenetic tree highlighting possible evolutionary relatedness of anaerobic thermophilic Archaea (modified from Eric Gaba, NASA Astrobiology Institute 2006).

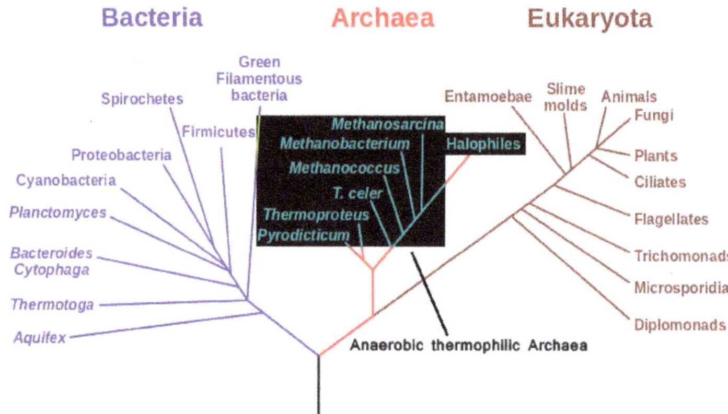

Although the first forms of life no longer exist, natural thermal environments are still abundant on Earth and some have properties similar to those environments in which life possibly first began. Many of these environments are characteristically anaerobic or have low levels of oxygen. The anaerobic feature can stem from a number of factors: remoteness of the environment from the atmosphere, low solubility of oxygen in water at elevated temperatures, hypersalinity, inputs of reducing gasses such as H_2S, or the consumption of oxygen by aerobic microorganisms on or near the water surface.

Natural environments for anaerobic thermophiles range from terrestrial volcanic sites (including solfatara fields) with temperatures slightly above ambient temperature, to submarine hydrothermal systems (sediments, submarine volcanoes, fumaroles and vents) with temperatures exceeding 300 °C, subterranean sites such as oil reservoirs, and solar heated surface soils with temperatures up to 65 °C (Figures 2 and 3). There are also human-made hot environments such as compost piles (usually around 60–70 °C but as high as 100 °C) slag heaps, industrial processes and water heaters [15].

Figure 2. Some environments where anaerobic thermophiles can be isolated: (**a**) A power plant in Iceland; (**b**) Terrestrial hot springs at Viterbo (Italy); (**c**) The hot pool of Bagno Vignoni (Italy).

Oil reservoirs, mines, and geothermal aquifers are examples of subsurface environments that thermophiles populate. Extreme thermophilic bacterial species of the genera *Geotoga* and *Petrotoga* (family *Thermotogaceae*) have so far only been found in deep subsurface oil reservoirs; on this basis, it has been proposed that these taxa represent typical indigenous *Bacteria* of this particular ecosystem. However, lately Thermotogales sequences have been found in mesobiotic environments [16] and novel species have been described [17]. Geothermal aquifers, such as the Great Artesian Basin of Australia, are considered to be markedly different from volcanically related hot springs in that they have low flow rates and long recharge times (around 1000 years) that affect the microbial populations therein. Besides natural thermal environments, thermophilic anaerobes are also found within anthropogenically heated environments, including coal refuse piles and compost heaps, and nuclear power plant effluent channels which contain not only spore-forming species, but also vegetative and active cells including *Bacteria* and *Archaea*.

Figure 3. Deep-sea hot ecosystems: (**a**) Hot sediment at the Guaymas Basin; (**b**,**c**) Drawings of black smokers located at a deep-sea hydrothermal vent area (courtesy of Focus Magazine and Jack Jones, respectively).

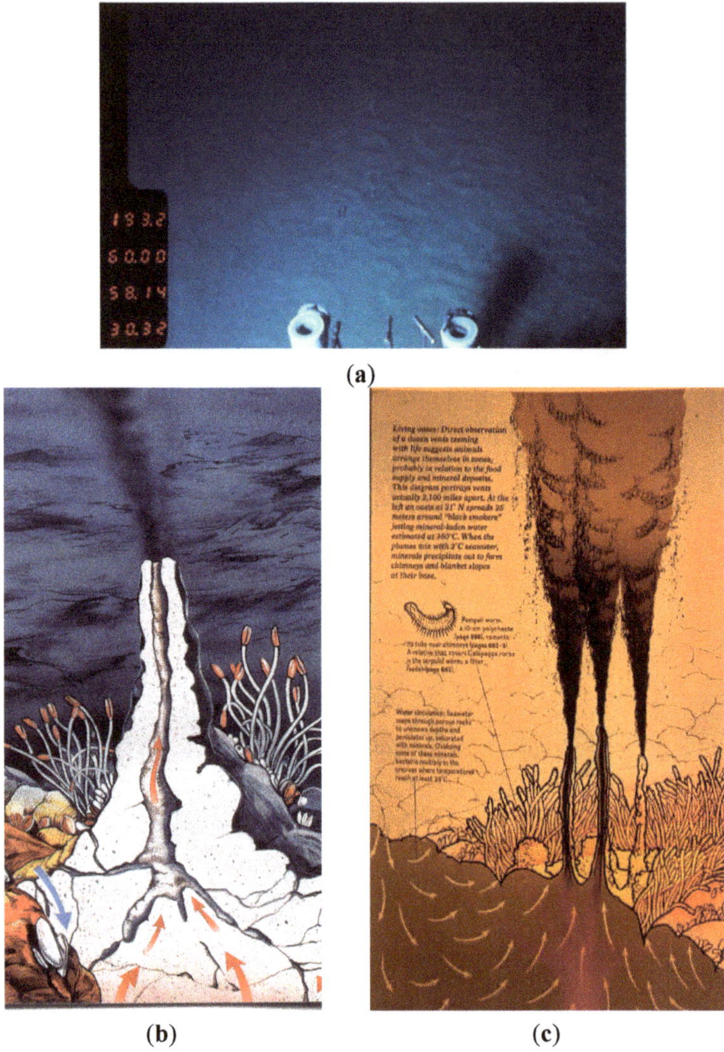

(a)

(b) (c)

Many environments are also temporarily hot, adaptation to which may be the reason some thermophiles are very fast-growing. Among the geothermally heated habitats are the alkaline, mainly carbonate-containing hot springs around a neutral pH, and acidic areas including some mud-holes. Most of the acidic high-temperature habitats contain elemental sulfur and metal sulfides and most isolates from these areas metabolize sulfur by either anaerobic respiration or fermentation. Ocean depths are under extreme pressures from the weight of the water column, and thus most anaerobic and thermophilic bacteria from these areas are piezotolerant, some are truly piezophilic, others such as *Pyrococcus* spp., *Thermococcus* spp., and *M. kandleri* show extensions of T_{max} under increased pressure [18–23] and all are at least halotolerant [24], and those isolated from solfataras generally acidophilic. Alike most described species of obligately aerobic thermophilic *Archaea* that are acidophilic, anaerobic thermophilic bacteria are generally unable to grow at acidic pH with some exemptions such as representatives of genera *Stygiolobus*, *Acidilobus*, and *Caldisphaera* [25–27]. On the other hand many anaerobic bacteria and some *Archaea* are capable of growing at an alkaline pH [28]. The anaerobic alkalithermophilic bacteria thus form an interesting group to study, and their relationships between temperature optimum and pH optimum for growth have been extensively investigated. This adaptability to high pH environments involves both cellular and biomolecular peculiar traits that are currently under investigation, particularly to exploit their potential biotechnological applications.

Among extreme environments, the deep sea is in general cold, but it is known to show areas of superheated water and widespread still-hot volcanic ocean crust beneath the flanks of the mid-ocean ridge and other rock structures, as well as geothermally heated shallower ocean waters.

A large group of anaerobic and thermophilic microorganisms have been isolated and studied from the deep-sea, particularly at both hydrothermal vents and sub-seafloor sites, either for their physiological properties or for their potential applications [29–34]. Representative deep-sea environments, if not in terms of geographical extension but certainly as the most spectacular, are the deep-sea hydrothermal vents. The highly dense and biologically diverse communities in the immediate vicinity of hydrothermal vent flows are in stark contrast to the surrounding bare seafloor. They comprise organisms with distinct metabolisms based on chemosynthesis and growth rates comparable to those from shallow water tropical environments, which have been rich sources of biologically active natural products. Fundamental discoveries in this regard will be accelerated by new cost-effective technologies in deep-sea research and more advanced molecular techniques.

Taxonomical and phylogenetic investigations have always been the main focus concerning research on deep-sea anaerobic thermophiles. Diversity and richness of deep-sea hydrothermal environments were particularly examined and shown to be as high as those in soil. As a matter of fact, sediments from deep-sea floors have always been great sources of novel bacterial isolates and recently new genera as well as species are being described from different sites in the ocean depths [35–37].

As far as concerns the diversity of sub-seafloor microorganisms, a "meta-enzyme approach" has been proposed as an ecological enzymatic method to explore the potential functions of microbial communities in extreme environments such as the deep marine habitats [38]. Detectable enzyme activities were used to predict the existence of a sizable population of viable anaerobic microorganisms even in deep sub-seafloor habitats. Moreover many microbial isolates produced a variety of extra-cellular enzymes such as proteases, amylases, lipases, chitinases, phosphatases, and deoxyribonucleases, giving them a great potential in terms of biotechnological applications.

A main topic in ecology and population dynamics of deep-sea anaerobic thermophiles is their colonization and distribution patterns along and around hydrothermal vent deposits. An approach based on the deployment of thermocouple arrays on two deep-sea hydrothermal vents at Guaymas Basin was adopted by Pagé *et al.* [39]. This aimed to measure *in situ* temperatures at which microorganisms colonize the associated mineral deposits. Spatial differences in archaeal diversity were observed in all deposits in relationship to *in situ* temperature. This study was the first direct assessment of *in situ* conditions experienced by microorganisms inhabiting actively forming hydrothermal deposits at different stages of structure development.

2. Growth Conditions

Microorganisms that grow optimally at elevated temperatures above 50 °C and can not use oxygen as terminal electron acceptor during electron transport phosphorylation are described as thermophilic anaerobes. They are of interest from basic and applied scientific perspectives and are studied to understand how life can thrive in environments previously considered inhospitable to life. Such environments include volcanic solfataras and hot springs high in sulfur and toxic metals, as well as abyssal hydrothermal vents with extremely high pressure and temperatures far above 100 °C [40].

Isolated species of thermophilic anaerobes include peculiar forms: for example, cells of the alkalithermophile *Clostridium paradoxum* become highly motile when sporulating, and *Moorella thermacetica*-like strains have exceptionally heat-resistant spores with D_{10} times of nearly 2 h at 121 °C. Also, *Pyrolobus fumarii* grows optimally at 106 °C, and the record-holder *Methanopyrus kandleri*-like strain grows at 122 °C under increased pressure [18]. *Thermobrachium celere* strains have doubling times of about 10 min while growing above pH 9.0 and above 55 °C [41] and the polyextremophilic *Natranaerobius* isolates simultaneously grow optimally up to 69 °C and above pH 9.5, and at a salt concentration above 4 M Na^+. They may be considered the most extremophile as they withstand the combination of multiple stressors. It will be of interest to evaluate whether those boundaries can be further extended by isolating other polyextremophiles [42–44].

The analyses of the biodiversity and patterns of biodiversity within thermal environments is an area of active research that continually expands as technology allows for novel approaches and more detailed analyses. Additionally, their thermostable enzymes, among other characteristics, make thermophilic anaerobes of significant interest for their biotechnological potential.

Contrary to any expectation, thermophilic anaerobes have also been isolated from mesobiotic and even psychrobiotic environments: two *Thermosediminibacter* species were isolated from ocean sediments of the Peru Margin at temperatures at or below 12 °C [45], uncharacterized *Thermoanaerobacter* species have been isolated from melted snow from Antarctica (unpublished results), alkalithermophiles have been isolated from many river sediments and wet meadows, and *Methanothermobacter thermoautotrophicus* and other thermophilic methanogens and chemolithoautotrophic acetogens can readily be found in lake sediments and rivers, streams, and ponds. Possible reasons for the presence of thermophilic anaerobes in environments where they were thought not to grow, considering their physiological properties, include (a) that the microorganisms are present but do not grow in these environments although they are able to carry out maintenance metabolism (e.g., as shown for *M. thermoautotrophicum* that is able to form methane at temperatures as low as 16 °C, although it is not able to multiply at temperatures below 22 °C (Wiegel unpl. results); (b) that they disperse only transiently from other thermal environments; (c) that they survive and multiply by taking advantage of temporary thermal piconiches that become available when proteinaceous biomass is degraded. The latter notion is further substantiated by observations that strains of *Calaromator* (Bas. *Thermobrachium) celer* isolated from mesobiotic environments show very short doubling times (between 10 min and 20 min), whereas the strains of the same species isolated from hot springs—which resemble a more constant thermobiotic environment—have doubling times of above 30 min (Wiegel, unpublished results) and also that the moderate thermophiles *C. paradoxum*- and the nonsporulating *C. thermophilus*-like cells are present in mesobiotic sewage sludge (<30 °C) at 1000 CFU/mL sludge [46]. However, so far, no direct molecular methods have been used to explain the growth of these taxa *in vivo*.

Thermophilic anaerobes in pure culture are characterized by a polyphasic approach, in which phenotypic and genotypic/phylogenetic properties are examined. Phenotypic characteristics of particular interest for this discussion include oxygen relationships and metabolic properties, such as energy production and carbon assimilation. Group-defining properties for extremophiles (also called marginal data set), such as temperature growth range (e.g., T_{min}, T_{opt}, and T_{max}) and pH growth range (e.g., pH_{min}, pH_{opt}, and pH_{max}), are particularly important. These values should be determined by measuring the doubling times over the range for growth and specifically noting where growth was obtained and where growth was not obtained. For other thermophilic extremophiles besides thermophilic alkali-/acido-thermophiles, the range and optima for their other characteristic properties, such as salt (halophiles) pressure (piezophiles), substrate concentration (oligophiles, osmophiles) and tolerance to metals or solvents are important when considering thermophilic anaerobes from habitats such as sun heated hypersaline lakes and deep-sea hydrothermal regions, deep oil wells/oil storage tanks, or heavily contaminated thermobiotic sites. Although genotypic characteristics such as G + C mol% of the genomic DNA and DNA-DNA relatedness between strains have been studied since the 1960s, in the past 20 years analysis of the 16S rRNA gene sequence (frequently backed up by DNA: DNA hybridization studies for taxa with 16S rRNA sequence similarity above 97%) has become standard, and the analysis of house keeping genes and more recently, the whole-genome sequencing of prokaryotes has become increasingly more common. As we enter deeper and deeper into the genomic era, genome sequencing will certainly become an essential part of the characterization and differentiation of novel taxa exceeding the importance of the 16S rRNA gene sequence analysis used today. In Table 1, classified thermophilic anaerobes with available sequenced genomes are reported, however, for most recent information, the reader should refer to the National Center for Biotechnology Information Taxonomy Database [47].

Table 1. Most representative thermophilic anaerobes with official nomenclature and genome sequenced.

Species	O_2-relationship and metabolism	Temperature range (optimum)	pH range (optimum)	Originally isolated from
Bacteria*; *Proteobacteria*; *Gammaproteobacteria*; *Chromatiales*; *Chromatiaceae*; Genus: *Thermochromatium				
Thermochromatium tepidum	AN PA/PH	34–57 (48–50)	(7)	Mammoth Hot Spring, Yellowstone National Park, USA
Bacteria*; *Bacteroidetes/Chlorobi* group; *Bacteroidetes*; *Bacteroidetes*; *Bacteroidales*; *Bacteroidaceae*; Genera: *Acetomicrobium*, *Anaerophaga				
Anaerophaga sp. strain HPS1	NR	NR	NR	Offshore hot spring sediment, China
Bacteria*; *Spirochaetes*; *Spirochaetes*; *Spirochaetales*; *Spirochaetaceae*; Genus: *Spirochaeta*, *Exilispira				
Spirochaeta thermophila	AN COH	40–73 (66–68)	5.9–7.7 (7.5)	Marine hot spring on the beach of an island of Kamchatka, also from a hot spring on Raoul Island, New Zealand
Bacteria*; *Firmicutes*; *Clostridia*; *Halanaerobiales*; *Halanaerobiaceae*; Genus: *Halothermothrix				
Halothermothrix orenii	AN COH	45–68 (60)	5.5–8.2 (6.5–7)	Chott El Guettar hypersaline lake, Tunisia
Bacteria*; *Firmicutes*; *Clostridia*; *Natranaerobiales*; *Natranaerobiaceae*; Genus: *Natranaerobius				
Natranaerobius thermophilus	AN COH	35–56 (53)	8.5–10.6 (9.5)	Sediment of alkaline, hypersaline lakes of the Wadi An Natrun

Table 1. *Cont.*

Species	O_2-relationship and metabolism	Temperature range (optimum)	pH range (optimum)	Originally isolated from
Bacteria*; *Firmicutes*; *Clostridia*; *Thermoanaerobacteriales*; *Thermoanaerobacteriaceae*; *Syntrophomonadaceae*; Genera: *Coprothermobacter*, *Gelria*, *Moorella*, *Thermacetogenium*, *Mahella*, *Thermoanaerobacterium*, *Thermoanaerobacter*, *Thermosediminibacter*, *Caldanaerobacter*, *Thermovenabulum*, *Tepidanaerobacter*, *Ammonifex*, *Thermanaeromonas*, *Thermhydrogenium*, *Caldanaerovirga*, *Fervidicola*, *Caldanaerobius				
Coprothermobacter proteolyticus	AN COH	35–70 (63)	5–8.5 (7.5)	Thermophilic digester fermenting tannery wastes and cattle manure
Moorella thermoacetica	AN COH	45–65 (55–60)	NR	Horse manure
Thermoanaerobacter ethanolicus	AN COH	37–78 (69])	4.4–9.9 (5.8–8.5)	Hot springs, Yellowstone National Park, USA
Thermoanaerobacter pseudoethanolicus	AN COH	(65)	NR	Hot Spring, Yellowstone National Park, USA
Caldanaerobacter subterraneus subsp. *tengcongensis*	AN COH	50–80 (75)	5.5–9 9 (7–7.5)	Hot spring, Tengcong, China
Ammonifex degensii	AN F-CLA	57–77 (70)	5–8 (7.5)	Kawah Candradimuka crater, Dieng Plateau, Java, Indonesia
Bacteria*; *Firmicutes*; *Clostridia*; *Clostridiales*; *Acidaminococcaceae*; Genus: *Thermosinus				
Thermosinus carboxydivorans	AN CLA	40–68 (60)	6.5–7.6 (6.8–7)	Norris Basin hot spring, Yellowstone National Park, USA
Bacteria*; *Firmicutes*; *Clostridia*; *Clostridiales*; *Peptococcaceae*; Genera: *Desulfotomaculum*, *Pelotomaculum*, *Carboxydothermus*, *Thermincola				
Pelotomaculum thermopropionicum	AN COH	45–65 (55)	6.7–7.5 (7)	Thermophilic upflow anaerobic sludge blanket reactor
Carboxydothermus hydrogenoformans	AN CLA	40–78 (70–72)	6.4–7.7 (6.8–7)	Freshwater hydrothermal springs, Kunashir Island, Kamchatka, Russia

Table 1. *Cont.*

Species	O_2-relationship and metabolism	Temperature range (optimum)	pH range (optimum)	Originally isolated from
Bacteria*; *Firmicutes*; *Clostridia*; *Clostridiales*; *Syntrophomonadaceae*; Genera: *Anaerobaculum*, *Syntrophothermus*, *Thermanaerovibrio*, *Carboxydocella*, *Anaerobranca*, *Thermosyntropha*, *Caldicellulosiruptor				
Caldicellulosiruptor lactoaceticus	AN COH	50–78 (68)	5.8–8.2 (7)	Hveragerði alkaline hot spring, Iceland
Caldicellulosiruptor owensensis	AN COH	50–80 (75)	5.5–9 (7.5)	Freshwater pond within the dry Owens Lake bed, California, USA
Caldicellulosiruptor kristjanssonii	AN COH	45–82 (78)	5.8–8 (7)	Hot spring, Iceland
Caldicellulosiruptor saccharolyticus	AN COH	45–80 (70)	5.5–8.0 (7.0)	Geothermal spring, Taupo, New Zealand
Caldicellulosiruptor bescii	AN COH			
Caldicellulosiruptor kronotskyensis	AN COH	(70)	(7)	Hot spring, Kamchatka, Russia
Caldicellulosiruptor hydrothermalis	AN COH	(65)	(7)	Hot spring, Kamchatka, Russia
Bacteria*; *Firmicutes*; *Clostridia*; *Clostridiales*; *Heliobacteriaceae*; Genus: *Heliobacterium				
Heliobacterium modesticaldum	AN PH & COH	25–56 (52)	(6–7)	Iceland, Yellowstone National Park, USA
Bacteria*; *Firmicutes*; *Clostridia*; *Clostridiales*; *Clostridiaceae*; *Caldicoprobacteraceae*; *Veillonellaceae*; Genera: *Alkaliphilus*, *Clostridium*, *Tepidibacter*, *Caloramator*, *Garciella*, *Caminicella*, *Caloranaerobacter*, *Thermobrachium*, *Thermohalobacter*, *Tepidimicrobium*, *Fervidicella*, *Caldicoprobacter*, *Sporolituus*, *Thermotalea*, *Lutispora				
Clostridium thermocellum	AN COH	28–69 (60)	(6.1–7.5)	Louisiana cotton bale and Compost heap
Clostridium stercorarium subsp. *stercorarium*	AN COH	(65)	(7.3)	Compost heap

Table 1. *Cont.*

Species	O_2-relationship and metabolism	Temperature range (optimum)	pH range (optimum)	Originally isolated from
Bacteria*; *Firmicutes*; *Bacilli*; *Bacillales*; *Bacillaceae*; Genera: *Anoxybacillus, Bacillus, Geobacillus, Vulcanibacillus				
Anoxybacillus flavithermus	FAE COH	30–72 (60–65)	5.5–9 (7)	A hot spring, New Zealand
Geobacillus thermodenitrificans	FAE COH	45–70	6–8	Sugar beet juice from extraction installations; Austria
Geobacillus thermoleovorans	FAE COH	35–78 (55–65)	(6.2–6.8)	Soil near hot water effluent, Bethlehem, PA, USA
Geobacillus thermoglucosidiasus	FAN COH	40–70 (60)	6–9 (7)	Japan soil
Bacteria*; *Proteobacteria*; delta/epsilon subdivisions; *Deltaproteobacteria*; *Desulfurellales*; *Desulfurellaceae*; Genera: *Desulfurella, Hippea				
Hippea maritima	AN COH	40–65 (52–54)	5.4–6.5 (5.8–6.2)	Shallow water hot vents, Bay of Plenty, New Zealand and Matupi Harbour, Papua New Guinea
Bacteria*; *Proteobacteria*; delta/epsilon subdivisions; *Epsilonproteobacteria*; *Nautiliales*; *Nautiliaceae*; Genera: *Nautilia, Lebetimonas, Caminibacter				
Caminibacter mediatlanticus	AN CLA	45–70 (55)	4.5–7.5 (5.5)	"Rainbow" deep-sea vent field, Mid-Atlantic Ridge
Bacteria*; *Deferribacteres*; *Deferribacteres*; *Deferribacterales*; *Deferribacteraceae*; Genera: *Deferribacter*; *Flexistipes* also, *Caldithrix* (unclassified *Deferribacteres*), *Calditerrivibrio				
Deferribacter desulfuricans	AN COH	40–70 (60–65)	5.0–7.5 (6.5)	From a black smoker vent from the hydrothermal fileds at the Suiyo Seamount in the Izu-Bonin Arc, Japan

Table 1. *Cont.*

Species	O_2-relationship and metabolism	Temperature range (optimum)	pH range (optimum)	Originally isolated from
Bacteria*; *Thermodesulfobacteria*; *Thermodesulfobacteria*; *Thermodesulfobacteriales*; *Thermodesulfobacteriaceae*; Genera: *Thermodesulfatator*, *Thermodesulfobacterium*, *Caldimicrobium*, *Thermosulfidibacter				
Thermodesulfatator indicus	AN CLA	55–80 (70)	6–6.7 (6.25)	The Kairei deep-sea hydrothermal vent field, Central Indian Ridge
Thermodesulfobacterium commune	AN COH	50–85 (70)	6.0–8.0	Ink Pot Spring, Yellowstone National Park, USA
Bacteria*; *Nitrospirae*; *Nitrospira*; *Nitrospirales*; *Nitrospiraceae*; Genus: *Thermodesulfovibrio				
Thermodesulfovibrio yellowstonii	AN COH	40–70 (65)	(6.8–7)	Thermal vent, Yellowstone National Park, USA
Bacteria*; *Dictyoglomi*; *Dictyoglomi*; *Dictyoglomales*; *Dictyoglomaceae*; Genus: *Dictyoglomus				
Dictyoglomus thermophilum	AN COH	50–80 (73–78)	5.9–8.3 (7)	Hot spring, Kumamoto Prefecture, Japan
Bacteria*; *Chloroflexi*; *Chloroflexi*; *Chloroflexales*; *Chloroflexaceae*; Genera: *Roseiflexus*, *Chloroflexus*, *Heliothrix				
Roseiflexus castenholzii	FAE PH (anaerobic)	45–55 (50)	7–9 (7.5–8)	Hot spring, Nakabusa, Japan
Chloroflexus aggregans	FAE PH (anaerobic)	(50–60)	7.0–9.0	Hot spring of the Okukinu Meotobuchi hot spring in Tochigi Perfecture, Japan
Chloroflexus aurantiacus	FAE PH (anaerobic)	(52–60)	(8)	Hot spring in the canyon at Sokokura, Hakone district, Japan
Heliothrix oregonensis	FAE PH	(40–55)	NR	Hot spring near Warm Springs River, Oregon, USA

Table 1. *Cont.*

Species	O_2-relationship and metabolism	Temperature range (optimum)	pH range (optimum)	Originally isolated from
Bacteria*; *Chloroflexi*; *Thermomicrobia*; *Thermomicrobiales*; *Thermomicrobiaceae*; Genus: *Thermomicrobium				
Thermomicrobium roseum	AN COH	(70–75)	6–9.4 (8.2–8.5)	Hot spring, Yellowstone National Park, USA
***Bacteria*; *Aquificae*; *Aquificae*; *Aquificales*; *Aquificaceae*; Genera: *Hydrogenivirga*, *Aquifex*, *Desulfurobacterium* (unclassified *Aquificales*), *Balnearium* (unclassified *Aquificales*), *Thermovibrio* (unclassified *Aquificales*)**				
Desulfurobacterium thermolithotrophum	AN CLA	40–75 (70)	4.4–7.5 (6)	"Snake Pit" vent field, Mid-Atlantic ridge
Thermovibrio ammonificans	AN CLA	60–80 (75)	5–7 (5.5)	Deep sea hydrothermal vent area, East Pacific Rise
Bacteria*; *Aquificae*; *Aquificae*; *Aquificales*; *Hydrogenothermaceae*; Genera: *Hydrogenothermus*, *Sulfurihydrogenibium*, *Persephonella				
Sulfurihydrogenibium azorense	FAE CLA	50–73 (68)	5.5–7 (6)	Near the Água do Caldeirão, Furnas, on São Miguel Island, Azores
Bacteria*; *Thermotogae*; *Thermotogae*; *Thermotogales*; *Thermotogaceae*; Genera: *Geotoga*, *Marinitoga*, *Petrotoga*, *Thermosipho*, *Thermotoga*, *Fervidobacterium*, *Thermococcoides*, *Kosmotoga				
Marinitoga camini	AN COH	25–65 (55)	5–9 (7)	Deep sea vent fields, Mid-Atlantic ridge
Petrotoga mobilis	AN COH	40–65 (58–60)	5.5–8.5 (6.5–7)	Oil reservoir production water from off-shore oil platforms, North Sea
Thermosipho melanesiensis	AN COH	45–80 (70)	3.5–9.5 (6.5–9.5)	Deep sea hydrothermal area, Lau Basin, southwest Pacific Ocean
Fervidobacterium nodosum	AN COH	41–79 (70)	6–8 (7)	Hot spring in New Zealand
Thermotoga lettingae	AN COH	50–75 (65)	6–8.5 (7)	Thermophilic, sulfate-reducing, slightly saline bioreactor

Table 1. *Cont.*

Species	O2-relationship and metabolism	Temperature range (optimum)	pH range (optimum)	Originally isolated from
Thermotoga maritima	**AN COH**	**55–90 (80)**	**5.5–9 (6.5)**	**Geothermally heated sea floors, Italy and the Azores**
Thermotoga petrophila	**AN COH**	**47–88 (80)**	**5.2–9 (7)**	**Production fluid of the Kubiki oil reservoir in Niigata, Japan**
Thermotoga neapolitana	**AN COH**	**55–90 (80)**	**5.5–9 (7)**	**Shallow submarine hot springs, Lucrino Bay, Naples, Italy**
Archaea*; *Crenarchaeota*; *Thermoprotei*; *Desulfurococcales*; *Desulfurococcaceae **Genera:** ***Acidilobus*, *Staphylothermus*, *Ignicoccus*, *Desulfurococcus Thermosphaera*, *Sulfophobococcus*, *Stetteria*, *Thermodiscus*,** (Also *Ignisphaera* of the ***Ignisphaera*** group)				
Thermosphaera aggregans	AN COH	65–90 (85)	5–7 (6.5)	"Obsidian Pool" Yellowstone National Park, USA
Staphylothermus marinus	AN COH	65–98 (92)	4.5–8.5 (6.5)	Vulcano Island, Italy, also a deep-sea black smoker of the East Pacific Rise
Archaea*; *Crenarchaeota*; *Thermoprotei*; *Desulfurococcales*; *Pyrodictiaceae*;** **Genera:** ***Pyrodictium*, *Hyperthermus*, *Pyrolobus				
Hyperthermus butylicus	AN COH	(95–107)	(7)	Hydrothermally heated flat-sea sediments off the coast of São Miguel Island, Azores
Archaea*; *Crenarchaeota*; *Thermoprotei*; *Thermoproteales*; *Thermofilaceae **Genus:** ***Thermofilum***				
Thermofilum pendens	AN COH	(85–90)	(5)	Icelandic solfataras

Table 1. *Cont.*

Species	O_2-relationship and metabolism	Temperature range (optimum)	pH range (optimum)	Originally isolated from
Archaea*; *Crenarchaeota*; *Thermoprotei*; *Thermoproteales*; *Thermoproteaceae ***Genera*: *Thermoproteus*, *Pyrobaculum*, *Thermocladium*, *Caldvirga***				
Thermoproteus neutrophilus	AN F-CLA	(85)	(6.8)	Hot spring, Iceland
Pyrobaculum arsenaticum	AN F-CLA	68–100 (81)	NR	Hot water pond, Pisciarelli Solfatara, Naples, Italy
Pyrobaculum islandicum	AN F-CLA	74–102 (100)	5–7 (6)	Boiling solfataras and geothermal waters, Iceland
Pyrobaculum calidifontis	FAE COH	75–100 (90–95)	5.5–8.0 (7.0)	Terrestrial hot spring Calamba, Laguna, the Philippines
Pyrobaculum aerophilum	FAE F-CLA	75–104 (100)	5.8–9 (7)	Boiling marine water hole, Maronti Beach, Ischia, Italy
Caldivirga maquilingensis	FAE COH	62–92 (85)	2.3–6.4 (3.7–4.2)	Acidic hot spring in the Philippines
Archaea*; *Euryarchaeota*; *Thermoplasmata*; *Thermoplasmatales*; *Thermoplasmataceae*;** **Genus: *Thermoplasma*; *Acidiplasma				
Thermoplasma acidophilum	FAE COH	45–63 (59)	0.5–4 (1–2)	Solfatara fields and self heated coal refuse piles
Thermoplasma volcanium	FAE COH	33–67 (60)	1–4 (2)	Submarine and continental solfataras at Vulcano Island, Italy; also from Java, Iceland and Yellowstone National Park, USA
Archaea*; *Euryarchaeota*; *Methanococci*; *Methanococcales*; *Methanocaldococcaceae **Genera: *Methanocaldococcus*, *Methanotorris***				
Methanocaldococcus jannaschii	AN CLA	50–86 (85)	5.2–7.0 (6.0)	"White smoker" chimney on the 20°N East Pacific Rise
Methanocaldococcus vulcanius	AN CLA	49–89 (80)	5.2–7 (6.5)	Deep-sea vent, 13°N thermal field, East Pacific Rise

Table 1. *Cont.*

Species	O_2-relationship and metabolism	Temperature range (optimum)	pH range (optimum)	Originally isolated from
Archaea*; *Euryarchaeota*; *Thermococci*; *Thermococcales*; *Thermococcaceae*; Genera: *Thermococcus*, *Pyrococcus*, *Palaeococcus				
Thermococcus barophilus	AN COH	48–100 (85)	(7)	"Snakepit" hydrothermal vent region of the Mid-Atlantic ridge
Thermococcus gammatolerans	AN COH	55–95 (88)	(6)	Guaymas Basin, Gulf of California
Thermococcus kodakarensis	AN COH	60–100 (85)	5–9 (6.5)	Solfatara on Kodakara Island, Kagoshima, Japan
Thermococcus sibiricus	AN COH	40–88 (78)	5.8–9 (7.5)	Samotlor oil reservoir, Western Siberia
Pyrococcus furiosus	AN COH	70–103 (100)	5–9 (7)	Shallow marine hydrothermal system at Vulcano Island, Italy
Pyrococcus horikoshii	AN COH	80–102 (98)	5–8 (7)	Hydrothermal fluid samples obtained at the Okinawa Trough vents in the NE Pacific Ocean, at a depth of 1395 m
Archaea*; *Euryarchaeota*; *Archaeoglobi*; *Archaeoglobales*; *Archaeoglobaceae*; Genera: *Archeoglobus*, *Geoglobus*, *Ferroglobus				
Archaeoglobus fulgidus	AN F-CLA	64–92 (83)	5.5–7.5	Marine hydrothermal systems at Vulcano island and at Stufe di Nerone, Naples, Italy
Archaea*; *Euryarchaeota*; *Methanopyri*; *Methanopyrales*; *Methanopyraceae*; Genus: *Methanopyrus				
Methanopyrus kandleri	AN CLA	84–110 (98)	5.5–7 (6.5)	Deep-sea sediment from the Guaymas Basin, Gulf of California, and from the shallow marine hydrothermal system of the Kolbeinsey ridge, Iceland

Table 1. *Cont.*

Species	O_2-relationship and metabolism	Temperature range (optimum)	pH range (optimum)	Originally isolated from
Archaea*; *Euryarchaeota*; *Methanobacteria*; *Methanobacteriales*; *Methanobacteriaceae*; Genera: *Methanobacterium*, *Methanothermobacter				
Methanothermobacter thermautotrophicus	AN CLA	40–75 (65–70)	6.0–8.8 (7.2–7.6)	Anaerobic sewage sludge digestor
Archaea*; *Euryarchaeota*; *Methanococci*; *Methanococcales*; *Methanococcaceae*; *Genus*: *Methanothermococcus				
Methanothermococcus thermolithotrophicus	AN CLA	30–70 (65)	6–8 (7)	Heated sea sediments near Naples, Italy
Archaea*; *Euryarchaeota*; *Methanomicrobia*; *Methanosarcinales*; *Methanosaetaceae*; *Methanocellales* Genus: *Methanothrix				
Methanothrix thermophila	AN COH	(55)	6.1–7.5 (6.7)	Mesophilic anaerobic sludge digestors
Archaea*; *Euryarchaeota*; *Methanomicrobia*; *Methanosarcinales*; *Methanosaetaceae*; *Methanocellales* Genus: *Methanocella				
Methanocella conradii	AN CLA	37–60 (55)	6.4–7.2 (6.8)	Rice field soil

3. Metabolism and Biotechnological Applications

In Table 2 potential applications for some of the described species are reported, particularly for the production of bioactive molecules and/or biocatalysts that may be important for industrial processes and biotechnologies.

Table 2. Biotechnological applications of major groups of extremophiles.

Enzymes, organic compounds and processes	Applications and products	Most representing Genera
Amylases and pullulanases	Glucose, fructose for sweeteners; polymer-degrading additives in detergents	*Pyrococcus*, *Thermococcus*, *Fervidobacterium*, *Dictyoglomus*, *Anaerobranca*
Cellulases and Xylanases	Paper bleaching	*Clostridium*, *Petrotoga*, *Thermotoga*, *Thermosypho*, *Moorella*, *Caldicoprobacter*, *Caldicellulosiruptor*
Proteases	Amino acid production from keratins, food processing, baking, brewing, detergents	*Thermoanaerobacter*, *Fervidobacterium*
DNA-polymerases and ligases	Genetic engineering	*Thermotoga*, *Pyrococcus*, *Thermococcus*, *Archaeoglobus*, *Thermoanaerobacter*
Ethanol	Chemical and food industries	*Clostridium*, *Thermoanaerobacter*; *Thermoanaerobacterium*, *Caldanaerobius*, *Caloramator*
Hydrogen and/or methane	Energy, fuels	*Clostridium*, *Carboxydocella*, *Thermincola*, *Thermosinus*, *Thermotoga*, *Carboxydothermus*, *Carboxydobrachium*, *Anaerobaculum*, *Methanotorris*, *Methanococcus*, *Methanothermococcus*, *Methanotermobacter*
Volatile fatty Acids	Chemical and food industries	*Clostridium*

The data reported here represent a summary of all that has been proposed and applied. A more exhaustive list of applications has been published by Vieille and Zeikus (2001).

Major metabolic possibilities can be observed in thermophiles, and there is no correlation between thermophily and metabolic properties, maybe with the exception of the reverse situation, *i.e.*, that the temperature limit for phototrophy is presently far below 70 °C. Amend and Shock [48] have previously described thermophilic and hyperthermophilic energenic reactions in depth, and their work is a key resource for the study of thermophilic metabolisms [49].

Chemoorganoheterotrophic metabolism (frequently in an incomplete form referred to as "heterotrophic") can be further divided into subcategories according to the substrates and include glycolytic, (hemi)cellulolytic, lipolytic, and proteo/peptidolytic metabolisms, amongst others. The Emden-Meyerhof and Entner-Doudoroff pathways are employed by glycolytic thermophilic anaerobes, but a variety of modifications have been discovered, predominantly within the *Archaea* [50]. Major fermentation products formed by glycolytic thermophilic anaerobes include acetate, butyrate, lactate, ethanol, CO_2, and H_2 and to a lesser degree the observed products propionate, propanol and butanol. Traces of various branched fatty acids from amino acid degradation are also detected since many glycolytic anaerobic thermophiles require yeast extract for growth and some even for metabolic activity.

The production of ethanol by glycolytic and cellulolytic taxa has been studied. Cellulose and hemicellulose are the most abundant renewable natural plant fibers, and their degradation, coupled with the production of "biofuels", such as ethanol by thermophilic anaerobes has been an intensely studied research area for the last 30 years, although research on fuel production leading to patents had already been done in the late 1920s, which includes the description and use of the oldest validly published anaerobic thermophile, *Clostridium thermocellum*. Recently, the focus has been shifting to butanol- and to H_2-production. An example for this is the use of the *Caldicellulosiruptor bescii* strain DSM 6725^T [51] and of similar anaerobic thermophilic bacteria [52].

As with cellulose-degrading thermophilic anaerobes, xylanolytic thermophilic anaerobes generate interest because the conversion of xylan—a component of plant hemicellulose and the second-most abundant renewable polysaccharide in biomass—to useful products might be coupled with an increasing efficiency of processing lignocellulose and the production of energy from renewable resources. Xylan is widely used as carbon and energy source among thermophilic anaerobic *Bacteria*, especially among members of the Firmicutes [53–56].

Among chemolithoautotrophic pathways, the methanogenic reaction $4H_2 + CO_2 \rightarrow CH_4 + 2H_2O$, is well characterized and used by thermophilic taxa within the *Methanobacteriaceae*, *Methanothermaceae*, *Methanocaldococcaceae*, and *Methanococcaceae*. Another, relatively recently described, interesting chemolithoautotrophic metabolism of anaerobic thermophiles makes use of CO, which occurs as a normal component of escaping volcanic gas of terrestrial and deep-sea hydrothermal origin. Several thermophilic anaerobes that grow lithotrophically on CO have indeed been isolated, performing the metabolic reaction $CO + H_2O \rightarrow CO_2 + H_2$ employed by the acetogens *Desulfotomaculum*, *Carboxydothermus*, *Hermolithobacter*, *Carboxydocella*, *Thermincola*, *Caldanaerobacter*, and *Thermosinus* [57].

The same CO-using reaction has also been observed within the *Archaea* in an isolate belonging to the genus *Thermococcus* (family *Thermococcaceae*). Another interesting chemolithotrophic strategy is employed by the acetogens using the Wood-Ljungdahl pathway (from the reaction: $3H_2 + CO_2 \rightarrow$ acetate). Both mesophilic and thermophilic taxa (e.g., *Moorella* species) are known to perform this reaction.

Chemolithotrophs generate energy chemolithotrophically and assimilate carbon heterotrophically. Thermophilic anaerobes with this metabolism include *Archaea* as *Archaeoglobus profundus* and *Stetteria hydrogenophila*, and *Bacteria* as *Desulfotomaculum alkaliphilum*, *Desulfotomaculum carboxydivorans*, *Thermincola carboxydiphila*, *T. ferriacetica* (which can also grow hemolithoautotrophically), *Caldithrix abyssi*, *Vulcanithermus ediatlanticus*, and *Oceanithermus profundus*.

Two mechanisms for collecting light energy and converting it into chemical energy are known: one depends on photochemical reaction centers containing (bacterio)-chlorophyll and the other employs rhodopsins. However, to the authors' knowledge, there are no rhodopsin-using thermophilic anaerobes yet described.

Within the phylum *Firmicutes* (family Heliobacteriaceae), *Heliobacterium modesticaldum* is an obligately anaerobic photoheterotroph that is also capable of growing chemoorganoheterotrophically [58]. *H. modesicaldum* is among the most recently discovered taxa containing (bacterio)-chlorophyll photochemical reaction centers; however, at present, it is not characterized in detail.

Many *Archaea* were initially described as being obligately dependent on S_0 reduction for the production of energy, but it has often been reported that some of the so-called "sulfur-dependent" *Archaea* grow well in co-culture with hydrogen-using thermophilic methanogens in the absence of sulfur. This is possible through interspecies hydrogen transfer, whereby growth-inhibiting hydrogen (from H^+ used as an electron acceptor) is removed without sulfur serving as the electron acceptor.

The *Ignicoccus*–"*Nanoarchaeum*" system has been described as a symbiotic relationship. It was discovered that small cocci were attached to the larger cells of a strain of *Ignococcus* isolated from the Kolbeinsey Ridge, in the north of Iceland [59]. These tiny cocci could be isolated from the larger cells and subsequently studied, but grew only when attached to their host. The genome sequence analysis of "*Nanoarchaeum*" showed that it was missing most of the enzymes required for nonparasitic growth.

The importance of sulfur in the metabolism of thermophilic anaerobes becomes evident when one considers that the majority of thermophiles (chemolithotrophs, as well as chemoheterotrophs) take advantage of the sulfur redox system. Amend and Shock [48] posed that the most common energy-yielding reaction under thermophilic conditions may be the reduction of elemental sulfur: $H_2 + S^\circ \rightarrow H_2S$.

Indeed, the diversity of known thermophilic anaerobic taxa that use this strategy is notable: the sulfur-reducing reaction has been reported within the *Pyrodictiaceae*, *Sulfolobaceae*, *Thermoanaerobacteriaceae*, *Thermoproteaceae*, *Aquificacea*, *Desulfurellaceae*, *Desulfurococcaceae*, *Thermococcaceae*, *Thermoplasmataceae*, *Thermofilaceae*, and *Thermotogaceae* genera. Thermophilic, sulfate-reducing *Bacteria* have been isolated from a wide range of environments, and many of these thermophiles belong to a phylogenetically coherent cluster of spore-forming *Desulfotomaculum* species (*Peptococcaceae* in the Phylum *Firmicutes*).

Thus, the role of sulfur in the metabolisms of thermophilic anaerobes can vary for different groups: it can be reduced, it can serve as an electron sink during fermentation, and it can function as a terminal electron acceptor to allow sulfur respiration.

Thermophilic anaerobic Fe(III)-reducing *Bacteria* and *Archaea* are found within nearly all thermobiotic environments and are usually diverse in terms of respiration, capable of growing chemoorganotrophically with fermentable substrates or chemolithoautotrophically with molecular hydrogen. Although only relatively recently described, a diverse set of thermophilic anaerobes is known to reduce Fe(III) [60]. Families of the *Bacteria* with taxa known to reduce Fe(III) include the *Bacillaceae*, *Peptococcaceae*, *Thermoanaerobacteriaceae*, *Acidaminococcaceae*, *Syntrophomonadaceae*, *Deferribacteraceae*, *Hydrogenothermaceae*, *Thermotogaceae*, and the *Thermodesulfobacteriaceae*.

Families of the *Archaea* with taxa known to reduce Fe(III) include the *Thermoproteaceae*, *Archaeoglobaceae*, and the *Thermococcaceae*. *Geoglobus ahangari*, of the *Archaeoglobaceae,* was reported as the first dissimilatory Fe(III)-reducing prokaryote obligately growing autotrophically on hydrogen. In some genera, such as *Thermoanaerobacter*, *Thermotoga*, and *Anaerobranca*, many of the species tested have been found to be capable of dissimilatory reduction of Fe(III), but overall it appears as though the ability to reduce Fe(III) does not correlate with an affiliation at the genus or species level. For example, although *Deferribacter abyssi* and *Deferribacter thermophilus* are closely related, having 98.1% 16S rRNA gene sequence similarity, *D. abyssi* is unable to reduce Fe(III) whereas it is a primary electron acceptor for *D. thermophilus.* The chemolithoautotrophic iron reducers are of special interest since they are believed to have been responsible for the Low Temperature Banded Iron Formations. Beside the dissimilatory iron reduction, several thermophiles are also able to use various other metals, sometimes in combination with iron, sometimes they only reduce other oxidized metal ions, either as soluble ions or even within specific minerals. *Pyrobaculum arsenaticum* has the ability to grow chemolithotrophically by arsenate reduction, and both *P. arsenaticum* and *Pyrobaculum aerophilum* can use selenate, selenite, or arsenate chemolithoorganotrophically. For some thermophiles it appears that the reduction of metal ions occurs partly or fully without energy formation through this process as a detoxification mechanism. *Thermoanaerobacter* strains isolated from the Piceance Basin in Colorado were able to reduce Co(III), Cr(VI), and U(VI), in addition to Mn(IV) and Fe(III) [61].

In addition to these described characteristics—O_2-relationship, temperature and pH profiles, and metabolic strategies—a number of additional physiological properties of thermophilic anaerobes should be examined and should, therefore, add to what is known about the diversity of thermophilic anaerobes. The NaCl optimum and tolerance of a prokaryote is often assessed. Thermophilic anaerobes of marine origin, for example, would be expected to grow best at marine salinity—around 3.5% (wt/vol) NaCl. Prokaryotes that grow optimally with high salinity are referred to as halophiles, and halophilic thermophilic anaerobes are known, as are halophilic alkalithermophiles [62].

Thermophilic anaerobes living at deep-sea hydrothermal vent sites must cope with the additional pressure exerted by the water column and are, therefore, piezotolerant or perhaps even piezophilic [63,64]. Both *Methanocaldococcus* (basonym *Methanococcus) jannaschii,* isolated from the 21 °N East Pacific Rise deep-sea hydrothermal vent site, and *Thermococcus barophilus*, obtained from the Snakepit region of the Mid-Atlantic Ridge, grow faster under increased hydrostatic pressure [22,23].

At its optimal growth temperature, the growth rate of *T. barophilus* was more than doubled at elevated hydrostatic pressure (40 MPa) compared with the growth rate at low pressure (0.3 MPa). Furthermore, *T. barophilus*, as well as "*Pyrococcus abyssi*" and *Pyrococcus* strain ES4, isolated from deep-sea hydrothermal vent sites, show an extension of their T_{max} with significant elevated hydrostatic pressure [21–23].

Representative genera of thermophilic anaerobes living at deep-sea hydrothermal vent sites include *Archaeoglobus*, *Thermodiscus*, *Thermoproteus*, *Acidianus*, *Pyrococcus*, *Thermococcus* and *Desulfurococcus*, which reduce sulfur or sulfate, *Sulfolobus* can oxidize H_2S or elemental sulfur, the methanogens *Methanothermus*, *Methanococcus* and *Methanopyrus*, and the nitrate reducers *Pyrobaculum* and *Pyrolobus*. *Sulfolobus* and *Acidianus* isolates can also oxidize ferrous iron, and with no doubt such a process plays a major role on the local environment and biogeochemical cycles. Examples of hyperthermophilic bacteria are included in the genera *Thermotoga* and *Aquifex*.

Some of the isolated thermophilic anaerobes also possess ionizing radiation resistance; for example, this characteristic is found in *Tepidimicrobium ferriphilum* (Order *Clostridiales*), which was isolated from a freshwater hot spring within the Barguzin Valley, Buryatiya, Russia [65]. The level of natural radioactivity at hydrothermal vents can be 100 times greater than that at Earth's surface because of the increased occurrence of elements, such as ^{210}Pb, ^{210}Po and ^{222}Rn. Indeed, *Archaea* of the family *Thermococcaceae*, *Thermococcus gammatolerans* and *Thermococcus* "*radiotolerans*" isolated from the Guaymas Basin, of the Gulf of California, and *Thermococcus* "*marinus*", isolated from the Snakepit hydrothermal site of the Mid-Atlantic Ridge have γ-irradiation resistance.

It is worth mentioning the moderate thermophiles and thermotolerant organisms, particularly for their potential applications as well as for their ecological roles. Among these are the cellulolytic *Clostridium thermocellum*, the acetogenic *Moorella thermoacetica/thermoautotrophica* and *Thermoanaerobacterium (former Clostridium) thermosaccharolyticum*, capable of growing in vacuum packed foods and thus known as the "can-swelling" organism [66–69]. The obligate mixotrophic *Thiomonas bhubaneswarensis*, the marine *Lutaonella thermophila* and *Thermophagus xiamenensis*, the cellulolytic bacteria *Clostridium clariflavum* and *Clostridium caenicola*, the faculatative microaerophilic *Caldinitratiruptor microaerophilus*, and a novel hydrogen-producing bacterium from buffalo-dung were described [70–75].

Novel isolates were isolated from waste disposal plants, methanogenic reactors and wetland systems. *Tepidanaerobacter acetatoxydans*, *Anaerosphaera aminiphila* and *Clostridium sufflavum* were isolated from two methanogenic processes [76–78], whereas *Anaerosalibacter bizertensis* and *Gracilibacter thermotolerans* were observed and described in artificial ecosystems [79,80].

4. Conclusions

Anaerobic thermophilic microorganisms have been known for a long time but it is always difficult to understand that some organisms do not only survive at high temperatures, but actually thrive in boiling water. They are one of the most interesting varieties of extremophilic organisms.

The main interest in anaerobic thermophiles during the last decades has mainly been on two issues dealing with basic and applied research: 1) the discovery of many novel hyperthermophilic *Archaea* (of which many can grow at 100 °C and above and a few even up to 121 °C), has attracted a great interest among the scientific community; 2) the realization that anaerobic thermophilic microorganisms can serve as excellent sources for thermostable biocatalysts was the driving force for implementing basic and applied research on thermophiles.

Due to the stress of living at such extreme temperatures, anaerobic thermophiles have evolved a variety of mechanisms that allow them to survive at temperatures other organisms cannot thrive at. These traits include unique membrane lipid composition, thermostable membrane proteins, and higher turnover rates for various protein enzymes. One of the most important attributes to the maintenance of homeostasis within the organism is that of the plasma membrane surrounding the organism. Aside from having to stabilize the plasma membrane at high temperatures, anaerobic thermophiles must also stabilize their proteins, DNA, RNA, and ATP. Study into how they manage thermostability at the protein and membrane structural level has elucidated many traits of protein, membrane and nucleic acid structure; however, there is not yet a full understanding of the principles of thermophily and thermostability of cell components. As a matter of fact, the process of heat stabilization for DNA, RNA, and ATP is not fully understood yet.

With no doubts anaerobic thermophiles are interesting from the viewpoint of the trend toward biotechnology as many chemical industrial processes employ high temperatures which would have to be lowered in order to use bioprocesses from mesophiles, and this could be avoided using enzymes of thermophiles.

One of the most interesting potential application of anaerobic thermophilic microorganisms is the production of biofuels that was particularly investigated in the last decades, mainly as research activities on the metabolism of pure or mixed cultures to produce biofuel, including methane and hydrogen, but also throughout extensive lab work with the aim to obtain ethanol from biomass by means of thermophilic biological processes.

Acknowledgments

Francesco Canganella truly acknowledges W. Jack Jones for having introduced him in the lab to the world of anaerobic thermophiles. Both authors sincerely acknowledge the support and friendships of many colleagues who have given fruitful advices and comments during the past years, so being part of their fortunate careers in microbiological research.

Author Contributions

Francesco Canganella is the main author of the paper and has been actively involved in research on anaerobic thermophiles for many years. Juergen Wiegel has been supervising both writing and editing of the paper. Moreover he has been a worldwide recognized leader on anaerobic thermophiles research.

Conflicts of Interest

The authors declare no conflict of interest.

References

1. Wiegel, J., Michael, A.W.W., Eds. *Thermophiles: The Keys to Molecular Evolution and the Origin of Life?* CRC Press: Boca Raton, FL, USA, 1998.
2. Stahl, D.A.; Lane, D.J.; Olsen, G.J.; Pace, N.R. Characterization of a Yellowstone hot spring microbial community by 5S rRNA sequences. *Appl. Environ. Microbiol.* **1985**, *49*, 1379–1384.

3. Sandbeck, K.A.; Ward, D.M. Temperature adaptations in the terminal processes of anaerobic decomposition of Yellowstone national park and icelandic hot spring microbial mats. *Appl. Environ. Microbiol.* **1982**, *44*, 844–851.
4. Friedmann, E.I. Endolithic microbial life in hot and cold deserts. *Orig. Life* **1980**, *10*, 223–235.
5. Brock, T.D. High temperature systems. *Annu. Rev. Ecol. Syst.* **1970**, *1*, 191–220.
6. Kempner, E.S. Upper temperature limit of life. *Science* **1963**, *142*, 1318–1319.
7. Averhoff, B.; Müller, V. Exploring research frontiers in microbiology: Recent advances in halophilic and thermophilic extremophiles. *Res. Microbiol.* **2010**, *161*, 506–514.
8. Basu, S.; Sen, S. Turning a mesophilic protein into a thermophilic one: A computational approach based on 3D structural features. *J. Chem. Inf. Model* **2009**, *49*, 1741–1750.
9. Kim, M.S.; Weaver, J.D.; Lei, X.G. Assembly of mutations for improving thermostability of *Escherichia coli* AppA2 phytase. *Appl. Microbiol. Biotechnol.* **2008**, *79*, 751–758.
10. Vieille, C.; Zeikus, G.J. Hyperthermophilic enzymes: Sources, uses, and molecular mechanisms for thermostability. *Microbiol. Mol. Biol. Rev.* **2001**, *65*, 1–43.
11. Daniel, R.M.; Toogood, H.S.; Bergquist, P.L. Thermostable proteases. *Biotechnol. Genet. Eng. Rev.* **1996**, *13*, 51–100.
12. Shaw, A.J.; Podkaminer, K.K.; Desai, S.G.; Bardsley, J.S.; Rogers, S.R.; Thorne, P.G.; Hogsett, D.A.; Lynd, L.R. Metabolic engineering of a thermophilic bacterium to produce ethanol at high yield. *Proc. Natl. Acad. Sci. USA* **2008**, *105*, 13769–13774.
13. Bustard, M.T.; Burgess, J.G.; Meeyoo, V.; Wright, P.C. Novel opportunities for marine hyperthermophiles in emerging biotechnology and engineering industries. *J. Chem. Technol. Biotechnol.* **2000**, *75*, 1095–1109.
14. Cai, G.; Jin, B.; Saint, C.; Monis, P. Genetic manipulation of butyrate formation pathways in *Clostridium Butyricum*. *J. Biotechnol.* **2011**, *155*, 269–274.
15. Oshima, T.; Moriya, T. A preliminary analysis of microbial and biochemical properties of high-temperature compost. *Ann. N. Y. Acad. Sci.* **2008**, *1125*, 338–344.
16. Nesbø, C.L.; Kumaraswamy, R.; Dlutek, M.; Doolittle, W.F.; Foght, J. Searching for mesophilic Thermotogales bacteria: "Mesotogas" in the wild. *Appl. Environ. Microbiol.* **2010**, *76*, 4896–4900.
17. Nesbø, C.L.; Bradnan, D.M.; Adebusuyi, A.; Dlutek, M.; Petrus, A.K.; Foght, J.; Doolittle, W.F.; Noll, K.M. *Mesotoga prima* gen. nov., sp. nov., the first described mesophilic species of the Thermotogales. *Extremophiles* **2012**, *16*, 387–393.
18. Takai, K.; Nakamura, K.; Toki, T.; Tsunogai, U.; Miyazaki, M.; Miyazaki, J.; Hirayama, H.; Nakagawa, S.; Nunoura, T.; Horikoshi, K. Cell proliferation at 122 degree C and isotopically heavy CH sub-production by a hyperthermophilic methanogen under high-pressure cultivation. *Proc. Natl. Acad. Sci. USA* **2008**, *105*, 10949–10954.
19. Summit, M.; Scott, B.; Nielson, K.; Mathur, E.; Baross, J. Pressure enhances thermal stability of DNA polymerase from three thermophilic organisms. *Extremophiles* **1998**, *2*, 339–345.

20. Pledger, R.J.; Crump, B.C.; Baross, J.A. A barophilic response by two hyperthermophilic, hydrothermal vent archaea: An upward shift in the optimal temperature and acceleration of growth rate at supra-optimal temperatures by elevated pressure. *FEMS Microbiol. Ecol.* **1994**, *14*, 233–242.
21. Di Giulio, M. A comparison of proteins from *Pyrococcus furiosus* and *Pyrococcus abyssi*: Barophily in the physicochemical properties of amino acids and in the genetic code. *Gene* **2005**, *346*, 1–6.
22. Marteinsson, V.T.; Birrien, J.-L.; Reysenbach, A.-L.; Vernet, M.; Dominique, M.; Gamacorta, A.; Messner, P.; Sleytr, U.B.; Prieur, D. *Thermococcus barophilus* sp. nov., a new barophilic and hyperthermophilic archaeon isolated under high hydrostatic pressure from a deep-sea hydrothermal vent. *Int. J. Syst. Bacteriol.* **1999**, *49*, 351–359.
23. Canganella, F.; Gonzalez, J.M.; Yanagibayashi, M.; Kato, C.; Horikoshi, K. Pressure and temperature effects on growth and viability of the hyperthermophilic archaeon *Thermococcus peptonophilus*. *Arch. Microbiol.* **1997**, *168*, 1–7.
24. Adams, M.W.W. The biochemical diversity of life near and above 100 °C in marine environments. *J. Appl. Microbiol.* **1999**, *85*, 108S–117S.
25. Segerer, A.; Trincone, A.; Gahrtz, M.; Stetter, K. *Stygiolobus azoricus* gen. nov., sp. nov. represents a novel genus of anaerobic, extremely thermoacidophilic archaebacteria of the order *Suvolobales*. *Int. J. Syst. Bacteriol.* **1991**, *41*, 495–501.
26. Prokofeva, M.I.; Kostrikina, N.A.; Kolganova, T.V.; Tourova, T.P.; Lysenko, A.M.; Lebedinsky, A.V.; Bonch-Osmolovskaya, E.A. Isolation of the anaerobic thermoacidophilic crenarchaeote *Acidilobus saccharovorans* sp. nov. and proposal of *Acidilobales* ord. nov., including *Acidilobaceae* fam. nov. and *Caldisphaeraceae* fam. nov. *Int. J. Syst. Evol. Microbiol.* **2009**, *59*, 3116–3122.
27. Itoh, T.; Suzuki, K.; Sanchez, P.C.; Nakase, T. *Caldisphaera lagunensis* gen. nov., sp. nov., a novel thermoacidophilic crenarchaeote isolated from a hot spring at Mt. Maquiling, Philippines. *Int. J. Syst. Evol. Microbiol.* **2003**, *53*, 1149–1154.
28. Wiegel, J. Anaerobic alkalithermophiles, a novel group of extremophiles. *Extremophiles* **1998**, *2*, 257–267.
29. Chung, D.; Cha, M.; Farkas, J.; Westpheling, J. Construction of a stable replicating shuttle vector for Caldicellulosiruptor species: Use for extending genetic methodologies to other members of this genus. *PLoS One* **2013**, doi:10.1371/journal.pone.0062881.
30. Fukui, K.; Bessho, Y.; Shimada, A.; Yokoyama, S.; Kuramitsu, S. Thermostable mismatch-recognizing protein MutS suppresses nonspecific amplification during Polymerase Chain Reaction (PCR). *Int. J. Mol. Sci.* **2013**, *14*, 6436–6453.
31. Lin, L.; Xu, J. Dissecting and engineering metabolic and regulatory networks of thermophilic bacteria for biofuel production. *Biotechnol. Adv.* **2013**, *31*, 827–837.
32. Frock, A.D.; Kelly, R.M. Extreme thermophiles: Moving beyond single-enzyme biocatalysis. *Curr. Opin. Chem. Eng.* **2012**, *1*, 363–372.

33. Merkel, A.Y.; Huber, J.A.; Chernyh, N.A.; Bonch-Osmolovskaya, E.A.; Lebedinsky, A.V. Detection of putatively thermophilic anaerobic methanotrophs in diffuse hydrothermal vent fluids. *Appl. Environ. Microbiol.* **2013**, *79*, 915–923.
34. Tsubouchi, T.; Shimane, Y.; Mori, K.; Usui, K.; Hiraki, T.; Tame, A.; Uematsu, K.; Maruyama, T.; Hatada, Y. Polycladomyces abyssicola gen. nov., sp. nov., a thermophilic filamentous bacterium isolated from hemipelagic sediment in Japan. *Int. J. Syst. Evol. Microbiol.* **2013**, *63*, 1972–1981.
35. Dang, H.; Luan, X.W.; Chen, R.; Zhang, X.; Guo, L.; Klotz, M.G. Diversity, abundance and distribution of amoA-encoding archaea in deep-sea methane seep sediments of the Okhotsk Sea. *FEMS Microbiol. Ecol.* **2010**, *72*, 370–385.
36. Canganella, F.; Wiegel, J. Extremophiles: From abyssal to terrestrial ecosystems and possibly beyond. *Naturwissenschaften* **2011**, *98*, 253–279.
37. Canganella, F.; Vettraino, A.M.; Trovatelli, L.D. The extremophilic bacteria: Ecology and agroindustrial applications. *Ann. Microbiol. Enzimol.* **1995**, *45*, 173–184.
38. Kobayashi, T.; Koide, O.; Mori, K.; Shimamura, S.; Matsuura, T.; Miura, T.; Takaki, Y.; Morono, Y.; Nunoura, T.; Imachi, H.; *et al.* Phylogenetic and enzymatic diversity of deep subseafloor aerobic microorganisms in organics- and methane-rich sediments off Shimokita Peninsula. *Extremophiles* **2008**, *12*, 519–527.
39. Pagé, A.; Tivey, M.K.; Stakes, D.S.; Reysenbach, A.L. Temporal and spatial archaeal colonization of hydrothermal vent deposits. *Environ. Microbiol.* **2008**, *10*, 874–884.
40. Blöchl, E.; Rachel, R.; Burggraf, S.; Hafenbradl, D.; Jannasch, H.W.; Stetter, K.O. *Pyrolobus fumarii*, gen. and sp. nov., represents a novel group of Archaea, extending the upper temperature limit for life to 113 degrees C. *Extremophiles* **1997**, *1*, 14–21.
41. Engle, M.; Li, Y.; Rainey, F.; DeBlois, S.; Mai, V.; Reichert, A.; Mayer, F.; Messmer, P.; Wiegel, J. *Thermobrachium celere* Y., gen. nov., sp. nov., a fast growing thermophilic, alkalitolerant, and proteolytic obligate anaerobe. *Int. J. Syst. Bacteriol.* **1996**, *46*, 1025–1033.
42. Abed, R.M.; Dobretsov, S.; Al-Fori, M.; Gunasekera, S.P.; Sudesh, K.; Paul, V.J. Quorum-sensing inhibitory compounds from extremophilic microorganisms isolated from a hypersaline cyanobacterial mat. *J. Ind. Microbiol. Biotechnol.* **2013**, doi:10.1007/s10295-013-1276-4.
43. Bowers, K.J.; Mesbah, N.M.; Wiegel, J. Biodiversity of poly-extremophilic Bacteria: Does combining the extremes of high salt, alkaline pH and elevated temperature approach a physico-chemical boundary for life? *Saline Syst.* **2009**, doi: 10.1186/1746-1448-5-9.
44. Mueller, D.R.; Vincent, W.F.; Bonilla, S.; Laurion, I. Extremotrophs, extremophiles and broadband pigmentation strategies in a high arctic ice shelf ecosystem. *FEMS Microbiol. Ecol.* **2005**, *53*, 73–87.
45. Lee, Y.J.; Wagner, I.D.; Brice, M.E.; Kevbrin, V.V.; Wiegel, J. *Thermosediminibacter oceani* gen. nov., sp. nov. and *Thermosediminibacter litoriperuensis* sp.nov., new anaerobic thermophilic bacteria isolated form Peru Margin. *Extremophiles* **2005**, *9*, 375–383.
46. Li, Y.; Mandelco, L.; Wiegel, J. Isolation and characterization of a moderately thermophilic anaerobic alkaliphile, *Clostridium paradoxum*, sp. nov. *Int. J. Syst. Bacteriol.* **1993**, *43*, 450–460.

47. National Center for Biotechnology Information Taxonomy Database. Available online: http://www.ncbi.nlm.nih.gov/genomes/MICROBES (accessed on 10 January 2014).
48. Amend, J.P.; Shock, E.L. Energetics of overall metabolic reactions of thermophilic and hyperthermophilic Archaea and Bacteria. *FEMS Microbiol. Rev.* **2001**, *25*, 175–243.
49. Amend, J.P.; Rogers, K.L.; Shock, E.L.; Gurrieri, S.; Inguaggiato, S. Energetics of chemolithoautotrophy in the hydrothermal system of Vulcano Island, southern Italy. *Geobiology* **2003**, *1*, 37–58.
50. Selig, M.; Xavier, K.B.; Santos, H.; Schönheit, P. Comparative analysis of Embden-Meyerhof and Entner-Doudoroff glycolytic pathways in hyperthermophilic archaea and the bacterium *Thermotoga. Arch. Microbiol.* **1997**, *167*, 217–232.
51. Yang, S.J.; Kataeva, I.; Wiegel, J.; Yin, Y.; Dam, P.; Xu, Y.; Westpheling, J.; Adams, M.W. Classification of *Anaerocellum thermophilum* strain DSM 6725 as *Caldicellulosiruptor bescii* sp. nov. *Int. J. Syst. Evol. Microbiol.* **2010**, *60*, 2011–2015.
52. Bhandiwad, A.; Guseva, A.; Lynd, L. Metabolic Engineering of *Thermoanaerobacterium thermosaccharolyticum* for increased n-Butanol production. *Adv. Microbiol.* **2013**, *1*, 46–51.
53. Su, X.; Han, Y.; Dodd, D.; Moon, Y.H.; Yoshida, S.; Mackie, R.I.; Cann, I.K. Reconstitution of a thermostable xylan-degrading enzyme mixture from the bacterium *Caldicellulosiruptor bescii*. *Appl. Environ. Microbiol.* **2013**, *79*, 1481–1490.
54. Han, Y.; Agarwal, V.; Dodd, D.; Kim, J.; Bae, B.; Mackie, R.I.; Nair, S.K.; Cann, I.K. Biochemical and structural insights into xylan utilization by the thermophilic bacterium Caldanaerobius polysaccharolyticus. *J. Biol. Chem.* **2012**, *287*, 34946–34960.
55. Sizova, M.V.; Izquierdo, J.A.; Panikov, N.S.; Lynd, L.R. Cellulose- and xylan-degrading thermophilic anaerobic bacteria from biocompost. *Appl. Environ. Microbiol.* **2011**, *77*, 2282–2291.
56. Shao, W.; Wiegel, J. Purification and characterization of two thermostable acetyl xylan esterases from Thermoanaerobacterium sp. strain JW/SL-YS485. *Appl. Environ. Microbiol.* **1995**, *61*, 729–733.
57. Sokolova, T.G.; Henstra, A.M.; Sipma, J.; Parshina, S.N.; Stams, A.J.; Lebedinsky, A.V. Diversity and ecophysiological features of thermophilic carboxydotrophic anaerobes. *FEMS Microbiol. Ecol.* **2009**, *68*, 131–141.
58. Kimble, L.K.; Mandelco, L.; Woese, C.R.; Madigan, M.T. *Heliobacterium modesticaldum*, sp. nov., a thermophilic heliobacterium of hot springs and volcanic soils. *Arch. Microbiol.* **1995**, *163*, 259–267.
59. Huber, H.; Hohn, M.J.; Rachel, R.; Fuchs, T.; Wimmer, V.C.; Stetter, K.O. A new phylum of Archaea represented by a nanosized hyperthermophilic symbiont. *Nature* **2002**, *417*, 63–67.
60. Liu, S.V.; Zhou, J.; Zhang, C.; Cole, D.R.; Gajdarziska-Josifovska, M.; Phelps, T.J. Thermophilic Fe(III)-reducing bacteria from the deep subsurface: The evolutionary implications. *Science* **1997**, *277*, 1106–1109.
61. Roh, Y.; Liu, S.V.; Li, G.; Huang, H.; Phelps, T.J.; Zhou, J. Isolation and characterization of metal-reducing *Thermoanaerobacter* strains from deep subsurface environments of the Piceance Basin, Colorado. *Appl. Environ. Microbiol.* **2002**, *68*, 6013–6020.

62. Mesbah, N.M.; Cook, G.M.; Wiegel, J. The halophilic alkalithermophile *Natranaerobius thermophilus* adapts to multiple environmental extremes using a large repertoire of $Na^+(K^+)/H^+$ antiporters. *Mol. Microbiol.* **2009**, *74*, 270–281.
63. Lucas, S.; Han, J.; Lapidus, A.; Cheng, J.F.; Goodwin, L.A.; Pitluck, S.; Peters, L.; Mikhailova, N.; Teshima, H.; Detter, J.C.; *et al.* Complete genome sequence of the thermophilic, piezophilic, heterotrophic bacterium Marinitoga piezophila KA3. *J. Bacteriol.* **2012**, *194*, 5974–5975.
64. Alain, K.; Marteinsson, V.T.; Miroshnichenko, M.L.; Bonch-Osmolovskaya, E.A.; Prieur, D.; Birrien, J.L. Marinitoga piezophila sp. nov., a rod-shaped, thermo-piezophilic bacterium isolated under high hydrostatic pressure from a deep-sea hydrothermal vent. *Int. J. Syst. Evol. Microbiol.* **2002**, *52*, 1331–1339.
65. Slobodkin, A.I.; Tourova, T.P.; Kostrikina, N.A.; Lysenko, A.M.; German, K.E.; Bonch-Osmolovskaya, E.A.; Birkeland, N.K. *Tepidimicrobium ferriphilum* gen. nov., sp. nov., a novel moderately thermophilic, Fe(III)-reducing bacterium of the order Clostridiales. *Int. J. Syst. Evol. Microbiol.* **2006**, *56*, 369–372.
66. Prevost, S.; Andre, S.; Remize, F. PCR detection of thermophilic spore-forming bacteria involved in canned food spoilage. *Curr. Microbiol.* **2010**, *61*, 525–533.
67. Canganella, F.; Wiegel, J. The Potential of Thermophilic Clostridia in Biotechnology. In *The Clostridia and Biotechnology*; Woods, D.R., Ed.; Butterworths Publishers: Stoneham, MA, USA, 1993; pp. 391–429.
68. Wiegel, J.; Ljungdahl, L.G. The importance of thermophilic bacteria in biotechnology. *Crit. Rev. Biotechnol.* **1986**, *3*, 39–107.
69. Kristjansson, J.K. *Thermophilic Bacteria*; CRC Press: Boca Raton, FL, USA, 1992.
70. Gao, Z.M.; Liu, X.; Zhang, X.Y.; Ruan, L.W. *Thermophagus xiamenensis* gen. nov., sp. nov., a moderately thermophilic and strictly anaerobic bacterium isolated from hot spring sediment. *Int. J. Syst. Evol. Microbiol.* **2013**, *63*, 109–113.
71. Fardeau, M.L.; Barsotti, V.; Cayol, J.L.; Guasco, S.; Michotey, V.; Joseph, M.; Bonin, P.; Ollivier, B. *Caldinitratiruptor microaerophilus*, gen. nov., sp. nov. isolated from a French hot spring (Chaudes-Aigues, Massif Central): A novel cultivated facultative microaerophilic anaerobic thermophile pertaining to the Symbiobacterium branch within the Firmicutes. *Extremophiles* **2010**, *14*, 241–247.
72. Romano, I.; Dipasquale, L.; Orlando, P.; Lama, L.; d'Ippolito, G.; Pascual, J.; Gambacorta, A. *Thermoanaerobacterium thermostercus* sp. nov., a new anaerobic thermophilic hydrogen-producing bacterium from buffalo-dung. *Extremophiles* **2010**, *14*, 233–240.
73. Arun, A.B.; Chen, W.M.; Lai, W.A.; Chou, J.H.; Shen, F.T.; Rekha, P.D.; Young, C.C. *Lutaonella thermophila* gen. nov., sp. nov., a moderately thermophilic member of the family *Flavobacteriaceae* isolated from a coastal hot spring. *Int. J. Syst. Evol. Microbiol.* **2009**, *8*, 2069–2073.

74. Panda, S.K.; Jyoti, V.; Bhadra, B.; Nayak, K.C.; Shivaji, S.; Rainey, F.A.; Das, S.K. *Thiomonas bhubaneswarensis* sp. nov., a novel obligately mixotrophic, moderately thermophilic, thiosulfate oxidizing bacterium. *Int. J. Syst. Evol. Microbiol.* **2009**, *59*, 2171–2175.
75. Shiratori, H.; Sasaya, K.; Ohiwa, H.; Ikeno, H.; Ayame, S.; Kataoka, N.; Miya, A.; Beppu, T.; Ueda, K. *Clostridium clariflavum* sp. nov. and *Clostridium caenicola* sp. nov., moderately thermophilic, cellulose-/cellobiose-digesting bacteria isolated from methanogenic sludge. *Int. J. Syst. Evol. Microbiol.* **2009**, *59*, 1764–1770.
76. Westerholm, M.; Roos, S.; Schnürer, A. Tepidanaerobacter acetatoxydans sp. nov., an anaerobic, syntrophic acetate-oxidizing bacterium isolated from two ammonium-enriched mesophilic methanogenic processes. *Syst. Appl. Microbiol.* **2011**, *34*, 260–266.
77. Nishiyama, T.; Ueki, A.; Kaku, N.; Ueki, K. Clostridium sufflavum sp. nov., isolated from a methanogenic reactor treating cattle waste. *Int. J. Syst. Evol. Microbiol.* **2009**, *59*, 981–986.
78. Ueki, A.; Abe, K.; Suzuki, D.; Kaku, N.; Watanabe, K.; Ueki, K. Anaerosphaera aminiphila gen. nov., sp. nov., a glutamate-degrading, Gram-positive anaerobic coccus isolated from a methanogenic reactor treating cattle waste. *Int. J. Syst. Evol. Microbiol.* **2009**, *59*, 3161–3167.
79. Rezgui, R.; Maaroufi, A.; Fardeau, M.L.; Ben, A.; Gam, Z.; Cayol, J.L.; Ben Hamed, S.; Labat, M. Anaerosalibacter bizertensis gen. nov., sp. nov., a halotolerant bacterium isolated from sludge. *Int. J. Syst. Evol. Microbiol.* **2012**, *62*, 2469–2474.
80. Lee, Y.J.; Romanek, C.S.; Mills, G.L.; Davis, R.C.; Whitman, W.B.; Wiegel, J. Gracilibacter thermotolerans gen. nov., sp. nov., an anaerobic, thermotolerant bacterium from a constructed wetland receiving acid sulfate water. *Int. J. Syst. Evol. Microbiol.* **2006**, *56*, 2089–2093.

Hot Spring Metagenomics

Olalla López-López, María Esperanza Cerdán and María Isabel González-Siso

Abstract: Hot springs have been investigated since the XIX century, but isolation and examination of their thermophilic microbial inhabitants did not start until the 1950s. Many thermophilic microorganisms and their viruses have since been discovered, although the real complexity of thermal communities was envisaged when research based on PCR amplification of the 16S rRNA genes arose. Thereafter, the possibility of cloning and sequencing the total environmental DNA, defined as metagenome, and the study of the genes rescued in the metagenomic libraries and assemblies made it possible to gain a more comprehensive understanding of microbial communities—their diversity, structure, the interactions existing between their components, and the factors shaping the nature of these communities. In the last decade, hot springs have been a source of thermophilic enzymes of industrial interest, encouraging further study of the poorly understood diversity of microbial life in these habitats.

Reprinted from *Life*. Cite as: López-López, O.; Cerdán, M.E.; González-Siso, M.I. Hot Spring Metagenomics. *Life* **2013**, *2*, 308-320.

1. Introduction

Currently there is a great interest in hot springs, which are the natural habitat of thermophilic and hyperthermophilic microorganisms with optimal growth temperatures of >55 °C and >80 °C, respectively. Enzymes obtained from them have been proved to be extremely valuable as biocatalysts for industrial and biotechnological purposes. A paradigm is Taq polymerase from *Thermus aquaticus* that led to the development of the polymerase chain reaction (PCR) technique [1].

The initial studies on hot springs focused only on their physicochemical properties and geological features, and it was not until the mid-XX century that the study of the microbiology of these ecosystems began [2]. The temperature in hot springs is usually over the limit of eukaryotic life (near to 60 °C), which limits the microbial life to Bacteria and Archaea (and their viruses). The earliest microbiological work was based on the isolation and identification of thermophilic microbial strains.

16S rRNA-based studies subsequently revealed that microbial diversity was much broader than suggested by culture-dependent techniques. In combination with the construction of metagenomic libraries, research on total environmental DNA produced a vast amount of information, providing detailed pictures of the microbial communities present in diverse thermal environments. Each hot spring differs from others in temperature, chemical composition and its gradients of temperature or light. Hot springs comprise several habitats, such as thermal fluids, microbial mats and sediments. This diversity of habitats provides a vast number of sites to sample, all with potential interest for metagenomic analysis. The increasing number of reports makes it easier to understand how physicochemical conditions and biological interactions have shaped these microbial communities within their specific environments. In this review, we will illustrate with several examples the usefulness of metagenomic techniques in expanding our knowledge about microbial communities in hot springs.

2. Bacteria and Archaea

The first studies on thermophilic microorganisms from hot springs were focused on isolation and characterization of thermophilic strains using culture-dependent approaches [2]. However, ~99% of the microorganisms within a particular environment proved uncultivable [3], although culture methods were improved to try to overcome the requirements of some of these reluctant strains that showed stringent growth conditions or inter-dependent microbial consortia to live [4]. The estimated number of microbial species that might be detected by further development of this methodology is still difficult to predict.

New molecular methodology gives support to the study of the whole biological diversity through a metagenomic approach that might help to characterize all the microorganisms living in an environment. The total present DNA, called "metagenome", is extracted and purified from environmental samples, which includes microorganisms that cannot be cultured. Molecular phylogenetic methods, based on the comparison of the sequences from the 16S rRNA genes, have revealed the hidden diversity of hot springs using the metagenomic approach. They have been extensively used to describe microbial communities in hot springs and identify novel thermophilic microorganisms since the mid-1990s. Early microbial diversity analysis combined PCR amplification of the 16S rRNA genes and their pattern analysis on denaturing gradient gel electrophoresis (DGGE). This provided the first insight into the true diversity of these environments, but only DNA sequencing of the amplified targets allowed that new microorganisms, most of them non-cultivable, could be identified and classified. More recently, the development of next-generation sequencing technologies allows metagenomic libraries to be rapidly constructed and sequenced. Therefore the sequences of the 16S rRNA genes or other relevant genes can easily be analyzed, providing a more comprehensive view of microbial diversity.

2.1. PCR Approach

The first PCR-based studies of 16S rRNA genes, carried out on organisms from the hot springs of Yellowstone National Park (YNP), revealed an unexpectedly high diversity in contrast with the estimate from culture-dependent studies. This finding led to a revolution in the understanding of phylogenetic relations between Archaea and Bacteria [5,6]. Since then, the same strategy has been used to analyze community profiles from hot springs all over the world. Phylogenetic diversity was greatly expanded with the finding of novel phylotypes, especially in the predominant phylogenetic groups from these hot springs, *i.e.*, the archaeal phylum *Crenarchaeota* and the bacterial division *Aquaficales* [6,7], and even to the discovery of a new archaeal phylum classified as the *Korarchaeota* [6]. However, the also new archaeal phylum *Nanoarchaeota* remained undetectable by conventional PCR-based studies until it was discovered using non-traditional culturing techniques. It is represented by *N. equitants,* a nanosized hypertermophilic symbiont from a submarine hot vent, whose rRNA gene sequence is unique, even in the highly conserved regions used as primer targets for PCR [8]. The newly found *Nanoarchaeota* and *Korarchaeota* are members of the deepest branch-offs of the rRNA phylogenetic tree [9]

Some approaches went far beyond by using additional biomolecular markers to analyze in parallel the microbial diversity. For instance, given the high abundance of the polysaccharide chitin in marine environments, chitinase genes were analyzed in coastal hot springs, that provide high diversity of valuable new chitinase genes [10]. From a hot spring in Bulgaria, the phylogeny of the archaeal community was analyzed using 16S RNA genes and genes of the glycoside hydrolase-4 family. The good correspondence between both affiliation assignment methods proved the usefulness of these gene encoding metabolic enzymes for phylogenetic studies in heterotrophic archaea [11]. The study also allowed the direct cloning of these genes of industrial interest [11].

Temperature is seen as the main factor that shapes microbial communities in hot springs with different locations [12]. Most hot springs limit the survival of eukaryotic organisms (their limiting temperature for thermophiles reaching 60 °C) and the presence of photosynthetic organisms (70 °C). A particular example is the unique coastal hot springs in Iceland, where the microbial populations are exposed to fluctuations of temperature (and salinity) as a consequence of periodic high tides. The areas suffering the longest hot-temperature periods showed a majority of terrestrial thermophilic bacteria, whereas the areas with the shortest hot-temperature periods showed predominance of moderately thermophilic microorganisms and the presence of mesophilic marine microorganisms and proteobacteria [10].

Although the structure of a bacterial community seems to be mainly determined by temperature, specific populations may be subjected to other factors. Geochemical features of hot springs might act as key determinants in community structure and diversity [7]. There is evidence that composition of an actinobacterial community is influenced by a combination of temperature and pH; this study also reveals the high degree of endemism in this group among different hot springs, even in those that are geographically distant [13].

Emerging pyrosequencing techniques overcome the otherwise limited sampling in library construction based on PCR amplified fragments and Sanger sequencing. Deep sampling achieves a better coverage of the microbial diversity, which is necessary to detect rare and minority microorganisms. Using these powerful techniques, diverse research groups have depicted a general view of thermal communities and have assigned abundance percentages to each identified phylum. The data confirm the predominance of the previously defined groups, but also show novel phylotypes related to specific features related to particular hot springs. The hot springs in Africa, which have been sampled hitherto, are dominated by phylotypes belonging to the *Proteobacteria* [14]. Groundwater in a thermal field in Russia shows that Archaea is dominated by a novel division in the phylum *Euryarchaeota* related to the order *Thermoplasmatales* (39% of all archaea) and by another abundant group (33% of all archaea) related to MCG1 lineage of the phylum *Crenarchaeota* [15]; both groups are widely spread in hot springs all over the world [15]. However, bacteria are dominated by thermoacidophilic methanotrophs and sulfur-oxidizing microorganisms that use inorganic substrates of volcanic origin [15]. The analysis of the taxonomic and metabolic features of the microbial community of a Colombian acidic hot spring showed that only a small proportion of the metagenomic sequences had matches against databases, possibly due to a high proportion of novel taxa; some groups potentially involved in the nitrogen and sulfur cycling in this environment have been described [16].

Temperature, as a shaping agent in a bacterial microbial mat of effluent channels from two hot springs in YNP, was analyzed by bar-coded pyrosequencing [12]. This technique entails the introduction of different codes in the primers used with the different samples so as to assign a sample origin to each sequence retrieved after pyrosequencing. Samples along the temperature gradient showed a loss of diversity and richness with increasing temperatures, probably as a consequence of the effect on primary producers affected by temperatures near the maximum that permits photosynthesis. Distribution of Cyanobacteria and *Chloroflexi* along the thermal gradient was the subject of special investigation. The results showed a general tendency to an alternative presence of the two groups in the abundance peak, which suggests that competition exists for physical space and/or limited nutrients. This explanation was proposed as more reliable than the longstanding hypothesis that co-adapted lineages of these bacteria maintain tightly co-occurring distributions along the gradient as a result of a producer-consumer relationship. In conclusion, temperature is revealed as the main factor in shaping bacterial microbial communities in hot springs of YNP, despite different taxon composition supported by specific solutes and physicochemical properties of water [12].

Differences in the importance of geographical distance and divergence of community structure were also found. A number of microbial species may be ubiquitous in hot springs. Some studies show the relevance of certain taxa distributed in hot springs world-wide, such as pJP89-related organisms and uncultured groups of the order *Thermoplasmatales* [15]. Nevertheless, other taxa seem to be endemic in certain regions [13,17]. The importance of geographic distance is controversial, but may be a significant factor that influences microbial communities according to some studies focused on actinobacteria and cyanobacteria populations [13,18], although this proved not to be significant in other studies [12]. The differences of diversity observed among bacterial populations cannot be attributable to the distance between hot springs, but to temperature variations in the hot springs [12].

2.2. Metagenomic Libraries

Metagenomic libraries yield comprehensive community profiles, and additional information on the lifestyles of the microorganisms present in environmental samples allow entire or partial community metabolic fluxes to be reconstructed. This approach leads to differences in community composition compared with PCR-based methods [19], and each one has advantages and disadvantages. PCR amplification of 16S rRNA may produce a bias due to the unequal amplification of species and chimeric sequences, whereas metagenomics may fail to detect rare species in a community. Uncultured bacteria can be detected by traditional studies of microbial diversity by PCR of 16S rRNA, but assignment to a metabolic type is difficult to achieve when there are not related cultivable microorganisms [7].

The identification of novel microbial species or phylotypes by analysis of 16S rRNA genes is a task not as simple to fulfil as is the species demarcation in animal and plants. There is a comprehensive review about this issue [20] that questions the reliability of selecting a cut-off of

97-98% for similarity in 16S rRNA sequences as a good criterion to discriminate new microbial species from closely related microorganisms. Those demarcated by the 97-98% cut-off might include different species-like ecotypes, which could potentially lead to an underestimate of the real microbial diversity in an environment. Tindall *et al.* [21] provide recommendations for analysis of the 16S rRNA genes in the field of prokaryote taxonomy. They indicate that when 16S rRNA gene similarity values are over 97% (over full pairwise comparisons) it is necessary to use other methods to discriminate new species, such as DNA-DNA hybridization or analysis of gene sequences with a greater resolution; these methods must also be correlated with phenotype characterization. At values above 95% of 16S rRNA gene similarity (over full pairwise comparisons), taxa should also be tested by other methods in order to establish whether separate genera are present.

Using metagenomics, genomes can be assembled and thereafter 16S rRNA can be analyzed along with other relevant genes, thereby more accurately demarcating new species. One example carefully analyzed is the cyanobacteria, *Synechococcus*, an inhabitant of microbial mats in hot springs along thermal gradients where ecologically different subpopulations have been found by exhaustive metagenomic analysis [22]. These subpopulations differ in certain metabolic capabilities and their genomes lack a conserved large-scale genomic order. This study highlights how useful metagenomics can be in understanding the heterogeneity of species in a natural microbial population. Metagenomic libraries are potent tools to identify and describe novel uncultured taxa, by assembly of their genomes or simply by exploring the metagenome, searching for functional genes to infer metabolic features and phylogenetic markers. This is a difficult task, given that often these uncultured members are minor components in a highly diverse community; besides, metagenomic information extracted from sequenced libraries can be partial and fragmented. Nevertheless, several new bacteria have been described in metagenomic libraries of microbial mats of hot springs, apart from the dominant populations of microbial mats (*Synechococcus*, *Chloroflexi* and *Roseiflexus* spp.-related strains), indicating that the microbial community is more phylogenetically and physiologically diverse than the studies of their 16S rRNA genes might indicate. For instance, discovery of a new member of anoxygenic chlorophototrops within the phylum *Chloroflexi*—where chlorotrophy was thought to be restricted to the class *Chloroflexi* [23], or the first phototrophic member of the phylum *Acidobacteria*, which was detected by BLAST search in metagenomic libraries of sequences from components of the type 1 reaction centre of chlorophototrophs [24]. The assignment of novel metabolic capabilities to these newly discovery taxa groups were only possible by analysis of available metagenomic information. In this way, an uncultured member of the phylum *Chlorobi*, provisionally named *Candidatus Thermochlorobacter aerophilum*, was found in the microbial mats of one alkaline siliceous hot springs at the YNP and was investigated through metagenomic and metatranscriptomic approaches. It was proposed to be an aerobic photoheterotroph that cannot oxidize sulfur compounds, cannot fix N_2, and does not fix CO_2 autotrophically. Metagenomic analyses suggest that it depends on other organisms in the mat, which provide fixed carbon and nitrogen, several amino acids, and other important nutrients [25]. Also, metagenomics of hydrothermal vents revealed the potential importance of H_2 as a key energy source in the deep ocean [26].

Recently, the partial assemble of the genome from a single cell became possible. In this way, the genome from one uncultured bacteria isolated from sediments of a hot spring in Oklahoma, which belongs to the world widely distributed candidate division OP11, provided important information about its lifestyle (metabolic capabilities, secretory pathways, cell wall structure, and defence mechanisms). These results should be taken with caution, since the genome of a single cell cannot be representative of the entire lineage [27].

There are many examples of metagenomic libraries used to investigate the diversity of microbial communities of hot springs in more complex studies and most of them were carried out in YNP: For instance, a detailed picture of community structure and ecology of a microbial mat sample from Mushroom hot spring was provided by comprehensive metagenomic studies of environmental DNA and RNA, which analyzed several phylogenetic and functional relationships between genes [23,28]. Metatranscriptome analysis at different time-points: during light-to-dark and during dark-to-light transitions showed different temporal pattern behavior in the different phototrophic populations [28].

One of the new issues addressed by metagenomics is an approach to understand the functional role of mobile genetic elements, e.g. insertion sequences (IS) [29]. These have the ability to produce rearrangements in the genome, making them a powerful force in genome evolution. Metagenomics provides a powerful method to gain an insight into the IS genome content and its location in a natural population, and to learn about the mechanisms of IS accumulation and survival against mutational forces. Clone ends usually show putative affiliation to different reference genomes in some metagenomic libraries. This suggests the existence of previous gene exchanges, even between distant lineages, many of them related with CRISPR (Clusters of Regularly Interspaced Short Palindromic Repeats)—associated proteins that have phage-host functions [23,30]. The thermophilic cyanobacterium, *Synechococcus,* shows many of such rearrangements in its genome by comparison among different strains, but which are dominant in a specific habitat. The abundance of IS in reference genomes of strains A and B' did not differ from natural communities, and provides evidence of recent lateral gene transfers between ecotype-like species. Furthermore, metagenomics provide a snapshot of the population, and allows the detection of deleterious mutations caused by IS that have not yet been selected against [29]. Comparative metagenomics of microbial communities inhabiting deep-sea hydrothermal vent chimneys showed a high proportion of transposases in the metagenomes, implying that horizontal gene transfer may be a common occurrence in the deep-sea vent chimney biosphere [31].

As well as with 16S rRNA based strategies, there are comprehensive studies where both temperature and geochemical composition were seen as factors that shape microbial communities from a metagenomic approach. A metagenomic study based on the Bison Pool hot spring (YNP) microorganism inhabitants in the effluent channel revealed a continuum of different microbial communities along a gradient of temperature and geochemical conditions, with different metabolic capabilities and lifestyles in accordance with these changing environmental conditions. Dominant taxa were identified in each sample location, and it was also possible to differentiate between specialist and generalist components in the microbial community [32].

3. Viruses

Viruses are the most abundant biological entities in every ecosystem, even among hot springs, although several estimates indicate a density lower than in other mesophilic aquatic systems [33]. Studies of viruses in this branch of environmental metagenomics do not only pursue the identification and classification of new types but also aim for a better understanding of the role of viruses in nature, and whether (and if so, how) they interact with other elements of the ecological systems. They are probably the only predators in these communities, and may be involved in the control of host mortality and the carbon cycle [33].

In the beginning, the only methodology available to study a virus was to culture its host, implying that studies were biased for lytic types in bacteriophages, which hindered further progress in this field. Metagenomic studies on viruses from thermal environments overcame this limitation by showing the enormous diversity and abundance of viruses in these ecosystems. For example, the consensus genome sequence of a novel GC-rich archaeal rudivirus recovered by iterative *de novo* read mapping and assembly from a hot spring metagenome was recently reported [34]. Nevertheless, enriched cultures continue to be used to identify new viruses with success. A bioreactor inoculated with a sample from a hot spring in YNP was maintained for two years at 85 °C and pH 6 and two new viruses were discovered by this approach [35].

Metagenomics is providing a vast amount of information about viruses in hot springs, but correct handling and understanding of new data remains in progress. Population dynamics are also being studied, but the data are poorly understood [36], the biggest problem being the correct virus classification. Viruses have been classified according to their host range and morphology, owing to the lack of universal genomic signatures like the 16S rRNA genes of prokaryotes. Failure to find homology to sequences coming from new viral metagenomes in GenBank using BLAST alignments made necessary the development of other tools. An alternative method is viral genome signature-based phylogenetic classification, using a database of oligonucleotide frequencies instead of sequence similarity, which gives a more reliable classification of the viral sequences and the assignment of a likely host [37].

Another problem is the bias in the studies. Most viruses being investigated are double-stranded DNA due to the methodology of library construction, but advances in this procedure made the study of RNA viruses possible. The first genome segments belonging to a putative positive-strand RNA virus replicating in archaeal hosts were recently isolated from several acidic hot springs of YNP. The virus might be related to the direct ancestor or eukaryotic viruses, whose origin remains unknown [38].

The first metagenomic study of viruses in geothermal environments was carried out relatively recently, with double-stranded DNA viruses of the Octopus and Bear Paw hot springs (YNP). The assembly of viral metagenomes indicated a high degree of heterogeneity, with the predominance of a lytic lifestyle. The occurrence of the appropriate machinery for lateral gene transfers and evidence of the replacement of cellular genes by non-orthologous viral genes (*i.e.*, the similarity between several proteins from cell and virus, such as helicases and DNA polymerases) suggested that viruses might play a critical role in the evolution of DNA and its replication mechanisms [33].

In the genomes of Bacteria and Archaea, there are loci called CRISPR that act as a molecular registration of phage attacks on the cell. They contain virus-derived sequences originating from previous viral infections or acquired by lateral transfers, which confer immunity against the phage that also contains one of them in its genome. Therefore, comparative analysis of viral sequences and cellular CRISPR might provide information about the archaeal or bacterial hosts of viruses [30,39]. A microarray assay has been designed to detect and analyze these viral sequences in the environmental samples, using metagenomic sequences from bacteria and archaea inhabiting acidic hot springs as probes. Effectively, it demonstrated its usefulness in detecting new viruses and monitoring changes in a viral population [36]. This method gives similar results to those obtained by the construction of metagenomic libraries, but it is less expensive and time-consuming.

Finally, metagenomic techniques show evidence of homologous recombination between viruses. The presence of capsid protein from RNA viruses in cirvovirus-like DNA suggest an event of gene transfer between unrelated RNA and DNA viruses, but the molecular mechanisms involved remain unclear. Further studies in metagenomic libraries might help to decipher virus evolution [40].

4. Novel Thermophilic Enzymes

Thermostable enzymes from thermophilic microorganisms are important biocatalysts for industrial and biotechnological purposes, given that they can work at high temperatures in which mesophilic enzymes would be denatured. There are now many thermophilic enzymes being used for biotechnological and industrial purposes. The classical example is Taq DNA polymerase from *Thermus aquaticus,* purified and isolated from hot springs [1], which made the development of the PCR amplification technique possible. Studies of the biodiversity in hot springs revealed the presence of complex communities containing novel microorganisms, which can be potential sources of novel enzymes with unique features of interest in industrial applications.

Thermophilic enzymes were primarily screened in a culture-based manner, but metagenomics of hot springs facilitate the search of new biocatalysts by functional screening for the desired activity or by shotgun sequencing and the search for the target enzyme in metagenomic libraries.

Sequence-based approaches are biased towards already-known families of enzymes, but some authors prefer this approach to circumvent problems ingetting the correct expression of the foreign gene coding in the desired enzyme in a heterologous host. This was the case with a thermostable Fe-superoxide dismutase that was identified from a partially sequenced metagenomic library using BLAST. The activity of this enzyme that eliminates toxic superoxide radicals has potential use as an additive in the cosmetic industry [41]. More recently, three novel genes conferring lipolytic and one gene conferring proteolytic activity were identified by mining a thermal spring volcanic metagenome. These genes were cloned into expression vectors and the recombinant proteins characterized showing special features [42].

By functional screening, entirely new classes of enzymes can be found, but the success of this approach requires the target protein to be compatible with transcription and translation machinery of the host. Although only 40% of the foreign proteins expressed in *E. coli* appear to be successfully expressed [43], several lipolytic enzymes, with potential industrial applications, have been isolated by functional screening from thermal springs in Thailand [44]. A curious case was a lipase isolated from soil in a hot spring area in India, which was thermo-labile but of thermophilic origin [45]. Thermostability of this protein was successfully improved by protein engineering, the mutated form being 144-fold more stable at 60 °C than the native enzyme [46].

Many DNA polymerases from thermophilic microorganisms have been exploited for biotechnological purposes since the discovery of the Taq polymerase. But the number of enzymes utilized for biotechnological purposes is strikingly low compared to the tremendous amount of diversity that exists in viral genes. A thermostable DNA polymerase of viral origin was isolated from a viral metagenomic library of Octopus hot spring in YNP [33] by identification of potential polymerase genes using BLAST alignment prior to functional screening. The most thermostable enzyme possesses reverse transcriptase and DNA polymerase activities. It has been modified to eliminate exonuclease activity, and the engineered enzyme has turned out to be a viable alternative to the traditional 2-enzyme systems employed in RT-PCR, with both higher specificity and sensitivity [47].

5. Conclusions

Less diversity and richness in hot springs than in other aquatic environments have been reported [12,33,48]. Analysis of thermal communities from all over the world has found the recurrent presence of certain groups, although in some hot springs there are specific strains that contribute significantly to the composition of the microbial community, and whose presence is correlated to the geochemical properties of hot springs. Apart from microbial community profiles, metagenomics assesses the effect of physicochemical conditions in community diversity from hot springs; although temperature seems to be the major factor, geochemical compositions and geographical distances are significant in some cases.

Discoveries in thermal environments have increased the knowledge about the evolution not only of bacteria and archaea, but also of viruses. As far as metagenomic studies are to be expanded, a more complete phylogeny of these groups could be drawn.

Hot springs are also a vast source of new and diverse thermophilic enzymes, many with potential uses in industry. Metagenomics is a powerful tool to identify yet unknown enzymes and provide industry with more cost-effective biocatalysts for specific purposes.

The reports summarized in this review and others not covered for brevity, but equally valuable, demonstrate the unequalled potential of metagenomics in the study of hot springs as well as the importance of continuing research in this field. We might expect that the findings already made probably represent only the tip of the iceberg.

Acknowledgments

The work of Olalla López was supported by a Maria Barbeito research contract from Xunta de Galicia. This research was supported by project grant 10MDS373027PR (Xunta de Galicia). General support to the laboratory was funded by Xunta de Galicia during 2008-11 (Consolidación D.O.G. 3-12-2008 EN:2008/008) and during 2012-13 (Consolidación D.O.G. 10-10-2012. EN: 2012/118); both co-financed by FEDER. The English presentation of the manuscript was improved by BioMedES (www.biomedes.co.uk).

References and Notes

1. Chien, A.; Edgar, D.B.; Trela, J.M. Deoxyribonucleic acid polymerase from the extreme thermophile *Thermus aquaticus*. *J. Bacteriol.* **1976**, *127*, 1550–1557.
2. Marsh, C.L.; Larsen D.H. Characterization of some thermophilic bacteria from the hot springs of Yellowstone National Park. *J. Bacteriol.* **1953**, *65*, 193–197.
3. Amann, R.I.; Ludwig, W.; Schleifer, K.H. Phylogenetic identification and *in situ* detection of individual microbial cells without cultivation. *Microbiol. Rev.* **1995**, *59*, 143–169.
4. Ghosh, D.; Bal, B.; Kashyap, V.K.; Pal, S. Molecular phylogenetic exploration of bacterial diversity in a Bakreshwar (India) hot spring and culture of *Shewanella*-related thermophiles. *Appl. Environ. Microbiol.* **2003**, *69*, 4332–4336.
5. Barns, S.M.; Fundyga, R.E.; Jeffries, M.W.; Pace, N.R. Remarkable archaeal diversity detected in a Yellowstone National Park hot spring environment. *Proc. Natl. Acad. Sci. USA* **1994**, *91*, 1609–1613.
6. Barns, S.M.; Delwiche, C.F.; Palmer, J.D.; Pace, N.R. Perspectives on archaeal diversity, thermophily and monophyly from environmental rRNA sequences. *Proc. Natl. Acad. Sci. USA* **1996**, *93*, 9188–9193.
7. Meyer-Dombard, D.R.; Shock, E.L.; Amend, J.P. Archaeal and bacterial communities in geochemically diverse hot springs of Yellowstone National Park, USA. *Geobiology* **2005**, *3*, 211–227.
8. Huber, H.; Hohn, M.J.; Rachel, R.; Fuchs, T.; Wimmer, V.C.; Stetter, K.O. A new phylum of *Archaea* represented by a nanosized hyperthermophilic symbiont. *Nature* **2002**, 417, 63–67.
9. Stetter, K.O. A brief history of the discovery of hyperthermophilic life. *Biochem. Soc. Trans.* **2013**, *41*, 416–420.
10. Hobel, C.F.V.; Marteinsson, V.T.; Hreggvidsson, G.O.; Kristjansson, J.K. Investigation of the microbial ecology of intertidal hot springs by using diversity analysis of 16S rRNA and chitinase genes. *Appl. Environ. Microbiol.* **2005**, *71*, 2771–2776.
11. Atanassov, I.; Dimitrova, D.; Stefanova, K.; Tomova, A.; Tomova, I.; Lyutskanova, D.; Stoilova-Disheva, M.; Radeva, G.; Danova, I.; Kambourova, M. Molecular characterization of the Archaeal diversity in Vlasa hot spring, Bulgaria, by using 16S rRNA and glycoside hydrolase family 4 genes. *Biotechnol. Biotecnol. Equip.* **2010**, *24*, 1979–1985.
12. Miller, S.R.; Strong, A.L; Jones, K.L; Ungerer, M.C. Bar-Coded pyrosequencing reveals shared bacterial community properties along the temperature gradients of two alkaline hot springs in Yellowstone National Park. *App. Environ. Microbiol.* **2009**, 4565–4572.
13. Valverde, A.; Tuffin, M.; Cowan, D.A. Biogeography of bacterial communities in hot springs: A focus on the actinobacteria. *Extremophiles* **2012**, *16*, 669–679.
14. Tekere, M.; Lötter, A.; Olivier, J.; Jonker, N.; Venter, S. Metagenomic analysis of bacterial diversity of Siloam hot water spring, Limpopo, South Africa. *Afr. J. Biotechnol.* **2011**, *10*, 18005–18012.

15. Mardanov, A.V.; Gumerov, V.M.; Beletsky, A.V.; Perevalova, A.A.; Karpov, G.A.; Bonch-Osmolovskaya, E.A.; Ravin, N.V. Uncultured archaea dominate in the thermal groundwater of Uzon Caldera, Kamchatka. *Extremophiles* **2011**, *15*, 365–372.
16. Jiménez, D.J.; Andreote, F.D.; Chaves, D.; Montaña, J.S.; Osorio-Forero, C.; Junca, H.; Zambrano, M.M.; Baena, S. Structural and functional insights from the metagenome of an acidic hot spring microbial planktonic community in the Colombian Andes. *PLoS One* **2012**, *7*, e52069.
17. Kanokratana, P.; Chanapan, S.; Pootanakit, K.; Eurwilaichitr, L. Diversity and abundance of Bacteria and Archaea in the Bor Khlueng Hot Spring in Thailand. *J. Basic Microbiol.* **2004**, *44*, 430–444.
18. Papke, R.T.; Ramsing, N.B.; Bateson, M.M.; Ward, D.M. Geographical isolation in hot spring cyanobacteria. *Environ. Microbiol.* **2003**, *5*, 650–659.
19. Shah, N.; Tang, H.; Doak, T.G.; Ye, Y. Comparing bacterial communities inferred from 16S rRNA gene sequencing and shotgun metagenomics. *Pac. Symp. Biocomput.* **2011**, 165–176.
20. Ward, D.M.; Cohan, F.M.; Bhaya, D.; Heidelberg, J.F.; Kühl, M.; Grossman, A. Genomics, environmental genomics and the issue of microbial species. *Heredity* **2008**, *100*, 207–219.
21. Tindall, B.J.; Rosselló-Móra, R.; Busse, H.-J.; Ludwig, W.; Kampfer, P. Notes on the characterization of prokaryote strains for taxonomic purposes. *Int. J. Syst. Evol. Micr.* **2010**, *60*, 249–266.
22. Bhaya, D.; Grossman, A.R.; Steunou, A.S.; Khuri, N.; Cohan, F.M.; Hamamura, N.; Melendrez, M.C.; Bateson, M.M.; Ward, D.M.; Heidelberg, J.F. Population level functional diversity in a microbial community revealed by comparative genomic and metagenomic analyses. *ISME J.* **2007**, *1*, 703–713.
23. Klatt, C.G.; Wood, J.M.; Rusch, D.B.; Bateson, M.M.; Hamamura, N.; Heidelberg, J.F.; Grossman, A.R.; Bhaya, D.; Cohan, F.M.; Kühl, M.; Bryant, D.A.; Ward, D.M. Community ecology of hot spring cyanobacterial mats: predominant populations and their functional potential. *ISME J.* **2011**, *5*, 1262–1278.
24. Bryant, D.A.; Costas, A.M.; Maresca, J.A.; Chew, A.G.; Klatt, C.G.; Bateson, M.M.; Tallon, L.J.; Hostetler, J.; Nelson, W.C.; Heidelberg, J.F.; Ward, D.M. *Candidatus Chloracidobacterium thermophilum*: An Aerobic Phototrophic Acidobacterium. *Science* **2007**, *317*, 523–526.
25. Liu, Z.; Klatt, C.G.; Ludwig, M.; Rusch, D.B.; Jensen, S.I.; Kühl, M.; Ward, D.M.; Bryant, D.A. Candidatus Thermochlorobacter aerophilum: An aerobic chlorophotoheterotrophic member of the phylum Chlorobi defined by metagenomics and metatranscriptomics. *ISME J.* **2012**, 6, 1869–1882.
26. Anantharaman, K.; Breier, J.A.; Sheik, C.S.; Dick, G.J. Evidence for hydrogen oxidation and metabolic plasticity in widespread deep-sea sulfur-oxidizing bacteria. *Proc. Natl. Acad. Sci. USA* **2013**, *110*, 330–335.
27. Youssef, N.H.; Blainey, P.C.; Quake, S.R.; Elshahed, M.S. Partial genome assembly for a candidate division OP11 single cell from an anoxic spring (Zodletone Spring, Oklahoma). *Appl. Environ. Microbiol.* **2011**, *77*, 7804–7814.

28. Liu, Z.; Klatt, C.G.; Wood, J.M.; Rusch, D.B.; Ludwig, M.; Wittekindt, N.; Tomsho, L.P.; Schuster, S.C.; Ward, D.M.; Bryant, D.A. Metatranscriptomic analyses of chlorophototrophs of a hot-spring microbial mat. *ISME J.* **2011**, *5*, 1279–1290.
29. Nelson, W.C.; Wollerman, L.; Bhaya, D.; Heidelberg, J.F. Analysis of insertion sequences in thermophilic cyanobacteria: exploring the mechanisms of establishing, maintaining, and withstanding high insertion sequence abundance. *Appl. Environ. Microbiol.* **2011**, *77*, 5458–5466.
30. Heidelberg, J.F.; Nelson, W.C.; Schoenfeld, T.; Bhaya, D. Germ warfare in a microbial mat community: CRISPRs provide insights into the co-evolution of host and viral genomes. *PLoS One* **2009**, *4*, e4169.
31. Xie, W.; Wang, F.; Guo, L.; Chen, Z.; Sievert, S.M.; Meng, J.; Huang, G.; Li, Y.; Yan, Q.; Wu, S.; Wang, X.; Chen, S.; He, G.; Xiao, X.; Xu, A. Comparative metagenomics of microbial communities inhabiting deep-sea hydrothermal vent chimneys with contrasting chemistries. *ISME J.* **2011**, *5*, 414–426.
32. Swingley, W.D.; Meyer-Dombard, D.R.; Shock, E.L.; Alsop, E.B.; Falenski, H.D.; Havig, J.R.; Raymond, J. Coordinating Environmental Genomics and Geochemistry Reveals Metabolic Transitions in a Hot Spring Ecosystem. *PLoS ONE* **2012**, *7*, e38108.
33. Schoenfeld, T.; Patterson, M.; Richardson, P.M.; Wommack, K.E.; Young, M.; Mead, D. Assembly of Viral Metagenomes from Yellowstone Hot Springs. *Appl. Environ. Microbiol.* **2008**, *74*, 4164–4174.
34. Servín-Garcidueñas, L.E.; Peng, X.; Garrett, R.A.; Martínez-Romero, E. Genome sequence of a novel archaeal rudivirus recovered from a mexican hot spring. *Genome Announc.* **2013**, *1*, e00040–12.
35. Garrett, R.A.; Prangishvili, D.; Shah, S.A.; Reuter, M.; Stetter, K.O.; Peng, X. Metagenomic analyses of novel viruses and plasmids from a cultured environmental sample of hyperthermophilic neutrophiles. *Environ. Microbiol.* **2010**, *12*, 2918–2930.
36. Snyder, J.C.; Bateson, M.M.; Lavin, M.; Young, M.J. Use of Cellular CRISPR (Clusters of Regularly Interspaced Short Palindromic Repeats) Spacer-Based Microarrays for Detection of Viruses in Environmental Samples. *Appl. Environ. Microbiol.* **2010**, *76*, 7251–7258.
37. Pride, D.T.; Schoenfeld, T. Genome signature analysis of thermal virus metagenomes reveals Archaea and thermophilic signatures. *BMC Genet.* **2008**, *9*, 420.
38. Bolduc, B.; Shaughnessy, D.P.; Wolf, Y.I.; Koonin, E.; Roberto, F.F.; Young, M. Identification of novel positive-strand RNA viruses by metagenomic analysis of archaea- 2 dominated Yellowstone Hot Springs. *J. Virol.* **2012**, *86*, 5562–5573.
39. Garcia Costas, A.M.; Liu, Z.; Tomsho, L.P.; Schuster, S.C.; Ward, D.M.; Bryant, D.A. Complete genome of *Candidatus Chloracidobacterium thermophilum*, a chlorophyll-based photoheterotroph belonging to the phylum *Acidobacteria*. *Environ. Microbiol.* **2012**, 14, 177–190.
40. Diemer, G.S.; Stedman, K.M. A novel virus genome discovered in an extreme environment suggests recombination between unrelated groups of RNA and DNA viruses. *Biol. Direct.* **2012**, *7*, 13.

41. He, Y.Z.; Fan, K.Q.; Jia, C.J.; Wang, Z.J.; Pan, W.B.; Huang, L.; Yang, K.Q.; Dong, Z.Y. Characterization of a hyperthermostable Fe-superoxide dismutase from hot spring. *Appl. Microbiol. Biotechnol.* **2007**, *75*, 367–376.
42. Wemheuer, B.; Taube, R.; Akyol, A.; Wemheuer, F.; Daniel, R. Microbial Diversity and Biochemical Potential Encoded by Thermal Spring Metagenomes Derived from the Kamchatka Peninsula. *Archaea* **2013**, *2013*, 136714.
43. Gabor, E.M.; Alkema, W.B.L.; Janssen, D.B. Quantifying the accessibility of the metagenome by random expression cloning techniques. *Environ. Microbiol.* **2004**, *6*, 879–886.
44. Tirawongsaroj, P.; Sriprang, R.; Harnpichamchai, P.; Thongaram, T.; Champreda, V.; Tanapongpipat, S.; Pootanakit, K.; Eurwilaichitr, L. Novel thermophilic and thermostable lipolytic enzymes from a Thailand hot spring metagenomic library. *J. Biotechnol.* **2008**, *133*, 42–49.
45. Sharma, P.K.; Singh, K.; Singh, R.; Capalash, N.; Ali, A.; Mohammad, O.; Kaur, J. Characterization of a thermostable lipase showing loss of secondary structure at ambient temperature. *Mol. Biol. Rep.* **2012**, *39*, 2795–2804.
46. Sharma, P.K.; Kumar, R.; Kumar, R.; Mohammad, O.; Singh, R.; Kaur, J. Engineering of a metagenome derived lipase toward thermal tolerance: Effect of asparagine to lysine mutation on the protein surface. *Gene* **2012**, *491*, 264–271.
47. Moser, M.J.; DiFrancesco, R.A.; Gowda, K.; Klingele, A.J.; Sugar, D.R.; Stocki, S.; Mead, D.A.; Schoenfeld, T.W. Thermostable DNA Polymerase from a Viral Metagenome Is a Potent RT-PCR Enzyme. *PLoS ONE* **2012**, *7*, e38371.
48. Kemp, P.F.; Aller, J.Y. Bacterial diversity in aquatic and other environments: What 16S rDNA libraries can tell us. *FEMS Microbiol. Ecol.* **2004**, *47*, 161–177.

Eukaryotic Organisms in Extreme Acidic Environments, the Río Tinto Case

Angeles Aguilera

Abstract: A major issue in microbial ecology is to identify the limits of life for growth and survival, and to understand the molecular mechanisms that define these limits. Thus, interest in the biodiversity and ecology of extreme environments has grown in recent years for several reasons. Some are basic and revolve around the idea that extreme environments are believed to reflect early Earth conditions. Others are related to the biotechnological potential of extremophiles. In this regard, the study of extremely acidic environments has become increasingly important since environmental acidity is often caused by microbial activity. Highly acidic environments are relatively scarce worldwide and are generally associated with volcanic activity or mining operations. For most acidic environments, low pH facilitates metal solubility, and therefore acidic waters tend to have high concentrations of heavy metals. However, highly acidic environments are usually inhabited by acidophilic and acidotolerant eukaryotic microorganisms such as algae, amoebas, ciliates, heliozoan and rotifers, not to mention filamentous fungi and yeasts. Here, we review the general trends concerning the diversity and ecophysiology of eukaryotic acidophilic microorganims, as well as summarize our latest results on this topic in one of the largest extreme acidic rivers, Río Tinto (SW, Spain).

Reprinted from *Life*. Cite as: Aguilera, A. Eukaryotic Organisms in Extreme Acidic Environments, the Río Tinto Case. *Life* **2013**, *3*, 363-374.

1. Introduction

Our ongoing exploration of Earth has led to continued discoveries of life in environments that have been previously considered uninhabitable. For example, we find thriving communities in the boiling hot springs of Yellowstone, the frozen deserts of Antarctica, the concentrated sulfuric acid in acid-mine drainages, and the ionizing radiation fields in nuclear reactors [1–4]. We find some microbes that grow only in brine and require saturated salts to live, and we find others that grow in the deepest parts of the oceans and require 500 to 1,000 bars of hydrostatic pressure. Life has evolved strategies that allow it to survive even beyond the daunting physical and chemical limits to which it has adapted to grow. To survive, organisms can assume forms that enable them to withstand freezing, complete desiccation, starvation, high levels of radiation exposure, and other physical or chemical challenges. We need to identify the limits for growth and survival and to understand the molecular mechanisms that define these limits. Biochemical studies will also reveal inherent features of biomolecules and biopolymers that define the physico-chemical limits of life under extreme conditions. Broadening our knowledge both of the range of environments on Earth that are inhabitable by microbes and of their adaptation to these habitats will be critical for understanding how life might have established itself and survived.

2. Eukaryotic Extremophiles

When we think of extremophiles, prokaryotes come to mind first. Thomas Brock's pioneering studies of extremophiles carried out in Yellowstone's hydrothermal environments, set the focus of life in extreme environments on prokaryotes and their metabolisms [3]. However, eukaryotic microbial life may be found actively growing in almost any extreme condition where there is a source of energy to sustain it, with the only exception of high temperature (>70 °C) and the deep subsurface biosphere [4]. The development of molecular technologies and their application to microbial ecology has increased our knowledge of eukaryotic diversity in many different environments [5]. This is particularly relevant in extreme environments, generally more difficult to replicate in the laboratory.

Recent studies based on molecular ecology have demonstrated that eukaryotic organisms are exceedingly adaptable and not notably less so than the prokaryotes, although most habitats have not been sufficiently well explored for sound generalizations to be made. In fact, molecular analysis has also revealed novel protist genetic diversity in different extreme environments [4]. Temperature is one of the main factors determining the distribution and abundance of species due to its effects on enzymatic activities [6]. All extremophiles that survive at high temperatures (95–115 °C) are microorganisms from the archaeal or bacterial domains. On the contrary, for eukaryotic microorganisms, the highest temperature reported is 62 °C, and most of the metazoans are unable to grow above 50 °C [7].

Surprisingly, photosynthetic prokaryotes, such as cyanobacteria, have never been found in hot acidic aquatic systems [8]. Instead, these ecological niches are usually profusely colonized by species of the order *Cyanidiales*, red unicellular algae [8]. Thus, species from the genera *Galdieria* and *Cyanidium* have been isolated from hot sulfur springs, showing an optimal growth temperature of 45 °C and a maximum growth temperature of 57 °C [9,10]. These extreme hot springs are usually acidic (pH 0.05–4) and frequently characterized by high concentrations of metals such as cadmium, nickel, iron or arsenic, which are highly toxic to almost all known organisms.

Additionally, phototrophic eukaryotic microorganisms have colonized environments characterized by temperatures at or below 0 °C. Some algal species bloom at the snow surface during spring, and complex microbial communities have been found on glaciers, probably the most widely studies environments after marine ice habitats. Aplanospores of *Chlamydomonas nivalis* are frequently found in high-altitude, persistent snowfields where they are photosynthetically active despite cold temperatures and high levels of ultraviolet radiation [11]. Distinct microbial communities composed of psychrophilic bacteria, microalgae and protozoa colonize and grow in melt pools on the ice surface, or in brine channels in the sub-ice platelet in the Arctic even during winter, at extremely low temperatures of −20 °C [12].

Additionally, a number of heterotrophic eukaryotes have also been reported to inhabit extremely acidic environments. Thus, the yeast *Rhodotorula* spp. is frequently encountered in acid mine drainage waters, and isolates belonging to other genera (e.g., *Candida, Cryptococcusor Purpureocillium* sp.) have also been described [13–15]. Among the filamentous fungi that have been isolated from acidic sites, are some of the most acidophilic of all microorganisms: *Acontium cylatium, Trichosporon cerebriae* and a *Cephalosporium* sp. have all been reported to grow at ca. pH 0 [16].

3. Acidic Environments. The Río Tinto (SW, Spain) Case

Highly acidic environments are relatively scarce worldwide and are generally associated with volcanic activity and mining operation [17]. The natural oxidation and dissolution of the sulfidic minerals exposed to oxygen and water results in acid production, and the process can be greatly enhanced by microbial metabolism [1,18]. At the same time, low pH facilitates metal solubility in water, particularly cationic metals (such as aluminum and many heavy metals), and therefore acidic water tends to have high concentrations of heavy metals [19] (Table 1).

Río Tinto (SW, Spain) is an unusual ecosystem due to its size (100 km long), rather constant acidic pH (mean value 2.3), high concentration of heavy metals (Fe, Cu, Zn, As, Mn, Cr, *etc.*) and high level of microbial diversity, mainly eukaryotic [20,21]. The river rises in Peña de Hierro, in the core of the Iberian Pyritic Belt, and reaches the Atlantic Ocean at Huelva (Figure 1). The Iberian Pyritic Belt is a geological entity of hydrothermal origin 250 km long and between 25 and 70 km wide, known to be one of the biggest deposits of metallic sulfides in the world [22,23]. One important characteristic of Río Tinto is the high concentration of ferric iron and sulfates found in its waters, products of the biooxidation of pyrite, the main mineral component of the system. Ferric iron is maintained in solution due to the acidic pH of the river and is responsible for the constant pH due to the buffer characteristics of this cation.

The combined use of conventional microbial ecology methods (enrichment cultures, isolation, phenotypic characterization) and molecular ecology, allowed most of the representative elements of the system to be identified. Eighty percent of the prokaryotic diversity in the water column corresponds to three bacterial genus: *Leptospirillum* spp., *Acidithiobacillus ferrooxidans* and *Acidiphilium* spp., all of them conspicuous members of the iron cycle [1].

Table 1. Physicochemical parameters at the most extreme sampling sites in Río Tinto (mean ± SD). Cond.—Conductivity (mS cm^{-1}); Redox.—redox potential (mV). Ions in mg L^{-1} except Fe in g L^{-1}.

Location	pH	Cond	Redox	Fe	Cu	As	Cd	Zn
Iz-Iz	1.8 ± 0.2	25.7 ± 2.3	569 ± 22	17 ± 4	12 ± 3	16 ± 4	43 ± 16	14 ± 3
ANG	1.5 ± 0.2	30.8 ± 3.4	471 ± 16	16 ± 3	132 ± 43	24 ± 3	30 ± 12	162 ± 5
UMA	1.6 ± 0.3	40.2 ± 8.3	473 ± 10	18 ± 7	85 ± 36	32 ± 5	40 ± 18	118 ± 4
RI	0.9 ± 0.3	38.9 ± 1.6	460 ± 30	22 ± 5	100 ± 36	48 ± 7	34 ± 11	94 ± 31
LPC	2.6 ± 0.3	3.70 ± 1.1	548 ± 70	0.2 ± 0.1	19 ± 7	0.2 ± 0.1	0.7 ± 0.1	50 ± 10

Figure 1. (**a**) General view of Río Tinto; (**b**) Photosynthetic biofilms formed by acidic *Klebsormidium* and *Zygnema*; (**c**) Photosynthetic biofilms formed by *Euglena mutabilis*.

4. Acidophilic Eukaryotic Diversity, an Ecological Paradox

Besides its extreme physico-chemical water characteristics, what makes Río Tinto a unique acidic environment is the unexpected degree of eukaryotic diversity found in its waters [20,21] and the fact that eukaryotic organisms are the principal contributors of biomass in the habitat (over 65% of the total biomass). Members of the phylum *Chlorophyta* such as *Chlamydomonas*, *Chlorella*, and *Euglena*, are the most frequent species followed by two filamentous algae belonging to the genera *Klebsormidium* and *Zygnemopsis* [24,25]. The most acidic part of the river, is inhabited by a eukaryotic community dominated by two species related to the genera *Dunaliella* and *Cyanidium* (*Rhodophyta*) well known for their high metal and acid tolerance [26]. Pennate diatoms are also present in the river forming large brown biofilms. These biofilms are usually clearly dominated by only one species related to the genus *Pinnularia*. Species belonging to these genera, especially *Pinnularia*, are fairly widespread at environments with pH values around 3.0 [27]. From all the environmental variables that affect freshwater diatoms, pH seems to be the most important and, most taxa show a preference for a narrow pH range [28]. The low diversity of diatoms present in Río Tinto in comparison with the diversity found in neighboring freshwaters, supports the idea that there is a threshold between pH 4.5 and 3.5 in which many species of diatoms are eliminated [27].

Molecular ecology techniques have identified algae closely related to those characterized phenotypically, emphasizing the high degree of eukaryotic diversity existing in the extreme conditions of Río Tinto [20,29]. Within the decomposers, fungi are very abundant and exhibit great diversity, including yeast and filamentous forms. A high percentage of the isolated hyphomycetes are able to grow in the extreme conditions of the river. Some of the isolated yeast species can also be found in less extreme aquatic environments, but the isolated dematiaceae seems to be specific to the extreme conditions of the habitat [30,31].

Additionally, fungal species such as *Hortaea werneckii* and *Acidomyces acidophilum* have been detected in Río Tinto by using molecular techniques [20]. The mixotrophic community is dominated by cercomonads and stramenopiles related to the genus *Bodo*, *Ochromonas*, *Labyrinthula* and *Cercomonas*. The protistan consumer community is characterized by two different species of ciliates tentatively assigned to the genera *Oxytrichia* and *Euplotes*. Amoebas related to the genus *Valhkampfia* and *Naegleria* can be found frequently even at the most acidic parts of the river (pH 1) and one species of heliozoan belonging to the genera *Actinophyris* seems to be the characteristic top predator of the benthic food chain in the river. We know from microscopic observations that rotifers also inhabit the river [21,24].

However, not only unicellular eukaryotic systems develop in the extreme conditions of the Tinto Basin. Different plants can be found growing in the acidic soils of the river banks (pH ca. 3, de la Fuente and Amils, personal communication). The strategies used by these plants to overcome the physiological problems associated to the extreme conditions of the habitat are diverse. Some are resistant to the heavy metals concentrated in the soils in which they grow while others specifically concentrate metals in different plant tissues. Recent analysis by XRD and Mossbauer spectroscopy of the iron minerals found in the rhizomes and leaves of *Imperata cylindrica*, an iron hyperaccumulator perennial grass growing in the Río Tinto banks, showed significant concentrations of jarosite and iron oxyhydroxides [32]. These results suggest that the management of heavy metals, in general, and iron, in particular, is much more complex and versatile in plants than has been reported to date [33]. Also, these results prove that multicellular complex systems can also develop in some extreme conditions, like those existing in Río Tinto.

As previously discussed, the prokaryotic diversity in Río Tinto water column is rather low, which corresponds to what should be expected from an extreme environment. In contrast, the unexpectedly high level of acidophilic eukaryotic diversity (Figure 2) poses an ecological paradox that is not well understood. It is obvious from these observations that adaptation to the extreme conditions of Río Tinto must be much easier than what we thought.

The extreme conditions of this ecosystem are rather recent (2 My) [34], so the adaptation of these complex organisms, which can be found in neutral aquatic environments nearby, to proton gradients between the inner (pH near neutrality) and outer part of the membranes (pH around 2) of five orders of magnitude and high concentrations of very toxic heavy metals (As, Cu, Zn, Cr, Al), must be relatively fast and efficient [29].

Figure 2. Light microscopy photographs of different eukaryotic species isolated from Río Tinto. (**a**) Filamentous green algae *Klebsormidium* sp.; (**b**) Diatoms; (**c**) Green algae *Chlamydomonas* spp.; (**d**) Heliozoa *Actinophrys* sp.

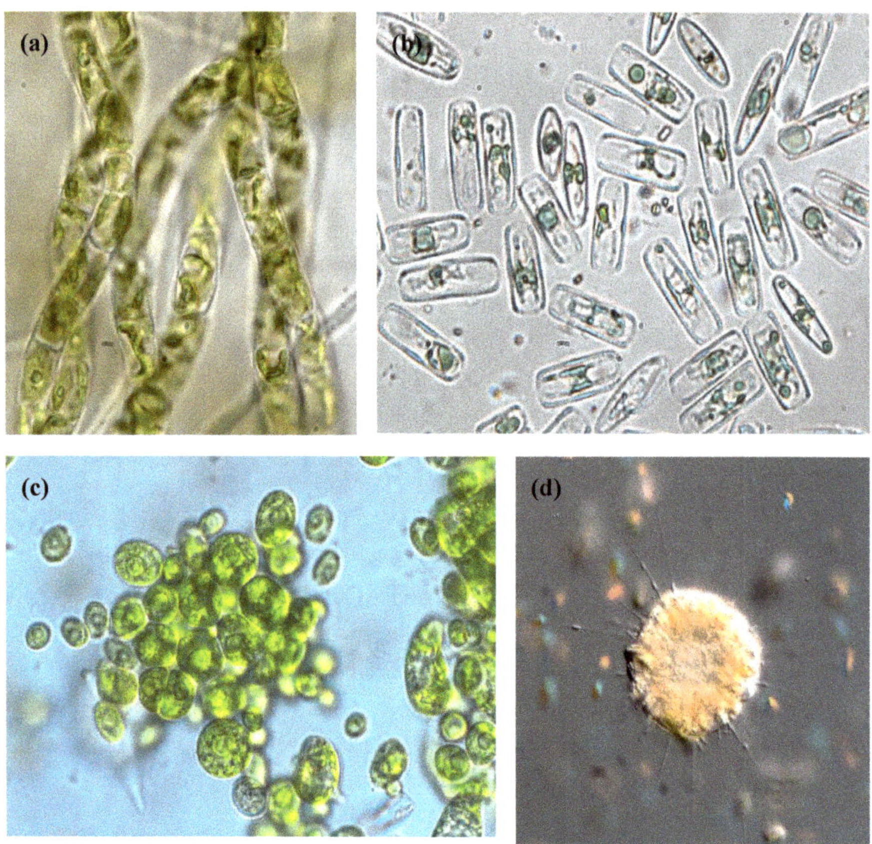

5. Photosynthesis in Acidic Environments

Photosynthesis is known to be particularly sensitive to stressful environmental conditions, such as salinity, pH or presence of toxicants. There are relatively few reports regarding photosynthesis in acidic environments in the literature, and most have focused on primary productivity measurements in acidic lakes. Thus, it has been reported that minimum primary productivity is mainly due to metal stress [35] or soluble reactive phosphate concentration [36]. However, low pH itself does not reduce photosynthetic activity [37]. Light is another limiting factor for primary productivity in acidic lakes. Adaptation to low light intensities has been reported for benthic biofilms of diatoms in acidic lakes [38].

In Río Tinto, the phototrophic biofilms analyzed exhibiting photoinhibition (*Euglena mutabilis*, *Pinnularia* sp. and *Chlorella* sp.) are usually located at the bottom of the river bed, covered by several centimetres of highly coloured red water. However, *Zygnemopsis* sp. showed a photosaturated behaviour probably because is a filamentous alga usually found at the water surface during the summer, when the sun irradiance is extremely high, and in this way can be considered a high light adapted species. In addition, the analyzed species can be considered as low light or shade adapted organisms due to their low Ic and Ik values (Table 2), which may be related to the fact that they develop under highly colored waters, that affect quantitatively and qualitatively the light available for phototrophic organisms [39]. Even at irradiances as low as 5 μmol m^{-2} s^{-1}, in the case of the diatom *Pinnularia* sp., photosynthetic activity was detected. These results are in agreement with previous data from sediments of acidic lakes, where photosynthetic ability at low light intensities (<1.2 μE m^{-2} s^{-1}) were found in a benthic biofilm of diatoms suggesting an efficient absorption of red light, the dominant wavelength available in these iron-rich acidic waters, by these organisms [38]. Maximum photosynthesis values (Pmax) were also low in comparison with other environments, in which rates higher than 200 μmol O_2 mg $Chla^{-1}$ h^{-1} are usually reached [40]. In acidic mining lakes, planktonic primary productivity is usually low probably due to the low phytoplankton biomass [41]. In our case, this cannot be the reason, since in Río Tinto, representing over 65% of the total biomass [30]. Another suggested reason for the low productivity in these extreme environments could be the lack of nutrients such as ammonium, phosphate or nitrate [36].

Table 2. Photosynthetic parameters of the different biofilms (*Chlorella*, *Euglena*, Diatom and *Zygnemopsis*) isolated from different locations at Río Tinto (AG, ANG, 3.1, NUR, SM and LPC). Compensation light intensity (Ic) and light saturation parameter (Ik) are expressed on photon basis (μmol photons m^{-2} s^{-1}). Photosynthetic efficiency (α) and photoinhibition factor (β) are expressed on Chl a basis (μmol O_2 mg $Chla^{-1}$ h^{-1}) [39].

Species	Ic	Ik	α	β
Chlo_AG	10.36 ± 3.26	59.65 ± 7.03	0.448 ± 0.13	0.0123 ± 0.01
Chlo_ANG	23.19 ± 3.24	120.34 ± 8.08	0.137 ± 0.02	0.0431 ± 0.01
Eug_3.1	18.93 ± 0.72	95.41 ± 9.23	0.448 ± 0.09	0.0450 ± 0.02
Eug_AG	18.44 ± 5.34	49.91 ± 5.89	0.278 ± 0.12	0.0441 ± 0.00
Eug_NUR	16.84 ± 0.76	96.23 ± 5.32	0.263 ± 0.02	0.0247 ± 0.02
Eug_SM	17.43 ± 4.59	48.53 ± 5.32	0.558 ± 0.05	0.0179 ± 0.00
Dia_NUR	5.06 ± 1.72	47.82 ± 6.47	1.423 ± 0.10	0.0426± 0.02
Zyg_LPC	38.89 ± 22.70	13.22 ± 3.23	0.249 ± 0.03	

All the photosynthetic parameters analyzed showed statistical significant differences among species and sampling locations. Thus, the *Euglena* biofilms isolated from different habitats of the Río Tinto (3.1, AG, NUR, and SM) showed different photosynthetic values despite they are mainly formed by the same phototrophic species. These three sampling locations showed different water environmental physochlemical characteristic [39]. These results could be explained by photoadaptation processes instead of photoacclimation procedures. Photoadaptation refers to changes in the genotype that arise either from mutations or from changes in the distribution of alleles within a gene pool, while photoacclimation refers to phenotypic adjustments that arise in response to variations of environmental factors.

Using proteomic analysis of global expression patterns of cellular soluble proteins in an acidophilic strain of *Chlamydomonas* sp. we found that several stress-related proteins are induced in the cells growing in natural river water, along with a complex battery of proteins involved in photosynthesis, primary and energy metabolism or motility [42]. When the 2-DE gels were compared, some of the most dramatic changes observed were related to proteins involved in the Calvin cycle and photosynthetic metabolism. In fact, three of the nine identified downregulated proteins found in cells grown in the presence of metals, were described from these metabolic pathways (Figure 3). The amount of the ribulose-1,5-bisphosphate carboxylase/oxygenase (RuBisCo) decreases significantly when cells grow in metal rich water. This decrease correlates with other proteins described from photosynthesis, such us cytrochrome C peroxidase, oxygen-evolving enhancer protein or photosystem I 11K protein precursor.

Figure 3. 2-DE preparative gels. The spots resolved by 2-DE from preparative gels were stained with (**A**) Cy3 cells growing under BG11/f2 artificial media at pH2. (**B**) Cy5 cells growing under natural metal-rich water NW/f2 at pH2.

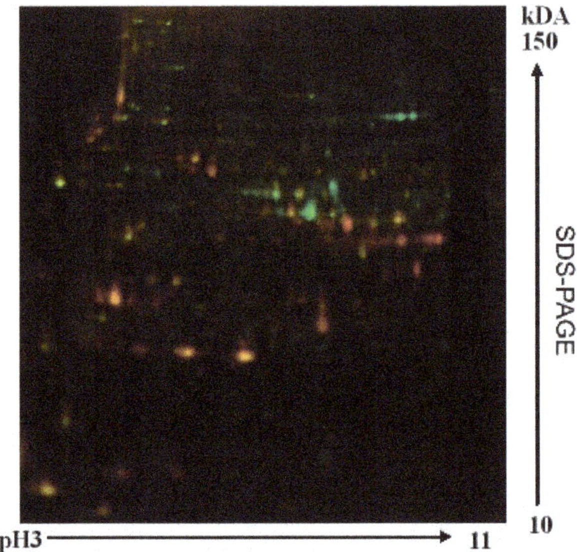

These results are closely related to the presence of high levels of heavy metals present in the natural acidic waters since, inhibition of photosynthetic activity is one of the most important cellular responses to metal stress conditions [43,44]. Similar results were found in *C. reinhardtii* in the presence of cadmium and copper [45,46], as well as in other photosynthetic organisms [47,48]. Although growth in extreme acidic environments is expected to require specific cellular adaptations of photosynthetic organisms, other studies have reported stress symptoms in acidophilic *Chlamydomonas* growing under acidic or metal-enriched natural water [36,49]. On the contrary, phytochrome B, phosphoribulokinase and phosphoglycerate kinase were up-regulated when cells were grown in metal rich acidic water. Phytochromes are a family of light-sensing proteins required for plant developmental responses to light [50]. Plants perceive the intensity, direction, and quality of light and use this information to optimize photosynthesis. Phytochrome is the best characterized of the photoreceptors involved in these light dependent responses. In our case, the induction of this protein in the cells under study could be due to the intense red color of Río Tinto water, caused by the high concentration of soluble ferric iron at the low pH of the river. This color has a marked effect on the quality and intensity of the light that reaches the cells. Experiments carried out in acidic mining lakes showed that only red light reaches the sediments of iron-rich water [38]. The increased levels of phytochrome could be an adaptation process to these environmental conditions. The remaining induced enzymes, phosphoglycerate kinase and phosphoribulokinase are similar results were found for *C. reinhardtii* under cadmium exposure suggesting a limitation of the photosynthetic electron transfer that might force the cell to reorganize its whole metabolism [46].

6. Conclusions

Extremophiles are not only important resources for developing novel biotechnological processes, but also ideal models for research in the ecological and molecular fields. An understanding of the versatility of life on Earth, as well as the mechanisms that allow some organisms to survive or develop in extreme conditions, will help to gain more insights into the evolution of life, the development of special ecosystems and in the search for life beyond Earth. Although most habitats have not been sufficiently well explored, colonization of extreme habitats is not usually limited to a unique taxonomic domain, and eukaryotic species are exceptionally adaptable comparable in this aspect to the prokaryotes, at least in this acidic environment. Among different extreme environments, acidic habitats are rather peculiar because in most cases they are the product of metabolism of active chemolithotrophic microorganisms. The microbial diversity characterization of Río Tinto allowed for detection of a high level of eukaryotic diversity, which contrasts with the rather low level of prokaryotic diversity found in the system. The highest concentration of biomass of the ecosystem corresponds to photosynthetic algae, although other protists, yeast and filamentous fungi have been isolated along the river. Most the photosynthetic eukaryotes in the Tinto ecosystem are forming biofilms. In this case, the distribution of these photosynthetic communities seems to be more influenced by the presence of heavy metals, than low pH. The high level of eukaryotic diversity found in the Tinto basin demonstrates that complex eukaryotic systems can thrive and dominate extremely acidic, heavy metal-laden environments. Since some of these acidophiles are closely related to cultured neutrophiles, we can conclude that eukaryotes must have the ability to adapt from neutral to acidic environments over relatively short periods of time. Thus, eukaryotic extremophiles are more widely distributed and phylogenetically diverse than previously thought.

Acknowledgments

This work has been supported by the Spanish MINECO CGL2011-02254/BOS Grant.

Conflicts of Interest

The author declares no conflict of interest.

References

1. González-Toril, E.; Llobet-Brossa, E.; Casamayor, E.O.; Amann, R.; Amils, R. Microbial ecology of an extreme acidic environment, the Tinto River. *Appl. Environ. Microbiol.* **2003**, *69*, 4853–4865.
2. Pikuta, E.V.; Hoover, R.B.; Tang, J. Microbial extremophiles at the limits of life. *Crit. Rev. Microbiol.* **2007**, *33*, 183–209.
3. Brock, T.D. *Thermophilic Microorganisms and Life at High Temperatures*; Springer-Verlag: Berlin, Germany, 1978; p. 245.
4. Roberts, D.M.L. Eukaryotic Cells under Extreme Conditions. In *Enigmatic Microorganisms and Life in Extreme Environments*; Seckbach, J., Ed.; Kluwer Academic Publication: London, UK, 1999; pp. 165–173.
5. Caron, D.A.; Countway, P.D.; Brown, M.V. The growing contributions of molecular biology and immunology to protistan ecology: Molecular signatures as ecological tools. *J. Euk. Microbiol.* **2004**, *51*, 38–48.
6. Alexandrof, V.Y. Conformational Flexibility of Macromolecules and Ecological Adaptations. In *Cells, Molecules and Temperature*; Springer-Verlag: Berlin, Germany, 1977; p. 342.
7. Rothschild, L.J.; Mancinelli, R.L. Life in extreme environments. *Nature* **2001**, *409*, 1092–1101.
8. Brock, T. Lower pH limit for the existence of blue-green algae: Evolutionary and ecological implications. *Science* **1973**, *179*, 480–483.
9. Seckbach, J. Evolutionary Pathways and Enigmatic Algae: *Cyanidium caldarium* (*Rhodophyta*) and Related Cells. In *Developments in Hydrobiology*; Kluwer Academic Publication: Dordrecht, Germany, 1994; p. 349.
10. Ciniglia, C.; Yoon, H.S.; Pollio, A.; Pinto, G.; Bhattacharya, D. Hidden biodiversity of the extremophilic Cyanidiales red algae. *Mol. Ecol.* **2004**, *13*, 1827–1838.
11. Stibal, M.; Elster, J.; Šabacká, M.; Kaštovská, K. Seasonal and diel changes in photosynthetic activity of the snow alga *Chlamydomonas nivalis* (*Chlorophyceae*) from Svalbard determined by pulse amplitude modulation fluorometry. *FEMS Microbiol. Ecol.* **2006**, *59*, 265–273.
12. Garrison, D.L.; Close, A.R. Winter ecology of the sea ice biota in Weddel Sea pack ice. *Mar. Ecol. Prog. Ser.* **1993**, *96*, 17–31.
13. Gross, S.; Robbins, E.I. Acidophilic and acid-tolerant fungi and yeasts. *Hydrobiologia* **2000**, *433*, 91–109.
14. Russo, G.; Libkind, D.; Sampaio, J.P.; VanBrock, M.R. Yeast diversity in the acidic Río Agrio-Lake Caviahue volcanic environment (Patagonia, Argentina). *FEMS Microbiol. Ecol.* **2008**, *65*, 415–424.
15. Oggerin, M.; Tornos, F.; Rodríguez, N.; del Moral, C.; Sánchez-Román, M.; Amils, R. Specific jarosite biomineralization by *Purpureocillium lilacinum*, an acidophilic fungi isolated from Río Tinto. *Environ. Microbiol.* **2013**, doi:10.1111/1462–2920.12094.
16. Schleper, C.; Puehler, G.; Kuhlmorgen, B.; Zillig, W. Life at extremely low pH. *Nature* **1995**, *375*, 741–742.

17. Baffico, G.D.; Díaz, M.M.; Wenzel, M.T.; Koschorreck, M.; Schimmele, M.; Neu, T.R.; Pedrozo, F. Community structure and photosynthetic activity of epilithon from a highly acidic (pH < 2) mountain stream in Patagania, Argentina. *Extremophiles* **2004**, *8*, 465–475.
18. Nordstrom, D.K.; Southam, G. Geomicrobiology of Sulphide Mineral Oxidation. In *Geomicrobiology: Interactions Between Microbes and Minerals*; Banfield, J.F., Nealson, K.H., Eds.; Mineralogical Society of America: Washington, DC, USA, 1997; Volume 35, pp. 361–390.
19. Johnson, D.B. Biodiversity and ecology of acidophilic microorganisms. *FEMS Microbiol. Ecol.* **1998**, *27*, 307–317.
20. Amaral, L.A.; Gómez, F.; Zettler, E.; Keenan, B.G.; Amils, R.; Sogin, M.L. Eukaryotic diversity in Spain's river of fire. *Nature* **2002**, *417*, 137.
21. Aguilera, A.; Manrubia, S.C.; Gómez, F.; Rodríguez, N.; Amils, R. Eukaryotic community distribution and their relationship to water physicochemical parameters in an extreme acidic environment, Río Tinto (SW, Spain). *Appl. Environ. Microbiol.* **2006**, *72*, 5325–5330.
22. Boulter, C.A. Did both extensional tectonics and magmas act as major drivers of convection cells during the formation of the Iberian Pyritic Belt massive sulfide deposits? *J. Geol. Soc. Lond.* **1996**, *153*, 181–184.
23. Leistel, J.M.; Marcoux, E.; Thieblemont, D.; Quesada, C.; Sanchez, A.; Almodovar, G.R.; Pascual, E.; Saez, R. The volcanic-hosted massive sulphidic deposits of the Iberian Pyritic Belt. *Miner. Deposita* **1998**, *33*, 2–30.
24. Aguilera, A.; Souza-Egipsy, V.; Gómez, F.; Amils, R. Development and structure of eukaryotic biofilms in an extreme acidic environment, Río Tinto (SW, Spain). *Microb. Ecol.* **2006**, *53*, 294–305.
25. Aguilera, A.; Gómez, F.; Lospitao, E.; Amils, R. A molecular approach to the characterization of the eukaryotic communities of an extreme acidic environment: Methods for DNA extraction and denaturing gradient electrophoresis analysis. *Syst. Appl. Microbiol.* **2006**, *29*, 593–605.
26. Visviki, I.; Santikul, D. The pH tolerance of *Chlamydomonas applanata* (Volvocales, Chlorophyta). *Arch. Environ. Cont. Toxicol.* **2000**, *38*, 147–151.
27. DeNicola, D.M. A review of diatoms found in highly acidic environments. *Hydrobiologia* **2000**, *433*, 111–122.
28. Battarbee, R.W.; Smol, J.P.; Meriläinen, J. Diatoms as Indicators of pH: An Historical Review. In *Diatoms and Lake Acidity*; Smol, J.P., Battarbee, R.W., Davis, R.B., Meriläinen, J., Eds.; Dr. W. Junk Publication: Dordrecht, Germany, 1986; pp. 5–14.
29. Aguilera, A.; Amils, R. Tolerance to cadmium in *Chlamydomonas* sp. (Chlorophyta) strains isolated from an extreme acidic environment, the Tinto River (SW, Spain). *Aquat. Toxicol.* **2005**, *75*, 316–329.
30. López-Archilla, A.I.; Marín, I.; Amils, R. Microbial community composition and ecology of an acidic aquatic environment: The Tinto river, Spain. *Microb. Ecol.* **2001**, *41*, 20–35.
31. López-Archilla, A.I.; González, A.E.; Terrón, M.C.; Amils, R. Diversity and ecological relationships of the fungal populations of an acidic river of Southwestern Spain: The Tinto River. *Can. J. Microbiol.* **2005**, *50*, 923–934.

32. Rodrıguez, N.; Menendez, N.; Tornero, J.; Amils, R.; de la Fuente, V. Internal iron biomineralization in *Imperata cilindrica*, aperennial grass: Chemical composition, speciation and plant localization. *New Phytol.* **2005**, *165*, 781–789.
33. Schmidt, W. Iron solutions: Acquisition strategies and signalling pathways in plants. *Trends Plant Sci.* **2003**, *8*, 188–193.
34. Fernandez-Remolar, D.C.; Morris, R.V.; Gruener, J.E.; Amils, R.; Knoll, A.H. The Río Tinto Basin, Spain: Mineralogy, sedimentary geobiology and implications for interpretation of outcrop rocks at Meridiani Planum, Mars. *Earth Plannet Sci. Lett.* **2005**, *240*, 149–167.
35. Niyogi, D.K.; Lewis, W.M.; McKnight, D.M. Effects of stress from mine drainage on diversity, biomass, and function of primary producers in mountain streams. *Ecosystems* **2002**, *5*, 554–567.
36. Spijkerman, E.; Barua, D.; Gerloff-Elias, A.; Kern, J.; Gaedke, U.; Heckathorn, S.A. Stress responses and metal tolerance of *Chlamydomonas acidophila* in metal-enriched lake water and artificial medium. *Extremophiles* **2007**, *11*, 551–562.
37. Guyre, R.A.; Konopka, A.; Brooks, A.; Doemel, W. Algal and bacterial activities in acidic (ph3) strip mine lakes. *Appl. Environ. Microbiol.* **1987**, *53*, 2069–2076.
38. Koschorrek, M.; Tittel, J. Benthic photosynthesis in acidic mining lake (pH 2.6). *Limnol. Oceanogr.* **2002**, *47*, 1197–1201.
39. Souza-Egipsy, V.; Altamirano, M.; Amils, R.; Aguilera, A. Photosynthetic performance of phototrophic biofilms in extreme acidic environments. *Environ. Microbiol.* **2011**, *13*, 2351–2358.
40. Ritchie, R.J. Fitting light saturation curves measured using modulated fluorometry. *Photosynth. Res.* **2008**, *96*, 201–215.
41. Nixdorf, B.; Krumbeeck, H.; Jander, J.; Beulker, C. Comparison of bacterial an phytoplankton productivity in extremely acidic mining lakes and eutrophic hard water lakes. *Acta Oecol.* **2003**, *24*, S281–S288.
42. Cid, C.; Garcia-Descalzo, L.; Casado-Lafuente, V.; Amils, R.; Aguilera, A. Proteomic analysis of the response of an acidophilic strain of *Chlamydomonas* sp. (Chlorophyta) to natural metal-rich water. *Proteomics* **2010**, *10*, 2026–2036.
43. Hanikenne, M. *C. reinhardtii* as a eukaryotic photosynthetic model for studies of heavy metal homeostasis and tolerance. *New Phytol.* **2003**, *159*, 331–340.
44. Pinto, E.; Sigaud-Kutner, T.; Leitão, M.; Okamoto, O. Heavy-metal induced oxidative stress in algae. *J. Phycol.* **2003**, *39*, 1008–1018.
45. Boswell, C.; Sharma, N.C.; Sahi, S.V. Copper tolerance and accumulation potential of *Chlamydomonas reinhardtii. Bull. Environ. Cont. Toxicol.* **2002**, *69*, 546–553.
46. Gillet, S.; Decottignies, P.; Chardonnet, S.; Maréchal, P. Cadmium response and redoxin targets in *Chlamydomonas reinhardtii*: A proteomic approach. *Photosynth. Res.* **2006**, *89*, 201–211.
47. Takamura, N.; Kasai, F.; Watanabe, M.M. Effects of Cu, Cd and Zn on photosynthesis of fresh water benthic algae. *J. Appl. Phycol.* **1989**, *1*, 39–52.

48. Wang, S.; Chen, F.; Sommerfeld, M.; Hu, Q. Proteomic analysis of molecular response to oxidative stress by the green alga *Haematococcus pluvialis* (Chlorophyceae). *Planta* **2004**, *220*, 17–29.
49. Langner, U.; Jakob, T.; Stehfest, K.; Wilhelm, C. An energy balance from absorbed protons to new biomass for *C. reinhardtii* and *C. acidophila* under neutral and extremely acidic growth conditions. *Plant Cell Environ.* **2009**, *32*, 250–258.
50. Furuya, M. Phytochromes: Their molecular species, gene families, and functions. *Annu. Rev. Plant Physiol. Plant Mol. Biol.* **1993**, *44*, 617–641.

Evolution of Microbial "Streamer" Growths in an Acidic, Metal-Contaminated Stream Draining an Abandoned Underground Copper Mine

Catherine M. Kay, Owen F. Rowe, Laura Rocchetti, Kris Coupland, Kevin B. Hallberg and D. Barrie Johnson

Abstract: A nine year study was carried out on the evolution of macroscopic "acid streamer" growths in acidic, metal-rich mine water from the point of construction of a new channel to drain an abandoned underground copper mine. The new channel became rapidly colonized by acidophilic bacteria: two species of autotrophic iron-oxidizers (*Acidithiobacillus ferrivorans* and "*Ferrovum myxofaciens*") and a heterotrophic iron-oxidizer (a novel genus/species with the proposed name "*Acidithrix ferrooxidans*"). The same bacteria dominated the acid streamer communities for the entire nine year period, with the autotrophic species accounting for ~80% of the micro-organisms in the streamer growths (as determined by terminal restriction enzyme fragment length polymorphism (T-RFLP) analysis). Biodiversity of the acid streamers became somewhat greater in time, and included species of heterotrophic acidophiles that reduce ferric iron (*Acidiphilium, Acidobacterium, Acidocella* and gammaproteobacterium WJ2) and other autotrophic iron-oxidizers (*Acidithiobacillus ferrooxidans* and *Leptospirillum ferrooxidans*). The diversity of archaea in the acid streamers was far more limited; relatively few clones were obtained, all of which were very distantly related to known species of euryarchaeotes. Some differences were apparent between the acid streamer community and planktonic-phase bacteria. This study has provided unique insights into the evolution of an extremophilic microbial community, and identified several novel species of acidophilic prokaryotes.

Reprinted from *Life*. Cite as: Kay, C.M.; Rowe, O.F.; Rocchetti, L.; Coupland, K.; Hallberg, K.B.; Johnson, D.B. Evolution of Microbial "Streamer" Growths in an Acidic, Metal-Contaminated Stream Draining an Abandoned Underground Copper Mine. *Life* **2013**, *3*, 189-210.

1. Introduction

Acidophilic microorganisms are defined as those that grow optimally, or exclusively, in low pH environments, and have been sub-divided into moderate (pH optima 3–5) and extreme (pH optima < 3) acidophiles. The latter comprise a wide range of physiologically- and phylogenetically-diverse prokaryotes and a more limited number of eukaryotic microorganisms (chiefly micro-algae and protozoa [1]). Occasionally, individual cells of acidophilic microorganisms aggregate and form gelatinous macroscopic growths that have been referred to as "acid streamers" (filamentous growths in flowing waters), "acid mats" (thick and often dense growths), and "pipes" / "snotites" (pendulous growths attached to mine roofs or other subterranean features [1]). These have been found in subterranean and surface environments, frequently in and around abandoned mine sites [2]. The microbial composition of such growths is highly variable, and is determined to a great extent by the physico-chemical nature of the waters in which they develop. In the extremely acidic (pH 0–2) and warm (35–45 °C) streams and pools within the abandoned Richmond mine at Iron Mountain, California, iron-oxidizing bacteria (*Leptospirillum* spp.) and archaea ("*Ferroplasma acidarmanus*") are often the dominant members of the acid streamer communities [3], whereas in higher pH (2–3) mine waters (e.g. at the Drei Konen und Ehrt pyrite mine in Germany, the Cae Coch pyrite mine in Wales) other iron-oxidizing bacteria (*Acidithiobacillus* spp. and "*Ferrovum myxofaciens*") tend to dominate in sites rich in pyrite and other sulfide minerals [4]. The Frasassi cave complex in Italy provides an interesting contrast to pyrite-rich mines in that the main energy source for the acidophiles that form snotites is reduced sulfur rather than ferrous iron, and consequently the dominant prokaryotes are sulfur-oxidizers (*Acidithiobacillus thiooxidans*, with smaller numbers of *Sulfobacillus* and a bacterium related to *Acidimicrobium* [5]). Most data on the microbial compositions of acid streamers and snotites are from a single or relatively limited number of analyses of such growths whose "age" (in terms of how long they have been established) is unknown. Consequently, little is known about how these communities establish and evolve with time.

Mynydd Parys (Parys Mountain) is an abandoned copper mine located in the north-west corner of the island of Anglesey, north Wales [6]. In the 18th century it was the world's largest copper-producing mine. Although extractive mining ended in around 1880, copper was still recovered from the metal-rich waters draining the site until the 1950s. The two mines at Mynydd Parys (the larger Parys and smaller Mona mines) were operated initially as deep underground mines though towards the end of their working lives they were converted to opencast operations. In the 20th century, controlled flooding and drainage of the underground adits was used to recover copper using "*in situ*" bioleaching, though this was before the role of bacteria in this process had been recognized. When this practice was abandoned, the valves controlling the release of the mine water were closed, and consequently groundwater accumulated within the underground shafts and adits, forming a large pond in the opencast void of the Parys mine. Between ~1955 and 2003, the water level within the underground mine was controlled by overflow through an open adit (the Mona adit), and mine water flowed from this outlet eastwards forming a narrow stream (the Afon Goch [7]). Concern of the risk that this relatively high-elevation impounded acidic water body posed to the nearby low-level coastal town of Amlwch led to the partial dewatering of the mine in 2003. Between April and July, *ca.* 274,000 m^3 of water was pumped from the mine into the Irish Sea, after which the retaining dam was destroyed, and the overflow water from the underground water body was constrained to flow northwards through a new channel which connected to an existing stream which had previously not received acid mine drainage (AMD) [8]. The diversion of mine water, draining Mynydd Parys into the new stream channel (the "Dyffryn Adda"), presented a unique opportunity to investigate how rapidly acidophilic microbial communities develop in a new drainage channel, and how the composition of these growths evolve with time. Here we report the results of a nine year study of the microbiology and geochemistry of the Dyffryn Adda from its inception as an AMD stream.

2. Results

2.1. Physico-Chemistry of Dyffryn Adda AMD

Prior to overflow AMD from Mynydd Parys being diverted into the Dyffryn Adda, water flowing through the channel was mostly run-off and groundwater from the neighboring fields (used for agricultural grazing) was moderately acidic (pH ~5) and not contaminated with metals (data not shown). Analysis of mine water samples taken from the upper end of the Dyffryn Adda since October 2003 have shown only relatively minor changes in many of the parameters measured over the nine year period (Table 1 and Figures 1–4). The subterranean water body is sufficiently remote from the land surface not to be affected by seasonal temperature fluctuations, and water draining from it into the Dyffryn Adda has a near constant temperature of *ca.* 11 °C. Figures 1–3 show changes in the various physico-chemical parameters measured between October 2003 and November 2007 when the stream was monitored on a monthly basis. Concentrations of copper showed some variation between months one and 12, and mine water conductivity increased sharply at the start of 2005 compared to pre-2005 values. Concentrations of DOC also varied from time to time though values were not seasonally-related and did not appear to correlate with any other measured parameter. Comparison of data obtained during the first four years and those obtained later (grouped as three time periods: 2003–2007, 2008–2009 samples, and 2011–2012 samples) show that the AMD discharged in the Dyffryn Adda has continued to remain fairly constant in its physico-chemical composition during this time (Figure 4). Measurements of the parameters listed in Table 1 in the Dyffryn Adda from the adit portal at regular intervals to the point at which it joined with a second stream, showed that any changes were either not detectable or minor (data not shown). This was probably due to the relatively rapid flow rate of the stream, which was recorded by personnel of the Environment Agency (UK) to have a mean value of 10 L s^{-1}.

2.2. Growth of Acid Streamers within the Dyffryn Adda

Macroscopic gelatinous microbial growths were found in the upper reaches of the Dyffryn Adda within three months of it receiving AMD from the abandoned copper mines. These rapidly colonized the entire length of the stream between the portal and its conjuncture with a second stream, forming long (up to 0.5 m) off-white filamentous growths that attached to rocks, branches and other structures in the stream. Occasionally, surfaces of some streamer growths were green-colored due to the presence of *Euglena mutabilis* [9] which had also previously been reported in the Afon Goch [7]. Figure 5 shows images of the acid streamers within the Dyffryn Adda taken during April 2005. The section of the stream from the adit portal to its juncture with a second stream has remained similarly fully colonized by acid streamer growths since that time. The conjoining stream has a circum-neutral pH, and the higher pH of the mixed waters causes ferric iron to hydrolyze and precipitate, enshrouding streamer growths downstream of the junction of the two streams.

Table 1. Mean physico-chemical data of mine water flowing from the Dyffryn Adda adit portal between October 2003 and November 2007. All concentrations shown are mg L^{-1}, except where indicated (n = 46).

Analyte	Mean value (standard error)
pH	2.53 (0.04)
Redox potential E_h (mV)	669 (2.6)
Conductivity (μS cm^{-1})	2506 (66)
Temperature (°C)	11.3 (0.1)
Oxygen (%)	12.7 (0.4)
Sulfate-S	800 (13)
Fe^{2+}	378 (14)
Fe_{total}	563 (11)
Zn	67 (1.6)
Cu	49 (2.2)
Mn	15.8 (0.7)
Al	2.15 (0.16)
DOC*	4.50 (0.29)

*dissolved organic carbon.

Figure 1. Changes in pH, temperature, dissolved oxygen and dissolved organic carbon (DOC) in acid mine drainage (AMD) at the Dyffryn Adda adit portal between October 2003 and November 2007.

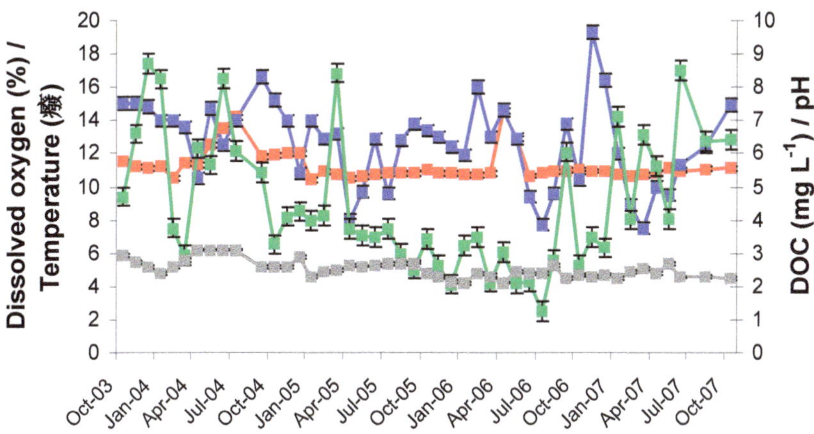

Key: ■, pH; ■, dissolved organic carbon (DOC); ■, temperature; ■, dissolved oxygen. (n = 3).

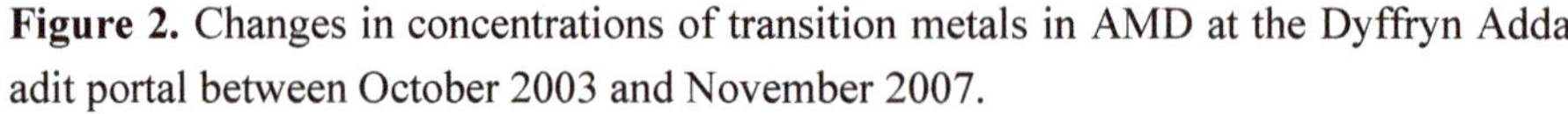

Figure 2. Changes in concentrations of transition metals in AMD at the Dyffryn Adda adit portal between October 2003 and November 2007.

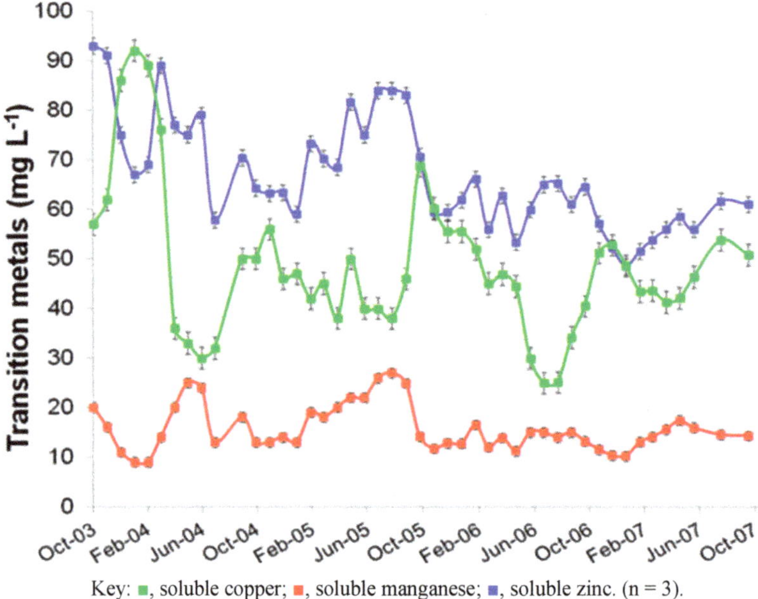

Key: ■, soluble copper; ■, soluble manganese; ■, soluble zinc. (n = 3).

Figure 3. Changes in iron, sulfate-S, redox potential and conductivity in AMD at the Dyffryn Adda adit portal between October 2003 and November 2007.

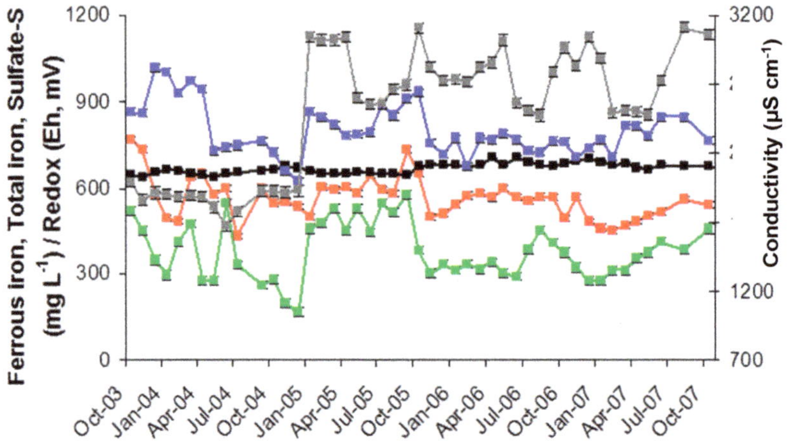

Key: ■, soluble ferrous iron; ■, total soluble iron; ■, redox potential; ■, sulfate-S; ■, conductivity. (n = 3)

Figure 4. Concentrations of metals and other physico-chemical parameters (temperature (T), pH, conductivity (E_c), redox potentials (E_h) and dissolved oxygen concentrations (DO)) in AMD at the Dyffryn Adda adit portal between October 2003 and November 2007 (P1), November 2008 and August 2009 (P2) and February 2011 and August 2012 (P3).

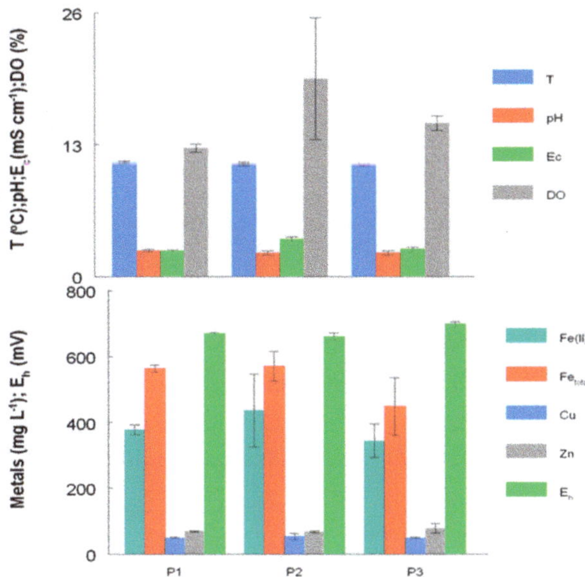

Figure 5. Acid streamer growths within the Dyffryn Adda, two years after acid mine drainage water from Mynydd Parys was diverted into this stream: (left) the screened portal at the upper end of the drainage channel; (right top) close up image of acid streamer growths close to the portal; (right middle) 15 m downstream of the portal, with *Euglena*-colonised surface streamers close to the stream bank; (right bottom) 90 m downstream of the portal, just before the Dyffryn Adda merges with a second stream.

2.3. Molecular Analysis of Acid Streamer Growths: Bacteria

Newly formed acid streamers sampled during 2003 showed that they were composed mainly of relatively few species of acidophilic bacteria (Figure 6). Streamers were sampled only at the Dyffryn Adda adit portal during October and November, while in December they were also sampled 90 m downstream of this point. terminal restriction enzyme fragment length polymorphism (T-RFLP) analysis indicated that the composition of the streamers was similar in all cases, with two species of autotrophic iron-oxidizing proteobacteria ("*Fv. myxofaciens*" and *Acidithiobacillus ferrivorans*) and a heterotrophic iron-oxidizing isolate (actinobacterium Py-F3) accounting for most of the biomass present.

Figure 6. Terminal restriction enzyme fragment length polymorphism (T-RFLP) profiles of bacterial 16S rRNA genes (digested with AluI) of acid streamers sampled at the Dyffryn Adda adit portal, and 90 m downstream of the portal (Dec (b)). Sampling period: October–December 2003. The 170 nt fragment was attributed to "*Fv. myxofaciens*", the 234 nt fragment to *At. ferrivorans* and the 235 and 478 (a pseudo T-RF) nt fragments to the actinobacterium isolate Py-F3. The 276 nt fragment was also thought to be a pseudo-T-RF, and the 417nt fragment was not identified.

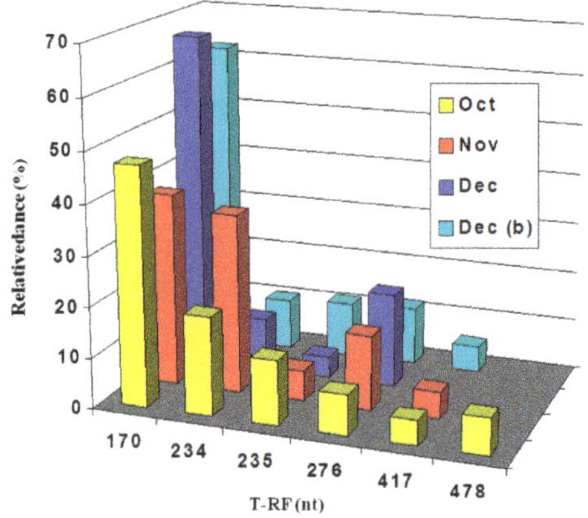

On the next three sampling occasions (April 2005, October 2007 and November 2008) acid streamers were sampled at the upper portal and at 15 m intervals downstream (to 90 m) to examine the heterogeneity of the microbial communities on a spatial as well as on a temporal basis. In general, acid streamer samples from different points within the Dyffryn Adda showed very similar bacterial compositions, as shown in the T-RFLP profiles for October 2007 (Figure 7a; other data not shown). Occasionally, amplified plastid 16S rRNA genes (mostly from micro-algae) skewed the bacterial data somewhat, but when these were removed, a similar picture emerged to the nascent streamers. The chemoautotrophic iron-oxidizing autotrophic "*Fv. myxofaciens*" and *At. ferrivorans* were the dominant bacteria present, with other terminally labeled restriction fragment lengths (T-RFs) identified corresponding mostly to heterotrophic acidophiles (of which actinobacterium Py-F3 was usually the dominant bacterium). There was also a notable increase in the biodiversity of the streamer growths, with several additional species being either firmly (gammaproteobacterium WJ2, *At. ferrooxidans* and *Leptospirillum ferrooxidans*) or tentatively (*Acidiphilium* and *Alicyclobacillus* spp.) identified compared to those detected during the first months of sampling, though some of these (in contrast to *At. ferrivorans*, "*Fv. myxofaciens*" and actinobacterium Py-F3) were only detected sporadically.

Acid streamers in the Dyffryn Adda were analyzed further on six occasions between March 2009 and August 2012, and a representative T-RFLP profile of 16S rRNA genes amplified with bacteria-specific primers (from August 2012) is shown in Figure 7b. A general pattern of bacterial diversity similar to that observed between 2005 and 2008, though with occasional subtle differences, was found during this period. In addition, eukaryotic microorganisms (identified from T-RFs of digested plastid 16S rRNA genes) were occasionally present in relatively large amounts, particularly in the summer months, in streamers sampled between 2009 and 2012.

Figure 7. T-RFLP profiles of bacterial 16S rRNA genes (digested with HaeIII) of acid streamers from the Dyffryn Adda: (**a**) sampled during October 2007; (**b**) sampled during August 2012. Color code (bars): yellow, adit portal; brown, +15 m; green, +30 m; red, +45 m; charcoal, +60 m; purple, +75 m; blue, +90 m. The major peaks identified corresponded to *At. ferrivorans* (70 nt) and "*Fv. myxofaciens*" (200 nt), with smaller and more sporadic T-RFs corresponding to heterotrophic acidophiles (actinobacterium Py-F3 (227 nt), gammaproteobacterium WJ2 (197 nt) and *Acidobacterium* sp. Thars 1 (214 nt)), the iron-oxidizing chemoautotrophs *L. ferrooxidans* (206 nt) and *At. ferrooxidans* (253 nt), or plastid 16S rRNA (383, 391 nt). The 219 and 234 nt fragments found as minor T-RFs in two streamer samples were tentatively identified as corresponding to *Alicyclobacillus* clone DAAP3B4 or *Acidiphilium* spp. (both obligate heterotrophs). Other minor peaks were not identified.

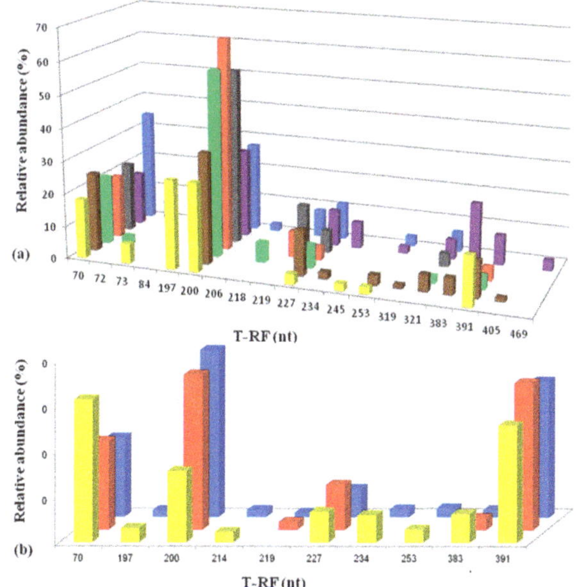

Figure 8a shows changes in the relative abundances of the bacteria identified in T-RFLP profiles over the nine year period covered by this study. Where multiple streamer samples were analyzed at any one time, average abundance data were calculated. Figure 8b shows the collated data for each of the acidophilic bacteria identified in the Dyffryn Adda streamers as mean values and standard deviations over the nine year sampling period. Data shown in Figure 8 have been corrected to take into account plastid genes which were also amplified by the bacterial primers used.

Figure 8. (**a**) Changes in the relative abundances of bacteria in acid streamers in Dyffryn Adda, determined by T-RFLP analysis of amplified 16S rRNA genes; (**b**) mean values and standard deviations of the relative abundance of different species of acidophilic bacteria in acid streams in the Dyffryn Adda, measured over a nine year sampling period. (n =42).

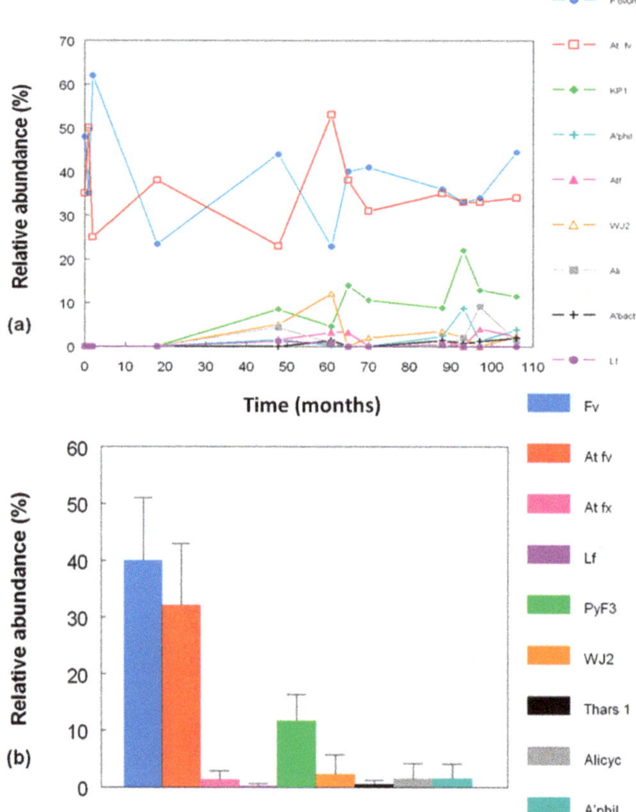

Key: •, "*Fv. myxofaciens*" (Fv); , *At. ferrivorans* (At fv); ▼, actinobacterium Py-F3; ▲ γ-proteobacterium WJ2; ▲, *At. ferrooxidans* (At fx); •, *L.ferrooxidans* (Lf); +, *Acidobacterium* sp. Thars 1. The identities of *Alicyclobacillus* clone DAAP3B4 (■) and *Acidiphilium* spp. (+) in T-RFLP profiles were inferred from fragment lengths produced using 1-2 restriction enzymes only.

2.4. Bacterial Isolates and Cloned Genes

The dominant bacteria identified in T-RFLP profiles of acid streamers were isolated on solid media and their identities confirmed by analysis of their 16S rRNA genes (Table 2). These were "*Ferrovum myxofaciens*" (strain P3G), *Acidithiobacillus ferrivorans* (strain Py-F1), and a novel actinobacterium (strain Py-F3). Other isolates included a second heterotrophic iron-oxidizing actinobacterium (strain Py-F2) which was 99.5% identical (16S rRNA gene) to the type strain of *Ferrimicrobium acidiphilum*, and two strains of iron-reducing alphaproteobacteria, Py-H1 (97% identity to *Acidocella aluminiidurans*) and Py-H3 (96% identity to *Acidiphilium rubrum*). The phylogenetic relationships of the Dyffryn Adda isolates to known species of acidophilic bacteria are shown in Figure 9.

Figure 9. Unrooted tree showing the phylogenetic relationship of bacteria isolated from acid streamers in the Dyffryn Adda with known species of acidophiles, based on comparison of their 16S rRNA genes (720 nucelotides). The scale bar represents 1% nucleotide sequence divergence and numbers at the nodes are bootstrap values out of 1000 trials. Type strains are identified by a superscript T. The acid streamer isolates are indicated in bold text.

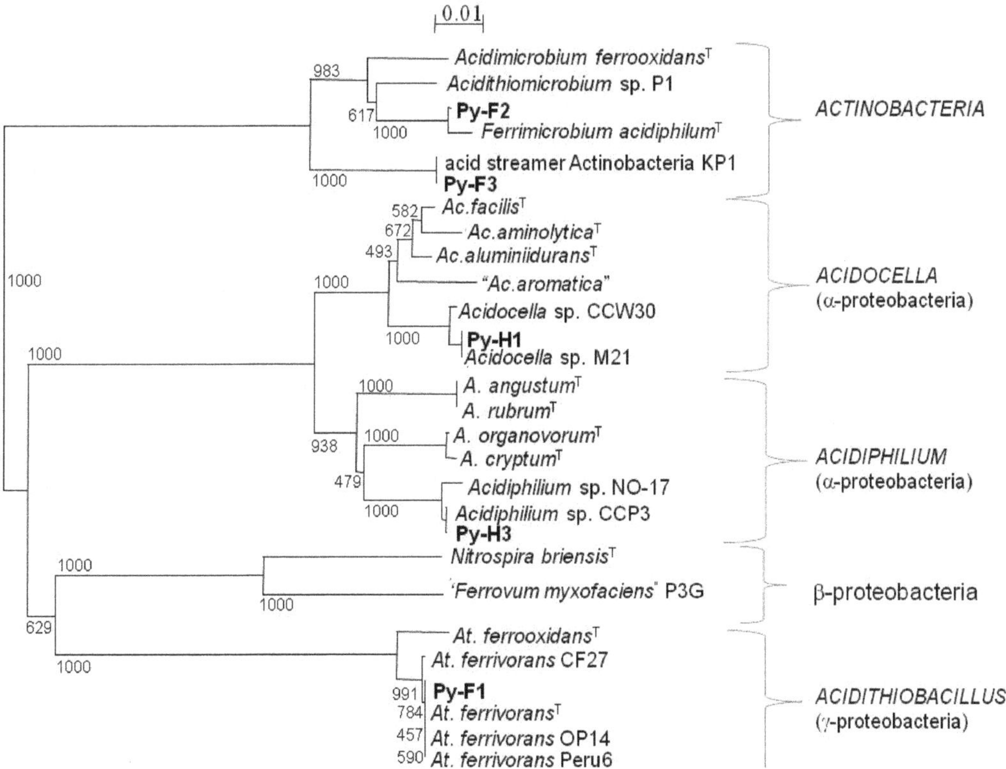

Table 2. Identities of microbial isolates and cloned genes obtained from acid streamers in the Dyffryn Adda.

Isolate/clone designation (GenBank Accession No.)	**Closest relative (GenBank Accession No.)**	**Identity (%) (16S rRNA gene)**	**Reference**
	Bacterial isolates		
P3G ("*Ferrovum myxofaciens*"^{T}) (HM044161)	"*Ferrovum myxofaciens*" PSTR (EF133508)	100	[10]
Py-H1 (KC208493)	*Acidocella* sp. strain M21(AY765998)	100	[10]
	Acidocella aluminiidurans (AB362219)	97.4	[11]
Py-H3 (KC208494)	*Acidiphilium* sp. CCP3(AY766000)	99.9	[10]
	Acidiphilium sp. NO-17 (AF376026)	99.5	[12]
	Acidiphilium rubrum^{T} (NR_025854)	95.8	[13]
Py-F1 (KC208495)	*Acidithiobacillus ferrivorans*^{T} (AF376020)	100	[14]
Py-F2 (KC208496)	*Ferrimicrobium acidiphilum*^{T} (NR_041798)	99.5	[15]
Py-F3 ("*Acidithrix ferrooxidans*") (KC208497)	Heterotrophic iron-oxidizing bacterium KP1(AY765991)	100	[10]
	Ferrimicrobium acidiphilum^{T} (NR_041798)	92.4	[15]
	Bacterial clones		
DAAP3B4 (KC208499)	*Alicyclobacillus* K23_bac (EF464642)	99.3	[16]
	Alicyclobacillus ferrooxydans^{T} (NR_044413)	90.0	[17]
PMC25 (KC208498)	*Actinobacterium* U2V-bac_a5 (JN982098)	95.9	[18]
	Ferrimicrobium acidiphilum^{T} (NR_041798)	94.8	[15]
	Isolate Py-F3	91.3	(this study)
	Archaeal clones		
DAAP3A2 (KC208501)	Clone from sediment in an acidic pit lake (FJ228391)	99.1	(GenBank entry)
	Methanomassiliicoccu luminyensis^{T} (HQ896499)	83.5	[19]

Table 2. *Cont.*

Isolate/clone designation (GenBank Accession No.)	Closest relative (GenBank Accession No.)	Identity (%) (16S rRNA gene)	Reference
	Archaeal clones		
DAAP3A1 (KC208500)	Clone from sediment in an acidic pit lake (FJ228392)	99.7	(GenBank entry)
	Methanomassiliicoccus luminyensis^{T} (HQ896499)	82.9	[19]
	Clone DAAP3A3	77.1	(this study)
	Clone DAAPA6	90.4	(this study)
DAAP3A3 (KC208502)	clone from coal-impacted forest wetland (AF523941)	98.5	[20]
	Thermogymnomonas acidicola^{T} (NR_041513)	90.6	[21]
	Clone DAAP3A6	76.8	(this study)
DAAP3A6 (KC2084503)	Clone from AMD stream (HE653789)	99.2	[22]
	Methanomassiliicoccus luminyensis^{T} (HQ896499)	80.6	[19]

2.5. Molecular Analysis of Acid Streamer Growths: Archaea

No archaeal 16S rRNA genes were amplified from DNA extracted from acid streamers sampled in 2003. In 2005 and 2007, 16S rRNA archaeal genes were amplified though subsequent T-RFLP analysis of these was not successful. Archaeal diversity in the acid streamers was, however, elucidated in analyses carried out between 2008 and 2012. T-RFLP analysis indicated that the diversity of archaea was relatively limited, with two dominant T-RF peaks and one minor peak in T-RFLP profiles generated by HaeIII digests (Figure 10). Clones corresponding to two of these peaks were obtained in libraries and identified by comparing gene sequences to those in GenBank (Table 2). A single clone (DAAP3A2) corresponding to the 180 nt T-RF was obtained, whereas three clones, all distantly related to each other, had T-RFs of 215 nt length (HaeIII digests). All of the clones were most closely related (of known species) to euryarchaeotes of the order *Thermoplasmatales* (Table 2). Changes in the relative abundances of archaea corresponding to the three T-RFs are shown in Figure 11.

2.6. Comparison of Acid Streamer and Planktonic Phase Prokaryotic Communities in the Dyffryn Adda

On two occasions (November 2008 and February 2011) both acid streamers and mine drainage waters were sampled at the same locations within the Dyffryn Adda and bacterial populations compared. Data from the November 2008 samples are shown in Figure 12. One major difference found at both times was that the iron-oxidizing autotroph *Leptospirillum ferrooxidans* accounted for a far greater part of the summated bacterial T-RFs in the mine water (average of 12.5% and

1.4% in 2008 and 2012, respectively) than in the acid streamers (corresponding mean values of 0.7 and 0.6%, and frequently not detected). The actinobacterium isolate Py-F3 showed the opposite trends (acid streamer abundance of 4.6% (2008) and 11.5% (2011) as opposed 0.7% (2008) and 1.5% (2011) in the mine water samples. In contrast to the bacterial communities, the relative abundances of the different archaeal T-RFs in the mine drainage water were found to be similar to those in the acid streamers on the single occasion (November 2008) when this was assessed (Figure 11).

Figure 10. T-RFLP profiles of archaeal 16S rRNA genes (digested with HaeIII) of acid streamers from the Dyffryn Adda adit portal (violet bars), and at 45 m (magenta bars) and 90 m (cream bars) downstream of the portal, sampled during March 2009.

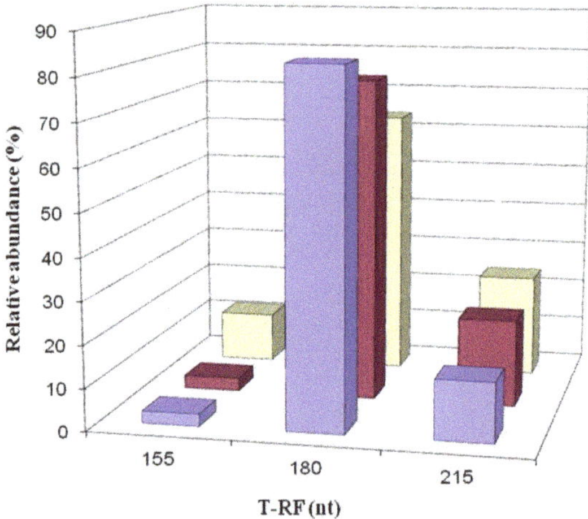

Figure 11. Changes in the relative abundances of archaea in acid streamers (and mine water in November 2008) in Dyffryn Adda, determined by T-RFLP analysis of amplified 16S rRNA genes. The red bars represent clone DAAP3A2 (180 nt), the green bars clones DAAP3A1, DAAP3A3 and DAAP3A6 (215 nt), and the blue bars represent unidentified archaea (155 nt). The gene fragment lengths referred to are from HaeIII digests and correspond to T-RFs shown in Figure 10.

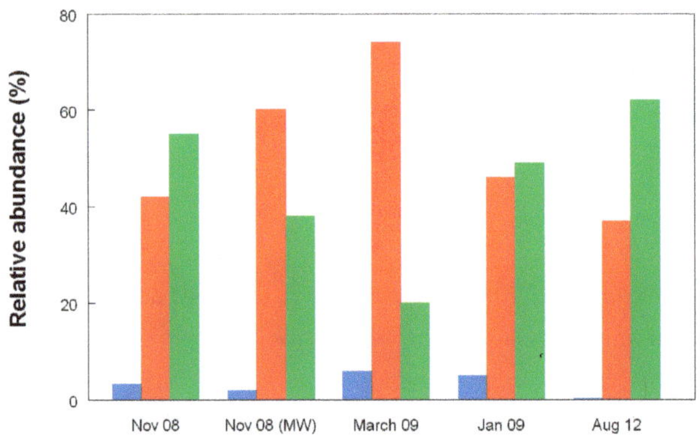

Figure 12. T-RFLP profiles of bacterial 16S rRNA genes (digested with HaeIII) of filtered mine water (cream and light blue bars) and acid streamers (orange and dark blue bars) from the Dyffryn Adda adit portal (cream and orange bars), and 90 m downstream of the portal (light and dark blue bars), sampled during November 2008. The major peaks identified corresponded to *At. ferrivorans* (70 nt), "*Fv. myxofaciens*" (200 nt), *L. ferrooxidans* (205 nt), actinobacterium Py-F3 (228 nt), and *At. ferrooxidans* (253 nt).

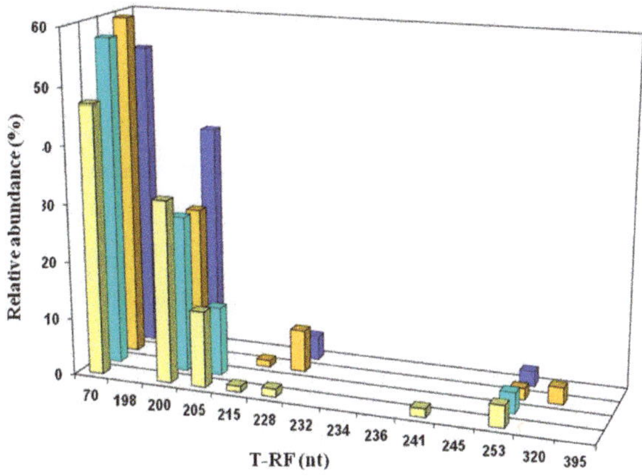

3. Discussion

This study has provided unique insights into how a newly-created extremely acidic environment becomes colonized by prokaryotic micro-organisms. The channel, which was excavated to divert subterranean water from the shafts and adits in the underground Mynydd Parys mines into an existing unpolluted stream, became heavily colonized with microbial "streamer" growths within a few months of its construction. These macroscopic growths have continued to occupy most of the volume of the AMD stream over the nine year study period. Analysis of the indigenous microbial populations has shown that, while the same three species of iron-oxidizing bacteria ("*Fv. myxofaciens*", *At. ferrivorans* (autotrophic species) and isolate Py-F3, which is, as described below, a novel heterotrophic species) have been the dominant members of the streamer communities, there has been a subtle but notable increase in biodiversity during the period of study. Obligately heterotrophic bacteria (e.g., gammaproteobacterium WJ2, *Acidobacterium*, *Acidiphilium* and *Alicyclobacillus* spp.) were detected more regularly by T-RFLP analysis six months after AMD first flowed through the Dyffryn Adda, though non iron-oxidizing heterotrophic acidophiles were always relatively minor, and often more transient, members of the streamer communities. A main reason why the streamer community is thought to be so stable is the near constant physico-chemical nature of the mine water. Its temperature varied by <1 °C (independent of season) over the study period, and its pH and content of ferrous iron (the main energy source in the stream water) and other dissolved solutes have also shown relatively small variations. It is interesting to note that, even though most of the bacteria in the acid streamers are iron-oxidizing acidophiles, only small changes in ferrous iron concentrations were detected between the adit portal and the point at which the Dyffryn Adda joined with the second stream. Water flows continuously from the underground mine, and the stream has never been observed to dry up, even during periods of prolonged lack of precipitation.

Most, if not all, of bacteria and archaea that comprise the acid streamer communities originate from the underground mine water body, the biogeochemistry of which has been described in detail elsewhere [2]. The microbial populations in the underground mine water were analyzed on several occasions during the de-watering of the mine, and bacteria identified included iron-oxidizing acidithiobacilli (*At. ferrivorans* and/or *At. ferrooxidans*), other iron-oxidizers (*L. ferrooxidans* and *Gallionella*-like bacteria (autotrophs) and the heterotrophic acidophile *Fm. acidiphilum*) and non iron-oxidizing heterotrophs (*Acidiphilium, Acidobacterium* and *Acidisphaera* [8]). When the water level within the mine was lowered, large microbial growths that had previously been submerged in the underground water body and which were described by enthusiasts exploring the mine as "drapes" were revealed. Analysis of these "drapes" showed that they were dominated by acidophilic heterotrophic bacteria (*Acidobacteria, Acidisphaera*, gammaproteobacterium WJ2 and species of actinobacteria) in contrast to the Dyffryn Adda acid streamers. The most surprising difference between the subterranean water and "drapes" and the acid streamers in the Dyffryn Adda was that "*Fv. myxofaciens*", the most abundant bacterium in the latter, was not detected in macroscopic growths or the water body in the underground mine. A similar scenario was reported at a pilot plant used to oxidize and precipitate iron from moderately acidic (pH 4.8) ferruginous ground-water in Nochten, Germany [23]. After a few months of continuous operation, "*Ferrovum*"-like bacteria were found to dominate the microbial communities in the pilot plant, even though other bacteria (*At. ferrooxidans* and *L. ferrooxidans*) had been used as inocula, and "*Ferrovum*" was not detected in the ground-water.

Autotrophic iron-oxidizing acidophiles have dominated the acid streamers in the Dyffryn Adda from its inception to the present day. An average of 72% of the bacterial community (as assessed by T-RFLP analysis) in the streamers over a nine year period was accounted for by two species, "*Fv. myxofaciens*" and *At. ferrivorans* (Figure 8b). Ferrous iron, continually supplied at ~600 mg/L in the inflowing mine water, represents the major energy source available (excluding sunlight) in the Dyffryn Adda. No reduced sulfur compounds were detected and acidophiles, such as *Acidithiobacillus thiooxidans*, that use only reduced forms of sulfur as electron donors, were also never detected. Dissolved organic carbon concentrations are relatively small, and how much of this is metabolized by acidophilic heterotrophs is unknown. The reasons for "*Fv. myxofaciens*" and *At. ferrivorans* being more successful "streamer" bacteria than other autotrophic iron-oxidizing acidophiles, such as *At. ferrooxidans* and *L. ferrooxidans*, probably relates to (i) their propensity to produce large amounts of extracellular polymeric substances, which causes them to grow (particularly in the case of "*Fv. myxofaciens*") as visible filaments and flocs *in vitro* [24], and (ii) to both species being psychro-tolerant (growing at 4 °C and above [18]), and therefore well adapted to the constant 11 °C mine water.

The ratio of bacteria to archaeal cells in the acid streamers was not determined in the present study, though analysis of streamers in the Afon Goch at Mynydd Parys [10] and at another acidic mine in north Wales [25] found that archaea accounted for only small fractions of the total microbial populations, at most. The current study indicated that diversity of archaea in the acid streamer growths was much more limited than that of the bacteria. No archaea were isolated, in contrast to bacteria, and all of the clones obtained were very distantly related to known species (though all appeared to be euryarchaeotes), precluding assigning physiological traits to the indigenous archaea.

The microbial composition of the acid streamers in the Dyffryn Adda contrasts greatly with some other streamer/mat growths that have been reported. In some cases, this is readily attributable to the contrasting physico-chemistries of the waters in which they form. For example, the much warmer (35–45 °C) and more acidic (pH 0–2) waters within the Richmond mine at Iron Mountain selects for more acidophilic and thermotolerant species (*Leptospirillum ferriphilum* and "*Ferroplasma acidarmanus*") while the extremely acidic (pH 0–1) cool (13 °C) sulfur-rich but iron-poor waters in the Frasassi cave select for sulfur-oxidizing bacteria, such as *At. thiooxidans* and *Sulfobacillus* spp. The abandoned Cantareras copper mine in south-west Spain, which has drainage water of similar chemical composition to that at the Dyffryn Adda, provides a more direct contrast [26]. Although the two major bacteria detected in the Dyffryn Adda streamers ("*Fv. myxofaciens*" and *At. ferrivorans*) were also found in the layered streamer/mat communities in a drainage channel at the Cantareras mine, heterotrophic bacteria (e.g. *Acidobacterium* and *Acidiphilium* spp., and novel species of acidophilic sulfate-reducers) were more abundant. This was considered to be due to extensive colonization of the streamer surfaces with acidophilic micro-algae, such as *Chlamydomonas acidophila*, which can provide the organic carbon used by heterotrophic bacteria [27]. Occasional surface growth of *Euglena* may also be the reason for the varying abundance and diversity of heterotrophic acidophiles found in the Dyffryn Adda. Why algal growths are far less extensive here than in the Cantareras mine is not known, but could be related to the lower temperature, regional differences in light intensity, and a generally higher ferric iron content of mine water at the Welsh site.

Based on comparisons of 16S rRNA gene sequences, two of the bacteria isolated from the Dyffryn Adda streamers, Py-H1 and Py-H3, appear to be novel species of *Acidocella* and *Acidiphilium*, respectively, while a third (Py-F3) appears to represent a novel genus. Isolate Py-F3 is an actinobacterium, 100% identical (16S rRNA gene sequence) to a bacterium that was previously also isolated from Dyffryn Adda streamers, KP1 [10]. Isolate Py-F3 appears to have similar physiological characteristics to an obligately acidophilic iron-oxidizing bacterium (CCH7) that was isolated from acid streamer growths within an abandoned pyrite mine [28] but not sequenced. Both Py-F3 and CCH7 grow as long filaments that entwine, forming gelatinous streamer-like growths *in vitro*. Both isolates Py-F3 require organic carbon and oxidize ferrous iron, though in the case of isolate CCH7 it appeared that energy from iron oxidation was not conserved. The nearest classified relative to isolate Py-F3 is *Ferrimicrobium acidiphilum*, which is also an obligately heterotrophic, iron-oxidizing acidophile, but which grows as single rods or short chains, and with a 16S rRNA gene identity of $<$ 93% is clearly a separate genus. The binomial "*Acidithrix ferrooxidans*" is proposed for isolate Py-F3 (denoting its ability for filamentous growth and to oxidize iron) and work on fully characterizing this novel actinobacterium is continuing. Interestingly, another apparently novel actinobacterium (clone PMC25) was detected, but has not yet been isolated, in the streamers. The 16S rRNA gene sequence identities to both *Fm.*

acidiphilum (95%) and "*Acidithrix ferrooxidans*" (< 92%) suggest that this is another novel species (or genus) of acidophilic actinobacteria.

4. Experimental Section

4.1. Site Description and Sampling Regime

Acid mine drainage from the underground Mynydd Parys mines (latitude 53.396784N; longitude-4.350629) following its partial de-watering in 2003, flows first through a buried pipe (*ca.* 50 m long) before entering the newly formed open channel in the upper reaches of the Dyffryn Adda stream, via a screened portal. Some 90 m downstream of the portal, the Dyffryn Adda converges with a relatively unpolluted stream, draining agricultural land, and the conjoined streams flow in a northerly direction for about 1.5 km before entering the Irish Sea. The chemistry and microbiology of the AMD stream at Dyffryn Adda was monitored for over nine years, from October 2003 (three months after the subterranean drainage from Mynydd Parys was diverted into this channel) to August 2012. Mine water chemistry was determined at monthly intervals between October 2003 and November 2007 and less frequently afterwards (only when acid streamers were also analyzed, a total of twelve sampling events).

4.2. Physico-Chemical Analyses

On-site measurements of temperature, conductivity, pH, dissolved oxygen (DO) and redox potentials (E_h; corrected to be relative to a standard hydrogen electrode reference electrode) were carried out using a YSI 556 MPS multi-meter (YSI instruments, Yellow Springs, Ohio). For laboratory analyses, water samples were filtered through 0.2 μm cellulose nitrate membrane filters into sterile sample tubes, with a sub-set of each sample acidified (by adding two drops of concentrated nitric acid) for subsequent analysis of transition metals (using ion chromatography [29]). Concentrations of sulfate were determined using a turbidometric technique [30] or by ion chromatography, ferrous iron using the ferrozine assay [31], and dissolved organic carbon (DOC) using a LABTOC DOC analyzer (Pollution & Process Monitoring Ltd., UK).

4.3. Molecular Analysis of Acid Streamer Growths

Samples of acid streamers (50 cm^3) were removed from the stretch of the Dyffryn Adda between the upper portal and the point where it joined the second stream, and put into sterile Falcon tubes. Samples were maintained at 4 °C and processed within 24 hours of collection.

Terminal restriction enzyme fragment length polymorphism (T-RFLP) was used routinely to determine the compositions of the microbial communities of acid streamer samples. This is a semi-quantitative, PCR-based technique that has been used successfully to assess similar growths at other mine sites (e.g., [25]). In brief, DNA was extracted from the acid streamer samples (0.1 g) using the MO-BIO Ultraclean Soil DNA Isolation kit (MO-BIO Laboratories Inc. USA) according to the manufacturer's instructions. Then 16S rRNA gene amplifications were performed in triplicate and products combined to avoid PCR bias [32] using Cy5-labelled fluorescent forward primers.

Amplification primers for bacterial 16S rRNA genes were the forward primer 27F (5'-AGAGTTTGATCCTGGCTCAG-3'), and the reverse amplification primers were either 1387R (5'-GGGCGGWGTGTACAAGGC-3'), 1492R (5'–TACGGYTACCTTGTTACGACT-3') or 536R (5'-CAGCSGCCGCGGTAAWC-3'). For Archaeal 16S rRNA gene amplifications the primers used were 20F (5'–TCCGGTTGATCCYGCCRG-3') and 915R (5'– GTGCTCCCCCGCCAATTCCT–3'). Bacterial PCR conditions were: 95 °C (5 min), followed by 30 cycles at 95 °C (30 s), 55 °C (30 s) and 72 °C (1.5 min), and a final extension at 72 °C (10 min). The conditions for archaeal PCR amplifications were 95 °C (5 min), followed by 30 cycles at 95 °C (30 s), 62 °C (30 s) and 72 °C (1.0 min), and a final extension at 72 °C (10 min).

The resulting PCR fragments were purified using SureClean (Bioline Ltd., UK), and digested separately with up to three restriction endonucleases; HaeIII, CfoI or AluI (Promega, UK) to differentiate the 16S rRNA gene fragments by producing digestion products of differing terminally labeled restriction fragment lengths (T-RFs). The digested products were analyzed by capillary electrophoresis (CEQ8000 genetic analysis system; Beckman-Coulter, UK) and sized by their mobility in comparison with fluorescently labeled size standards. Total relative abundances (%) of the individual T-RFs were calculated from peak areas which are directly related to the fluorescent peak intensity, e.g., large peak intensity indicates a greater total abundance in the environmental samples. Identification of the major micro-organisms represented by individual T-RFs was facilitated by comparison of the latter against a database maintained at the authors' laboratory, which was compiled from extensive T-RF data obtained from acidophilic bacterial and archaeal isolate and clone sequences. In cases where T-RFs from all three enzyme digests corresponded to a known micro-organism, identification was classed as "firm", whereas if only two of the three digests corresponded to a known micro-organism, identification was regarded as "tentative".

4.4. Molecular Analysis of Planktonic-Phase Bacteria

On two occasions (November 2008 and February 2011) the bacteria present in the AMD flowing through the Dyffryn Adda were also analyzed using T-RFLP of extracted DNA. Approximately, 500 mL of mine water was filtered (sterile 0.22 µm cellulose nitrate filters) at the adit site and the filters were stored at −20 °C until DNA extraction. DNA extraction used the same kit and protocol as stated previously except whole filters were aseptically cut into pieces and placed into the bead beating tube. The extracted DNA was used as template for bacterial and archaeal 16S rRNA T-RFLP PCR using the fluorophore-labeled primers and conditions as stated above.

4.5. Clone Library Construction

For the construction of a clone library of 16S rRNA genes, unlabeled PCR fragments were generated as described above, purified and ligated into the pGem-T-Easy vector system (Promega, UK) according to the manufacturer's instructions. Insert-vector ligation mixes (3:1 ratio) were used to transform *Escherichia coli* DH5α [33] and cells plated on selective media. The 16S rRNA gene inserts in the clones were amplified by PCR, differentiated via restriction enzyme fragment length polymorphism (RFLP), and visualized using agarose gel electrophoresis. Plasmid inserts that generated distinct RFLP patterns were further screened using T-RFLP to ascertain whether these corresponded to unknown T-RFs in streamer profiles. These were isolated with Strataprep plasmid mini-prep kits (Agilent technologies, USA) and sequenced (Macrogen, Inc., Korea). Chromatogram files were visualized using Chromas Lite, sequences edited to generate a contiguous gene sequence for each isolate, and then these were aligned using the BlastN on-line software (NCBI). Gene sequences were compared to those contained in the Genbank database.

4.6. Cultivation Analysis of Acid Streamer Bacteria

Small fragments (*ca.* 0.5 cm^3) of acid streamer samples were removed from the bulk samples using sterile tweezers, placed in 1 ml of sterile acidic (pH 2.5) water, and bacterial cells were dislodged by vortexing for 20 s. The streamer remnants were removed by gentle centrifugation and the cell suspensions streak-inoculated onto a range of solid overlay and non-overlay media formulated to promote the growth of autotrophic and heterotrophic acidophiles [34]. Plates were incubated for 3–4 weeks at 30 °C and inspected on a regular basis. Preliminary identification of isolates that grew on the different media was based on their colony morphologies (e.g. ferric iron-stained colonies that grew on ferrous iron media were identified as acidophilic iron-oxidizers) and their cell morphologies (e.g. highly motile curved rods that formed ferric iron-stained colonies were identified as *Leptospirillum* spp.). To confirm their identities, bacterial isolates were purified by repeated single colony isolation on solid media, and PCR amplification of 16S rRNA genes carried out using either single colonies (from plates) or pellets (from liquid cultures). These were sequenced by Macrogen using the 27F and 1387R primer pair.

5. Conclusions

A long-term study of microbial communities in a new drainage channel receiving acidic (pH 2.5) water from an abandoned copper mine found that acid streamer growths rapidly colonized the drain channel and displayed changes in community profiles with time. Three iron-oxidizing bacteria dominated the streamer communities, two of which (*Acidithiobacillus ferrivorans* and "*Ferrovum myxofaciens*") were obligate chemolithotrophs and the third (a novel genus/species with the proposed name "*Acidithrix ferrooxidans*") a heterotrophic acidophile. With time, the biodiversity of the acid streamers increased, though other bacteria that were identified were always present in relatively low abundance. Archaea were only detected in the streamer growths some time after the drainage stream was established, and showed limited diversity. This study provides rare insights into how microbial communities colonize and evolve in a newly-created extreme environment.

Acknowledgements

We wish to thank Kathryn Wakeman and Stewart Rolfe for their help with some areas of the work described. KC acknowledges the financial support given by the Natural Environment Research Council (U.K.) and Rio Tinto Technical Services Ltd. (Industrial CASE studentship NER/S/C/2001/06450). OFR is grateful to the Natural Environment Research Council (U.K.) and Paques b.v. (The Netherlands) for the provision of an industrial CASE research studentship

(NER/S/C/2003/11752). Part of this work was carried out within the European Union-sponsored projects "BioMinE" project (FP6 contract NMP1-CT-500329-1) and "Umbrella" (FP7 contract 226870).

References

1. Johnson, D.B. Extremophiles: acidic environments. In *Encyclopaedia of Microbiology*, 2nd Edition; Schaechter, M. Ed.; Elsevier: Oxford, UK, 2009; pp.107-126.
2. Johnson, D.B. Geomicrobiology of extremely acidic subsurface environments. *FEMS Microbiol. Ecol.* **2012**, *81*, 2–12.
3. Edwards, K.J.; Gihring, T.M.; Banfield, J.F. Seasonal variations in microbial populations and environmental conditions in an extreme acid mine drainage environment. *Appl. Environ. Microbiol.* **1999**, *65*, 3627–3632.
4. Ziegler, S.; Ackermann, S.; Majzlan, J.; Gescher, J. Matrix composition and community structure analysis of a novel bacterial leaching community. *Environ. Microbiol.* **2009**, *11*, 2329–2338.
5. Jones, D.S.; Albrecht, H.L.; Dawson, K.S.; Schaperdoth, I.; Freeman, K.H.; Pi, Y.; Pearson, A.; Macalady, J.L. Community genomic analysis of an extremely acidophilic sulfur-oxidizing biofilm. *ISME J.* **2012,** *6*,158–170.
6. Bevins, R.E. *A Mineralogy of Wales*; Geological Series No. 16; National Museum of Wales Press: Cardiff, UK, 1994.
7. Walton, K.C.; Johnson, D.B. Microbiological and chemical characteristics of an acidic stream draining a disused copper mine. *Environ. Pollut.* **1992**, *76*, 169–175.
8. Coupland, K.; Johnson, D.B. Geochemistry and microbiology of an impounded subterranean acidic water body at Mynydd Parys, Anglesey, Wales. *Geobiol.* **2004**, *2*, 77–86.
9. Valente, T.M.; Gomez, C.L. The role of two acidophilic algae as ecological indicators of acid mine drainage sites. *J. Iberian Geol.* **2007**, *33*, 283–294.
10. Hallberg, K.B.; Coupland, K.; Kimura, S.; Johnson, D.B. Macroscopic "acid streamer" growths in acidic, metal-rich mine waters in north Wales consist of novel and remarkably simple bacterial communities. *Appl. Environ. Microbiol.* **2006**, *72*, 2022–2030.
11. Kimoto, K.; Aizawa, T.; Urai, M.; Bao Ve,N.; Suzuki, K.; Nakajima, M.; Sunairi, M. *Acidocella. aluminiidurans* sp. nov., an aluminium-tolerant bacterium isolated from *Panicum. repens* grown in a highly acidic swamp in actual acid sulfate soil area of Vietnam. *Int. J. Syst. Evol. Microbiol.* **2010**, *60*, 764–768.
12. Johnson, D.B.; Rolfe, S.; Hallberg, K.B.; Iversen, E. Isolation and phylogenetic characterisation of acidophilic microorganisms indigenous to acidic drainage waters at an abandoned Norwegian copper mine. *Environ. Microbiol.* **2001**, *3*, 630–637.
13. Kishimoto, N.; Kosako, Y.; Wakao, N.; Tano, T.; Hiraishi, A. Transfer of *Acidiphilium facilis* and *Acidiphilium aminolytica* to the genus *Acidocella* gen. nov., and emendation of the genus *Acidiphilium*. *Syst. Appl. Microbiol.* **1995**, *18*, 85–91.
14. Hallberg, K.B.; González-Toril, E.; Johnson, D.B. *Acidithiobacillus ferrivorans* sp. nov.; facultatively anaerobic, psychrotolerant, iron- and sulfur-oxidizing acidophiles isolated from metal mine-impacted environments. *Extremophiles* **2010**, *14*, 9–19.

15. Johnson, D.B.; Bacelar-Nicolau, P.; Okibe, N.; Thomas, A.; Hallberg, K.B. Characteristics of *Ferrimicrobium acidiphilum* gen. nov., sp. nov., and *Ferrithrix thermotolerans* gen. nov., sp. nov.: heterotrophic iron-oxidizing, extremely acidophilic Actinobacteria. *Int. J. Syst. Evol. Microbiol.* **2009,** *59*, 1082–1089.
16. Winch, S.; Mills, H.J.; Kostka, J.E.; Fortin, D.; Lean, D.R. Identification of sulfate-reducing bacteria in methylmercury-contaminated mine tailings by analysis of SSU rRNA genes. *FEMS Microbiol. Ecol.* **2009**, *68*, 94–107.
17. Jiang, C.Y.; Liu, Y.; Liu, Y.Y.; You, X.Y.; Guo, X.; Liu, S.J. *Alicyclobacillus ferrooxydans* sp. nov., a ferrous-oxidizing bacterium from solfataric soil. *Int. J. Syst. Evol. Microbiol.* **2008**, *58*, 2898–2903.
18. Urbieta, M.S.; Gonzalez Toril, E.; Aguilera, A.; Giaveno, M.A.; Donati, E. First prokaryotic biodiversity assessment using molecular techniques of an acidic river in Neuquen, Argentina. *Microb. Ecol.* **2012**, *64*, 91–104.
19. Dridi, B.; Fardeau, M.L.; Ollivier, B.; Raoult, D.; Drancourt, M. *Methanomassiliicoccus luminyensis* gen. nov., sp. nov., a methanogenic archaeon isolated from human faeces. *Int. J. Syst. Evol. Microbiol.* **2012**, *62*, 1902–1907.
20. Brofft, J.E.; McArthur, J.V.; Shimkets, L.J. Recovery of novel bacterial diversity from a forested wetland impacted by reject coal. *Environ. Microbiol.* **2002**, *4*, 764–769.
21. Itoh, T.; Yoshikawa, N.; Takashina, T. *Thermogymnomonas acidicola* gen. nov., sp. nov., a novel thermoacidophilic, cell wall-less archaeon in the order *Thermoplasmatales*, isolated from a solfataric soil in Hakone, Japan. *Int. J. Syst. Evol. Microbiol.* **2007**, *57*, 2557–2561.
22. Volant, A.; Desoeuvre, A.; Casiot, C.; Lauga, B.; Delpoux, S.; Morin, G.; Personne, J.C.; Hery, M.; Elbaz-Poulichet, F.; Bertin, P.N.; *et al.* Archaeal diversity: temporal variation in the arsenic-rich creek sediments of Carnoulès Mine, France. *Extremophiles.* **2012**, *16*, 645–657.
23. Heinzel, E.; Janneck, E.; Glombitza, F.; Schlömann, M.; Seifert, J. Population dynamics of iron-oxidizing communities in pilot plants for the treatment of acid mine waters. *Environ. Sci. Technol.* **2009**, *43*, 6138–6144.
24. Hedrich, S.; Schlömann, M.; Johnson, D.B. The iron-oxidizing *Proteobacteria. Microbiol.* **2011**, *157*, 1551–1564.
25. Kimura, S.; Bryan, C.G.; Hallberg, K.B.; Johnson, D.B. Biodiversity and geochemistry of an extremely acidic, low temperature subterranean environment sustained by chemolithotrophy. *Environ. Microbiol.* **2011**, *13*, 2092–2104.
26. Rowe, O.F.; Sánchez-España, J.; Hallberg, K.B.; Johnson, D.B. Microbial communities and geochemical dynamics in an extremely acidic, metal-rich stream at an abandoned sulfide mine (Huelva, Spain) underpinned by two functional primary production systems. *Environ. Microbiol.* **2007**, *9*, 1761–1771.
27. Ňancucheo, I.; Johnson, D.B. Acidophilic algae isolated from mine-impacted environments and their roles in sustaining heterotrophic acidophiles. *Front. Microbiol.* **2012**, *3*, 325.
28. Johnson, D.B.; Ghauri, M.A.; Said, M.F. Isolation and characterisation of an acidophilic heterotrophic bacterium capable of oxidizing ferrous iron. *Appl. Environ. Microbiol.* **1992**, *58*, 1423–1428.

29. Ňancucheo, I.; Johnson, D.B. Selective removal of transition metals from acidic mine waters by novel consortia of acidophilic sulfidogenic bacteria. *Microb. Biotechnol.* **2012**, *5*, 34–44.
30. Kolmert, A.; Wikstrom, P.; Hallberg, K.B. A fast and simple turbidometric method for the determination of sulfate in sulfate-reducing bacterial cultures. *J. Microbiol. Meth.* **2000**, *41*, 179–184.
31. Lovley, D.R.; Phillips, E.J.P. Rapid assay for microbially reducible ferric iron in aquatic sediments. *Appl. Environ. Microbiol.* **1987**, *53*, 1536–1540.
32. Suzuki, M.T.; Giovannoni, S.J. Bias caused by template annealing in the amplification of mixtures of 16S rRNA genes by PCR. *Appl. Environ. Microbiol.* **1996**, *62*, 625–630.
33. Sambrook, J; Fritsch E.F.; Maniatis, T. *Molecular Cloning: A Laboratory Manual*, 2nd Ed.; Cold Spring Harbor Laboratory Press: Cold Spring Harbor, NY, USA, 1989.
34. Johnson, D.B.; Hallberg, K.B. Techniques for detecting and identifying acidophilic mineral-oxidizing microorganisms. In *Biomining*; Rawlings, D.E., Johnson, D.B., Eds; Springer-Verlag: Heidelberg, Germany, 2007; pp. 237–262.

Quorum Sensing in Some Representative Species of *Halomonadaceae*

Ali Tahrioui, Melanie Schwab, Emilia Quesada and Inmaculada Llamas

Abstract: Cell-to-cell communication, or quorum-sensing (QS), systems are employed by bacteria for promoting collective behaviour within a population. An analysis to detect QS signal molecules in 43 species of the *Halomonadaceae* family revealed that they produced *N*-acyl homoserine lactones (AHLs), which suggests that the QS system is widespread throughout this group of bacteria. Thin-layer chromatography (TLC) analysis of crude AHL extracts, using *Agrobacterium tumefaciens* NTL4 (pZLR4) as biosensor strain, resulted in different profiles, which were not related to the various habitats of the species in question. To confirm AHL production in the *Halomonadacea*e species, PCR and DNA sequencing approaches were used to study the distribution of the *luxI*-type synthase gene. Phylogenetic analysis using sequence data revealed that 29 of the species studied contained a LuxI homolog. Phylogenetic analysis showed that sequences from *Halomonadaceae* species grouped together and were distinct from other members of the *Gammaproteobacteria* and also from species belonging to the *Alphaproteobacteria* and *Betaproteobacteria.*

Reprinted from *Life.* Cite as: Tahrioui, A.; Schwab, M.; Quesada, E.; Llamas, I. Quorum Sensing in Some Representative Species of *Halomonadaceae*. *Life* **2013**, *3*, 260-275.

1. Introduction

The definition of extreme environments has been the subject of considerable controversy and as yet, due to its complexity, no consensus of opinion has been reached. In 1979 Brock defined extreme environments as those habitats with low species diversity [1]. It has been suggested, for example, that in hypersaline environments, a typical extreme habitat, environmental factors such as a high salt concentration along with low oxygen concentrations, high or low temperatures, basic pH and solar radiation may contribute to limiting their biodiversity [2]. Nevertheless, the results of a large number of ecological studies conducted recently in extreme environments indicate that in fact they contain a fairly high diversity of species [3–6].

According to 16S rDNA gene-sequence analysis, the family *Halomonadaceae* [7], within the order *Oceanospirillales,* forms a separate phylogenetic lineage within the class *Gammaproteobacteria. Halomonadaceae* contains the largest number of halophilic species described to date [8], including 10 genera of halophilic and halotolerant bacteria [9]: *Aidingimonas* (one species), *Carnimonas* (one species), *Chromohalobacter* (nine species), *Cobetia* (two species), *Halomonas* (80 species), *Kushneria* (five species), *Modicisalibacter* (one species) and *Salinicola* (three species); as well as two genera of non-halophilic bacteria: *Halotalea* (one species) and *Zymobacter* (one species). Many members of the *Halomonadaceae* family are moderately halophilic since they grow best in media containing from 0.5 M to 2.5 M NaCl [10], although some can grow over a very broad range of salt concentrations due to their ability to accumulate organic compounds to adapt themselves to changes in environmental osmolarity. Compatible solutes such as betaine can be taken up from the external medium or others, such as ectoine, synthesized by the cells themselves [8,11,12]. *Halomonadaceae* species have been isolated from very different habitats, including the sea, salterns, saline soils and endorheic lakes, and have also been found in some seafoods, marine invertebrates and even in a mural painting [9,13].

Extensive studies carried out over the last 30 years into this group of halophiles have led us to a better understanding of their biodiversity, phylogenetic relationships, physiological and haloadaptative mechanisms and, more recently, their biotechnological applications. Some moderate halophiles are recognised for their potential use in biotechnology because of their capacity to produce exopolysaccharides, enzymes and compatible solutes such as ectoine—used as a stabilizer for enzymes—as well as for their active role in the process of denitrification and the degradation of aromatic compounds [6,13,14].

Within the *Halomonadaceae*, *Halomonas* is one of the genera most frequently isolated from hypersaline waters and soils by conventional-culture techniques [11,15]. Molecular-ecology techniques based on 16S rDNA sequence analysis suggest that *Halomonas ventosae* is the predominant species in these habitats [16]. Nevertheless, the ecological role that *Halomonas* species play in these habitats and their relationships with other halophilic and non-halophilic microorganisms are still unknown.

Bacteria have evolved sophisticated mechanisms to co-ordinate gene expression, such as quorum sensing (QS) [17–19], which involves the production of signal molecules known as autoinducers. Autoinducers include, among others, *N*-acyl homoserine lactones (AHLs), produced by the *Proteobacteria*, oligopeptides, produced by the *Firmicutes*, and furanosylborate diester (AI-2), which is produced by both *Proteobacteria* and *Firmicutes* and used for interspecies communication [20,21]. The classical AHL-based systems contain a *luxI* homologue gene, which is responsible for the synthesis of AHLs, and a *luxR* homologue gene, which is an AHL-dependent transcriptional regulator. AHL-based systems involve the accumulation of AHL molecules in the extracellular medium until a critical concentration is reached, at which point the AHLs bind a transcriptional activator, which triggers the expression of target genes, including the *luxI* gene, leading in turn to the production of more AHLs [22–24] and the expression of virulence factors and exoenzymes, conjugal DNA transfer, control of plasmid-copy number, production of and susceptibility to antibiotics, biofilm formation and exopolysaccharide production [20,21].

AHL-dependent systems have been reported in many genera belonging to the phylum *Proteobacteria*. In addition, studies based on genome sequencing have revealed that many bacteria contain possible *luxI/luxR* homologues and, in some cases, multiple coexisting QS systems [25]. Nevertheless, little is known about the QS systems in halophilic microorganisms. In our experiments we have found that four exopolysaccharide-producing species of the genus *Halomonas* produce AHL-autoinducer molecules [26]. We have also described the structure of AHL molecules produced by *Halomonas anticariensis* (C_4-HSL, C_6-HSL, C_8-HSL and C_{12}-HSL) [26] and identified and characterized the QS genes *hanR/hanI* involved in their production [27]. In addition, we have proved that the QS system is regulated by a GacS/GacA two-component system, suggesting its integral involvement in the intercellular communication strategies of this bacterium [28].

In this present study we have detected and identified quorum-sensing systems that rely upon the production of AHLs in 43 species belonging to the *Halomonadaceae* family. Using PCR and DNA sequencing approaches, we have studied the distribution of the LuxI-type synthase in 29 of these species and constructed a phylogenetic tree, that was based on a partial sequence of this protein.

2. Results and Discussion

2.1. Detection of AHLs in the Halomonadaceae Family

We have investigated the existence of AHL-dependent QS systems in 43 species of the *Halomonadaceae* family. These 43 bacteria, the type strains of their respective species, were isolated from very different saline habitats, including salterns, saline soils, marshes and seawater, among others (Table 1). They include 12 species discovered by our research group during the course of ecological and taxonomic studies conducted in hypersaline environments in Spain, Chile and Morocco [29–41]. The strains used in this study include representatives of the following genera: *Chromohalobacter* (one species), *Cobetia* (one species), *Halomonas* (33 species), *Halotalea* (one species), *Kushneria* (three species), *Modicisalibacter* (one species) and *Salinicola* (two species). So far AHL signal molecules have only been detected in four exopolysaccharide-producing species of *Halomonas* described in our laboratory: *H. eurihalina, H. maura, H. ventosae* and *H. anticariensis* [26].

To overcome the limitation of the "cross-streak" method [42], in which each couple of sensor test strains must be cultured under optimum conditions without interfering with each other, we extracted AHLs from all the strains assayed and added them to agar plates upon which the biosensor strain had already been spread (see Experimental Section). Our choice of biosensor strains was ultimately based on previous experience in our laboratory [26]. Thus we chose the *Chromobacterium violaceum* strain CV026, a mutant which cannot synthesize its own quorum-sensing signal molecules and responds to exogenously added short-chain AHLs (C_4-C_6-HSLs), producing a pigment called violacein [43], and also *Agrobacterium tumefaciens* NTL4 (pZLR4), a sensitive, broad-spectrum AHL-responsive reporter that is unable to produce its own AHLs and contains a *lacZ* fusion to the quorum-sensing regulated gene *traG*. This latter strain is sensitive to AHLs with medium-to-long acyl chains that, when added exogenously, activate *lacZ* fusion, which is detectable by the appearance of a blue stain in the presence of X-Gal [44]. All the strains tested synthesized signal molecules to activate the biosensor *A. tumefaciens* NTL4 (pZLR4) (see Figure 1 for some examples). No signal was detected when a sample from an uninoculated cultured medium of MY 7.5% (w/v) was tested as negative control (data not shown). These results initially suggested that most strains were able to produce AHLs and therefore probably possess at least one AHL-QS system. Nevertheless, only *Halomonas rifensis* HK31^T and *H. anticariensis* FP35^T, used as control, produced AHLs in sufficient quantities to activate *C. violaceum* CV026 under our assay conditions. As is demonstrated below, these two strains produce about five times more AHL than the rest of the species tested (Figure 2). We have in fact already described how some species of *Halomonas*, such as *H. anticariensis* FP35^T, synthesize much greater quantities of AHLs than others, such as *H. eurihalina, H. maura* and *H. ventosae* [26].

Figure 1. *N*-acyl homoserine lactone (AHL) production by *Halomonas salina* F8-11^T, *H. eurihalina* F9-6^T, *H. pacifica* DSM 4742^T and *H. variabilis* DSM 3051^T. A volume of 5μL of AHLs previously extracted from the bacterial cultures were visualized on agar plate diffusion assay by means of the indicator strain *A. tumefaciens* NTL4 (pZLR4).

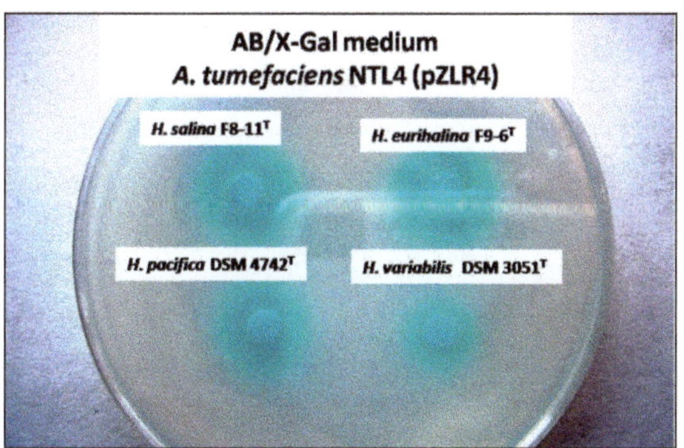

2.2. Characterization of the AHLs

The use of the indicator organisms in combination with thin-layer chromatography (TLC) provides a simple, rapid way of determining the number and nature of the AHLs produced by a particular strain [44]. We analysed the culture extracts of the 43 *Halomonadaceae* strains (Table 1) using TLC in combination with the biosensor *A. tumefaciens* NTL4 (pZLR4). This analysis showed the production of different AHL profiles amongst the various genera e.g. *Chromohalobacter salexigens* (Figure 2, lane 2), *Cobetia marina* (Figure 2, lane 3), *Halomonas anticariensis* (Figure 2, lane 6), *Halotalea alkalilenta* (Figure 2, lane 37), *Kushneria marisflavi* (Figure 2, lane 40) and *Salinicola halophilus* (Figure 2, lane 42) and also among certain species such as *Halomonas alimentaria*YKJ-16^T (Figure 2, lane 4), *H. anticariensis* FP35^T (Figure 2, lane 6), *H. desiderata* FB2^T (Figure 2, lane 11) and *H. eurihalina* F9-6^T (Figure 2, lane 14). Similarly, in a previous study [26] we found that strains belonging to the same species showed the same AHL profiles whilst different species showed different profiles. In just the same way, significant differences have been identified in the AHL profiles of the marine species *Vibrio salmonicida* [45] and *V. anguillarum* [46].

The AHL patterns from *Halomonadaceae* species contain from one to three spots with mobilities similar to those of the C_8-HSL, C_6-HSL and 3-oxo-C_6-HSL standards. Differences were also observed in the quantities of AHLs synthesized, *Halomonas anticariensis* FP35^T and *Halomonas rifensis* HK31^T synthesizing about five times more AHLs than any of the other 41 species examined (Figure 2. lanes 6 and 29). *Halomonas variabilis* DSM 3051^T (Figure 2, lane 35) and *Salinicola salarius* M27^T (Figure 2, lane 43) produce as low an amount of signal as that from the uninoculated cultured medium MY 7.5% (w/v) [26], although, their AHL-extracts did activate the *A. tumefaciens* NTL4 (pZLR4) indicator strain when the diffusion plate assay was carried out (data not shown). The most predominant AHL molecule was C_6-HSL. In *H. anticariensis* FP35^T this AHL had been previously identified by gas chromatography/mass spectrometry (GM/MS) and electrospray ionization tandem mass spectrometry (ESI MS/MS) [26], suggesting that this signal molecule may well be biologically active in intercellular communication strategies within the *Halomonadaceae* family. The synthesis of short-chain-acyl AHLs, such as C_6-HSL and C_8-HSL, is also very common among the species belonging to the *Vibrionaceae* family, which are ubiquitous in marine environments [45,46].

Figure 2. Thin-layer chromatography (TLC) analysis of the AHLs produced by the 43 species of *Halomonadaceae*: lane 1, *Carnimonas nigrificans* CTCBS1^T, lane 2, *Chromohalobacter salexigens* DSM3043^T; lane 3, *Cobetia marina* 219^T; lane 4, *Halomonas alimentaria* YKJ-16^T; lane 5, *H. almeriensis* M8^T; lane 6, *H. anticariensis* FP35^T; lane 7, *H. aquamarina* 558^T; lane 8, *H. campaniensis* 5AG^T; lane 9, *H. cerina* SP4^T; lane 10, *H. denitrificans* M29^T; lane 11, *H. desiderata* FB2^T; lane 12, *H. elongata* 1H9^T; lane 13, *H. eurihalina* F9-6^T, lane 14, *H. fontilapidosi* 5CR^T, lane 15, *H. gudaonensis* SL014B-69^T; lane 16, *H. halmophila* ACAM 71^T; lane 17, *H. halodenitrificans* ATCC 13511^T; lane 18, *H. halodurans* DSM 5160^T; lane 19, *H. koreensis* SS20^T; lane 20, *H. magadiensis* 21 MI^T; lane 21, *H. maura* S-31^T; lane 22, *H. meridiana* ACAM 246^T; lane 23, *H. mongoliensis* Z-7009^T; lane 24, *H. nitroreducens* 11-S^T; lane 25, *H. organivorans* G-16.1^T; lane 26, *H. pacifica* DSM 4742^T; lane 27, *H. pantelleriensis* AAP^T; lane 28, *H. ramblicola* RS-16^T; lane 29, *H. rifensis* HK31^T; lane 30, *H. saccharevitans* AJ275^T; lane 31, *H. salina* F8-11^T; lane 32, *H. shengliensis* SL014B-85^T; lane 33, *H. stenophila* N12^T; lane 34, *H. subglaciescola* ACAM 12^T; lane 35, *H. variabilis* DSM 3051^T; lane 36, *H. ventosae* Al-12^T, lane 37, *Halotalea alkalilenta* AW-7^T; lane 38, *Kushneria avicenniae* MW2a^T; lane 39, *K. indalinina* CG2.1^T; lane 40, *K. marisflavi* SW32^T; lane 41, *Modicisalibacter tunisiensis* LIT2^T; lane 42, *Salinicola halophilus* CG4.1^T; lane 43, *S. salarius* M27^T. Each lane contains 10 μL of AHL crude extract except for lanes 6 and 29, which contain 5 μL. Lane S; synthetic AHL standards: oxo-C_6-HSL (4.7 pmol), C_6-HSL (804 pmol), C_8-HSL (31.6 pmol), C_{10}-HSL (2 nmol), C_{12}-HSL (4.8 nmol).

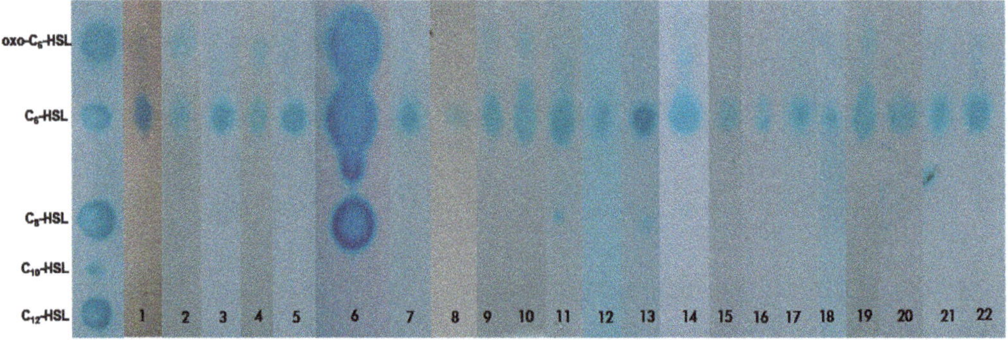

Figure 2. *Cont.*

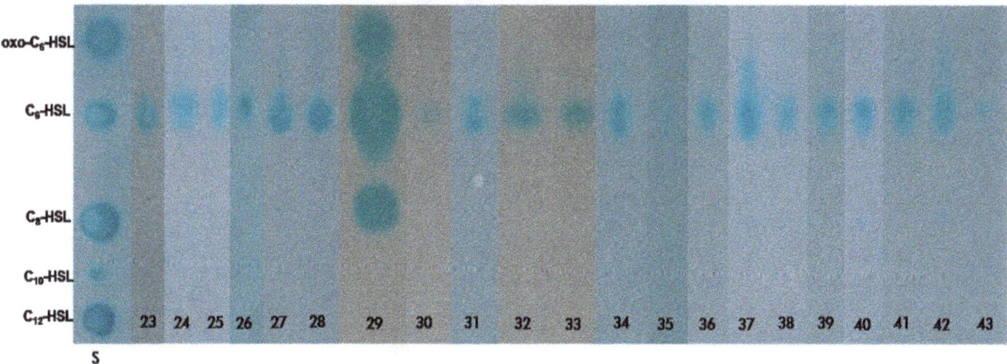

To carry out a phylogenetic analysis of the LuxI synthase in the 29 positive species of the *Halomonadaceae* family we conducted a multiple sequence alignment, including amino-acid sequences of LuxI synthase that had been experimentally determined in other members of the phyla *Alphaproteobacteria, Betaproteobacteria* and *Gammaproteobacteria.* The phylogenetic tree constructed according to the neighbour-joining method showed that all the amino-acid sequences from the *Halomonadaceae* family grouped together and were distinct from the rest of the *Gammaproteobacteria* analysed, and also from the species belonging to the *Alphaproteobacteria* and *Betaproteobacteria* (Figure 3a). The clustering of the 29 *Halomonadaceae* in which the *luxI* fragment was detected was not related to the habitat from which they were isolated (Table 1). The distribution of the *Halomonadaceae* family with respect to the rest of species analysed in the phylogenetic tree based on the 16S rDNA sequences (Figure 3b) was similar to that obtained from the LuxI sequence (Figure 3a). This result indicates that the LuxI amino-acid partial region used in this study is conserved among the family *Halomonadaceae* members.

Figure 3. Phylogenetic trees based on LuxI sequences (a) and 16S rDNA sequences (b) found in some members of *Alphaproteobacteria*, *Betaproteobacteria* and *Gammaproteobacteria*, including the species studied here belonging to the *Halomonadaceae* family. Abbreviations for bacterial genus names: *A, Aliivibrio; Ac, Acidithiobacillus; Ae, Aeromonas; Ag, Agrobacterium; Az, Azospirillium; B, Burkholderia; Car, Carnimonas; C, Chromobacterium; Er, Erwinia; En, Enterobacter; H, Halomonas; Halot, Halotalea; K, Kushneria; M, Modicisalibacter; Me, Mesorhizobium; Ps, Pseudomonas; Rh, Rhizobium; Ra, Ralstonia; Rho, Rhodobacter; Si, Sinorhizobium; S, Salinicola; Se, Serratia; V, Vibrio; Y, Yersinia.* The scale bar indicates the mean number of substitutions per site. Bootstrap values were obtained from 1,000 replicates via neighbour-joining algorithms using the MEGA program. Only branches with values >50% are shown. The branches highlighted in red are sequences from *Alphaproteobacteria*, in green from *Betaproteobacteria* and in blue from *Gammaproteobacteria*. The sequence accession numbers are given in brackets.

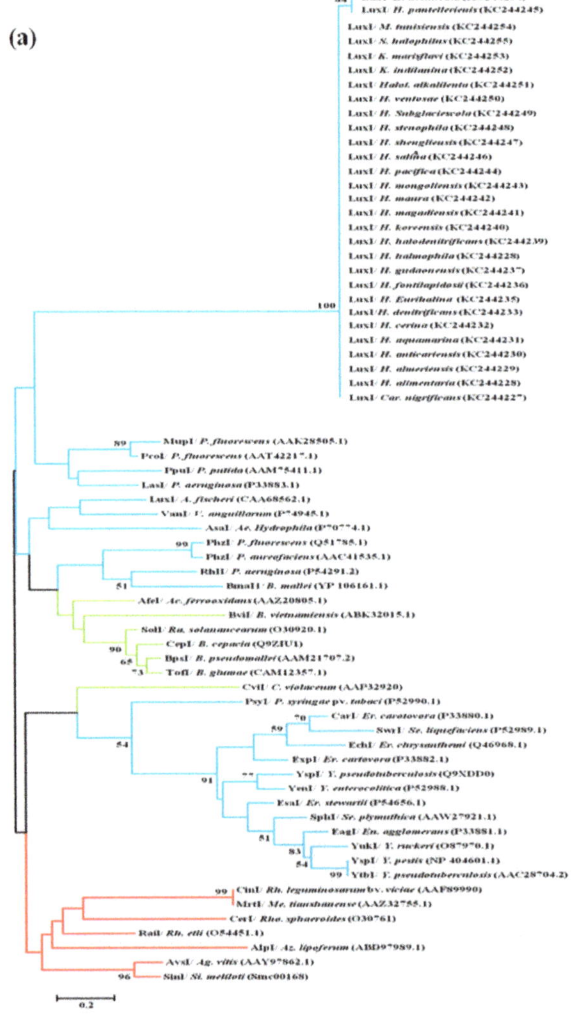

Figure 3. *Cont.*

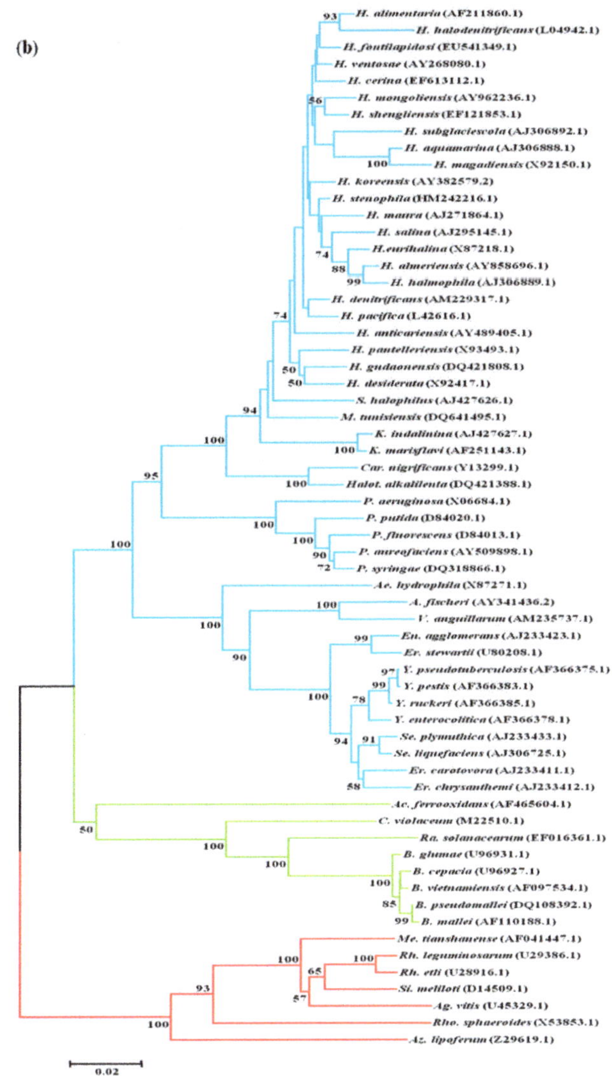

3. Experimental Section

Bacterial strains. We used the type strains of 43 species belonging to the *Halomonadaceae* family (Table 1). Strains were cultured at 32 °C in MY medium (3 g malt extract, 3 g yeast extract, 10 g glucose and 5 g peptone per litre) [52,53] modified with a balanced mixture of sea salts [54].

Table 1. Species of the *Halomonadaceae* family included in this study [9].

Species	Strain	Ecological Niches
1. *Carnimonas nigrificans*	CTCBS1^T	Cured meat, Spain
2. *Chromohalobacter salexigens*	DSM 3043^T	Solar saltern, Netherlands
3. *Cobetia marina*	219^T	Sea water, USA
4. *Halomonas alimentaria*	YKJ-16^T	Jeotgal, a traditional Korean fermented seafood, Korea
5. *H. almeriensis*	M8^T	Solar saltern, south-east Spain
6. *H. anticariensis*	FP35^T	Saline wetland, southern Spain
7. *H. aquamarina*	558^T	Pacific ocean
8. *H. campaniensis*	5AG^T	Mineral pool, Italy
9. *H. cerina*	SP4^T	Saline soil, Spain
10. *H. denitrificans*	M29^T	Saline water, Korea
11. *H. desiderata*	FB2^T	Municipal sewage works, Germany
12. *H. elongata*	1H9^T	Solar saltern, Netherlands
13. *H. eurihalina*	F9-6^T	Saline soil, Spain
14. *H. fontilapidosi*	CR-5^T	Saline soil, southern Spain
15. *H. gudaonensis*	SL014B-69^T	Saline soil contaminated by crude oil, China
16. *H. halmophila*	ACAM 71^T	Dead Sea, Israel
17. *H. halodenitrificans*	ATCC 13511^T	Meat in brine
18. *H. halodurans*	DSM 5160^T	Great Bay estuary, USA
19. *H. koreensis*	SS20^T	Solar saltern, Korea
20. *H. magadiensis*	21 MI^T	Soda lake, East-African Rift Valley
21. *H. maura*	S-31^T	Saltern, Morocco
22. *H. meridiana*	ACAM 246^T	Saline lake, Antarctic
23. *H. mongoliensis*	Z-7009^T	Soda lake, Mongolia
24. *H. nitroreducens*	11-S^T	Solar saltern, Chile
25. *H. organivorans*	G-16.1^T	Hypersaline habitats contaminated by aromatic organic compounds, southern Spain
26. *H. pacifica*	DSM 4742^T	Pacific ocean
27. *H. pantelleriensis*	AAP^T	Hard lake sand, Pantelleria island, Italy
28. *H. ramblicola*	RS-16^T	Saline soil, south-east Spain
29. *H. rifensis*	HK-31^T	Solar saltern, Morocco
30. *H. saccharevitans*	AJ275^T	Salt lake and a subterranean saline well, China
31. *H. salina*	F8-11^T	Saline soil, Spain
32. *H. shengliensis*	SLO14B-85^T	Saline soil contaminated with crude oil, China
33. *H. stenophila*	N12^T	Saline soil, Spain
34. *H. subglaciescola*	ACAM 12^T	Antarctic hypersaline, meromictic lake
35. *H. variabilis*	DSM 3051^T	Great Salt Lake, USA

Table 1. *Cont.*

Species	Strain	Ecological Niches
36. *H. ventosae*	Al-12^T	Saline soil, south-eastern Spain
37. *Halotalea alkalilenta*	AW-7^T	Alkaline olive-mill waste (alpechin), Greece
38. *Kushneria avicenniae*	MW2a^T	Salty leaves of *Avicennia germnans* trees growing near solar salterns, Puerto Rico
39. *K. indalinina*	CG2.1^T	Solar saltern, south-east Spain
40. *K. marisflavi*	SW32^T	Water from the Yellow Sea, Korea
41. *Modicisalibacter tunisiensis*	LIT2^T	Oilfield-water injection sample, southern Tunisia
42. *Salinicola halophilus*	CG4.1^T	Solar saltern, south-east Spain
43. *S. salarius*	M27^T	Saline water, Korea

Note: Species shaded in grey indicates that the LuxI homolog has been detected by PCR.

Chromobacterium violaceum CV026 was cultured at 30 °C in LB medium supplemented with 2.5 mM $CaCl_2$ and 2.5 mM $MgSO_4$ (LB/MC) and containing 50 μg kanamycin per mL [43]. *Agrobacterium tumefaciens* NTL4 (pZLR4) was cultured at 30 °C in LB medium supplemented with 2.5 mM $CaCl_2$ and 2.5 mM $MgSO_4$ (LB/MC), in MGM minimal medium (11 g Na_2HPO_4, 3 g KH_2PO_4, 0.5 g NaCl, 1 g glutamate, 10 g mannitol, 1 mg biotin, 27.8 mg $CaCl_2$ and 246 mg $MgSO_4$ per litre) containing 50 μg gentamycin per ml, and in AB medium [44,55].

Extraction and detection of AHLs. AHL molecules were extracted following the technique described in our previous studies [56,57]. Briefly, 20 mL cultures were grown until the early stationary phase (optical density of approximately 2.8 at 600 nm) and then extracted twice with equal volumes of dichloromethane. The extracts were dried and suspended in 40 μL of 70% v/v methanol.

To detect AHLs, an overnight culture of one of the AHL indicator strains [*Chromobacterium violaceum* CV026 or *Agrobacterium tumefaciens* NTL4 (pZLR4)] was diluted 1:100 in 5ml of the corresponding medium and poured onto LB/MC and AB supplemented with 80 μg of 5-bromo-4-chloro-3-indolyl-β-D-galactopyranoside (X-Gal) per ml agar plates. Once the plates were dry, paper disks 5 mm in diameter were placed onto them and the AHL samples applied. The assay plates were incubated overnight at 32 °C to allow the indicator organisms to grow and surround the paper disks with either purple or blue haloes.

Thin-layer chromatography analysis of AHLs. To characterize the AHLs, the samples were subjected to analytical and preparative thin-layer chromatography (TLC). AHL samples and standards were spotted onto a TLC plate and developed with 70% v/v methanol in water. The plate was air-dried and overlaid with top agar containing the *A. tumefaciens* NTL4 (pZLR4) indicator strain before being incubated at 32 °C. For the *A. tumefaciens* NTL4 (pZLR4) overlay, a 6–8 h culture in MGM medium was mixed with an equal volume of fresh medium, 1.5% w/v Bacto Agar and 80 μg of X-Gal per mL [26].

The standard AHLs used were: *N*-(β-ketocaproyl)-dl-homoserine lactone (3-oxo-C_6-HSL), *N*-hexanoyl-dl-homoserine lactone (C_6-HSL), *N*-octanoyl-dl-homoserine lactone (C_8-HSL) and *N*-decanoyl-dl-homoserine lactone (C_{10}-HSL) (Sigma®).

Chromosomal DNA extraction, autoinducer synthase gene amplification and sequencing. Chromosomal DNA was isolated and purified according to Marmur's protocol [58], modified by Martín-Platero and co-workers [59]. The purified DNA was dissolved in 50 μL doubly distilled water and checked by agarose gel electrophoresis [60] An internal segment of the autoinducer synthase gene was amplified from approximately 100 ng of chromosomal DNA by using the primers *luxI*-F: 5'-GGGAGATATATACTGTAA-3' and *luxI*-R: 5'-TGAGGTATTATTCTGCAA-3'. These primers were designed to target a highly conserved region of the *hanI* autoinducer synthase gene of *Halomonas anticariensis* FP35^{T}. The *hanI* gene is about 645 bp and the primer pair amplified approximately 386 bp of the conserved active site of the enzyme which contains the three conserved amino acids Arg71 (R71), Glu101 (E101) and Arg104 (R104) [27]. PCR entailed 30 cycles of 30 s at 95 °C, 30 s at 50 °C and 30 s at 72 °C. The annealing temperature was determined by PCR with a temperature gradient from 40 °C to 60 °C. All of the PCRs were run in a T100™ thermal cycler (Bio-Rad).

The PCR fragments were purified and sequenced with *luxI*-F or *luxI*-R primers using a BigDye Terminator Cycle Sequencing Kit in an ABI 3100 DNA sequencer (Applied Biosystems). The DNA sequences thus obtained were analysed using a BLAST search of the GenBank database [61] to align homologous regions of autoinducer synthase gene sequences from different isolates.

Phylogenetic analysis. A phylogenetic tree was constructed using version 4 of the MEGA (Molecular Evolutionary Genetics Analysis) software [62] after multiple alignments of the data by CLUSTALW [63] and the alignments were checked manually. Distances and clustering were determined according to the neighbour-joining method and bootstrap values were measured on the basis of 1,000 replications.

Nucleotide sequence accession number. The autoinducer synthase DNA sequences reported here have been deposited in the GenBank database under accession numbers from KC244227 to KC244255.

4. Conclusions

Screening for AHL signal molecules in 43 species belonging to the *Halomonadaceae* family revealed that the AHL-QS system is widespread within this group of bacteria. We did however find diversity within the AHL-profile signalling molecules produced by the different genera, and even between the molecules produced by different species from the same genus. Such variety would seem to be consistent with the ecological, physiological, metabolic and taxonomic diversity among them. The role of QS signalling in these extremophilic microorganisms remains to be elucidated and further work needs to be done to explore this bacterial cell-cell communication process in the multispecies communities.

Acknowledgments

This research was supported by grants from the Spanish Ministry of Education and Science (CGL2008-02399/BOS; AGL2009-07656), the Andalucian Government Council for Education, Science and Business (P07-CVI-03150) and from the Andalucian Research Project. Ali Tahrioui was supported by a postgraduate grant from the Junta de Andalucía. We thank our colleague Rafael de la Haba of the University of Seville for providing some of the type strains used in this work and J. Trout for revising and editing our English text.

References

1. Brock, T. Halophilic-blue-green algae. *Arch. Microbiol.* **1976**, *107*, 109–111.
2. Rodríguez-Valera, F. Characteristics and microbial ecology of hypersaline environments. In *Halophilic Bacteria*; Rodríguez-Valera, F., Ed.; CRC Press: Boca Raton, Florida, USA, 1988; Volume 1, pp. 3–30.
3. Cifuentes, A.; Antón, J.; De Wit, R.; Rodríguez-Valera, F. Diversity of *Bacteria* and *Archaea* in sulphate-reducing enrichment cultures inoculated from serial dilution of *Zostera noltii* rhizosphere samples. *Environ. Microbiol.* **2003**, *5*, 754–764.
4. González-Toril, E.; Llobet-Brossa, E.; Casamayor, E.O.; Amann, R.; Amils, R. Microbial ecology of an extreme acidic environment, the Tinto River. *Appl. Environ. Microbiol.* **2003**, *69*, 4853–4865.
5. Horikoshi, K.; Grant, W.D. *Extremophiles: Microbial Life in Extreme Environments*; Wiley-Liss: New York, USA, 1998.
6. Ventosa, A. Unusual micro-organisms from unusual habitats: Hypersaline environments. In *Prokaryotic Diversity: Mechanisms and Significance*; Logan, N.A., Lappin-Scott, H.M., Oyston, P.C.F., Eds.; Cambridge University Press: New York, USA, 2006; pp. 223–253.
7. Franzmann, P.D.; Wehmeyer, U.; Stackebrandt, E. *Halomonadaceae* fam. nov., a new family of the Class *Proteobacteria* to accommodate the genera *Halomonas* and *Deleya*. *Syst. Appl. Microbiol.* **1988**, *11*, 16–19.
8. Ventosa, A.; Nieto, J.J.; Oren, A. Biology of moderately halophilic aerobic bacteria. *Microbiol. Mol. Biol. Rev.* **1998**, *62*, 504–544.
9. Euzéby, J.P. List of Prokaryotic Names with Standing in Nomenclature. **2012**, Available online: http://www.bacterio.cict.fr/ (accessed on 15 November 2012).
10. Kushner, D.; Kamekura, M. Physiology of halophilic eubacteria. *Halophilic Bacteria*; Rodríguez-Valera, F. Ed.; CRC Press: Boca Raton, Florida, USA, 1988; Volume 1, pp. 87–103.
11. Ventosa, A.; Mellado, E.; Sánchez-Porro, C.; Márquez, M.C. Halophilic and halotolerant micro-organisms from soils. In *Microbiology of Extreme Soils*; Dion, P., Nautiyal, C., Eds.; Springer-Verlag: Heidelberg, Germany, 2008; Volume 13, pp. 87–115.
12. Nieto, J.J.; Carmen, V.M. Synthesis of osmoprotectants by moderately halophilic bacteria: Genetic and applied aspects. In *Recent Research and Development in Endocrinology*; Transworld Research Network: Kerala, India, 2002; pp. 403–418.
13. De la Haba, R.R.; Sánchez-Porro, C.; Márquez, M.C.; Ventosa, A. Taxonomy of Halophiles. In *Extremophiles Handbook*; Horikoshi, K., Ed.; Springer: New York, NY, USA, 2011.
14. Oren, A. Industrial and environmental applications of halophilic microorganisms. *Environ. Technol.* **2010**, *31*, 825–834.

15. Kaye, J.Z.; Baross, J.A. High incidence of halotolerant bacteria in Pacific hydrothermal-vent and pelagic environments. *FEMS Microbiol. Ecol.* **2000**, *32*, 249–260.
16. Oueriaghli, N.; González-Domenech, C.M.; Martínez-Checa, F.; Muyzer, M.; Quesada, E.; Béjar, V. Estudio molecular de la diversidad del género *Halomonas* en Rambla Salada mediante DGGE, CARD-FISH y análisis multivariable. Presented at the XIV Reunión del Grupo de Taxonomía Filogenia y Biodiversidad Microbiana (SEM), Granada, España, 10–11 May 2012.
17. Parker, C.T.; Sperandio, V. Cell-to-cell signalling during pathogenesis. *Cell. Microbiol.***2009**, *11*, 363–369.
18. Williams, P. Quorum sensing, communication and cross-kingdom signalling in the bacterial world. *Microbiology* **2007**, *153*, 3923–3938.
19. Jung, K.; Fried, L.; Behr, S.; Heermann, R. Histidine kinases and response regulators in networks. *Curr. Opin. Microbiol.* **2012**, *15*, 118–124.
20. González, J.E.; Marketon, M.M. Quorum sensing in nitrogen-fixing rhizobia. *Microbiol. Mol. Biol. Rev.* **2003**, *67*, 574–592.
21. Whitehead, N.A.; Barnard, A.M.; Slater, H.; Simpson, N.J.; Salmond, G.P. Quorum-sensing in Gram-negative bacteria. *FEMS Microbiol. Rev.* **2001**, *25*, 365–404.
22. Eberhard, A.; Longin, T.; Widrig, C.A.; Stranick, S.J. Synthesis of the *lux* gene autoinducer in *Vibrio fischeri* is positively autoregulated. *Arch. Microbiol.* **1991**, *155*, 294–297.
23. Fuqua, W.C.; Winans, S.C.; Greenberg, E.P. Quorum sensing in bacteria: the LuxR-LuxI family of cell density-responsive transcriptional regulators. *J. Bacteriol.* **1994**, *176*, 269–275.
24. Swift, S.; Williams, P.; Stewart, G.S.A.B. *N*-acyl homoserine lactones and quorum sensing in proteobacteria. In *Cell-Cell Signaling in Bacteria.*; Dunny, G.M., Winans, S.C., Eds.; American Society of Microbiology Press: Washington, DC, USA, 1999; pp. 291–314.
25. Case, R.J.; Labbate, M.; Kjelleberg, S. AHL-driven quorum-sensing circuits: their frequency and function among the *Proteobacteria. The ISME J.* **2008**, *2*, 345–349.
26. Llamas, I.; Quesada, E.; Martínez-Cánovas, M.J.; Gronquist, M.; Eberhard, A.; González, J.E. Quorum sensing in halophilic bacteria: Detection of *N*-acyl-homoserine lactones in the exopolysaccharide-producing species of *Halomonas*. *Extremophiles.* **2005**, *9*, 333–341.
27. Tahrioui, A.; Quesada, E.; Llamas, I. The *hanR/hanI* quorum-sensing system of *Halomonas anticariensis*, a moderately halophilic bacterium. *Microbiology* **2011**, *157*, 3378–3387.
28. Tahrioui, A.; Quesada, E.; Llamas, I. Genetic and phenotypic analysis of the GacS/GacA system in the moderate halophile *Halomonas anticariensis*. *Microbiology* **2013**, *159*, 461–473.
29. Amjres, H.; Béjar, V.; Quesada, E.; Abrini, J.; Llamas, I. *Halomonas rifensis* sp. nov., an exopolysaccharide-producing, halophilic bacterium isolated from a solar saltern. *Int. J. Syst. Evol. Microbiol.* **2011**, *61*, 2600–2605.
30. Dobson, S.J.; Franzmann, P.D. Unification of the genera *Deleya* (Baumann et al. 1983), *Halomonas* (Vreeland et al. 1980), and *Halovibrio* (Fendrich 1988) and the Species *Paracoccus halodenitrificans* (Robinson and Gibbons 1952) into a single genus, *Halomonas*, and placement of the Genus *Zymobacter* in the family *Halomonadaceae*. *Int. J. Syst. Bacteriol.* **1996**, *46*, 550–558.

31. González-Domenech, C.M.; Béjar, V.; Martínez-Checa, F.; Quesada, E. *Halomonas nitroreducens* sp. nov., a novel nitrate- and nitrite-reducing species. *Int. J. Syst. Evol. Microbiol.* **2008**, *58*, 872–876.
32. González-Domenech, C.M.; Martínez-Checa, F.; Quesada, E.; Béjar, V. *Halomonas cerina* sp. nov., a moderately halophilic, denitrifying, exopolysaccharide-producing bacterium. *Int. J. Syst. Evol. Microbiol.* **2008**, *58*, 803–809.
33. González-Domenech, C.M.; Martínez-Checa, F.; Quesada, E.; Béjar, V. *Halomonas fontilapidosi* sp. nov., a moderately halophilic, denitrifying bacterium. *Int. J. Syst. Evol. Microbiol.* **2009**, *59*, 1290–1296.
34. Llamas, I.; Béjar, V.; Martínez-Checa, F.; Martínez-Cánovas, M.J.; Molina, I.; Quesada, E. *Halomonas stenophila* sp. nov., a halophilic bacterium that produces sulphate exopolysaccharides with biological activity. *Int. J. Syst. Evol. Microbiol.* **2011**, *61*, 2508–2514.
35. Luque, R.; Béjar, V.; Quesada, E.; Martínez-Checa, F.; Llamas, I. *Halomonas ramblicola* sp. nov., a moderately halophilic bacterium from Rambla Salada, a Mediterranean hypersaline rambla in south-east Spain. *Int. J. Syst. Evol. Microbiol.* **2012**, *62*, 2903–2909.
36. Martínez-Cánovas, M.J.; Béjar, V.; Martínez-Checa, F.; Quesada, E. *Halomonas anticariensis* sp. nov., from Fuente de Piedra, a saline-wetland, wildfowl reserve in Málaga, Southern Spain. *Int. J. Syst. Evol. Microbiol.* **2004**, *54*, 1329–1332.
37. Martínez-Cánovas, M.J.; Quesada, E.; Llamas, I.; Béjar, V. *Halomonas ventosae* sp. nov., a moderately halophilic, denitrifying, exopolysaccharide-producing bacterium. *Int. J. Syst. Evol. Microbiol.* **2004**, *54*, 733–737.
38. Quesada, E.; Valderrama, M.J.; Bejar, V.; Ventosa, A.; Gutierrez, M.C.; Ruiz-Berraquero, F.; Ramos-Cormenzana, A. *Volcaniella eurihalina* gen. nov., sp. nov., a moderately halophilic nonmotile Gram-negative rod. *Int. J. Syst. Bacteriol.* **1990**, *40*, 261–267.
39. Valderrama, M.J.; Quesada, E.; Béjar, V.; Ventosa, A.; Gutierrez, M.C.; Ruiz-Berraquero, F.; Ramos-Cormenzana, A. *Deleya salina* sp. nov., a moderately halophilic Gram-negative bacterium. *Int. J. Syst. Bacteriol.* **1991**, *41*, 377–384.
40. Martínez-Checa, F.; Béjar, V.; Martínez-Cánovas, M.J.; Llamas, I.; Quesada, E. *Halomonas almeriensis* sp. nov., a moderately halophilic, exopolysaccharide-producing bacterium from Cabo de Gata, Almería, south-east Spain. *Int. J. Syst. Evol. Microbiol.* **2005**, *55*, 2007–2011.
41. Mellado, E.; Moore, E.R.B.; Nieto, J.J.; Ventosa, A. Phylogenetic inferences and taxonomic consequences of 16S ribosomal DNA sequence comparison of *Chromohalobacter marismortui*, *Volcaniella eurihalina*, and *Deleya salina* and reclassification of *V. eurihalina* as *Halomonas eurihalina* comb. nov. *Int. J. Syst.Bacteriol.* **1995**, *45*, 712–716.
42. Steindler, L.; Venturi, V. Detection of quorum-sensing *N*-acyl homoserine lactone signal molecules by bacterial biosensors. *FEMS Microbiol. Lett.* **2007**, *266*, 1–9.
43. McClean, K.H.; Winson, M.K.; Fish, L.; Taylor, A.; Chhabra, S.R.; Cámara, M.; Daykin, M.; Lamb, J.H.; Swift, S.; Bycroft, B.W.; *et al.* Quorum sensing and *Chromobacterium violaceum*: Exploitation of violacein production and inhibition for the detection of *N*-acyl homoserine lactones. *Microbiology* **1997**, *143*, 3703–3711.

44. Shaw, P.D.; Ping, G.; Daly, S.L.; Cha, C.; Cronan, J.E., Jr.; Rinehart, K.L.; Farrand, S.K. Detecting and characterizing *N*-acyl-homoserine lactone signal molecules by thin-layer chromatography. *Proc. Natl. Acad. Sci. USA* **1997**, *94*, 6036–6041.
45. García-Aljaro, C.; Eberl, L.; Riedel, K.; Blanch, A. Detection of quorum-sensing-related molecules in *Vibrio scophthalmi*. *BMC Microbiol.* **2008**, *8*, 138.
46. Yang, Q.; Han, Y.; Zhang, X.H. Detection of quorum sensing signal molecules in the family *Vibrionaceae*. *J. Appl. Microbiol.* **2011**, *110*, 1438–1448.
47. Parsek, M.R.; Greenberg, E.P. Acyl-homoserine lactone quorum sensing in Gram-negative bacteria: A signaling mechanism involved in associations with higher organisms. *Proc. Natl. Acad. Sci. USA* **2000**, *97*, 8789–8793.
48. Parsek, M.R.; Schaefer, A.L.; Greenberg, E.P. Analysis of random and site-directed mutations in rhlI, a *Pseudomonas aeruginosa* gene encoding an acylhomoserine lactone synthase. *Mol. Microbiol.* **1997**, *26*, 301–310.
49. Hanzelka, B.L.; Parsek, M.R.; Val, D.L.; Dunlap, P.V.; Cronan, J.E.; Greenberg, E.P. Acylhomoserine lactone synthase activity of the *Vibrio fischeri* AinS protein. *J. Bacteriol.* **1999**, *181*, 5766–5770.
50. Milton, D.L.; Chalker, V.J.; Kirke, D.; Hardman, A.; Cámara, M.; Williams, P. The LuxM Homologue VanM from *Vibrio anguillarum* directs the synthesis of *N*-(3-hydroxyhexanoyl)-homoserine lactone and *N*-hexanoyl-homoserine lactone. *J. Bacteriol.* **2001**, *183*, 3537–3547.
51. Laue, B.E.; Jiang, Y.; Chhabra, S.R.; Jacob, S.; Stewart, G.S.A.B.; Hardman, A.; Downie, J.A.; O'Gara, F.; Williams, P. The biocontrol strain *Pseudomonas fluorescens* F113 produces the Rhizobium small bacteriocin, N-(3-hydroxy-7-cis-tetradecenoyl) homoserine lactone, via HdtS, a putative novel *N*-acylhomoserine lactone synthase. *Microbiology* **2000**, *146*, 2469–2480.
52. Haynes, W.C.; Wickerham, L.J.; Hesseltine, C.W. Maintenance of cultures of industrially important microorganisms. *Appl. Microbiol.* **1955**, *3*, 361–368.
53. Moraine, R.A.; Rogovin, P. Kinetics of polysaccharide B-1459 fermentation. *Biotechnol. Bioeng.* **1966**, *8*, 511–524.
54. Rodríguez-Valera, F.; Ruíz-Berraquero, F.; Ramos-Cormenzana, A. Characteristics of the heterotrophic bacterial populations in hypersaline environments of different salt concentrations. *Microb. Ecol.* **1981**, *7*, 235–243.
55. Cha, C.; Gao, P.; Chen, Y.-C.; Shaw, P.D.; Farrand, S.K. Production of acyl-Homoserine lactone quorum-sensing signals by Gram-negative plant-associated bacteria. *Mol. Plant. Microbe In.* **1998**, *11*, 1119–1129.
56. Marketon, M.M.; Gronquist, M.R.; Eberhard, A.; González, J.E. Characterization of the *Sinorhizobium meliloti sinR/sinI* locus and the production of novel *N*-acyl homoserine lactones. *J. Bacteriol.* **2002**, *184*, 5686–5695.
57. Llamas, I.; Keshavan, N.; González, J.E. Use of *Sinorhizobium meliloti* as an indicator for specific detection of longchain *N*-acyl homoserine lactones. *Appl. Environ. Microbiol.* **2004**, *70*, 3715–3723.

58. Marmur, J. A procedure for the isolation of deoxyribonucleic acid from micro-organisms. *J. Mol. Biol.* **1961**, *3*, 208–218.
59. Martín-Platero, A.M.; Valdivia, E.; Maqueda, M.; Martínez-Bueno, M. Fast, convenient, and economical method for isolating genomic DNA from lactic acid bacteria using a modification of the protein "salting-out" procedure. *Anal. Biochem.* **2007**, *366*, 102–104.
60. Sambrook, J.; Russel, D.W., *Molecular Cloning: A Laboratory Manual* 3rd Ed.; Cold Spring Harbor Laboratory Press: New York, USA, 2001.
61. National Center for Biotechnology Information, N. Available online: http://www.ncbi.nlm.nih.gov/ (accessed on 15 November 2012).
62. Tamura, K.; Dudley, J.; Nei, M.; Kumar, S. MEGA4: Molecular Evolutionary Genetics Analysis (MEGA) Software Version 4.0. *Mol.Biol. Evol.* **2007**, *24*, 1596–1599.
63. Thompson, J.D.; Gibson, T.J.; Plewniak, F.; Jeanmougin, F.; Higgins, D.G. The CLUSTAL_X windows interface: Flexible strategies for multiple sequence alignment aided by quality analysis tools. *Nucleic Acids Res.* **1997**, *25*, 4876–4882.

Heterotrophic Protists in Hypersaline Microbial Mats and Deep Hypersaline Basin Water Columns

Virginia P. Edgcomb and Joan M. Bernhard

Abstract: Although hypersaline environments pose challenges to life because of the low water content (water activity), many such habitats appear to support eukaryotic microbes. This contribution presents brief reviews of our current knowledge on eukaryotes of water-column haloclines and brines from Deep Hypersaline Anoxic Basins (DHABs) of the Eastern Mediterranean, as well as shallow-water hypersaline microbial mats in solar salterns of Guerrero Negro, Mexico and benthic microbialite communities from Hamelin Pool, Shark Bay, Western Australia. New data on eukaryotic diversity from Shark Bay microbialites indicates eukaryotes are more diverse than previously reported. Although this comparison shows that eukaryotic communities in hypersaline habitats with varying physicochemical characteristics are unique, several groups are commonly found, including diverse alveolates, strameonopiles, and fungi, as well as radiolaria. Many eukaryote sequences (SSU) in both regions also have no close homologues in public databases, suggesting that these environments host unique microbial eukaryote assemblages with the potential to enhance our understanding of the capacity of eukaryotes to adapt to hypersaline conditions.

Reprinted from *Life*. Cite as: Edgcomb, V.P.; Bernhard, J.M. Heterotrophic Protists in Hypersaline Microbial Mats and Deep Hypersaline Basin Water Columns. *Life* **2013**, *3*, 346-362.

1. Introduction

Hypersaline waters (generally >10% NaCl; [1]) have salinities that exceed the 3.5% total salt of most oceans. These include salt or soda lakes, salterns, coastal lagoons, and deep hypersaline anoxic brines, and contain Bacteria, Archaea and Eukarya [2]. Hypersaline environments are characterized by a low water content or water activity (a_w) because of the high-salt concentrations [1]; this presents challenges for organisms living in these habitats. With increasing salinity, halophilic (requiring salinities greater than normal seawater salinity) and halotolerant (able to survive salinities greater than normal seawater) eukaryotes appear to comprise smaller fractions of communities. Typically, growth does not occur at a_w below 0.72 because, below this, there is not enough water available for dissolving nutrients, general metabolic processes, and for hydrating proteins and nucleic acids [3]. Other stresses may also be present in hypersaline habitats, such as intense solar radiation and elevated temperatures in shallow lakes, solar salterns and salt production plants, intense pressures in deep hypersaline anoxic basins, and vastly different ionic compositions of the salts themselves. For instance, if the salt in the brine originates from seawater (thalassohaline) it will be dominated by sodium chloride, however, if the salts originate from other sources (athalassohaline), other ions will dominate, and the ionic composition can vary widely [1].

To prevent loss of cellular water to the environment, microbial eukaryotes (and other halotolerant and halophilic organisms) require a way to balance the osmotic pressure created by their hypersaline habitat. Some halophiles accumulate solutes within the cytoplasm and others use sodium pumps to expel sodium ions out of the cell while concentrating potassium ions inside the cell to balance osmotic pressure [4]. Some halotolerant algae are known to balance osmotic pressure by producing or taking up organic molecules from the environment, such as glycerol [1]. A full discussion of adaptive mechanisms of microorganisms to hypersaline habitats is outside the scope of this review. Extensive literature exists describing the adaptive mechanisms for Bacteria and Archaea and, to a lesser extent, Eukarya to hypersaline conditions, including high guanine to cytosine ratios in DNA, high concentrations of acidic residues on exteriors of proteins, and unique lipids, cellular architectures, pigments, physiologies and metabolisms (e.g., [1]; and extensively discussed in [5]).

Protists are an essential component of microbial food webs that play a central role in global biogeochemical cycles, thus making them key players in sustaining the healthy functioning of any ecosystem. Protists include autotrophic (capable of making organic molecules from inorganic sources via photosynthesis, e.g., algae) and heterotrophic (those that prey on preformed organic carbon, including other microbes, and hence contribute to pools of dissolved organic carbon through release of metabolic wastes and "sloppy feeding", e.g., ciliates and flagellates) microbial eukaryotes. Microscopical observations of hypersaline habitats long ago revealed heterotrophic protists to be present (e.g., [6–10] and more recently, [11–13]). However, studies indicated protists were rare or nonexistent in extremely hypersaline (over 30%) environments (e.g., [13–18]) (for contrasting viewpoint see [19]). Autotrophs and heterotrophs of moderately hypersaline habitats (6–15%) are suspected of being euryhaline representatives of marine forms that have adapted to life in extremely salty conditions (e.g., [10–13,20,21]. Such studies based primarily upon microscopic observations can sometimes underestimate diversity, particularly with respect to heterotrophic nanoflagellates and naked amoebae (discussed in [22]). The heterotrophic nanoflagellates, ciliates and amoebae appear to be the major groups of protists adapted to life in hypersaline environments (e.g., [12]). Most investigations of halophilic ciliate and flagellated protists have examined habitats with 10–20% salinities, and some of the taxa identified appear ubiquitous in hypersaline environments (e.g., the marine green flagellate *Dunaliella* spp. in hypersaline lakes; [16]), while some appear uniquely adapted to specific ratios of particular ions (e.g., [23]). The physiological underpinnings of this remain to be determined.

Heterotrophic bicosoecids and non pigmented chrysomonads belonging to stramenopiles are protists known to be important components of many aquatic microbial communities [24]. Halotolerant and halophilic bicosoecids have been isolated from hypersaline locations that can grow in both normal seawater and 17.5% salinity; these have not been previously reported in marine or freshwater environments using culturing or PCR-based approaches [22]. At higher salinities (up to 18%), the diversity of stramenopiles appears to be significantly less than in typical marine conditions, consisting primarily of *Halocafeteria*, and other "Cafeteria" species [22]. Investigations of higher salinities (e.g., deep brine basins with salinity of 28%) did not reveal stramenopiles [25]. Dinoflagellates are another group of heterotrophic protists frequently documented in environments with salinities between 6 and 30% (e.g., [11,26–28]).

Our view of heterotrophic protist diversity in hypersaline environments is expanding, particularly with additional data from molecular-based surveys of these habitats. There is increasing evidence of extremely halophilic heterotrophs that are distinct from marine or freshwater forms, such as the ciliate *Trimyema koreanum* sp. nov. in solar salterns with salinity of 29% [29], and a diversity of other heterotrophs from nearly saturated brines (30% or more salinity) that could not grow at salinities less than 7.5% [29,30–32]. Additionally, there are molecular and microscopic observations of heterotrophic protists in habitats up to 36% salinity (e.g., [25,33,34]). This paper compares heterotrophic protists in two shallow hypersaline habitats: hypersaline microbial mats in solar salterns and microbialites in Shark Bay, Australia, with deep anoxic brines at the bottom of the Mediterranean Sea. We focus on heterotrophs because autotrophic communities in deep-sea hypersaline habitats are not comparable to shallow water communities within the photic zone. There are many physicochemical differences between these habitats that likely contribute to variations in microbial communities between locations, including water depth, temperature, salinity, salt ionic composition, concentrations of sulfide and methane, *etc*. A comparison of protist populations in these habitats reveals that protists are present in hypersaline waters and sediments, and that physicochemical differences in habitats select for unique protist populations in different hypersaline habitats.

2. Views of Eukaryotic Diversity in Selected Hypersaline Habitats

2.1. Microbialites in Hypersaline Shark Bay, Australia

Hamelin Pool, in Shark Bay, Western Australia, is one of the few sites where microbialites, or lithifying microbial mats, are forming today (Figure 1). These microbialites, are found along the margins of Hamelin Pool [35], and form by trapping and binding of carbonates by extracellular polymeric substances (EPS), produced by filamentous cyanobacteria and other bacteria (e.g., [36]). Because fossilized microbialites comprise the earliest visible record of life on Earth (e.g., [37,38]), the modern Hamelin Pool microbialites have been intensively studied. Waters of Hamelin Pool are typically 6–7% salinity.

Figure 1. Hamelin Pool, Shark Bay, W. Australia microbialites. (**A**) smooth mat; (**B**) colloform mat.

Tong [39] reported 41 different species of heterotrophic nanoflagellates based on light microscopic observations of hypersaline waters of Shark Bay. These included apusomonads, cercomonads, choanoflagellates, cryptomonads, euglenids, heteroloboseids, stramenopiles, and several groups of uncertain taxonomic affiliation. When sediments from four different sites in the Western Australia Shark Bay area with differing salinities were examined, fewer species were observed to overlap with previously documented species from normal marine environments in the most hypersaline samples (four times the salinity of seawater) than samples from waters that were only two times the salinity of seawater [11]. This finding is consistent with the generally accepted idea that eukaryotic microbial diversity declines as hypersalinity increases, as discussed above.

Examination of eukaryotic microbial diversity has been more limited than investigations of prokaryotic diversity in Shark Bay microbialites. Al-Qassab *et al.* [21] identified flagellate protists in modern stromatolites in Shark Bay, and foraminiferal tests (shells) were observed in some thrombolites [40]. Different microbialite types have varied degrees of lamination, and groups of eukaryotes such as the foraminifera, which are known bioturbators of sediments, have the potential to influence sediment fabric (discussed in Bernhard *et al.*, 2013 [41]).

Using a combination of next generation and Sanger-based sequencing approaches we gathered as comprehensive a picture of eukaryotic diversity as possible in the 0–1 cm and 1–2 cm intervals of different microbialite samples collected in Hamelin Pool, Shark Bay in June 2011. Salinity is reported in Practical Salinity Units (PSU), which is a measure of salt content of water based upon electrical conductivity of a sample relative to reference standard of seawater. Normal seawater at 15 °C has a salinity of 3.5%, or 35 PSU. The salinity of the overlying water at the Shark Bay sampling site was 66–72 PSU (Table 1). We sampled pustular mats, which are irregular, clotted mats; colloform mats, which are coarse, laminoid wavy mats; and smooth mats, which are fine, laminoid structures (nomenclature according to [35,42]). Total RNA was extracted using the FastRNA Pro Soil-Direct Kit according to the manufacturer's instructions, with the exception that a Turbo DNase step (Ambion) was included prior to the RNA Matrix cleanup. Total RNA was converted to cDNA and used as a template for all PCR reactions using a Superscript One-Step RT-PCR kit (Invitrogen) and eukaryotic small subunit ribosomal RNA V4 hypervariable region PCR primers [43] or general primers for foraminifera (S14F1/S17, [44]) to focus on the more active fraction of the community. Foraminifera-specific amplifications were necessary because general V4 primers do not detect most foraminifera. Foraminiferal PCR products were cloned into pCR4-TOPO using the TOPO TA Cloning Kit (Invitrogen) for Sanger sequencing (one 96-well plate per microbialite sample), and pyrotags were submitted for Titanium pyrosequencing. Sequence data were processed for quality control, clustered into operational taxonomic units (OTUs) at 97% sequence identity in QIIME [45], and taxonomic assignments were made using JAguc [46].

In contrast to previous analyses on microbial eukaryotes in Shark Bay microbialites, we found diverse communities of eukaryotes in all microbialite samples examined (Figure 2). Eukaryotic signatures were dominated by Alveolata (10–50% of OTUs), unclassified eukaryotes (5–45%), and stramenopiles (10–30%). The dominant alveolates differed in different microbialite types. In colloform mats, alveolates were dominated by Heterocapsaceae and *Protodinium* (Dinophyceae), and more diversity was observed in the 0–1 cm depth interval than in the 1–2 cm depth. In pustular mats, the greatest diversity of alveolates was observed, with OTUs from both ciliates and dinoflagellates. Alveolates in smooth mats were dominated by dinoflagellate OTUs affiliating with Gymnodiniales (40%) in the 0–1 cm interval, and in the 1–2 cm interval, 90% of OTUs affiliated with ciliates (Litostomatea). Stramenopile OTUs were dominated in all mat samples by representatives of labyrinthulids affiliating with Thraustrochytridae (60–85% of stramenopile OTUs). The high number of OTUs that could not be assigned a taxonomy based on BLASTn (<80% sequence similarity to GenBank sequences) suggests the presence of novel eukaryotic lineages in Hamelin Pool microbialites.

Table 1. Physicochemical data for several Eastern Mediterranean deep hypersaline anoxic basins and Hamelin Pool, Shark Bay. In this context, "interface" equates to the halocline.

Sample	Coordinates		Water Depth (m)	Total Salinity (PSU)	Oxygen (mL/L)
Discovery Interface [2]	35°19'N	21°41'E	3,580	70 [1]	0.50
Thetis Interface [3]	34°40'N	22°08'E	3,259	80	0.68
Thetis Brine [3]	34°40'N	22°08'E	3,415	340	0
Bannock Brine [2]	34°17'N	20°00'E	3,790	280	0
Bannock Interface [2]	34°17'N	20°00'E	3,300	246	0.50
Atalante Upper Interface [5]	35°18'N	21°23'E	3,499	39	0.44
Atalante Lower Interface [5]	35°18'N	21°23'E	3,501	365	0
Urania Interface [6]	35°13'N	21°28'E	3,467	63	1.22
Hamelin Pool microbialites [7]	26°15'S	114°14'E	0–3	66–72	supersaturated at 0–1 cm, 0 at 1–2 cm
Guerrero Negro saltern mats [4]	27°41'N	113°55'W	1–2	90	na

[1] Using the conventional sensor mounted on the Niskin rosette, the measurement of conductivity is not reliable in athalassohaline brines enriched by divalent cations. na = not available. [2] Edgcomb *et al.*; 2009 [47]; [3] Stock *et al.* 2011 [48]; [4] Bebout *et al.* 2002 [49]; [5] Alexander *et al.* 2009 [25]; [6] Orsi *et al.* 2012 [50]; [7] this study.

The highest number of rhizarian OTUs in the surface 0–1 cm samples was recovered from pustular mat samples. In all other mat samples, differences were observed in foraminiferal OTUs in different microbialite types, and OTUs were detected from within Rotaliida, Textulariida, Milliolina, and Allogromiida (thecate, non-mineralized forms). The latter group is of particular interest because they are modern representatives of basal foraminifera that likely evolved in the Precambrian (e.g., [51]), when microbialites dominated Earth's biosphere.

Figure 2. Stacked histogram of eukaryotic operational taxonomic units (out) composition of (97% sequence similarity, weighted data presentation) in Hamelin Pool, Australia microbialite and water samples based on eukaryote sequences (SSU) rRNA signatures (cDNA template). Y-axis corresponds to fraction of OTUs affiliating with each grouping out of 100%. S = smooth mat, C = colloform, P = Pustular.

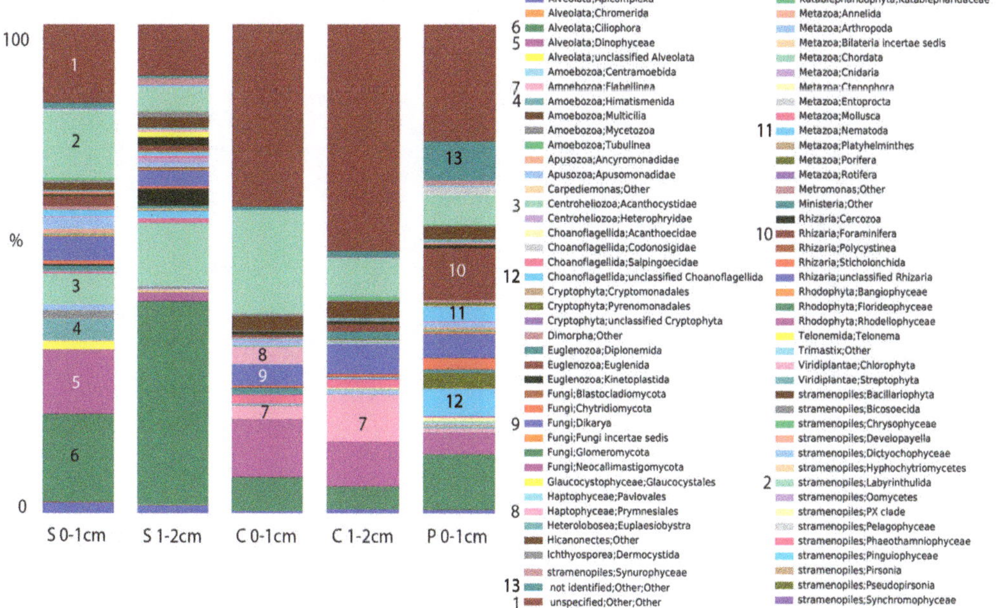

As an independent means to assess eukaryotic presence in microbialites, we performed microscopic studies using a viability indicator. CellTracker® Green CMFDA (CTG; Invitrogen) was used as an indicator of active hydrolytic esterase activity in marine populations. The Fluorescently Labeled Embedded Core (FLEC) method [52], which preserves the life positions of microbes in a sample, was also used to assess the living positions of eukaryotes within microbialites by collecting syringe cores of 1.5-cm inner diameter from the same microbialites sampled for sequence analysis. At least five replicate cores were collected per microbialite structure. Along with overlying seawater, cores were incubated with 1 μM CTG at ambient light and temperature for approximately 6–8 hours, after which 3% glutaraldehyde in 0.1 M cacodylate sodium salt buffer was introduced into each core for approximately one hour. Cores were then rinsed three times in buffer and transported to the laboratory for further processing. Processing steps through polymerization followed Bernhard *et al.* [52] except that cores were embedded in LR White rather than the typical Spurrs' resin. Polymerized cores were sectioned coronally with an Isomet low speed rock saw. Both sides of each section were scanned with a Leica FLIII stereomicroscope equipped with epifluorescence capabilities to identify fluorescent objects suggestive of eukaryotes (size, shape). Promising targets were imaged with an Olympus Fluoview 300 Laser Scanning Confocal Microscope (LSCM). FLEC analysis supports sequence analysis indicating that microbial eukaryotes inhabit the microbialites (Figure 3). For example, foraminifera are easily identified in many FLEC sections (e.g., Figure 3B,C). Other picoeukaryotes (0.2–2 μm) are more difficult to identify, but putative protists can be quite common (Figure 3).

Figure 3. LSCM images of microbialites in FLEC sections from Hamelin Pool. (**A**) Sediment-water interface showing masses of cyanobacteria comprising pustular mat; (**B**) Foraminifer (*) inhabiting smooth mat; (**C**) Foraminifera (*) inhabiting colloform mat. Arrows = indeterminate protists; * = foraminifer; + = ooid with concentric layering. Scales = 200 μm.

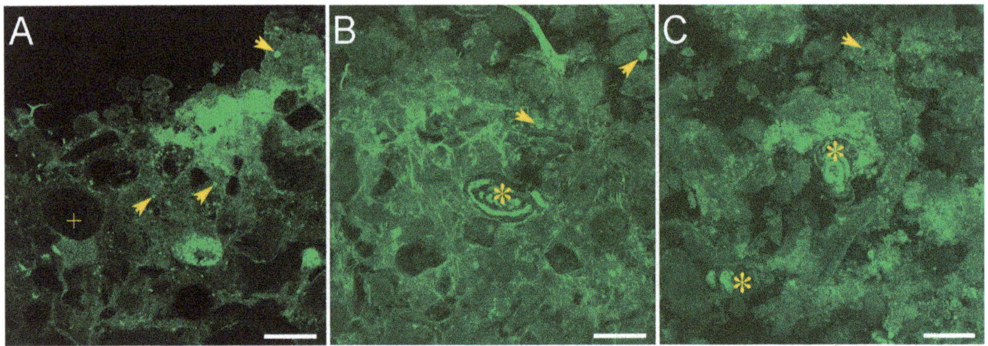

2.2. Microbial Mats in Hypersaline Solar Salterns

Extensive microbial mats grow within the hypersaline lagoons of the Exportadora de Sal SA saltern in Guerrero Negro, Baja California Sur, Mexico. Salterns, or salt works, involve a series of sun-baked lagoons where seawater is gradually evaporated, to the point where salts including sodium chloride precipitate, and are harvested. Feazal *et al.* [28] examined eukaryotic diversity in Guerrero Negro mat samples collected from a pond with 90 PSU salinity [44]. The mat cores from which they extracted DNA were 6 cm thick, and were sectioned into 1 mm intervals for 0–6 mm and ~1 cm intervals downcore. The extent of eukaryotic diversity these authors detected using Sanger sequencing of small subunit ribosomal RNA genes (SSU rRNA) was extremely low, with only 15 ribotypes identified among 890 clones analyzed [28]. There are several possible explanations for the low diversity relative to the microbialite samples, including potential primer biases, swamping of clone libraries by metazoan (377 nematode signatures from two taxa out of 890 clones analyzed) SSU rRNA signatures in their DNA-based analysis, and screening clones based on Terminal Restriction Fragment Length Polymorphism (T-RFLP), which may have underestimated diversity. An RNA-based analysis of several of these salterns coupling next generation molecular methods with microscopy would provide further insights into eukaryotic diversity in this hypersaline habitat.

2.3. Deep Hypersaline Anoxic Basins in the Eastern Mediterranean Sea (DHABs)

Deep brines on the seafloor exist in different locations, including Orca Basin in the Gulf of Mexico, the Red Sea, and the Eastern Mediterranean Sea. DHAB, which are found in many locations in the Eastern Mediterranean Sea are discussed here, and are thought to have formed through the dissolution of buried Messinian evaporitic deposits, followed by accumulation of brines in sea floor depressions ([53] and references therein).

Several studies of E. Mediterranean DHAB water columns have extended our knowledge of the environmental factors that define the limits of life for microbial eukaryotes and have provided insights into novel eukaryotic diversity in these planktic habitats (e.g., [25,33,42,43,54,55]). Recovery of sequences of many taxonomic groups in these studies with no known homologues in public databases suggests these pelagic habitats harbor organisms with possibly novel metabolic/physiological characteristics.

E. Mediterranean DHABs such as Discovery Basin (Figure 4) are found more than 3,000 m below sea level. There is very little mixing of these brines with the overlying seawater due to their high density (typically ranging from 1.13 to 1.35 × 10^3 kg m^{-3} relative to Mediterranean seawater 1.03 × 10^3 kg m^{-3}). The steep halocline that results, can be only a few meters or less in thickness, oxygen concentrations drop to undetectable at the base of the halocline, and salinity increases dramatically, often to around 10 times normal seawater [56]. The chemical compositions of the different basins are distinct, with widely varying concentrations of sulfide, methane, and various cations and anions ([56] Supplementary Material) (Table 1). Some brines, such as that found in Discovery Basin, are athallasohaline, with Mg^{2+} concentrations up to 5,000 mM compared with 300–650 mM in other basins, and ca. 60 mM in regular seawater. All other basins reported here are thalassohaline. Sodium concentrations can vary widely in the different basins. For example, sodium is 70 mM in Discovery Basin brine, and almost 4,700 mM in Atalante Basin brine [56]. Combined with the very high pressures associated with the depths of these basins, DHABs represent some of the most extreme habitats on Earth.

Figure 4. Discovery Basin, Eastern Mediterranean Sea (3,582 m depth). Image taken with ROV *Jason*, showing the Deep Hypersaline Anoxic Basins (DHAB) "beach" (white zone where the halocline intersects the seafloor) at the edge of the brine pool (right). Note floating garbage in the brine pool. ©Woods Hole Oceanographic Institution.

In the water column, abundant chemosynthetic bacterial communities along DHAB oxyclines (e.g., [56–60]) appear to support active pelagic protist communities there [25,33,42,43]. Moderately hypersaline systems are known to sustain rich and diverse communities of mostly halotolerant eukaryotes [13]. Habitats with salinities in excess of 30% are not thought to harbor significant protist diversity [14–18] (but see opposing viewpoint in [19]). Although previous studies suggested habitats with salinities in excess of 30% did not harbor significant protist diversity (see above), initial investigations into protist diversity in several Eastern Mediterranean DHAB haloclines and brines using DNA-based [42] and RNA-based [25] molecular approaches suggested that these pelagic habitats not only harbor diverse protistan communities, but that these communities are largely unique to the water columns of these basins and share little overlap with overlying waters with typical marine salinity and oxygen tension.

The first indication that haloclines and brines of DHABs support active microbial eukaryotes came from two studies in 2009 that presented profiles of SSU rRNA genes in clone libraries, one based on RNA extracted from halocline samples from Atalante basin (upper halocline 39 PSU, lower halocline 365 PSU) [25] and one based on DNA extracted from haloclines and brines of Bannock (halocline 246 PSU and brine 280 PSU) and Discovery basins (halocline 70 PSU) [42]. Similar to Hamelin Pool microbialite samples, abundant signatures of alveolates were recovered from Discovery and Bannock samples (75% of OTUs at 98% sequence similarity), most of which were from dinoflagellates (62%) and ciliates (12%) [42]. Fungi were the third most abundant group (17% of OTUs), particularly in brine samples of both basins. Signatures were also recovered from stramenopiles, euglenozoans, metazoans, plants, cercozoans and kinetoplastids. Due to the steep density gradient typical of haloclines, organic material (dead cells and detritus) accumulates at the top of haloclines and likely is partly responsible for fueling the active chemosynthetic prokaryotic communities on which these protists and fungi feed. Phagotrophic protists are known to be quite successful along oxyclines where prokaryotes are abundant [61–63] and so it follows that they are also successful within these haloclines. Fungi are active remineralizers of organic material that accumulates at these haloclines, and some which sinks into the brine. The presence of plant and metazoan signatures within the DNA-based clone libraries reflects their likely detrital origin. Nonetheless, the dominant protist signatures from Bannock and Discovery Basin haloclines and brines [42] are distinct from the picture of open-ocean pelagic communities in the photic zone that are usually dominated by stramenopiles and pigmented picoplankton taxa [64–66].

The density of protists that halocline and brine habitats can support is significant. For example, in a study of Thetis basin, which has one of the highest salt concentrations reported for DHABs (340 PSU), protist counts of ca. 0.6×10^4 per liter were reported in the anoxic brine [43]. The study of Thetis was based on RNA extracted from water samples, which is less likely than DNA to originate from dead or inactive cells, and identified fungi as the most diverse taxonomic group of eukaryotes in the brine (38% of OTUs based on 98% sequence similarity), followed by ciliates and stramenopiles, each accounting for 20% of phylotypes. Ciliate OTUs detected in the Thetis study were closely related to sequences detected in surveys of other DHABs, suggesting specific adaptations of ciliates to these habitats [43]. In addition to OTUs from dinoflagellates, haptophytes, choanoflagellates and jakobids, OTUs affiliating with marine stramenopiles (MAST) were detected in the brine samples, expanding the known salinity range of these taxa. Beta-diversity analyses supported the uniqueness of brine *vs.* halocline communities [43].

The RNA-based study of SSU rRNA gene signatures from Atalante basin halocline by Alexander *et al.* [25] also supported the presence of active protists in these hypersaline habitats. Almost the same number of OTUs (99% sequence similarity) were recovered from the upper halocline as from the extremely hypersaline lower halocline (43 and 42, respectively). In that study, alveolates also dominated the protist community, and ciliates were the most common group of alveolates in both the upper and lower halocline (18 and 21 OTUs, respectively). Only 12 OTUs (including seven ciliate, two choanoflagellate, and one each fungal, radiolarian, and jakobid OTUs) were shared between the two samples that were only separated by ~1.5 m water depth but differed in salinity by 324 PSU. Although salinity differences may be the primary driver behind observed differences in protist community composition, other environmental factors, including oxygen and ammonia concentrations likely also play a role [25]. Fungal and radiolarian OTUs were common in the upper halocline (39 PSU), but only a single representative of each group was detected in the lower halocline (365 PSU). Stramenopile, haptophyte, rhizarian and chlorophyte signatures were detected exclusively in the upper halocline, while cryptophyte and diverse dinoflagellate OTUs were detected exclusively in the lower halocline [25]. Comparisons of protist communities found in Bannock, Discovery, and Atalante using Jaccard indices support the notion that unique basin chemistries select for unique protist communities (e.g., [42] comparison of Bannock and Discovery). This is supported by a recent broad comparison of eukaryotic communities in many different DHAB haloclines and brines using T-RFLP by Filker *et al.* [55].

Confirmation of active/living protists in different DHAB samples was obtained using scanning electron microscopy (SEM) and fluorescence *in situ* hybridization (FISH) to visualize intact cells on filters. SEM images of intact ciliates and flagellates (Figure 5) showed that most ciliates on halocline and brine filters hosted prokaryotic epibionts. The role of these symbioses in DHAB habitats is under investigation. In the course of examining SEM filters from Discovery Basin halocline water samples, it was noted that abundant kinetoplastid cells were present. Kinetoplastids have been reported previously from anoxic and high-salt environments [13]. Kinetoplastid-specific PCR primers were used to amplify kinetoplastid signatures from brine and halocline water samples from six basins with differing chemistries [33]. Amplifications were successful only for halocline samples from three of six basins (and not for normal seawater controls). FISH probes were developed using the SSU rRNA gene sequence of an "unidentified clade" of kinetoplastid signatures that was observed to dominate the Discovery Basin halocline clone library. It was revealed using FISH that this clade represented up to 10% of the total protist community in the Discovery Basin halocline (6.4×10^3 kinetoplastids per liter belonging to this clade). This clade most likely represents a new genus [33]. Finding signatures of this "unidentified clade" of kinetoplastids from only halocline filters from three of six basins lends further support to the notion that unique basin chemistries are driving protist community composition.

Figure 5. Scanning electron micrographs of microbial eukaryotes from the water-column haloclines of Urania and Discovery Basins in the Eastern Mediterraean Sea. (**A**) A scuticociliate morphotype consistently associated with epibiotic bacteria (**B**) that has been found to be the most abundant eukaryotic morphotype in the Urania halocline [45]; (**B**) A flagellate in Urania halocline. (**C**) A larger ciliate associated with long (10–20 µm), thin, filamentous bacteria, which was the most abundant eukaryotic morphotype in the Discovery halocline [45]. Scale bars = 2 µm. Photos by William Orsi.

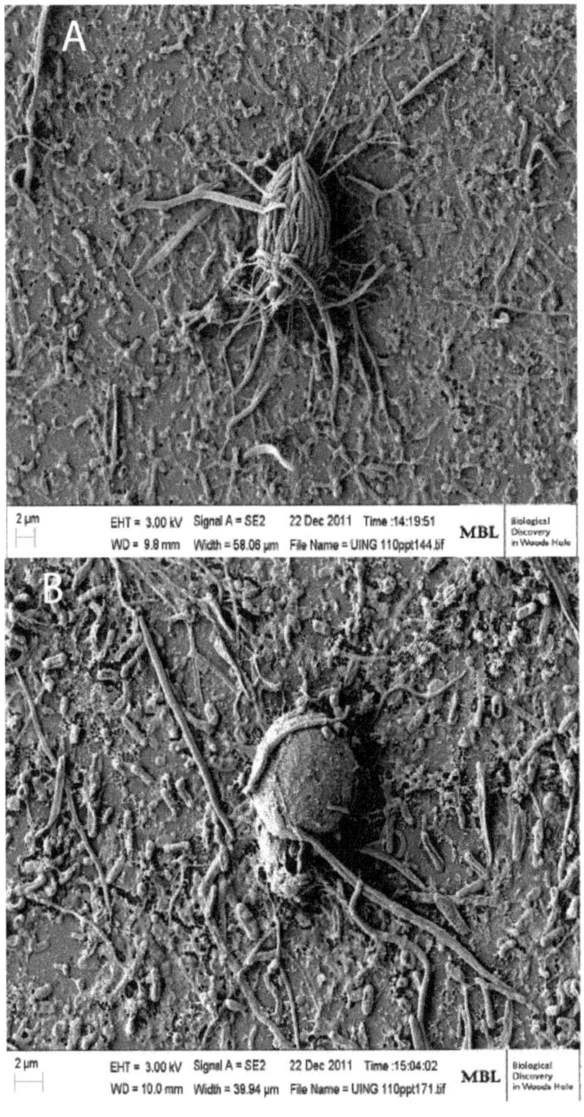

Figure 5. *Cont.*

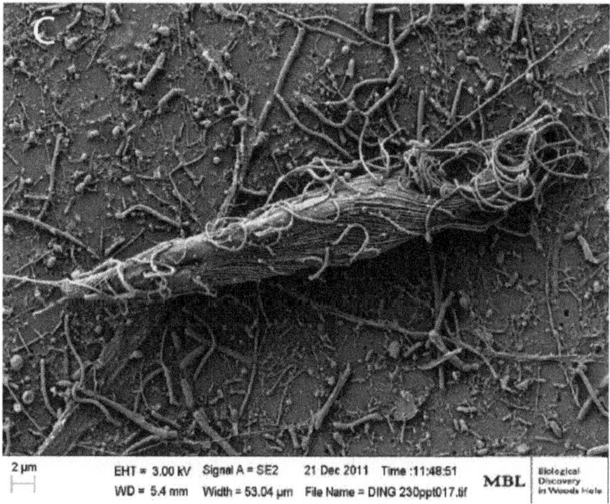

3. Commonalities between Protist Communities in DHABs and Hypersaline Shallow Water Mat Communities (Hamelin Pool and Guerrero Negro)

Direct comparisons between protist communities detected in DHAB halocline and brine waters with those in shallow water hypersaline sedimentary microbial mat communities is difficult for a number of reasons. First, communities in benthic and aquatic habitats are normally quite distinct, and second, the sampling and methodological approaches used in each of the studies discussed here are different. Nonetheless there are groups of protists and fungi that are common to both shallow hypersaline mat and DHAB environments, including diverse alveolates, stramenopiles, and fungi, as well as radiolaria. Within alveolates, groups that were common included ciliates (including signatures of Oligohymenophorea, Spirotrichea, Heterotrichea, Plagiopylea, and Phyllopharyngea) and uncultured Group I and Group II alveolates, as well as diverse dinoflagellates (including signatures of Gonyaulacales, Prorocentrales, Peridiniales, Gymnodiniales, Syndiniales). Stramenopile signatures shared include those from Labyrinthulidae, Bicosoecida, and Marine Stramenopile Group 3 (MAST-3). Radiolarian sequences shared between DHABs and Hamelin Pool microbialite samples included signatures of Polycystinea and Acantharia. In both Hamelin Pool microbialites and DHAB haloclines and brines many sequences of novel eukaryotes with no close homologues in public databases were recovered, suggesting these habitats host unusual protist and fungal communities. The taxonomic identification of the eukaryotic cells behind those signatures, as well as the determination of the environmental role of those eukayotes and their impact on carbon and other nutrient cycling in hypersaline habitats, is a fascinating avenue for future investigation. Furthermore, groups of protists are identified here that are common to different hypersaline habitats. By studying cultured representatives of these groups we will gain a better understanding of the physiological underpinnings behind their ability to adapt to hypersaline conditions. Future studies of heterotrophic grazing impacts in hypersaline water columns and sediments will enhance our understanding of carbon and other nutrient cycling in these habitats.

Acknowledgments

We thank the captains and crews of the R/V *Oceanus*, R/V *Walton Smith*, and R/V *Atlantis* for their hard work to assure the success of our sampling objectives, as well as the owners/operators of Carbla Station,Western Australia, and Ross Mack, World Heritage Site Ranger. Shark Bay work was a collaboration with Roger Summons (MIT) and Pieter Visscher (UConn); the DHAB study was a collaboration with Konstantinos Kormas (U. Thessaly) and Thorsten Stoeck (U. Kaiserslautern). This research was funded by NSF OCE-0849578 to VE and JMB, and NSF OCE-0926421 to JMB and VE.

Conflict of Interest

The authors declare no conflict of interest.

References

1. Litchfield, C.D. Survival strategies for microorganisms in hypersaline environments and their relevance to life on early Mars. *Meteorites Planet Sci.* **1998**, *33*, 813–819.
2. Baxter, B.K.; Litchfield, C.D.; Sowers, K.; Griffith, J.D.; DasSarma, P.A.; DasSarma, S. Microbial Diversity of Great Salt Lake. In *Adaptation to Life in High Salt Concentrations in Archaea, Bacteria, and Eukarya*; Gunde-Cimerman, N., Oren, A., Plemenitas, A., Eds.; Springer: Dordrecht, The Netherlands, 2005; Volume 9, pp. 9–25.
3. Brock, T.D.; Madigan, M.T.; Martinko, J.M. *Biology of Microorganisms*, 7th ed.; Benjamin Cummings: San Francisco, CA, USA, 1994; p. 909.
4. Galinsky, E.A. Compatible solutes of halophilic bacteria: Molecular principles, water-soluble interactions, stress protection. *Experientia* **1993**, *49*, 487–496.
5. Oren, A. *Halophilic Microorganisms and Their Environments*; Kluwer Academic: Dordrecht, The Netherlands, 2003; p. 575.
6. Ruinen, J.; Raas Becking, L.G.M. Rhizopods living in unusual environments. *Arch. Néerl Zool.* **1938**, *3*, 183–198.
7. Volcani, B.E. The Microorganisms of the Dead Sea. In *Papers Collected to Commemorate the 70th Anniversary of Dr. Chaim Weizmann*; Daniel Sieff Research Institute: Rehovoth, Israel, 1944; pp. 71–85.
8. Namyslowski, B. Adaptation of zooflagellates to higher salinity. *Biol. Vnutr. Vod. Inform Bull.* **1913**, *61*, 21–24.
9. Kirby, H. Two protozoa from brine. *Trans. Am. Microsc. Soc.* **1932**, *51*, 8–15.
10. Ruinen, J. Notizen über Salzflagellaten. II Über die Verbereitung der Salzflagellaten. *Arch. Protistenkd.* **1938**, *90*, 210–258.
11. Patterson, D.J.; Simpson, A.G. Heterotrophic flagellates from coastal marine and hypersaline sediments in Western Australia. *Eur. J. Protistol.* **1996**, *32*, 423–448.
12. Post, F.J.; Borowitzka, L.J.; Borowitzka, M.A.; Mackay, B.; Moulton, T. The protozoa of a Western Australian hypersaline lagoon. *Hydrobiologia* **1983**, *105*, 95–113.

13. Hauer, G.; Rogerson, A. Heterotrophic Protozoa from Hypersaline Environments. In *Adaptation to Life at High Salt Concentrations in Archaea, Bacteria, and Eukarya*; Gunde-Cimerman, N., Oren, A., Plemenitaš, A., Eds.; Springer: Dordrecht, The Netherlands, 2005; pp. 519–540.
14. Por, F. A classification of hypersaline waters, based on trophic criteria. *Mar. Ecol.* **1980**, *1*, 121–131.
15. Ramos-Cormenzana, A. Halophilic Organisms and Their Environment. In *General and Applied Aspects of Halophilic Microorganisms*; Rodriguez-Valera, F., Ed.; Plenum Press: New York, NY, USA, 1991; pp. 15–24.
16. Oren, A. Diversity of halophilic microorganisms: Environments, phylogeny, physiology, and applications. *J. Indust. Microbiol. Biotechnol.* **2002**, *28*, 56–63.
17. Pedros-Alió, C.; Calderón-Paz, J.I.; MacLean, M.H.; Medina, G.; Marrasé, C.; Gasol, J.M.; Guixa-Boixereu, N. The microbial food web along salinity gradients. *FEMS Microbiol. Ecol.* **2000**, *32*, 143–155.
18. Elloumi, J.; Carrias, J.-F.; Ayadi, H.; Sime-Ngando, T.; Boukhris, M.; Bouain, A. Composition and distribution of planktonic ciliates from ponds of different salinity in the solar saltwork of Sfax, Tunisia. *Estuar. Coast. Shelf Sci.* **2006**, *67*, 21–29.
19. Finlay, B.J. Physiological ecology of free-living protozoa. *Adv. Microbiol. Ecol.* **1990**, *11*, 1–34.
20. McLachlan, J. The culture of *Dunaliella tertiolecta* Butcher—A euryhaline organism. *Can. J. Microbiol.* **1960**, *6*, 367–379.
21. Al-Qassab, S.; Lee, W.J.; Muray, S.; Simpson, A.G.B.; Patterson, D.J. Flagellates from stramatolites and surrounding sediments in Shark Bay, Western Australia. *Acta Protozool.* **2002**, *41*, 91–144.
22. Park, J.S.; Simpson, A.G.B. Characterization of halotolerant Bicosoecida and Placididea (Stramenopila) that are distinct from marine forms, and the phylogenetic pattern of salinity preference in hetertrophic stramenopiles. *Environ. Microbiol.* **2010**, *12*, 1173–1184.
23. Park, J.S. Effects of different ion compositions on growth of obligately halophillic protozoan *Halocafeteria seosinensis*. *Extremophiles* **2012**, *16*, 161–164.
24. Fenchel, T. Ecology of heterotrophic microflagellates. IV. Quantitative occurrence and importance as bacterial consumers. *Mar. Ecol. Prog. Ser.* **1982**, *9*, 35–42.
25. Alexander, E.; Stock, A.; Breiner, H.W.; Behnke, A.; Bunge, J.; Yakimov, M.M.; Stoeck, T. Microbial eukaryotes in the hypersaline anoxic L'Atalante deep-sea basin. *Environ. Microbiol.* **2009**, *11*, 360–381.
26. Ayadi, H.; Toumi, N.; Abid, O.; Medhioub, K.; Hammami, M.; Sime-Ngando, T.; Amblard, C.; Sargos, D. Qualitative and quantitative study of phyto- and zooplankton communities in the saline ponds of Sfax, Tunisia. *Revue Des. Sci. L'Eau* **2002**, *15*, 123–135.
27. Laybourn-Parry, J.; Quayle, W.; Henshaw, T. The biology and evolution of Antarctic saline lakes in relation to salinity and trophy. *Polar. Biol.* **2002**, *25*, 542–552.

28. Feazel, L.M.; Spear, J.R.; Berger, A.B.; Harris, J.K.; Frank, D.N.; Ley, R.E.; Pace, N.R. Eucaryotic diversity in a hypersaline microbial mat. *Appl. Environ. Microbiol.* **2007**, *74*, 329–332.
29. Cho, B.C.; Park, J.S.; Xu, K.; Choi, J.K. Morphology and molecular phylogeny of *Trimyema koreanum* n. sp., a ciliate from the hypersaline water of a solar saltern. *J. Eukaryot. Microbiol.* **2008**, *55*, 417–426.
30. Park, J.S.; Cho, B.C.; Simpson, A.G.B. *Halocafeteria seoinensis* gen. et sp. nov. (Bicosoecida), a halophilic bacterivorous nanoflagellate isolated from a solar saltern. *Extremophiles* **2006**, *10*, 493–504.
31. Park, J.S.; Simpson, A.G.B.; Brown, S.; Cho, B.C. Ultrastructure and molecular phylogeny of two heterolobosean amoebae, *Euplaesiobystra hypersalinica* gen. et sp. nov. and *Tulamoeba peronaphora* gen. et sp. nov., isolated from an extremely hypersaline habitat. *Protist* **2009**, *160*, 265–283.
32. Park, J.S.; Simpson, A.G.B.; Lee, W.J.; Cho, B.C. Ultrastructure and phylogenetic placement within Heterolobosea of the previously unclassified, extremely halophilic heterotrophic flagellate *Pleurostomum flabellatum* (Ruinen 1938). *Protist* **2007**, *158*, 397–413.
33. Edgcomb, V.P.; Orsi, W.; Breiner, H.-W.; Stock, A.; Filker, S.; Yakimov, M.M.; Stoeck, T. Novel kinetoplastids associated with hypersaline anoxic lakes in the Eastern Mediterranean deep-sea. *Deep Sea Res.* **2011b**, *58*, 1040–1048.
34. Edgcomb, V.P.; Orsi, W.; Taylor, G.T.; Vdacny, P.; Taylor, C.; Suarez, P.; Epstein, S. Accessing marine protists from the anoxic Cariaco Basin. *ISME J.* **2011**, *5*, 1237–1241.
35. Jahnert, R.L.; Collins, L.B. Significance of subtidal microbial deposits in Shark Bay, Australia. *Mar. Geol.* **2011**, *286*,106–111.
36. Dupraz, C.; Visscher, P.T. Microbial lithification in marine stromatolites and hypersaline mats. *Trends Microbiol.* **2005**, *13*, 429–438.
37. Allwood, A.C.; Walter, M.R.; Kamber, B.S.; Marshall, C.P.; Burch, I.W. Stromatolite reef from the Early Archaean era of Australia. *Nature* **2006**, *441*, 714–718.
38. Grotzinger, J.P.; Knoll, A.H. Stromatolites in Precambrian carbonates: Evolutionary mileposts or environmental dipsticks? *Ann. Rev. Earth Planet Sci.* **1999**, *27*, 313–358.
39. Tong, S.M. Heterotrophic flagellates from the water column in Shark Bay, Western Australia. *Mar. Biol.* **1997**, *128*, 517–536.
40. Papineau, D.; Walker, J.J.; Mojzsis, S.J.; Pace, N.R. Composition and structure of microbial communities from stromatolites of Hamelin Pool in Shark Bay, Western Australia. *Appl. Environ. Microbiol.* **2005**, *71*, 4822–4832.
41. Bernhard, J.M.; Edgcomb, V.P.; Visscher, P.T.; McIntyre-Wressnig, A.; Summons, R.E.; Bouxsein, M.; Louis, L.; Jeglinski, M. Microbialites at Highborne Cay, Bahamas: insights on foraminiferal inhabitants and influence on their microfabric. *Proc. Natl. Acad. Sci USA* **2013**, in press.
42. Logan, B.W. Cryptozoon and associate stromatolites from the Recent, Shark Bay, Western Australia. *J. Geol.* **1961**, *69*, 517–533.

43. Stoeck, T.; Bass, D.; Nebel, M.; Christen, R.; Jones, M.D.; Breiner, H.W.; Richards, T.A. Multiple marker parallel tag environmental DNA sequencing reveals a highly complex eukaryotic community in marine anoxic water. *Mol. Ecol.* **2010**, *19*, 21–31.
44. Pawlowski, J. Introduction to the molecular systematics of foraminifera. *Micropaleontology* **2000**, *46*, 1–12.
45. Caporaso, J.G.; Kuczynski, J.; Strombaugh, J.; Bittinger, K.; Bushman, F.D.; Costello, E.K.; Fierer, N.; Gonzalez Pena, A.; Goodrich, J.K.; Gordon, J.I.; *et al.* QIIME allows analysis of high-throughput community sequencing data. *Nat. Meth.* **2010**, *7*, 335–336.
46. Nebel, M.E.; Wild, S.; Holzhauser, M.; Reitzig, R.; Sperber, M.; Stoeck, T. Jaguc—A software package for environmental diversity analyses. *J. Bioinf. Comp. Biol.* **2011**, *9*, 749–773.
47. Edgcomb, V.; Orsi, W.; Leslin, C.; Epstein, S.S.; Bunge, J.; Jeon, S.; Yakimov, M.M.; Behnke, A.; Stoeck, T. Protistan community patterns within the brine and halocline of deep hypersaline anoxic basins in the eastern Mediterranean Sea. *Extremophiles* **2009**, *13*, 151–167.
48. Stock, A.; Breiner, H.-W.; Pachiadaki, M.; Edgcomb, V.; Filker, S.; LaCono, V.; Yakimov, M.M.; Stoeck, T. Microbial eukaryote life in the new hypersaline deep-sea basin Thetis. *Extremophiles* **2011**, *16*, 21–34.
49. Bebout, B.M.; Carpenter, S.P.; Des Marais, D.J.; Discipulo, M.; Embaye, T.; Garcia-Pichel, F.; Hoehler, T.M.; Hogan, M.; Jahnke, L.L.; Keller, R.M.; *et al.* Long-term manipulations of intact microbial mat communities in a greenhouse collaboratory: Simulating earth's present and past field environments. *Astrobiology* **2002**, *2*, 383–402.
50. Orsi, W.; Charvet, S.; Bernhard, J.; Edgcomb, V.P. Prevalence of partnerships between bacteria and ciliates in oxygen-depleted marine water columns. *Front. Ext. Microbiol.* **2012**, *3*, 341.
51. Bosak, T.; Lahr, D.J.G.; Pruss, S.B.; Macdonald, F.A.; Gooday, A.J.; Dalton, L.; Matys, E.D. Possible early foraminiferans in post-Sturtian (716–635 Ma) cap carbonates. *Geology* **2012**, *40*, 67–70.
52. Bernhard, J.M.; Visscher, P.T.; Bowser, S.S. Submillimeter life positions of bacteria, protists, and metazoans in laminated sediments of the Santa Barbara Basin. *Limnol. Oceanogr.* **2003**, *48*, 813–828.
53. Cita, M.B. Exhumation of Messinian evaporites in the deep-sea and creation of deep anoxic brine filled collapsed basins. *Sed. Geol.* **2006**, *188–189*, 357–378.
54. Danovaro, R.; Dell'Anno, A.; Pusceddu, A.; Gambi, C.; Heiner, I.; Kristensen, R.M. The first metazoa living in permanently anoxic conditions. *BMC Biol.* **2010**, *8*, 30.
55. Filker, S.; Stock, A.; Breiner, H.W.; Edgcomb, V.P.; Orsi, W.; Yakimov, M.M.; Stoeck, T. Environmental selection of protistan communities in hypersaline anoxic deep-sea basins, Eastern Mediterranean Sea. *Microbiologyopen* **2013**, *2*, 54–63.
56. Van der Wielen, P.W.; Bolhuis, H.; Borin, S.; Daffonchio, D.; Corselli, C.; Giuliano, L.; D'Auria, G.; de Lange, G.J.; Huebner, A.; Varnavas, S.P.; *et al.* The enigma of prokaryotic life in deep hypersaline anoxic basins. *Science* **2005**, *307*, 121–123.

57. Daffonchio, D.; Borin, S.; Brusa, T.; Brusetti, L.; van der Wielen, P.W.; Bolhuis, H.; Yakimov, M.M.; D'Auria, G.; Giuliano, L.; Marty, D.; *et al.* Stratified prokaryote network in the oxic-anoxic transition of a deep-sea halocline. *Nature* **2006**, *440*, 203–207.
58. Hallsworth, J.E.; Yakimov, M.M.; Golyshin, P.N.; Gillion, J.L.; D'Auria, G.; de Lima Alves, F.; La Cono, V.; Genovese, M.; McKew, B.A.; Hayes, S.L.; *et al.* Limits of life in $MgCl_2$-containing environments: Chaotropicity defines the window. *Environ. Microbiol.* **2007**, *9*, 801–813.
59. Van der Wielen, P.W.; Heijs, S.K. Sulfate-reducing prokaryotic communities in two deep hypersaline anoxic basins in the Eastern Mediterranean deep sea. *Environ. Microbiol.* **2007**, *9*, 1335–1340.
60. Yakimov, M.M.; Giuliano, L.; Cappello, S.; Denaro, R.; Golyshin, P.N. Microbial community of a hydrothermal mud vent underneath the deep-sea anoxic brine lake Urania (eastern Mediterranean). *Orig. Life Evol. Biosph.* **2007**, *37*, 177–188.
61. Bernhard, J.M.; Buck, K.R.; Farmer, M.A.; Bowser, S.S. The Santa Barbara Basin is a symbiosis oasis. *Nature* **2000**, *403*, 77–80.
62. Taylor, G.T.; Scranton, M.L.; Iabichella, M.; Ho, T.-Y.; Thunell, R.C.; Muller-Karger, F.; Varela, R. Chemoautotrophy in the redox transition zone of the Cariaco Basin: A significant midwater source of organic carbon production. *Limnol. Oceanogr.* **2001**, *46*, 148–163.
63. Edgcomb, V.; Orsi, W.; Bunge, J.; Jeon, S.O.; Christen, R.; Leslin, C.; Holder, M.; Taylor, G.T.; Suarez, P.; Varela, R.; *et al.* Protistan microbial observatory in the Cariaco Basin, Caribbean. I. Pyrosequencing *vs.* Sanger insights into species richness. *ISME J.* **2011**, *5*, 1344–1356.
64. Countway, P.D.; Gast, R.J.; Dennett, M.R.; Savai, P.; Rose, J.M.; Caron, D.A. Distinct protistan assemblages characterize the euphotic zone and deep sea (2,500 m) of the western North Atlantic (Sargasso Sea and Gulf Stream). *Environ. Microbiol.* **2007**, *9*, 1219–1232.
65. Massana, R.; Castresana, J.; Balagué, V.; Guillou, L.; Romari, K.; Groisillier, A.; Valentin, K.; Pedrós-Alió, C. Phylogenetic and ecological analysis of novel marine stramenopiles. *Appl. Environ. Microbiol.* **2004**, *70*, 3528–3534.
66. Not, F.; Gausling, R.; Azam, F.; Heidelberg, J.F.; Worden, AZ. Vertical distribution of picoeukaryotic diversity in the Sargasso Sea. *Environ. Microbiol.* **2007**, *9*, 1233–1252.

Properties of *Halococcus salifodinae*, an Isolate from Permian Rock Salt Deposits, Compared with Halococci from Surface Waters

Andrea Legat, Ewald B. M. Denner, Marion Dornmayr-Pfaffenhuemer, Peter Pfeiffer, Burkhard Knopf, Harald Claus, Claudia Gruber, Helmut König, Gerhard Wanner and Helga Stan-Lotter

Abstract: *Halococcus salifodinae* BIp^T DSM 8989^T, an extremely halophilic archaeal isolate from an Austrian salt deposit (Bad Ischl), whose origin was dated to the Permian period, was described in 1994. Subsequently, several strains of the species have been isolated, some from similar but geographically separated salt deposits. *Hcc. salifodinae* may be regarded as one of the most ancient culturable species which existed already about 250 million years ago. Since its habitat probably did not change during this long period, its properties were presumably not subjected to the needs of mutational adaptation. *Hcc. salifodinae* and other isolates from ancient deposits would be suitable candidates for testing hypotheses on prokaryotic evolution, such as the molecular clock concept, or the net-like history of genome evolution. A comparison of available taxonomic characteristics from strains of *Hcc. salifodinae* and other *Halococcus* species, most of them originating from surface waters, is presented. The cell wall polymer of *Hcc. salifodinae* was examined and found to be a heteropolysaccharide, similar to that of *Hcc. morrhuae*. Polyhydroxyalkanoate granules were present in *Hcc. salifodinae*, suggesting a possible lateral gene transfer before Permian times.

Reprinted from *Life*. Cite as: Legat, A.; Denner, E.B.M.; Dornmayr-Pfaffenhuemer, M.; Pfeiffer, P.; Knopf, B.; Claus, H.; Gruber, C.; König, H.; Wanner, G.; Stan-Lotter, H. Properties of *Halococcus salifodinae*, an Isolate from Permian Rock Salt Deposits, Compared with Halococci from Surface Waters. *Life* **2013**, *3*, 244-259.

1. Introduction

Halococcus salifodinae BIp^T DSM 8989^T was obtained as a viable isolate from Permian rock salt deposits of a mine in Bad Ischl, Austria [1,2]. The strain grew optimally at a salinity of 20%–25%, a pH value of 7.4 and at 40 °C. Subsequently, several halococcal strains were isolated from similar sites in England and Germany, which had identical 16S rRNA gene sequences and numerous similar properties as the Bad Ischl strain BIp^T [3].

The genus *Halococcus* [4], emended by Oren *et al.* [5] currently comprises seven formally described species, which are listed here with their sites of isolation and reference in brackets: *Hcc. morrhuae* (seawater, saline lakes, salterns and salted products, [6]), *Hcc. saccharolyticus* (marine salterns, [7]), *Hcc. salifodinae* (rock salt from mines in Germany and Austria, also from brine in a salt mine in England, [1]), *Hcc. dombrowskii* (bore core from a salt mine in Austria, [8]), *Hcc. hamelinensis* (stromatolites of Shark Bay, Hamelin Pool in Western Australia, [9]), *Hcc. qingdaonensis* (crude sea-salt sample collected near Qingdao in Eastern China, [10]) and *Hcc. thailandensis* (fermented fish sauce produced in Thailand) [11]. Thus, two species—*Hcc.*

salifodinae and *Hcc. dombrowskii* - were isolated from Permo-Triassic salt sediments, whereas the other five species can be regarded as inhabitants of hypersaline surface waters or heavily salted products.

A study by Wright [12] using 16S rRNA gene sequences of 61 haloarchaeal taxa, revealed that the mean genetic divergence over all possible pairs of halophilic archaeal 16S rRNA gene sequences was 12.4 ± 0.38%, indicating close relatedness. In comparison, the greatest genetic divergence within methanogenic archaea was 34.2% [12]. Within the halophilic archaea, *Halococcus* species form an even closer related group (see Figure 1), with 16S rRNA gene sequence similarities of 98.2%–98.7% between *Hcc. thailandensis*, *Hcc. morrhuae*, *Hcc. qingdaonensis* and *Hcc. dombrowskii*, and somewhat lower similarities of 93.7%–94.1% between *Hcc. hamelinensis*, *Hcc. saccharolyticus* and *Hcc. salifodinae* [11].

Figure 1. Distance-matrix neighbor-joining tree, showing the phylogenetic relationships of *Halococcus* type strains. The tree is based on an alignment of 16S rRNA gene sequences. Bootstrap values higher than 70 out of 1000 subreplicates are indicated at the respective bifurcations. The tree was constructed using the neighbour-joining method of Saitou and Nei [13]. The bar represents the scale of estimated evolutionary distance (1 % substitutions at any nucleotide) from the point of divergence. *Halobacterium noricense* was used as an outgroup.

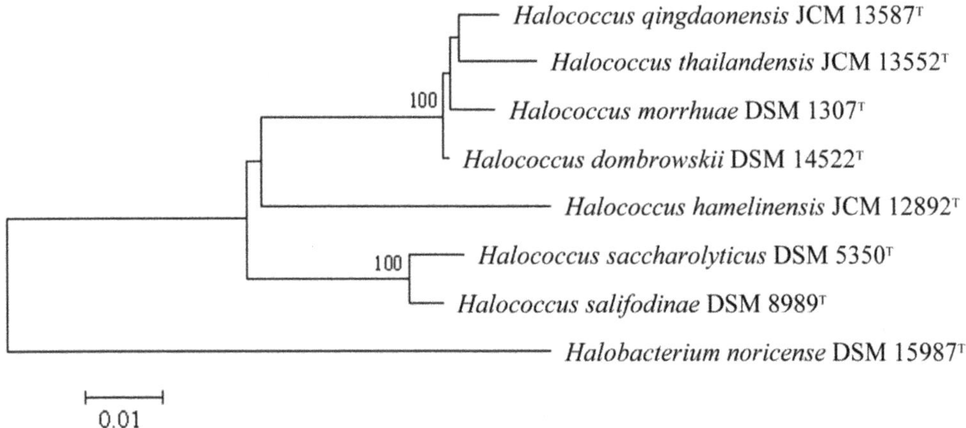

Our notions on prokaryotic evolution and evolution in general have been shaped by the concept of a molecular clock, which suggests an approximately uniform rate of molecular evolution among species and duplicated proteins over time [14]. Although subject to various criticisms, molecular-clock techniques still remain the only way to infer the timing of gene duplications and speciation events in the absence of fossil or biogeographical records [14]. The concept was applied previously to date the sequence divergences of halophilic archaeal protein-encoding genes, compared to the divergence of homologous non-halophilic eubacterial protein-encoding genes, assuming a point of haloarchaeal species diversion of 600 million years before present [15]. However, modern results from genome sequences revealed a much more complex history of life than can be depicted in bifurcating trees [16]. Widespread horizontal gene transfer—although occurring to different extents—, endosymbioses, gene losses and other processes cause the presence of different molecules with different histories in a species, and members of the same species were found to differ dramatically in gene content, leading to the suggestion of a fuzzy species concept in prokaryotes [16].

Some of these problems and uncertainties might be resolvable when viable microorganisms from well-dated ancient geological sites would be compared on a molecular basis with contemporary species. A crucial issue is the proof that microorganisms from ancient materials, like million year old deep subseafloor sediments, or Permian salt evaporites, are as old as the geological sites from which they were obtained (see [17–19] for discussions). The determination of the age of a single average bacterium is not possible with currently available methods, since its mass is only about a picogram. Thus, claims of ancient microorganisms were often dismissed as being due to laboratory contaminations.

Recently, small particles of about 0.4 μm in diameter were imaged by microscopy directly within fluid inclusions of 22,000–34,000 year old salt bore cores and, following successful culturing, identified as haloarchaea [18,20]. Embedding of halophilic microorganisms in fluid inclusions upon formation of salt crystals is well known, and fluid inclusions have been suggested as sites for preservation of microbial life [21–23]. In addition, Gramain *et al.* [24] reported isolation of haloarchaea from well-dated salt bore cores of Pliocene age (5.3 to 1.8 million years). Thus there is a growing body of evidence that haloarchaea survive for great lengths of time [24].

Here we review the properties of coccoid haloarchaea isolated from Permo-Triassic salt sediments, and relate them to those of halococci, which were isolated from surface waters. In addition, new data on *Halococcus salifodinae* concerning the chemical composition of its cell wall are included as well as DNA-DNA hybridization experiments between several strains of the species. Recently, the first genome sequence of a halococcus, *Hcc. hamelinensis* 100A6^T, became available [25] and therefore information for several genes (*phaC* synthases; subunit A of the rotary A-ATPase) is examined here for their potential use in delineating the evolution of haloarchaeal cocci.

2. Results and Discussion

2.1. General Description of halococci [26]

Halococci are cells of 0.8-1.5 μm diameter, occurring in pairs, tetrads, sarcina packets, or large clusters [1,26]; see Figure 2, left panel. A striking difference to other genera of the *Halobacteriaceae* is their resistance to lysis in water (or generally hypotonic solutions). They are non-motile, strictly aerobic and extremely halophilic, requiring at least 2.5 M NaCl for growth and 3.5–4.5 M NaCl for optimum growth [26]. Their optimum growth temperature is between 30–40 °C but most strains can grow up to 50 °C.

2.2. Properties of Isolates from Permo-Triassic Salt Sediments and Surface Waters

Following the formal description of *Halococcus salifodinae* BIp^T DSM 8989^T as a novel species from a Permian salt deposit [1], a detailed comparison with similar isolates from a British halite formation (strain Br3) and from a bore core of the salt mine in Berchtesgaden, Germany (strain BG2/2) was undertaken [3]. In addition, two further isolates (strains H2, N1) from the Bad Ischl salt mine were similar enough to *Hcc. salifodinae* BIp^T to consider them strains of the same species, obtained 8 years after the initial rock salt samples were taken [3]. The sequences of the 16S rRNA genes of all five strains were identical, as were their polar lipid composition, colonial and cellular morphology, cell size, cellular arrangement, and pigmentation. Strong similarities were found between whole-cell protein patterns, G+C contents, growth characteristics, enzyme content and susceptibility to antibiotics. Table 1 provides a comparison of biochemical characteristics of the five strains of *Hcc. salifodinae* (numbered 1–5) with the other presently known six halococcal species (numbers 6–11). All strains of *Hcc. salifodinae* were positive for alkaline phosphatase, esterase (C4), esterase lipase (C8), oxidase and catalase (Table 1). Variable reactions among strains were observed for acid phosphatase, N-acetyl-β-glucosaminidase, nitrate reduction, hydrolysis of Tween 20 and gelatine liquefaction (Table 1). Starch and Tween 80 were hydrolysed by strains BIp^T, BG2/2 and Br3, but casein was not. Goh *et al.* [9] reported that the two isolates of *Hcc. hamelinensis* were negative for oxidase activity, whereas *Hcc. morrhuae* NRC 16008, *Hcc. saccharolyticus* ATCC 49257^T and *Hcc salifodinae* DSM 8989^T were all positive. The API ZYM strips revealed that the two isolates of *Hcc. hamelinensis* were positive for leucine arylamidase, but negative for trypsin, as were all other halococci.

The results confirmed the assignment of strains 1–5 from salt deposits to the same species, *Hcc. salifodinae*. *Hcc. salifodinae* is distinct from other halococci, but, based on 16S rRNA sequences, appeared phylogenetically more closely related to *Hcc. saccharolyticus*. The similarity of the 16S rDNA sequence of *Hcc. salifodinae* BIp^T DSM 8989^T to that of *Hcc. saccharolyticus* DSM 5350^T was 98.9%. A similarity value of >97% necessitates the determination of the DNA-DNA homology by hybridization experiments, in order to delineate the identity of species [27]. Therefore, DNA-DNA hybridization between *Hcc. salifodinae* BIp^T DSM 8989^T and *Hcc. saccharolyticus* DSM 5350^T was performed and showed a value of 63.6%. DNA-DNA hybridization data confirmed that the two strains represent two different *Halococcus* species, since it is accepted that strains of a single species exhibit $\geq$ 70% DNA relatedness [28]. DNA-DNA hybridization was also carried out among the five *Hcc. salifodinae* strains and revealed values in the range of 82.6% to 95.0%, corroborating the assignment of the strains to a single species. Thus it was demonstrated that in geographically separated halite deposits—located in Austria, Germany and England—of similar geological age, identical species of halococci are present. It can therefore be speculated that their native environment may have been the ancient Zechstein sea, spreading over large parts of what is now Europe [29] and it is tempting to suggest that they might be marker organism for salt deposits from certain geological periods [30].

Table 1. Characteristics of five independently isolated strains of *Halococcus salifodinae* from three different locations [3] and other *Halococcus* species. 1, *Hcc. salifodinae* BIp^T DSM 8989^T, type strain; 2, *Hcc. salifodinae* BG2/2 (DSM 13045); 3, *Hcc. salifodinae* Br3 (DSM 13046); 4, *Hcc. salifodinae* H2 (DSM 13071); 5, *Hcc. salifodinae* N1 (DSM 13070); 6, *Hcc. saccharolyticus* DSM 5350^T (data from [26]); 7, *Hcc. hamelinensis* JCM 12892^T (data from [9]); 8, *Hcc. thailandensis* (data from [11]); 9, *Hcc. dombrowskii DSM* 14522^T (data from [8]); *10, Hcc. qingdaonensis* JCM 13587^T (data from [10]); *11, Hcc. morrhuae* DSM 1307^T (data from [8,26]).

Characteristic*	1	2	3	4	5	6	7	8	9	10	11
Oxidase	+	+	+	+	+	+	-	+	+	-	+
Catalase	+	+	+	+	+	+	+	+	+	+	+
Alkaline phosphatase	(+)	+	+	+	+						
Esterase (C4)	+	+	+	+	+						
Lipase esterase (C8)	+	+	+	+	+						
Lipase (C14)	-	-	-	-	-						
Leucine arylamidase	-	-	-	-	-		+				
Trypsin	-	-	-	-	-		-				
Acid phosphatase	-	-	+	+	+				+		-
Cystine arylamidase	-	-	-	-	-				+		-
Nitrate reduction	+	+	+	-	+	+		+	+		+
Gelatin liquefaction	+	-	-			v	-	-	+	-	v
Hydrolysis of starch	+	+	+			-	+	-			v
casein	-	-	-			-		-			
Tween 20	-	+	+					-			
Tween 80	+	+	+			-				-	+
Sensitivity to anti-biotics: Tetracycline	+	+	+	+	+	-	-	-	-	-	-
" : Chloramphenicol	+	+	+	+	+	-		-	-	+	-
" : Novobiocin	+	+	+	+	+	-	+	+	+		+

* +, positive reaction; -, negative reaction; (+) weak reaction; v, variable; empty box: no data available.

2.3. Cell Wall of Hcc. salifodinae

The cell wall of *Halococcus* species is very prominent as seen in TEM micrographs (Figure 2, right panel; [1,3]) with a thickness of 50–60 nm reported for *Hcc. morrhuae* [31]. The material appears amorphous and the formation of septa is visible (white arrows).

Figure 2. Left panel: Scanning electron micrograph of *Halococcus salifodinae* Br3 (DSM 13046), grown in liquid culture medium [1]. Bar, 500 nm. Right panel: Transmission electron micrograph of an ultrathin section of *Halococcus salifodinae* BIp^T DSM 8989^T. Cells are surrounded by an amorphous layer of wall material. Septum formation is visible (white arrows). Bar, 760 nm.

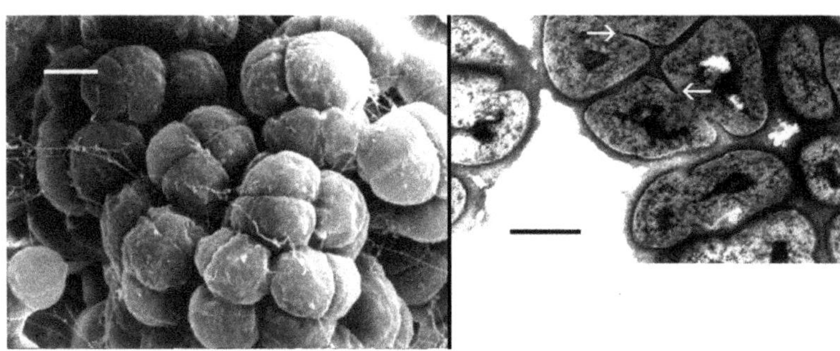

So far extremely halophilic Archaea are known to have developed three different cell wall types: (a) S-layers [32], (b) a heteropolysaccharide [33] and (c) a glutaminylglycan [34]. Layers of (glyco-) protein subunits (S-layers) represent the most common cell wall structures among Archaea [35,36]. The occurrence of S-layers, which were found in extremely halophilic archaeal genera such as *Halobacterium* and *Haloferax* [32] or several methanogenic genera such as *Methanococcus* [37] and *Methanothermus* [38] in the Euryachaeota branch, was not detected in *Halococcus salifodinae*. The chemical composition of only two cell walls from coccoid haloarchaea has been reported to date: *Hcc. morrhuae* CCM 859 possesses a heteropolysaccharide [33] and the haloalkalophile *Natronococcus occultus* contains a glutaminylglycan [34]. No molecular structures of these heteropolysaccharides are available yet. Since *Hcc. salifodinae* BIp^T originated from an ancient habitat existing for about 250 million years, the cell surface structure of the organism was of special interest and the chemical composition of its cell wall was analyzed.

No protein, as in S-layers, was found but instead, different monosaccharides were identified as constituents of the cell wall polymer, as was the case with the closely related species *Halococcus morrhuae* [33,39–41]. The results of the chemical analysis are summarized in Table 2. These data showed that the cell wall composition of *Halococcus salifodinae* is very similar to that of *Halococcus morrhuae*. Both cell wall polymers are composed of the neutral sugars galactose, mannose and glucose, the amino sugars glucosamine and galactosamine and the uronic acids glucuronic acid and galacturonic acid. Glycine and lysine were the only amino acids which could be detected in small amounts. However, the molar ratios of the cell wall constituents differed significantly between *Hcc. morrhuae* und *Hcc. salifodinae*.

These results allow us to speculate that a heteropolysaccharide forms the main cell wall polymer of *Halococcus salifodinae,* as was described for *Halococcus morrhuae* [33]. The carbohydrates of the cell wall sacculi of *Halococcus morrhuae* are arranged in three domains. These three domains are partly linked by N-glycyl-glucosaminyl bridges [41]. The data suggested that *Halococcus salifodinae*, a viable isolate from Permian rock salt deposits, has not developed a novel cell wall type, but possesses most likely a similarly structured, but modified heteropolysaccharide as the closely related species *Halococcus morrhuae*.

Table 2. Chemical composition of the cell wall polymer of *Halococcus morrhuae* and *Halococcus salifodinae*.

Cell wall constituents[a]	*Hcc. morrhuae* CCM 859[b]	*Hcc. salifodinae* BIp^T DMS 8989^T
Glucose	440	470
Mannose	350	220
Galactose	270	360
Ribose	n.d.	60
Glucosamine	380	180
Galactosamine	200	80
Glucuronic acid	470	60
Galacturonic acid	200	20
Gulosaminuronic acid	110	n.d.
Acetate	620	660
Sulfate	1470	1580
Phosphate	120	130
Glycine	100	7
Lysine	n.d.	1

[a] nmol/mg cell wall (dry weight); [b] data from [33,42]; n.d., not determined

The coccus is the simplest of possible cell shapes. The coccoid morphology has been suggested as the first (bacterial) cell morphology [43], but later, arguments for a rod-shaped morphology were viewed as more likely, mainly a better surface to volume ratio, providing an increased area for uptake of nutrients [44]. Still, this issue has not been solved yet, and it will be interesting to see a comparison of the synthesis of peptidoglycan, the nearly ubiquitous cell wall polymer of bacteria, which is lacking in halococci [39], with that of heteropolysaccharides. Peptidoglycan in Gram-positive cells consists of a thick layer of several nm width, similar to the heteropolysaccharide layer of halococci [39]. Both polymers appear functionally identical, providing stability to the cell, both form septa during cell division (in contrast to constrictions of most Gram-negative bacteria), but, judging from their different compositions, their biochemical pathways must involve quite a different set of enzymes.

2.4. Production of polyhydroxyalkanotes (PHA)

Due to the considerable biotechnological and medical potential, the pathways of bacterial synthesis of polyhydroxyalkanotes (PHA) have been examined intensely (for a review see [45]). Accumulation of PHAs by haloarchaea was first reported by Fernandez-Castillo *et al.* [46]). So far, the best PHA producer of the family *Halobacteriaceae* is *Haloferax mediterranei* (see [47]). Recently, evidence for the production of polyhydroxyalkanoates by halococci was published ([48]), which included staining and electron microscopy of PHA granules as well as chemical identification by NMR from *Hcc. morrhuae* DSM 1307^T, *Hcc. saccharolyticus* DSM 5350^T, *Hcc.*

salifodinae BIp^T DSM 8989^T, *Hcc. dombrowskii* DSM 14522^T, *Hcc. hamelinensis* JCM 12892^T and *Hcc. qingdaonensis* JCM 13587^T. Genetic information about haloarchaeal PHA synthases is still sparse. Of interest is the finding that high homologies exist to the bacterial enzymes as reported by several authors [47,49–52]. In a phylogenetic tree of PHA polymerases (*phaC* synthases), bacterial and haloarchaeal sequences clustered together, and the most closely affiliated microorganisms shared habitats of marine origin [47]. These observation suggested horizontal gene transfer [53]. It was even proposed that *phaC* synthases belonging to Class III of halophilic and non-halophilic microorganisms may have had a common ancestor [47].

2.5. BLAST Search of Genes in the Genome of Hcc. hamelinensis.

The genome sequence of *Hcc. hamelinensis* 100A6^T has recently been published [25]. This allows preliminary comparisons with the genome content of other halococci, since as noted above, all halococci appear to be closely related. BLAST (Basic Local Alignment Search Tool [54]) searches were carried out with two examples, *phaC* (see 2.4.) and subunit A of the haloarchaeal ATPsynthase.

2.5.1. Polyhydroxyalkanaote Synthase (*phaC*)

The gene coding for *phaC* (polyhydroxyalkanaote synthase) from *Haloferax mediterranei*, accession number ACB10370 [52] was used with the program TBLASTN. The identities for the (translated) protein (492 amino acids) were 56%, similarities were 71%. A search with the nucleotide sequence for the related *phaC* (1425 nucleotides) of *Haloarcula hispanica* ATCC 33960 [52] resulted in 77% identities in the genome sequence of *Hcc. hamelinensis* 100A6^T.

2.5.2. Subunit A (*atpA*) of the Archaeal ATP Synthase

Fundamental enzyme complexes in all cells are the rotary ATP synthases/ATPases, which catalyze the synthesis of ATP at the expense of a proton or ion gradient and include three related members (for a recent review see [55]). The A-ATP synthase is present in archaea, which is similar to the eukaryotic V-ATPases. Preliminary information of the occurrence of A-ATPases in isolates from Permian salt sediments was obtained by immunological and biochemical properties with strain 54R, a close relative of the rod-shaped *Halorubrum saccharovorum* [56], whose ATPase has been characterized in detail [57,58]. However, no information exists yet about rotary ATPases from halococci. A BLAST search with the nucleotide sequence of subunit A of the A-ATPase from *Halobacterium salinarum* strain NRC-1 (NC_002607.1; length of 1758) revealed an identity of 81% in the genome sequence of *Hcc. hamelinensis* 100A6^T.

2.6. Are Permo-Triassic Isolates Suitable for Evolutionary Studies?

2.6.1. What Type of Results Can be Expected?

Using 16S rRNA sequences as a chronometer, Dennis and Shimmin [15] estimated that *Halobacterium*, *Haloferax*, and *Haloarcula* diverged from a common ancestor about

600×10^6 years ago. This calculation was based on the assumption of a constant and uniform rate of sequence diversions of 1% per 50×10^6 years [59]. This time frame appears rather short, since evidence (although disputed) for haloarchaeal DNA was found in Silurian salt sediments (416–429 million years old), some from *Halobacterium* species and some of unknown affiliation [60], perhaps from older and now extinct microorganisms. This type of questions may be answerable, *i.e.*, the molecular clock concept could perhaps be verified, at least with highly conserved genes, when genome sequences of strains of Permo-Triassic origin are available. The evolution of biochemical pathways for cell wall synthesis or production of PHAs could be clarified (see above). Also, insights into the evolution of very complex cellular systems, e.g., ATP synthases, could be gained. Other expected results should be information about gene losses (see [16]), which could be detected with sequenced genomes of "ancient" subsurface prokaryotes. This could then explain unexpected phylogenetic results, for example, distribution of genes which do not fit a tree [16]. More information about horizontal gene transfer should probably become available.

2.6.2. Which Strains Should be Used for Comparative Studies?

Comparative 16S rRNA gene sequence analyses showed a similarity of 98.9% between *Hcc. salifodinae* BIp^T DSM 8989^T and *Hcc. saccharolyticus* DSM 5350^T. Thus, *Hcc. saccharolyticus* DSM 5350^T would appear to be an appropriate counterpart for *Hcc. salifodinae* BIp^T DSM 8989^T to carry out comparisons between a "contemporary" and a "Permian" genome, although the strains belong to different species, due to their DNA-DNA hybridization values of $<$ 70% and several different phenotypic properties. A description of isolates from surface waters of salterns in Goa, India, showed other suitable candidates: halococcoid isolates were found with 98–99% similarities of 16S rRNA genes to *Hcc. salifodinae* and *Hcc. saccharolyticus* [61]. From salt mines in Turkey, two halococcoid strains were obtained with 16S rRNA similarities of 99.8% and 99.3%, respectively, to *Hcc. dombrowskii*, and one pleomorphic strain with 99.7% similarity to *Hbt. noricense* [62]. Both *Hcc. dombrowskii* and *Hbt. noricense* originated also from Permo-Triassic salt sediments [8,63]. Detailed descriptions of all strains and whole genome sequences should yield meaningful comparisons.

3. Experimental Section

Growth of microorganisms: *Halococcus salifodinae* strains Blp^T (DSM 8989^T), BG2/2 (DSM 13045), Br3 (DSM 13046), N1 (DSM 13070), H2 (DSM 13071), *Hcc. morrhuae* DSM 1307^T, *Hcc. saccharolyticus* DSM 5350^T and *Hcc. dombrowskii* DSM 14522^T were grown in modified M2 medium as described by Denner *et al.* [1], which contained (g/l): yeast extract (5.0), casamino acids (5.0), NaCl (200.0), Tris/HCl (12.1), $MgCl_2 \times 6\ H_2O$ (20.0), $CaCl_2 \times 2\ H_2O$ (2.0) and KCl (0.2) at pH 7.4. For testing the presence of polyhydroxyalkanoates, strains were also grown in synthetic medium with 1% (w/v) glucose [64], except that KBr was used instead of NaBr. *Hcc. hamelinensis* JCM 12892^T and *Hcc. qingdaonensis* JCM 13587^T were grown in complex medium (DSM no. 372, http://www.dsmz.de/microorganisms/medium/pdf/DSMZ_Medium372.pdf). All cultures were incubated at 37 °C with shaking at 180 rpm in an Innova 4080 incubator (New Brunswick Scientific).

Preparation and analysis of cell walls: For preparation of cell walls, cultures were harvested by centrifugation at 8000 rpm for 20 min. The cell pellet was washed three times with deionized water and cells were disrupted with glass beads (Ø 0.5–0.7 mm) in a Braun cell homogenisator (model MKS) for 20 min. The crude cell wall preparation was incubated overnight with trypsin (0.5 mg/ml; Merck) in a 0.05 M (pH 7.8) potassium phosphate buffer at 37 °C. Cell wall preparations were washed four times with deionized water and freeze-dried (Lyovac, Leybold-Heraeus). Neutral sugars and uronic acids were released from cell walls under vacuum with 2 M HCl at 100 °C for 2 h and 3 h, respectively; amino sugars with 4 M HCl at 100 °C for 16 h. After removal of the acid the aqueous uronic acid solution was adjusted to pH 9 with an ammonia solution (1 M) and incubated for 2 h at room temperature to delactonize. The compounds were identified and quantified by HPLC using a CarboPac®PA1 column (Dionex) and pulsed amperometric detection. Monosaccharides and amino sugars were separated with a gradient reaching from 16 to 300 mM NaOH; for separation of uronic acids a gradient consisting of (A) 100 mM NaOH and (B) a solution of 100 mM NaOH with 1 M sodium acetate was used. Acetate [65], sulfate [66] and phosphate [67] were determined by the described methods.

Enzyme tests: The API ZYM system (bioMerieux) was used for the identification of 19 enzymatic activities [68]. Test strips were inoculated with cells in the exponential phase of growth and were incubated at 37–39 °C for up to 24 h [8]. Standard tests (oxidase, catalase, nitrate reduction, gelatin liquefaction, hydrolysis of starch, casein and Tween) were performed as described previously [8] or as recommended by Oren *et al.* [69]. Tests were performed at least in triplicate.

DNA-DNA hybridization. DNA was isolated as described by Cashion *et al.* [70]. Levels of DNA-DNA hybridization were determined spectrophotometrically by the renaturation method of De Ley *et al.* [71], with the modifications by [72] and [73]. Renaturation rates were computed by the program TRANSFER.BAS [74]. These experiments were carried out by the Identification Service of the DSMZ, Braunschweig, Germany.

Other methods: For comparative phylogenetic analyses 16S rRNA gene sequences from validly described *Halococcus* spp. were obtained from the European Molecular Biology Laboratory (EMBL) web interface and fitted in a subset of aligned archaeal sequences obtained from the Ribosomal Database Project II [75]. Phylogenetic relationships of the sequences were constructed by using distance-matrix methods (corrections as in Jukes and Cantor [76]). The web-based software MEGA 2 (http://www.megasoftware.net; [77]) and Clustal X [78] were used for sequence analysis and for construction of the phylogenetic tree, including maximum-likelihood and maximum parsimony methods. Confidence of the branching patterns was assessed by bootstrap analysis (1000 replicates). Scanning and transmission electron microscopy was performed as described previously [1,3].

4. Conclusions

From considerations of close genetic relatedness and origin from ancient geological materials it is concluded that *Halococcus salifodinae* strains can be considered as living fossils and constitute a promising source of novel evolutionary information.

Acknowledgments

This work was supported in part by the Austrian Science Funds (FWF), projects P16260 and P18256 (to HSL). We thank Christiane Grünewald for excellent technical assistance.

References

1. Denner, E.B.M.; McGenity, T.J.; Busse, H.-J.; Wanner, G.; Grant, W.D.; Stan-Lotter, H. *Halococcus salifodinae* sp. nov., an archaeal isolate from an Austrian salt mine. *Int. J. System. Bacteriol.* **1994**, *44*, 774–780.
2. Radax, C.; Gruber, G.; Stan-Lotter, H. Novel haloarchaeal 16S rRNA gene sequences from Alpine Permo-Triassic rock salt. *Extremophiles* **2001**, *5*, 221–228.
3. Stan-Lotter, H.; McGenity, T.J.; Legat, A.; Denner, E.B.M.; Glaser, K.; Stetter, K.O.; Wanner, G. Very similar strains of *Halococcus salifodinae* are found in geographically separated Permo-Triassic salt deposits. *Microbiol.* **1999**, *145*, 3565–3574.
4. Schoop, G. *Halococcus litoralis*, ein obligat halophiler Farbstoffbildner. *Dtsch Tierärztl Wochens* **1935**, *43*, 817–820.
5. Oren, A.; Arahal, D.R.; Ventosa, A. Emended descriptions of genera of the family *Halobacteriaceae*. *Int. J. Syst. Evol. Microbiol.* **2009**, *59*, 637–642.
6. Kocur, M.; Hodgkiss, W. Taxonomic status of the genus *Halococcus* Schoop. *Int. J. Syst. Bacteriol.* **1973**, *23*, 151–156.
7. Montero, C.G.; Ventosa, A.; Rodriguez-Valera, F.; Kates, M.; Moldoveanu, N.; Ruiz-Berraquero, F. *Halococcus saccharolyticus* sp. nov., a new species of extremely halophilic non-alkaliphilic cocci. *Syst. Appl. Microbiol.* **1989**, *12*, 167–171.
8. Stan-Lotter, H.; Pfaffenhuemer, M.; Legat, A.; Busse, H.-J.; Radax, C.; Gruber, C. *Halococcus dombrowskii* sp. nov., an archaeal isolate from a Permian alpine salt deposit. *Int. J. Syst. Evol. Microbiol.* **2002**, *52*, 1807–1814.
9. Goh, F.; Leuko, S.; Allen, M.A.; Bowman, J.P.; Kamekura, M.; Neilan, B.A.; Burns, B.P. *Halococcus hamelinensis* sp. nov., a novel halophilic archaeon isolated from stromatolites in Shark Bay, Australia. *Int. J. Syst. Evol. Microbiol.* **2006**, *56*, 1323–1329.
10. Wang, Q.-F.; Li, W.; Yang, H.; Liu, Y.-L.; Cao, H.-H.; Dornmayr-Pfaffenhuemer, M.; Stan-Lotter, H.; Guo, G.-Q. *Halococcus qingdaonensis* sp. nov., a halophilic archaeon isolated from a crude sea-salt sample. *Int. J. Syst. Evol. Microbiol.* **2007**, *57*, 600–604.
11. Namwong, S.; Tanasupawat, S.; Visessanguan, W.; Kudo, T.; Itoh, T. *Halococcus thailandensis* sp. nov., from fish sauce in Thailand. *Int. J. Syst. Evol. Microbiol.* **2007**, *57*, 2199–2203.

12. Wright, A.-D.G. Phylogenetic relationships within the order Halobacteriales inferred from16S rRNA gene sequences. *Int. J. Syst. Evol. Microbiol.* **2006**, *56*, 1223–1227.
13. Saitou, N.; Nei, M. The neighbor-joining method: A new method for reconstructing phylogenetic trees. *Mol. Biol. Evol.* **1987**, *4*, 406–425.
14. Kumar, S. Molecular clocks: four decades of evolution. *Nat. Rev. Genetics* **2005**, *6*, 654–662.
15. Dennis, P.P.; Shimmin, L.C. Evolutionary divergence and salinity-mediated selection in halophilic Archaea. *Microb. Mol. Biol. Rev.* **1997**, *61*, 90–104.
16. Gogarten, J.P.; Townsend, J.P. Horizontal gene transfer, genome innovation and evolution. *Nat. Rev. Microbiol.* **2005**, *3*, 679–687.
17. McGenity, T.J.; Gemmell, R.T.; Grant, W.D.; Stan-Lotter, H. Origins of halophilic microorganisms in ancient salt deposits. *Environ. Microbiol.* **2000**, *2*, 243–250.
18. Schubert, B.A.; Lowenstein, T.K.; Timofeeff, M.N.; Parker, M.A. Halophilic Archaea cultured from ancient halite, Death Valley, California. *Environ. Microbiol.* **2010**, *12*, 440–454.
19. Fendrihan, S.; Dornmayr-Pfaffenhuemer, M.; Gerbl, F.W.; Holzinger, A.; Grösbacher, M.; Briza, P.; Erler, A.; Gruber, C.; Plätzer, K.; Stan-Lotter, H. Spherical particles of halophilic Archaea correlate with exposure to low water activity - implications for microbial survival in fluid inclusions of ancient halite. *Geobiology* **2012**, *10*, 424–433.
20. Schubert, B.A.; Lowenstein, T.K.; Timofeeff, M.N. Microscopic identification of prokaryotes in modern and ancient halite, Saline Valley and Death Valley, California. *Astrobiology* **2009**, *9*, 467–482.
21. Norton, C.F.; Grant, W.D. Survival of halobacteria within fluid inclusions in salt crystals. *J. Gen. Microbiol.* **1988**, *134*, 1365–1373.
22. Mormile, M.R.; Biesen, M.A.; Gutierrez, M.C.; Ventosa, A.; Pavlovich, J.B.; Onstott, T.C.; Fredrickson, J.K. Isolation of *Halobacterium salinarum* retrieved directly from halite brine inclusions. *Environ. Microbiol.* **2003**, *5*, 1094–1102.
23. Fendrihan, S.; Legat, A.; Pfaffenhuemer, M.; Gruber, C.; Weidler, G.; Gerbl, F.; Stan-Lotter, H. Extremely halophilic archaea and the issue of long-term microbial survival. *Rev. Environ. Sci. Biotech.* **2006**, *5*, 203–218.
24. Gramain, A.; Chong Díaz G.C.; Demergasso, C.; Lowenstein, T.K.; McGenity, T.J. Archaeal diversity along a subterranean salt core from the Salar Grande (Chile). *Environ. Microbiol.* **2011**, *13*, 2105–2121.
25. Burns, B.P.; Gudhka, R.K.; Neilan, B.A. Genome sequence of the halophilic archaeon *Halococcus hamelinensis*. *J. Bacteriol.* **2012**, *194*, 2100–2101.
26. Grant, W.D.; Genus, I.V. *Halococcus* Schoop 1935a, 817[AL]. In *Bergey's Manual of Systematic Bacteriology*, 2nd ed.; Boone D.R., Castenholz, R.W., Garrity, G.M., Eds.; Springer-Verlag: New York, NY, USA, 2001; Volume 1, pp. 311–314.
27. Stackebrandt, E.; Goebel, B.M. Taxonomic note: a place for DNA-DNA reassociation and 16S rRNA sequence analysis in the present species definition in bacteriology. *Int. J. Syst. Bacteriol.* **1994**, *44*, 846–849.

28. Wayne, L.G.; Brenner, D.J.; Colwell, R.R.; Grimont, P.A.D.; Kandler, O.; Krichevsky, M.I.; Moore, L.H.; Moore, W.E.C.; Murray, R.G.E.; Stackebrandt, E.; Starr, M.P.; Trüper, H.G. International Committee on Systematic Bacteriology. Report of the ad hoc committee on reconciliation of approaches to bacterial systematics. *Int. J. Syst. Bacteriol.* **1987**, *37*, 463–464.
29. Zharkov, M.A. *History of Paleozoic Salt Accumulation*; Springer Verlag: Berlin, Germany, 1981.
30. Stan-Lotter, H.; Radax, C.; McGenity, T.J.; Legat, A.; Pfaffenhuemer, M.; Wieland, H.; Gruber, C.; Denner, E.B.M. From intraterrestrials to extraterrestrials - viable haloarchaea in ancient salt deposits. In *Halophilic Microorganisms*; Ventosa A., Ed.; Springer Verlag: New York, NY, USA, 2004; pp. 89–102.
31. Kocur, M.; Smid, B.; Martinec, T. The fine structure of extreme halophilic cocci. *Microbios* **1972**, *5*, 101–107.
32. Sumper, M.; Berg, E.; Mengele, R.; Strobel, I. Primary structure and glycosylation of the S-Layer protein of *Haloferax volcanii*. *J. Bacteriol.* **1990**, *172*, 7111–7118.
33. Schleifer, K.H.; Steber, J.; Mayer, H. Chemical composition and structure of the cell wall of *Halococcus morrhuae*. *Zbl. Bakt. Hyg. 1. Abt. Orig.* **1982**, *3*, 171–178.
34. Niemetz, R.; Kärcher, U.; Kandler, O.; Tindall, B.J.; König, H. The cell wall polymer of the extremely halophilic archaeon *Natronococcus occultus*. *Eur. J. Biochem.* **1997**, *249*, 905–911.
35. Kandler, O.; König, H. Cell envelopes of Archaebacteria. In *The Bacteria vol. VII*; Woese, C.R., Wolfe, R.S., Eds.; Academic Press: New York, NY, USA, 1985; pp. 413–457.
36. König, H.; Rachel, R.; Claus, H. Proteinaceous surface layers of *Archaea*: ultrastructure and biochemistry. In *Archaea. Molecular Cell Biology*; Cavicchioli, R., Ed.; ASM Press: Washington, DC, USA, 2007; pp. 315–340.
37. Claus, H.; Akca, E.; Debaerdemaeker, T.; Evrard, C.; Deqlercq, J.P.; Harris, J.R.; Schlott, B.; König, H. Molecular organization of selected prokaryotic S-layer proteins. *Can. J. Microbiol.* **2005**, *51*, 731–743.
38. Kärcher, U.; Schröder, H.; Haslinger, E.; Allmeier, G.; Schreiner, R.; Wieland, F.; Haselbeck, A.; König, H. Primary structure of the heterosaccharide of the surface glycoprotein of *Methanothermus fervidus*. *J. Biol. Chem.* **1993**, *268*, 26821–26826.
39. Brown, A.D.; Cho, K.Y. The walls of extremely halophilic cocci: Gram-positive bacteria lacking muramic acid. *J. Gen. Microbiol.* **1970**, *62*, 267–270.
40. Reistad, R. Cell wall of an extremely halophilic coccus. Investigation of ninhydrin-positive compounds. *Arch. Microbiol.* **1972**, *82*, 24–30.
41. Steber, J.; Schleifer, K.H. N-Glycyl-glucosamine, a novel constituent in the cell wall of *Halococcus morrhuae*. *Arch. Microbiol.* **1979**, *123*, 209–212.
42. Steber, J. Untersuchungen zur chemischen Zusammensetzung und Struktur der Zellwand von *Halococcus morrhuae*. PhD thesis, Technical University, Munich, Germany, 1976.
43. Koch, A.L. What size should a bacterium be? A question of scale. *Annu. Rev. Microbiol.* **1996**, *50*, 317–334.
44. Koch, A.L. Were Gram-positive rods the first bacteria? *Trends Microbiol.* **2003**, *11*, 166–170.

45. Rehm, H.A. Biogenesis of microbial polyhydroxyalkanoate granules: A platform technology for the production of tailor-made bioparticles. *Curr. Issues Mol. Biol.* **2007**, *9*, 41–62.
46. Fernandez-Castillo, R.; Rodriguez-Valera, F.; Gonzales-Ramos, J.; Ruiz-Berraquero, F. Accumulation of poly(β-hydroxybutyrate) by halobacteria. *Appl. Environ. Microbiol.* **1986**, *51*, 214–216.
47. Quillaguamán J.; Guzmán, H.; Van-Thuoc, D.; Hatti-Kaul, R. Synthesis and production of polyhydroxyalkanoates by halophiles: current potential and future prospects. *Appl. Microbiol. Biotechnol.* **2010**, *85*, 1687–1696.
48. Legat, A.; Gruber, C.; Zangger, K.; Wanner, G.; Stan-Lotter, H. Identification of polyhydroxyalkanoates in *Halococcus* and other haloarchaeal species. *Appl. Microbiol. Biotechn.* **2010**, *87*, 1119–1127.
49. Baliga, N.S.; Bonneau, R.; Facciotti, M.T.; Pan, M.; Glusman, G.; Deutsch, E.W.; Shannon, P.; Chiu, Y.; Weng, R.S.; Gan, R.R.; Hung, P.; Date, S.V.; Marcotte, E.; Hood, L.; Ng, W.V. Genome sequence of *Haloarcula marismortui*: A halophilic archaeon from the Dead Sea. *Genome Res.* **2004**, *14*, 2221–2234.
50. Bolhuis, H.; Palm, P.; Wende, A.; Falb, M.; Rampp, M.; Rodriguez-Valera, F.; Pfeiffer, F.; Oesterhelt, D. The genome of the square archaeon *Haloquadratum walsbyi*: life at the limits of water activity. *BMC Genomics* **2006**, *7*, 169.
51. Han, J.; Lu, Q.; Zhou, L.; Zhou, J.; Xiang, H. Molecular characterization of the $phaEC_{Hm}$ genes, required for biosynthesis of poly(3-hydroxybutyrate) in the extremely halophilic archaeon *Haloarcula marismortui*. *Appl. Environ. Microbiol.* **2007**, *73*, 6058–6065.
52. Lu, Q.; Han, J.; Zhou, L.; Zhou, J.; Xiang, H. Genetic and biochemical characterization of the poly(3-hydroxybutyrate-*co*-3-hydroxyvalerate) synthase in *Haloferax mediterranei*. *J. Bacteriol.* **2008**, *190*, 4173–4180.
53. Kalia, V.C.; Lal, S.; Cheema, S. Insight into the phylogeny of polyhydroxyalkanoate biosynthesis: horizontal gene transfer. *Gene* **2007**, *389*, 19–26.
54. Altschul, S.A.; Madden, T.L.; Schäffer, A.A.; Zhang, J.; Zhang, Z.; Miller, W.; Lipman, D.J. Gapped BLAST and PSI-BLAST: a new generation of protein database search programs. *Nucleic Acids Res.* **1997**, *25*, 3389–3402.
55. Muench, S.P.; Trinick, J.; Harrison, M.A. Structural divergence of the rotary ATPases. *Q Rev. Biophys.* **2011**, *44*, 311–356.
56. Stan-Lotter, H.; Sulzner, M.; Egelseer, E.; Norton, C.F.; Hochstein, L.I. Comparison of membrane ATPases from extreme halophiles isolated from ancient salt deposits. *Origins Life Evol. Biosphere.* **1993**, *23*, 53–64.
57. Hochstein, L.I.; Kristjansson, H.; Altekar, W. The purification and subunit structure of a membrane-bound ATPase from the archaebacterium *Halobacterium saccharovorum*. *Biochem. Biophys. Res. Commun.* **1987**, *147*, 295–300.
58. Stan-Lotter, H.; Hochstein, L.I. A comparison of an ATPase from the archaebacterium *Halobacterium saccharovorum* with the F_1 moiety from the *Escherichia coli* ATP synthase. *Eur. J. Biochem.* **1989**, *179*, 155–160.

59. Ochman, H.; Wilson, A. Evolution in bacteria: evidence for a universal rate in cellular genomes. *J. Mol. Evol.* **1987**, *26*, 74–86.
60. Park, J.S.; Vreeland, R.H.; Cho, B.C.; Lowenstein, T.K.; Timofeeff, M.N.; Rosenzweig, W.D. Haloarchaeal diversity in 23, 121 and 419 MYA salts. *Geobiology* **2009**, *7*, 515–523.
61. Mani, K.; Salgaonkar, B.B.; Braganca, J.M. Culturable halophilic archaea at the initial and crystallization stages of salt production in a natural solar saltern of Goa, India. *Aquat. Biosyst.* **2012**, *8*, 15.
62. Yildiz, E.; Ozcan, B.; Caliskan, M. Isolation, characterization and phylogenetic analysis of halophilic Archaea from a salt mine in central Anatolia (Turkey). *Polish J. Microbiol.* **2012**, *61*, 111–117.
63. Gruber, C.; Legat, A.; Pfaffenhuemer, M.; Radax, C.; Weidler, G.; Busse, H.-J.; Stan-Lotter, H. *Halobacterium noricense* sp. nov., an archaeal isolate from a bore core of an alpine Permian salt deposit, classification of *Halobacterium* sp. NRC-1 as a strain of *H. salinarum* and emended description of *H. salinarum*. *Extremophiles* **2004**, *8*, 431–439.
64. Lillo, J.; Rodriguez-Valera, F. Effects of culture conditions on poly (ß -hydroxybutyric acid) production by *Haloferax mediterranei*. *Appl. Environ. Microbiol.* **1990**, *56*, 2517–2521.
65. Bergmeyer, H.U. *Methoden der enzymatischen Analyse*; Verlag Chemie: Weinheim, Germany, 1974.
66. Dodgston, K.S.; Price, R.G. A note on the determination of the ester sulphate content of sulphated polysaccharides. *Biochem. J.* **1962**, *84*, 106–110.
67. Chen, P.S.; Toribara, T.Y.; Warner, H. Microdetermination of phosphorus. *Analyt. Chem.* **1956**, *28*, 1756–1758.
68. Humble, M.W.; King, A.; Phillips, I. API ZYM: A simple rapid system for the detection of bacterial enzymes. *J. Clin. Pathol.* **1977**, *30*, 275–277.
69. Oren, A.; Ventosa, A.; Grant, W.D. Proposed minimal standards for description of new taxa in the order *Halobacteriales*. *Int. J. Syst. Bacteriol.* **1997**, *47*, 233–238.
70. Cashion, P.; Hodler-Franklin, M.A.; McCully, J.; Franklin, M. A rapid method for the base ratio determination of bacterial DNA. *Anal. Biochem.* **1977**, *81*, 461–466.
71. De Ley, J.; Cattoir, H.; Reynaerts, A. The quantitative measurement of DNA hybridisation from renaturation rates. *Eur. J. Biochem.* **1970**, *12*, 133–142.
72. Huß, V.A.R.; Festl, H.; Schleifer, K.H. Studies on the spectrometric determination of DNA hybridisation from renaturation rates. *System. Appl. Microbiol.* **1983**, *4*, 184–192.
73. Escara, J.F.; Hutton, J.R. Thermal stability and renaturation of DNA in dimethylsulphoxide solutions: acceleration of renaturation rate. *Biopolymers* **1980**, *19*, 1315–1327.
74. Jahnke, K.-D. BASIC computer program for evaluation of spectroscopic DNA renaturation data from GILFORD System 2600 spectrometer on a PC/XT/AT type personal computer. *J. Microbiol. Meth.* **1992**, *15*, 61–73.
75. Maidak, B.L.; Cole, J.R.; Lilburn, T.G.; Parker, C.T., Jr.: Saxman, P.R.; Farris, R.J.; Garrity, G.M.; Olsen, G.J.; Schmidt, T.M.; Tiedje, J.M. The RDP-II (Ribosomal Database Project). *Nucleic Acids Res.* **2001**, *29*, 173–174.

76. Jukes, T.H.; Cantor, R.R. Evolution of protein molecules. In *Mammalian Protein Metabolism*; Munro, H.N, Ed.; Academic Press: New York, NY, USA, 1969; Volume 3, pp. 21–132.
77. Kumar, S.; Dudley, J.; Nei, M.; Tamura, K. MEGA: A biologist-centric software for evolutionary analysis of DNA and protein sequences. *Brief Bioinform.* **2008**, *9*, 299–306.
78. Thompson, J.D.; Gibson, T.J.; Plewniak, F.; Jeanmougin, F.; Higgins, D.G. The CLUSTAL_X windows interface: Flexible strategies for multiple sequence alignment aided by quality analysis tools. *Nucleic Acids Res.* **1997**, *25*, 4876–4882.

Molecular Mechanisms of Adaptation of the Moderately Halophilic Bacterium *Halobacillis halophilus* to Its Environment

Inga Hänelt and Volker Müller

Abstract: The capability of osmoadaptation is a prerequisite of organisms that live in an environment with changing salinities. *Halobacillus halophilus* is a moderately halophilic bacterium that grows between 0.4 and 3 M NaCl by accumulating both chloride and compatible solutes as osmolytes. Chloride is absolutely essential for growth and, moreover, was shown to modulate gene expression and activity of enzymes involved in osmoadaptation. The synthesis of different compatible solutes is strictly salinity- and growth phase-dependent. This unique hybrid strategy of *H. halophilus* will be reviewed here taking into account the recently published genome sequence. Based on identified genes we will speculate about possible scenarios of the synthesis of compatible solutes and the uptake of potassium ion which would complete our knowledge of the fine-tuned osmoregulation and intracellular osmolyte balance in *H. halophilus*.

Reprinted from *Life*. Cite as: Hänelt, I.; Müller, V. Molecular Mechanisms of Adaptation of the Moderately Halophilic Bacterium *Halobacillis halophilus* to Its Environment. *Life* **2013**, *3*, 234-243.

1. Introduction

Salt marshes are costal ecosystems in the upper intertidal zone between land and open sea water. Soil and water of this area face drastic changes in salinities since the land is regularly flooded by tides. In addition, water evaporates in summer leading to dryness and extremely high salinities of up to 3 M NaCl, while extensive rainfalls can decrease the salinity to fresh water concentrations. Organisms from all kingdoms of life that live in these areas have to adapt to such changing salinities by various strategies of osmoadaptation.

A well-studied model organism for osmoadaptation is the rod-shaped, endospore-forming, Gram-positive bacterium *Halobacillus halophilus*, which was isolated from a salt marsh on the North Sea coast of Germany and was originally described as *Sporosarcina halophila* [1]. Based on 16S rRNA homologies, it is now phylogenetically classified within the order *Bacillales*, Class *Bacilli*, Phylum *Firmicutes* and has been renamed to *Halobacillus halophilus* [2]. Being moderately halophilic, *H. halophilus* grows optimally between 0.5 and 2.0 M NaCl but can tolerate NaCl concentrations of up to 3.0 M NaCl with a growth rate of 38% of the optimum [3]. Strategies of osmoadaption of *H. halophilus* have been studied extensively in the past decades demonstrating a highly salinity- and growth phase-dependent adaption. This review will summarize the molecular mechanisms of osmoadaption in context of the recently published genome.

2. Hybrid Strategy for Long-Term Adaptation to Saline Environments

In general, two strategies are known to cope with changing or constantly high salinities of the environment. Organisms that grow optimally in the presence of extremely high salinities of up to 5 M NaCl, termed halophiles, accumulate intracellular KCl in concentrations higher than the external NaCl concentration to maintain a turgor pressure. This so called 'salt-in' strategy is found in the *Halobacteriales* (archaea) and the bacterium *Salinibacter ruber* [4,5]. Cellular processes and machineries of organisms following this 'salt-in' strategy are adapted to high internal KCl, such that in general these halophiles are restricted to growth at high salt. A more flexible strategy is found in moderately halophilic bacteria that grow over a wide range of salinities (typically 0.5–1 to 3 M NaCl) [6]. This strategy, the 'low-salt-in' strategy, relies on the accumulation of high concentrations of organic compatible solutes. Compatible solutes are small, mainly neutral but polar compounds (sugars, amino acids and derivates as well as polyols) which are highly soluble in water and do not interfere with the cellular metabolism. Thus, other than for KCl, a broad variation of the intracellular concentration of those compounds is possible without effecting cellular processes. The uptake or synthesis of compatible solutes retains a cytoplasm iso-osmotic with or slightly hyperosmotic compared to its surroundings.

H. halophilus has originally been described as a bacterium that amasses compatible solutes to establish a cellular turgor [2]. However, later it was shown that it also accumulates molar concentrations of chloride in the cytoplasm [3]. This survival strategy is now seen as a unique hybrid strategy of the moderately halophilic *H. halophilus* to cope with changing salinities of the environment.

The hypothesis of this hybrid strategy arose from the fact that *H. halophilus* amasses compatible solutes as well as chloride in the cytoplasm as we will discuss below. However, similar conclusions resulted from the recently published analysis of the proteome of *H. halophilus* deduced from the genome sequence [7]. Comparisons of different genome sequences revealed a clear distinction of proteomes from extreme halophiles and non-halophiles with a higher number of acidic proteins in the extreme halophiles. Within this comparison, the proteome of *H. halophilus* takes an intermediate position as the averaged isoelectric point of all proteins of 6.6 is slightly acidic. This trend was seen for soluble as well as for membrane proteins and supports an intermediate strategy in osmoadaptation [7].

3. The Chloride Modulon

One of the outstanding, or even to our knowledge, unique physiological features of *H. halophilus* is its chloride dependence of growth, gene expression and enzymatic activity. Growth of *H. halophilus* was shown to strictly depend on the presence of chloride, in absence of chloride the strain does not grow. As a consequence of high external salinities, Cl^- was shown to accumulate in the cytoplasm following the 'salt-in' strategy. While the internal Cl^- concentration is negligible at low external Cl^- concentrations, it increases to 50% of the external Cl^- concentration at higher salt concentrations [3]. Moreover, *H. halophilus* not only depends on Cl^- to compensate for increasing external salinities but Cl^- also regulates cellular processes. Increasing external chloride concentrations and not the salinity in general were determined to regulate the germination of endospores [8] and the motility of the vegetative cell [9] but also to control the expression of genes and the enzymatic activity of proteins needed for the halophilic life style of *H. halophilus* [10]. Most of these regulated proteins are key players of the biosynthesis of compatible solutes which are essential to cope with high and changing salinities. All known processes regulated by chloride are summarized in the chloride modulon (Figure 1).

Figure 1. The chloride regulon of *Halobacillus halophilus*. All known processes that were identified to be influenced by the presence of chloride are summarized. Both, on transcriptional and enzyme activity level, chloride was found to have stimulating effects [10]. For further explanations see text.

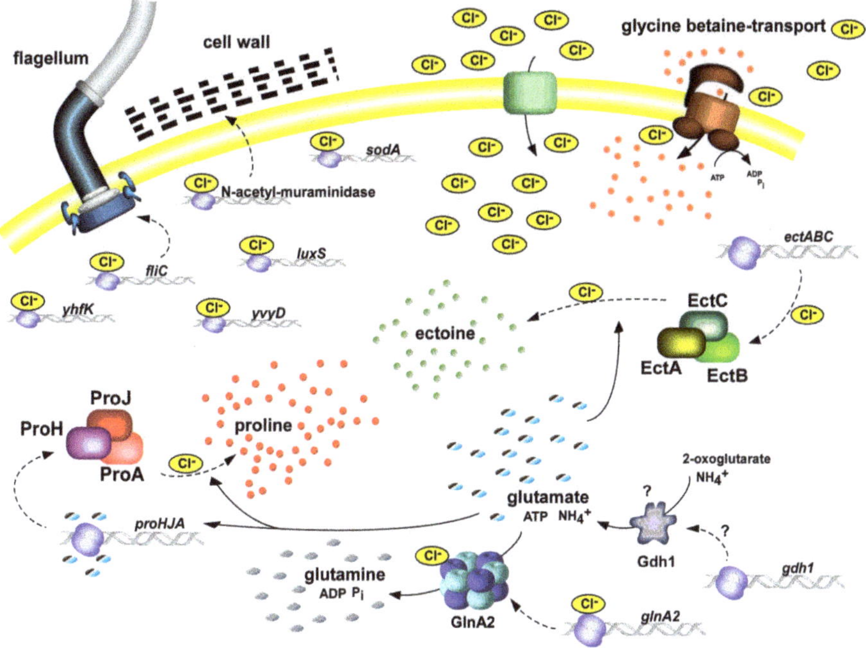

An open question is how chloride accumulates in the cytoplasm. Since the measured internal chloride concentrations are too high to be in equilibrium with the membrane potential, it has been suggested that chloride is actively accumulated into the cytoplasm of *H. halophilus* [3]. However, so far no gene could be identified to encode for a chloride transporter or even a simple channel [7]. Several genes are annotated to encode for potential symporters but their substrate specificity remains to be elucidated. To completely understand the chloride modulon, the chloride transporter has to be found and characterized in detail.

Also the counterion for chloride is not known but is likely to be potassium as shown for many other organisms. However, it is not yet understood why potassium is the main monovalent intracellular cation and why it is preferred compared to Na^+ in this function. In bacteria, K^+ might modulate activity and correct folding of proteins more effectively than Na^+. Another explanation for the accumulation of K^+ and the concomitant active extrusion of Na^+ is that this situation enables the cells to establish an inwardly directed electrochemical transmembrane Na^+ gradient, which in consequence is used for energy-consuming processes like secondary transport or flagella movement [11]. Interestingly, the genome of *H. halophilus* encodes for Ktr-type potassium transporters (Hbhal_1246, Hbhal_2758, Habl_4548) and two potassium channels (Hbhal_3837, Hbhal_3881) only while high-affinity, ATP-dependent transporters like the Kdp system were not identified [7]. Further studies are needed to elucidate the role of potassium and identify involved uptake systems.

4. Biosynthesis of Compatible Solutes

Next to accumulating internal Cl^-, *H. halophilus* synthesizes a cocktail of different solutes to combat external salinity. The major solutes are glutamate, glutamine and proline but also ectoine, N^{δ}-acetyl ornithine and N^{ε}-acetyl lysine are produced [7,10]. The biosynthesis pathways of glutamate, glutamine, proline and ectoine are predicted based on the genome sequence and studied biochemically (Figure 2) while the pathways of N^{δ}-acetyl ornithine and N^{ε}-acetyl lysine are deduced from the genome sequence only.

Three main enzymatic reactions are known for the biosynthesis of glutamate and glutamine [12]. The biosynthesis of glutamate can either be accomplished by the action of a sequence of glutamine synthetase and glutamate synthase (GOGAT) or by the action of a glutamate dehydrogenase (GDH) while glutamine is synthesized by a glutamine synthetase (Gln) only. The genome contains two putative open reading frames encoding a glutamate synthase (*gdh1* and *gdh2*), only one open reading frame encoding the large subunit of a glutamate synthase (*gltA*) and two open reading frames each potentially encoding the small subunit of a glutamate synthase (*gltB1* and *gltB2*). The glutamine synthetase Gln is encoded by two open reading frames (*glnA1* and *glnA2*). While the former (*glnA1*) clusters with a gene (*glnR*) encoding the regulatory protein GlnR which is known to be essential in nitrogen metabolism from *Bacillus subtilis*, the latter (*glnA2*) lies solitary and is predicted to be regulated by a promotor recognized by σ^B, the general stress σ-factor.

Proline is synthesized from glutamate by a sequence of a glutamate 5-kinase (ProJ), a glutamate 5-semialdehyde dehydrogenase (ProA) and a pyrroline-5-carboxylate reductase (ProH) [13]. The enzymes are encoded by a cluster of three genes—*proH, proJ* and *proA*, which are organized in an operon.

Also the three biosynthetic genes (*ectABC*) for the production of ectoine are arranged on one operon [14]. Ectoine is produced from aspartate semialdehyde by a sequence of a diaminobutyrate-2-oxoglutarate transaminase (EctB), a diaminobutyric acid acetyltransferase (EctA) and an ectoine synthase (EctC).

Figure 2. Proposed biochemical pathways of the main compatible solutes in *Halobacillus halophilus*. The pathways for the synthesis of glutamate, glutamine, proline and ectoine are shown in black, the involved enzymes (and coding genes) in gray. Stimulating effects of chloride and glutamate on gene transcription and enzyme activity are indicated [10].

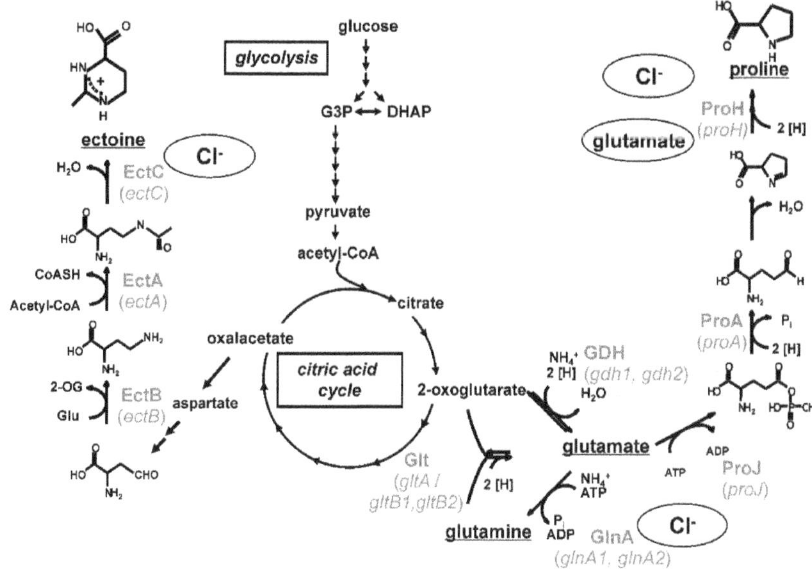

The pathway for the production of N^{δ}-acetyl ornithine still has to be elucidated but presumably ornithine is the direct precursor, for which several pathways are known. All necessary genes encoding the corresponding enzymes were identified in the genome of *H. halophilus* and possible pathways were described in detail by Saum and colleagues recently (Figure 3) [7].

The last potential compatible solute that is synthesized by *H. halophilus* is N^{ε}-acetyl lysine. However, its role as compatible solute still has to be confirmed by further studies. Based on the now available genome sequence it was assumed that *H. halophilus* is capable of synthesizing lysine by use of the classical diaminopimelate pathway [7]. In this pathway aspartate is initially activated by an aspartate kinase. *H. halophilus* possesses two copies of this enzyme (*dapG1*, Hbhal_3090, *dapG2*, Hbhal_3465) which might be, similar to the glutamine synthetase, subject to different modes of regulation. The activated aspartyl moiety in consequence gets reduced to the corresponding semialdehyde [catalysed by an aspartate semialdehyde dehydrogenase (*asd*, Hbhal_3089)] which consequently undergoes a condensation reaction with one molecule of pyruvate resulting in the formation of 2,3-dihydrodipicolinate. This reaction, which is the first that differs from the ectoine biosynthesis pathway, is catalyzed by the dihydrodipicolinate synthase. Three copies of the corresponding gene were identified (*dapA1*, Hbhal_2387, *dapA2*, Hbhal_3091, *dapA3*, Hbhal_5017) which may indicate its critical role in the biosynthesis. It is likely that the different genes are controlled by different demands such as the need for lysine as a compound in protein biosynthesis, the need of N^{ε}-acetyl lysine as osmoprotectant or the need to provide precursors for the biosynthesis of peptidoglycan. The following sequence of reactions leading to lysine is then catalyzed by a dihydrodipicolinate reductase (Hbhal_3261), a tetrahydrodipicolinate N-acetyltransferase (Hbhal_2790), an acetyltransferase, an N-acetyldiaminopimelate deacetylase (Hbhal_2791), a diaminopimelate epimerase (*dapF*, Hbhal_2627) and finally a diaminopimelate decarboxylase (*lysA*, Hbhal_3343). The final acetylation at the ε-amino group requires an acetyltransferase (Hbhal_3877).

Figure 3. Putative biosynthetic pathways of N^{δ}-acetyl ornithine. Most likely, N^{δ}-acetyl ornithine is formed from ornithine catalysed by a so far unidentified N^{δ}-acetyltransferase. Based on the genome sequence, the pool of ornithine could be replenished by the conversion of glutamate, proline or arginine catalysed by the reactions depicted in the diagram. The asterisk '*' indicates genes for which no gene name has yet been assigned [7].

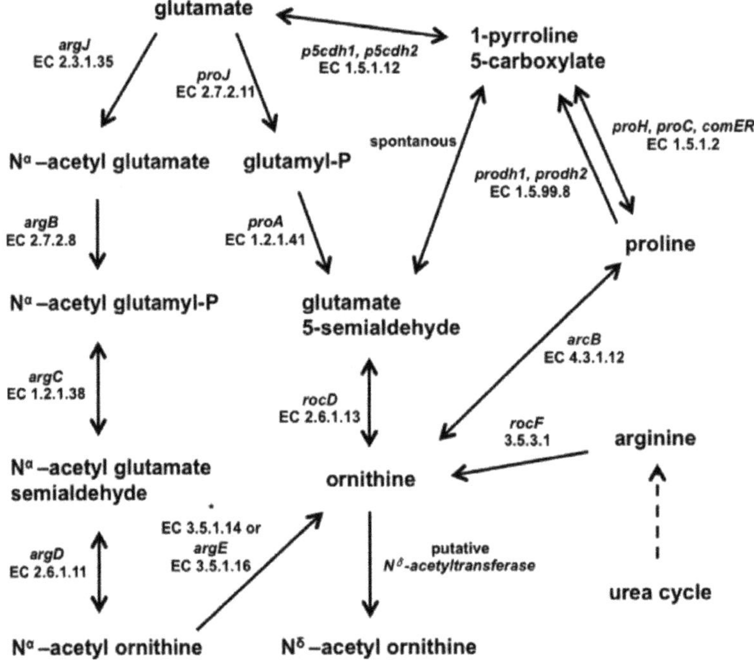

5. Salinity- and Growth Phase-dependent Adaptation of the Solute Pool

Interestingly, the compatible solutes listed above do not appear all at the same time but *H. halophilus* switches its osmolyte content depending on salinity (Figure 4) [10]. At intermediate salinities of around 1.5 M NaCl, glutamate and glutamine are the major solutes. Transcription analyses after an osmotic upshock from 0.8 to 2 M NaCl have shown that one of the putative glutamate dehydrogenase genes (*gdh1*) was induced and the mRNA level increased within 1.5 hours to about four-fold compared to the level before the upshock. This enables an increased production of glutamate from 2-oxoglutarate and NH^{4+}. In contrast, the transcript levels of the second glutamate dehydrogenase gene (*gdh2*) were close to the detection limit likely being involved in nitrogen metabolism rather than osmoregulation. So far, also the glutamate synthase gene (*gltA*) did not seem to be involved in osmoregulation. Glutamine is synthesized by the action of a glutamine synthetase which is encoded by two genes (*glnA1* and *glnA2*) in *H. halophilus*. On a transcriptional level only the expression of *glnA2* was shown to be upregulated at increasing salt concentrations with a maximal increase of transcripts of about 4-fold (compared to the value at 0.4 M NaCl) at 1.5 M NaCl or higher. The expression of *glnA1* was not affected. Moreover, the expression of *glnA2* and especially the glutamine synthetase activity were shown to be chloride-dependent being increased with increasing chloride concentrations in the surrounding media. The maximal enzymatic activity was found at 2.5 M NaCl or higher [12]. This is in line with the chloride dependence of growth and the accumulation of Cl into the cytoplasm as described above. However, it is unknown how chloride modulates the enzymatic activity. Both, a direct interaction of chloride with the glutamine synthetase or the participation of a regulatory protein that senses and mediates the concentration of chloride, are possible.

Figure 4. Accumulation of glutamine, glutamate, proline and ectoine is dependent on the NaCl concentration of the medium. Cells of *H. halophilus* were cultivated in mineral salt medium (G10) in the presence of the NaCl concentrations indicated. They were harvested in the exponential growth phase (OD_{578}. 0.6 to 0.8), compatible solutes were extracted, and the concentrations of glutamine (black), glutamate (dark gray), proline (light gray) and ectoine (white) were measured by HPLC [14].

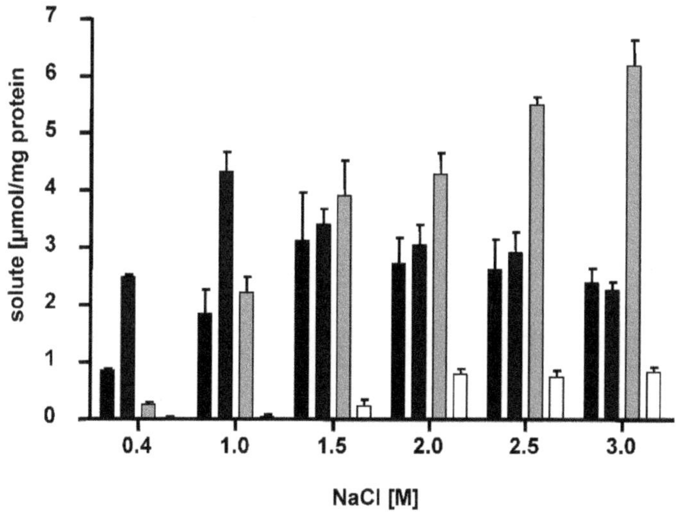

At high salinities (2.0 M NaCl or higher), glutamine and glutamate pools stay rather constant but proline is produced in addition, and becomes the dominant solute. It was shown that the transcription of the *pro* operon was increased with increasing salinities with a maximum at 2.5 M. The mRNA level reached a maximum 1.5 hours after an osmotic upshock while the maximal concentration of proline was determined after 6 hours. Consequently, the increased amount of enzymes led to an increased production of proline [13]. However, not only NaCl but also

Na-glutamate had an effect on gene expression. Compared to NaCl Na-glutamate was shown to dramatically increase the *proHJA* mRNA concentration. Since proline is produced from glutamate and Na-glutamate had a more stimulating effect on gene expression than NaCl, NaCl now is supposed to be the initial signal for the proline production only. Glutamate instead acts as 'second messenger' that further regulates the pro operon expression by its internal concentration which increases with increasing salinity [13].

In addition to the salinity-dependent solute regulation, another layer of regulation is active: proline contents are maximal in exponentially growing cultures but reduced in stationary phase cultures. Under these conditions, ectoine is synthesized. To resolve the time-dependent kinetics of ectoine production *H. halophilus* cells were subjected to an osmotic upshock from 0.8 to 2.0 M NaCl and the biosynthesis of ectoine was measured at the levels of transcription, translation and solute accumulation. Transcripts were readily detectable directly after the upshock, but increased dramatically with time and reached a maximum 3 hours later. Most important, the expression of ect genes was preceded by expression of genes responsible for glutamine, glutamate or proline biosynthesis. The signal leading to *ect* gene transcription is therefore assumed to be an indirect one mediated by one or more yet to be identified factors rather than by the presence of the osmolyte. The production of the ectoine synthase EctC nicely corresponds to the increase of *ectC* transcript. Both were found to increase 2- fold. Surprisingly, 4 hours after upshock the EctC concentration again decreased with time and the level reached a value only slightly above the value at the beginning, although the external stress was still present. This decrease, however, was not reflected in the ectoine concentration, which steadily increased and reached a maximum 18 hours after upshock. Again, this demonstrates a great delay in accumulation compared to proline that reached its maximum already 6 hours after upshock and hints to a role of ectoine not only in the immediate response to osmotic upshock but to a function as a more general protectant in the cell [14].

To quickly adapt to changing salinities and growth phases *H. halophilus* regulates the synthesis of solutes by both increasing expression of the enzymes involved and activating the produced enzymes in a chloride-dependent manner. *H. halophilus* also has a gene encoding a potential ectoine hydroxylase, but hydroxyectoine has not been detected yet [7]. *H. halophilus* does not have the genetic capacity for *de novo* biosynthesis of glycine betaine, but can take up choline from the environment and oxidize it to glycine betaine [15].

6. Catabolic Traits and Nutritional Versatility

The observed preference of specific compatible solutes is also reflected by possible catabolic traits and nutritional versatility of *H. halophilus*. *H. halophilus* is a chemoorganoheterotrophic, strictly aerobic bacterium with great nutritional versatility. It is able to hydrolyze complex substrates such as casein, gelatin, DNA, starch and pullulan [1]. Genes encoding two extracellular proteases (Hbhal 5155 and Hbhal 4449), one amylase (Hbhal 4101) and one pullulanase (Hbhal 2962) were identified in the genome. In addition, *H. halophilus* is able to grow on hexoses such as glucose or fructose (by way of glycolysis) and on amino acids. Noteworthy is the use of carbon sources that are also used as compatible solutes such as glutamate and proline. Proline and glutamate, major compatible solutes of *H. halophilus*, are good growth substrates and may also be used as nitrogen sources, indicating a sophisticated regulatory network balancing different cellular needs. For example, proline is degraded by proline dehydrogenase (ProDH) and Δ^1-pyrroline-5-carboxylate dehydrogenase (P5CDH) to glutamate via Δ^1-pyrroline-5-carboxylate of which *H. halophilus* has two isogenes each for *prodh* and *p5cdh*. *prodh2* and *p5cdh2* form an operon (put operon) that is involved in the utilization of proline as carbon and energy source whereas ProDH1 and P5CDH1 may be involved in supplying the cell with nitrogen from proline [7].

In contrast to *Halomonas elongata*, the genome of *H. halophilus* does not encode for ectoine utilization genes, consistent with the observation that ectoine is only a minor solute in *H. halophilus*.

7. Concluding Remarks

H. halophilus is the first moderate halophilic bacterium shown to use a hybrid strategy for osmoadaptation by accumulating both molar concentrations of chloride and compatible solutes. This unique feature enables *H. halophilus* to grow over a broad range of salinities and to adapt sufficiently to rapidly changing environments. Other than hyperosmotic organisms, *H. halophilus* can survive at low NaCl concentrations but also copes with relatively high salt concentrations of up to 3 M. The sophisticated salinity- and growth phase-dependent adaptation of the accumulated solutes is remarkable and probably demonstrates a long lasting evolution being optimally prepared for its changing environment. This adaptation is also reflected by the use of a dominant compatible solute, such as carbon and nitrogen, to guarantee energy optimization.

Acknowledgement

Financial support by the Deutsche Forschungsgemeinschaft is gratefully acknowledged.

References

1. Claus, D.; Fahmy, F.; Rolf, H.J.; Tosunoglu, N. *Sporosarcina halophila* sp. nov., an obligate, slightly halophilic bacterium from salt marsh soils. *Syst. Appl. Microbiol.* **1983**, *4*, 496–506.
2. Spring, S.; Ludwig, W.; Marquez, M.C.; Ventosa, A.; Schleifer, K.-H. *Halobacillus* gen. nov., with descriptions of *Halobacillus litoralis* sp. nov. and *Halobacilus trueperi* sp. nov., and transfer of *Sporosarcina halophila* to *Halobacillus halophilus* comb. nov. *Int. J. Syst. Bacteriol.* **1996**, *46*, 492–496.
3. Roessler, M.; Müller, V. Quantitative and physiological analyses of chloride dependence of growth of *Halobacillus halophilus*. *Appl. Environ. Microbiol.* **1998**, *64*, 3813–3817.
4. Galinski, E.A.; Trüper, H.G. Microbial behavior in salt-stressed ecosystems. *FEMS Microbiol. Rev.* **1994**, *15*, 95–108.
5. Ventosa, A.; Nieto, J.J.; Oren, A. Biology of moderately halophilic aerobic bacteria. *Microbiol. Mol. Biol. R* **1998**, *62*, 504–544.
6. Roessler, M.; Müller, V. Osmoadaptation in bacteria and archaea: Common principles and differences. *Environ. Microbiol.* **2001**, *3*, 743–754.

7. Saum, S.; Pfeiffer, F.; Palm, P.; Rampp, M.; Schuster, S.; Müller, V.; Oesterhelt, D. Chloride and organic osmolytes: A hybrid strategy to cope with elevated salinities by the moderately halophilic, chloride-dependent bacterium *Halobacillus halophilus*. *Environ. Microbiol.* **2012**, doi: 10.1111/j.1462-2920.2012.02770.x.
8. Dohrmann, A.B.; Müller, V. Chloride dependence of endospore germination in *Halobacillus halophilus*. *Arch. Microbiol.* **1999**, *172*, 264–267.
9. Roessler, M.; Müller, V. Chloride, a new environmental signal molecule involved in gene regulation in a moderately halophilic bacterium, *Halobacillus halophilus*. *J. Bacteriol.* **2002**, *184*, 6207–6215.
10. Saum, S.H.; Müller, V. Regulation of osmoadaptation in the moderate halophile *Halobacillus halophilus*: Chloride, glutamate and switching osmolyte strategies. *Saline Syst.* **2008**, *4*, 4.
11. Bakker, E.P. *Cellular K^+ and K^+ Transport Systems in Prokaryotes*; CRC Press: Boca Raton, FL, USA, 1992; pp. 205–224.
12. Saum, S.H.; Sydow, J.F.; Palm, P.; Pfeiffer, F.; Oesterhelt, D.; Müller, V. Biochemical and molecular characterization of the biosynthesis of glutamine and glutamate, two major compatible solutes in the moderately halophilic bacterium *Halobacillus halophilus*. *J. Bacteriol.* **2006**, *188*, 6808–6815.
13. Saum, S.H.; Müller, V. Salinity-dependent switching of osmolyte strategies in a moderately halophilic bacterium: Glutamate induces proline biosynthesis in *Halobacillus halophilus*. *J. Bacteriol.* **2007**, *189*, 6968–6975.
14. Saum, S.H.; Müller, V. Growth phase-dependent switch in osmolyte strategy in a moderate halophile: Ectoine is a minor osmolyte but major stationary phase solute in *Halobacillus halophilus*. *Environ. Microbiol.* **2008**, *10*, 716–726.
15. Burkhardt, J.; Sewald, X.; Bauer, B.; Saum, S.H.; Müller, V. Synthesis of glycine betaine from choline in the moderate halophilic *Halobacillus halophilus*: Co-regulation of two divergent, polycistronic operons. *Environ. Microbiol. Rep.* **2009**, *1*, 38–43.

Chapter 3:
Structure and Function of Extremophilic Biomolecules

Surface Appendages of Archaea: Structure, Function, Genetics and Assembly

Ken F. Jarrell, Yan Ding, Divya B. Nair and Sarah Siu

Abstract: Organisms representing diverse subgroupings of the Domain Archaea are known to possess unusual surface structures. These can include ones unique to Archaea such as cannulae and hami as well as archaella (archaeal flagella) and various types of pili that superficially resemble their namesakes in Bacteria, although with significant differences. Major advances have occurred particularly in the study of archaella and pili using model organisms with recently developed advanced genetic tools. There is common use of a type IV pili-model of assembly for several archaeal surface structures including archaella, certain pili and sugar binding structures termed bindosomes. In addition, there are widespread posttranslational modifications of archaellins and pilins with N-linked glycans, with some containing novel sugars. Archaeal surface structures are involved in such diverse functions as swimming, attachment to surfaces, cell to cell contact resulting in genetic transfer, biofilm formation, and possible intercellular communication. Sometimes functions are co-dependent on other surface structures. These structures and the regulation of their assembly are important features that allow various Archaea, including thermoacidophilic, hyperthermophilic, halophilic, and anaerobic ones, to survive and thrive in the extreme environments that are commonly inhabited by members of this domain.

Reprinted from *Life*. Cite as: Jarrell, K.F.; Ding, Y.; Nair, D.B.; Siu, S. Surface Appendages of Archaea: Structure, Function, Genetics and Assembly. *Life* **2013**, *3*, 86-117.

1. Introduction

The study of Archaea has led to great advancements in many fields of biology [1,2], with discoveries that have aided the understanding of processes common to Eucarya and/or Bacteria. In addition, other findings resulted in reports that highlight the novelty of the organisms representing the third Domain of life, such as their unusual and often unique surface appendages [3–6]. Like their bacterial counterparts, archaeal cells can possess a variety of surface structures that are critical in many aspects of their interactions with the environment. These structures are either (a) entirely unique to the Domain Archaea with no equivalent in either of the other Domains (*i.e.*, hami [7] and cannulae [8]) (b) similar to known bacterial structures but with archaeal-specific twists (pili [9–11]) or (c) they are structures which only superficially resemble appendages found in the bacterial domain with fundamental variations (archaella, formerly known as archaeal flagella [12–17]). Throughout this review, we will use the term archaellum which has been proposed as a new designation for the structure formerly called the archaeal flagellum. This designation was suggested since the archaeal structure does not resemble the bacterial flagellum in structure or assembly although both function in swimming. Discussion of the merits of the new term within the scientific community continues [18,19].

Given the limited number of Archaea for which tractable genetic tools are available [20], it is no surprise that the studies of archaeal surface appendages are limited in most instances to a few model organisms, such as *Halobacterium, Haloferax, Sulfolobus* and *Methanococcus.* Most recently, structural and genetic studies of pili and archaella in Archaea have focused on *Methanococcus maripaludis* and *Sulfolobus* species like *S. acidocaldarius* (Figure 1). The appearance of hami and cannulae is limited thus far to reports in a single genus that lacks genetic systems. Thus data on these unique structures are confined to biochemical and physiological analyses coupled with exquisite electron microscopic studies.

Genetic studies have revealed a preference of Archaea to utilize a bacterial type IV pili model to assemble many of their surface appendages, although the structures formed from these subunits are themselves unique [9–11,21]. This assembly mechanism is characterized, among other things, by the presence of class III signal peptides on the major structural proteins, requiring their removal by a dedicated signal peptidase [22–26]. In addition, another very common feature of the major structural subunits of various archaeal surface appendages is attachment of glycan, with N-linked glycan attachment being the best studied [5,8,9,27–30].

Interactions of archaeal cells with their environment through their surface appendages can include such functions as swimming, swarming, attachment to abiotic and biotic surfaces, aggregation, intercellular communication, DNA uptake, virus attachment, nutrient uptake, and biofilm formation. Functional analysis of the various archaeal appendages has confirmed their role in many of these important processes and often multiple functions have been attributed to a single structure or the co-operation of more than one structure is needed in carrying out a particular function [7,9,26,31–38]. The first reports of regulation of surface structure biosynthesis [39] indicate that, in *S. acidocaldarius*, regulation of archaella and pili biosynthesis are linked allowing the cells to adapt to changing environments by simultaneously increasing expression of one structure while repressing synthesis of the other.

In addition to genetic analysis of several archaeal appendages, several diverse archaeal surface structures (archaella, pili and Iho670 fibres) with the common type IV pili-like assembly model, have now been studied by various electron microscopic and imaging techniques revealing in all cases unusual and unique features [9–11,21,40].

In this contribution, we have reviewed the available data on surface structures from various Archaea from structural, functional, genetic and assembly aspects.

Figure 1. Appendages on the well-studied Archaea *M. maripaludis* and *S. acidocaldarius*. (**A**) Electron micrograph of *M. maripaludis* showing thin pili (arrows) with thicker and more numerous archaella. Bar = 0.5 µm. Courtesy of S.I. Aizawa. Prefectural University of Hiroshima, Japan. (**B**) Electron micrograph of *S. acidocaldarius* showing the presence of three different appendages namely archaella (14nm diameter, black arrow), Aap pili (10–12 nm, white arrow) and threads (5 nm, grey arrow). Bar = 0.5 µm. Courtesy of A.-L. Henche and S.V. Albers, Max Planck Institute for Terrestrial Microbiology, Marburg Germany.

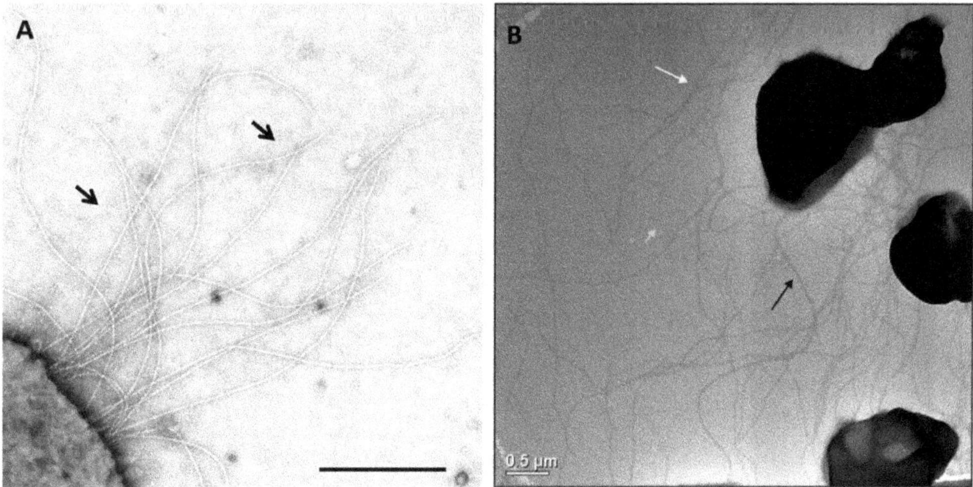

2. Widespread Use of the Bacterial Type IV Pili System for Archaeal Surface Structures

Archaeal type IV pilus-like structures include archaella [12,17,41,42], type IV-like pili [9,28,34,36], the sugar-binding structure termed the bindosome [43], and the unusual, brittle Iho670 fibers from *Ignicoccus hospitalis* [10] . Bacterial type IV pilins are synthesized as precursor proteins with a class III signal peptide, which is cleaved by a prepilin peptidase (PilD in *Pseudomonas aeruginosa*) [44]. Unlike signal peptidase I or II, whose cleavage site is on the periplasmic side of the cytoplasmic membrane, PilD cleaves the signal peptide from the cytoplasmic side, leaving the hydrophobic N terminal α-helix as part of the mature pilin. Similarly, structural proteins in archaeal type IV pilus-like structures are also synthesized as precursor proteins with usually short signal peptides (often 6–12 amino acids in length; Table 1) and processed by a unique signal peptidase homologous to the bacterial prepilin peptidase [22,23,25,45].

Table 1. Seleted archaeal proteins with type IV prepilin-like signal petides.

Archaellins	Gene	Signal peptide	Start of mature protein
Halobacterium salinarum R1	FlgA1	MFEFITDEDERG	QVGIGTLIVFIAMVLVAAIA
	FlgA2	MFEFITDEDERG	QVGIGTLIVFIAMVLVAAIA
	FlgB1	MFEFITDEDERG	QVGIGTLIVFIAMVLVAAIA
	FlgB2	MFEFITDEDERG	QVGIGTLIVFIAMVLVAAIA
	FlgB3	MFEFITDEDERG	QVGIGTLIVFIAMVLVAAIA
	FlgX	MESMRR	QVGIGTLVVFMAMILVAAMA
Haloferax volcanii DS2	FlgA1	MFENINEDRG	QVGIGTLIVFIAMVLVAAIA
	FlgA2	MFNNITDDDRG	QVGIGTLIVFIAMVLVAAIA
Methanococcus maripaludis S2	FlaB1	MKIKEFLKTKKG	ASGIGTLIVFIAMVLVAAVA
	FlaB2	MKITEFMKNKKG	ASGIGTLIVFIAMVLVAAVA
	FlaB3	MVKKFMKSKKG	AVGIGTLIIFIAMVLVAAIA
Sulfolobus solfataricus P2	FlaB	(MNSKK)MLKEYNKKVKRKG	LAGLDTAIILIAFIITASVL
Pilins			
Adhesion pilins			
Sulfolobus acidocaldarius MW001	AapA	MYNKITMISRYRYDKRRIRA	LSGAIVALILVIAGVIIATA
	AapB	MNIEVKKSKKKNMRA	LSGAIVALILVIAGVIIAIA
UV-induced pilins			
Sulfolobus solfataricus P2	UpsA	MMWLKA	ISSIFSTLIVVMITLSLIVP
	UpsB	MLQLMMKGGYKLKKRKG	LSSILGTVIVLAITLVLGGL
EppA-dependent pilins			
Methanococcus maripaludis S2	EpdA	MFKRFNRG	QISFEFSIIVLSILLISTIT
	EpdB	MSKG	QVSVEFIVLFLALLVAVVVS
	EpdC	MIKMLQLPFNKKG	QVSFDFIIAMLFLLLIFAFM
	MMP1685	MKFLEKLTSKKG	QIAMELGILVMAAVAVAAIA
	MMP1283	MSVALKKFFSKRG	QLSLEFSVLVLAVITAAILL
Others			
Iho670 fibers			
Ignicoccus hospitalis KIN4/I[T]	Igni_0670	MKIARKG	VSPVIATLLLILIAVAAAVL
Bindosome components			
Sulfolobus solfataricus P2	GlcS	MKRKYPYSLAKG	LTSTQIAVIVAVIVIVIIIG
	AraS	MSRRRLYKA	ISRTAIIIIVVVIIIAAIAG
	BasA	MRG	ISEAITVVFLILVTLIAIAI
	BasB	MKG	QASVIAMVVIFFLIIATIGL

The various archaeal type IV pili-like systems also share homologues of ATPases and a conserved cytoplasmic membrane protein needed for assembly with the bacterial system [46,47].

3. Archaella

The first appendage in Archaea that was hypothesized to use the type IV pilus-like assembly mechanism was the archaellum [42,48] and it remains the best-studied archaeal type IV pilus-like appendage. Archaella are found commonly on the surfaces of diverse Archaea including methanogens, extreme halophiles, thermoacidophiles and hyperthermophiles [49]. The number and location of archaella varies enormously among different species [49]. Archaella diameters are typically between 10-14 nm [49], although thicker filaments have been reported [50]. Although the major function of this appendage is swimming via filament rotation, like the bacterial flagellum, its structure and assembly are remarkably much more closely related to type IV pili [5,15–17,49,51].

3.1. Fla Operon–Genetic Location of Archaellar Associated Genes

3.1.1. Archaellins

In Archaea, most genes involved in archaellation are typically found clustered in the *fla* operon, which usually begins with several archaellin genes (*flaA* and/or *flaB)* encoding the major filament structural proteins, followed by either a complete set of *fla*-associated genes *flaC* to *J* (Euryarchaeaota) or a subset of these genes (Crenarchaeota), as shown in Figure 2 [15]. Genetic studies show that all of the *fla*-associated genes successfully deleted are essential for archaellation [17,52,53]. While some of the *fla* associated genes (*flaI*, *flaJ* and *flaK*) are homologues of genes found in the type IV pili systems of bacteria, others appear to be archaeal specific with no homologues in the bacterial domain [15]. Genes required for posttranslational modification of the archaellin structural proteins (*i.e.* the *agl* genes for N-linked glycosylation, some of which are essential for archaella formation) are located elsewhere in the genome, sometimes scattered in several loci [27,54–56].

Figure 2. The *fla* operons responsible for archaella formation in a representative euryarchaeote, *M. maripaludis,* and crenarchaeote *S. acidocaldarius*. Homologues in the two systems are shown in identical colours. The prepilin peptidase-like enzymes, typically located outside the operon, are also shown (FlaK and PibD).

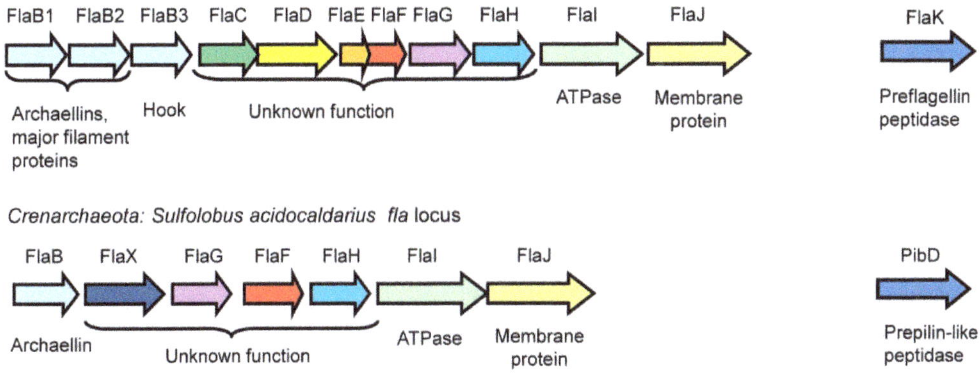

The first archaellin genes were identified in *Halobacterium salinarum* where five archaellin genes are located in two operons: *flgA1-A2* in locus A, and *flgB1-B3* in locus B, while the *fla*-associated genes are transcribed separately [57,58]. All five gene products were identified in archaella samples. Later, complete genome analysis revealed the presence of a predicted sixth archaellin gene, designated *flgX*, located at a distance from the other *flg* genes. The two archaellins in locus A are sufficient to form archaellar filaments [59]. In another halophile, *Haloferax volcanii*, genes encoding the major archaellin FlgA1 and minor archaellin FlgA2 are also transcribed separately from the *fla* associated genes [26]. In *Haloarcula marismortui*, two archaellin genes are found but one is on the chromosome and the second on a plasmid. In different phenotypes of *H. marismortui*, either archaellin can be the major filament protein, with the other found only in very minor amounts [50]. In *Methanococcus voltae*, the major archaellins FlaB1, FlaB2 and the minor archaellin FlaA compose the archaellar filament and the minor archaellin FlaB3 forms the hook region [60,61]. The gene for *flaA* is located immediately upstream of the *fla* operon and transcribed separately [60]. In a related methanogen, *M. maripaludis* S2, the archaellar filament is composed of the major archaellins FlaB1 and FlaB2, while FlaB3 is responsible for the hook region [52]. Deletion of *flaB3* in *M. maripaludis* resulted in functional, but hookless, archaellar filaments [52,61], consistent with the hypothesis that filament assembly occurs before addition of hook subunits at the base (opposite of the bacterial flagella assembly mechanism [62,63]). Interestingly, in the genome sequences of three other *M. maripaludis* strains (C5, C6 and C7), there is a fourth, much longer, archaellin gene in the *fla* operon. The short hook-like filaments in mutant *H. salinarum* cells are also comprised of an archaellin [64]. However, there are many other Archaea in which archaellar hook regions have never been observed, including species that carry a single archaellin gene [12,50]. There are only rare occasions where Archaea with a single archaellin gene have been shown to be archaellated. Studied examples include *Sulfolobus* species [25] and most recently *Halorubrum lacusprofundi* [65], proving that functional helical archaella could be formed from a single archaellin.

3.1.2. FlaCDEFG

In *H. salinarum*, FlaC and E are fused as one protein FlaCE. FlaCE/FlaD proteins were found to have indirect interactions with the chemotaxis proteins via three new identified proteins, indicating that FlaCE/FlaD might be involved in the switch of archaella motor (see below) [66]. In *M. voltae*, FlaC, FlaD and FlaE are all membrane proteins with as yet unidentified functions. Interestingly, both FlaD and a truncated C-terminal version of FlaD have been found in *M. voltae*, *M. maripaludis* and *Methanocaldococcus jannaschii*. The truncated version has high sequence similarity to FlaE [52,67,68]. In *M. maripaludis*, deletions of *flaC* are nonarchaellated; however, deletions of *flaD* or *flaE* could not be created [52]. In light of the halophile evidence, these mutants could be very exciting as they might still assemble archaella but be impaired in rotation switching.

In Crenarchaeota, *flaCDE* are missing, but *flaX*, absent from Euryarchaeota *fla* operons, is present. Switching of rotation of archaella has not been reported in crenarchaeotes, which would be consistent with a role for *flaCDE* in motor switching in euryarchaeotes. Bioinformatics analysis shows that regions in FlaX are homologous with methyl-accepting proteins, implying this protein might be involved in signal transduction [16]. In *S. acidocaldarius*, FlaX is essential for archaella assembly, and FlaH, FlaI and FlaJ are needed to maintain the stability of this protein, suggesting an interaction between these proteins [17]. Recently, the interaction between FlaX and FlaI, the ATPase motor, was confirmed. FlaX forms ring-like oligomers with a diameter of ~30 nm, which is about twice the diameter of the archaellar filament (~14 nm [4]). Both functions are dependent on the presence of the C terminus of FlaX [69].

So far little is known about FlaF and FlaG except the fact that they are essential for archaellation [17,52,68].

3.1.3. FlaHIJ

FlaH, -I and -J are thought to form a secretory complex in archaella assembly [16,68]. *In silico* analysis shows that FlaH is a potential ATPase-like protein containing a typical Walker A motif but an incomplete Walker B motif, which begs the question whether FlaH actually has ATPase function [16]. So far direct biochemical data addressing this issue is lacking [16]. FlaJ is predicted to be an integral membrane protein containing seven to nine transmembrane domains with two highly charged cytoplasmic loops [16,68], and likely forms a crucial component of the central core complex for archaella assembly [16]. It is a homologue of the conserved membrane component of bacterial type IV pili systems (PilC in *P. aeruginosa*) [47] and likely interacts directly with FlaI [16].

FlaI belongs to the "secretion superfamily ATPase" or "T2S/T4S ATPase" family involved in bacterial type II secretion, type IV secretion and type IV pili assembly, as well as archaella assembly [47]. Numerous conserved motifs (such as Walker A and B boxes and P-loop motifs involved in nucleotide binding; aspartate box and histidine box) are shared by members of this superfamily.

Recently, detailed biochemical studies on FlaI from *S. acidocaldarius* were published [70]. FlaI was shown to be a Mn^{2+}-dependent ATPase with an optimum pH of 6.5 and temperature of 75 °C. Mutations of key amino acids in the conserved motifs reveal that the Walker A motif is involved in ATP binding, and the Walker B motif is involved in ATP hydrolysis. FlaI ATPase activity is strongly activated by archaeal tetraether lipids but not *E. coli* lipid extracts. FlaI also undergoes an ATP-dependent hexamerization in solution. It is still unclear whether the energy generated by FlaI is used for the archaella assembly by archaellin translocation, or for driving the archaella rotation, or even for both [70].

3.2. FlaK-Signal Peptidase for Archaellin Maturation

Archaellins are synthesized as preproteins with a short signal peptide similar to bacterial type IV pilins [22,23,71,72]. Genetic studies show that the removal of the signal peptide is essential for the archaella assembly [23,26]. The most extensively studied archaeal type IV prepilin-like peptidases are FlaK in *M. voltae* and *M. maripaludis* [22,23,73] and PibD in *Sulfolobus solfataricus* [25,72]. Site-directed mutagenesis studies revealed that FlaK/PibD belong to an unusual family of aspartic acid proteases in which two aspartic acid residues, one located within a conserved GxGD motif, are critical for the peptidase activity [23,72,73]. The recently solved *M. maripaludis* FlaK crystal structure confirmed the presence of six transmembrane helices and demonstrated that the enzyme must undergo a conformational change in order to bring the two catalytic aspartic acid residues, located in transmembrane helix 1 and 4, into close proximity [73] .

The typical length of the signal peptide on archaellins is 6–12 amino acids [15]. Site-directed mutagenesis studies investigated the importance of various amino acid positions in the signal peptide of archaellins. The highly conserved glycine at the -1 position (the cleavage site is +1) was shown to be critical for peptidase cleavage, with the usually basic amino acids at positions -2 and -3 and the conserved +3 glycine also playing important roles [74]. Similar studies conducted on the glucose binding protein precursor, a substrate for PibD in *Sulfolobus* indicated PibD was more flexible in accepting amino acid substitutions around the cleavage site [25]. In *M. maripaludis*, FlaK specifically processes pre-archaellins while the type IV prepilins are processed by another type IV prepilin-like peptidase, EppA (see below) [24]. The *Sulfolobus* PibD, on the other hand, has a more flexible substrate diversity, ranging from archaellins and type IV pilins to the sugar binding proteins that comprise the bindosome [25]. Recent study has indicated that PibD can also process the archaellins of *M. voltae* [75] . In that report, PibD was shown to cleave engineered archaellin signal peptides as short as three and four amino acids whereas FlaK needed a minimal signal peptide length of five amino acids for cleavage. This further supports the more flexible nature of the PibD enzyme. Recently, the prepilin peptidase in *Hfx. volcanii*, also designated PibD, was found to be responsible for the processing of both archaellin FlgA2 and two other type IV pilin-like proteins [26] . The signal peptide cleavage sites of selected archaellins are included in Table 1 [5].

3.3. N-Glycosylation Modifications of Archaellin

N-glycosylation is a significant and likely widespread post-transcriptional modification for archaellins. All the archaellins in *M. voltae*, *M. maripaludis*, *H. salinarum* and *Hfx. volcanii* are modified with N-linked glycans [29,30,76,77]. Analysis of available archaellin sequences only rarely reveals ones that lack potential N-linked glycosylation sites (such as both archaellins of *H. marismortui* [65]) while some archaellins show a remarkable number of such sites: *Methanothermococcus thermolithotrophicus* FlaB2 has the most with 16 predicted N-glycosylation sites. For most organisms, further work is needed to confirm whether these sequons are actually occupied by N-glycan [15]. Using archaellin as the reporter protein, an N-glycosylation model has been established in *Methanococcus* spp, and its role in the archaellum assembly model in *M. maripaludis* is depicted in Figure 3. In the cytoplasm, the N-glycan precursor is assembled on an unknown lipid carrier by sequential addition of sugar monomers by the corresponding glycosyltransferases, followed by flipping of the glycan across the plasma membrane by an as yet unidentified enzyme. Finally, the N-glycan is transferred *en bloc* to the N-X-S/T (X=/P) motifs in the archaellins by the oligosaccharyltransferase AglB [54,55]. In *M. voltae*, both the archaellins and S-layer protein are N-glycosylated with a trisaccharide β-ManNAcA6Thr-(1–4)-β-GlcNAc3NAcA-(1–3)-β-GlcNAc [54]. AglH and AglA are glycosyltransferases responsible for the 1st and the 3rd sugar, respectively, and AglC and AglK are either the glycosyltransferases or involved in the biosynthesis of the 2nd sugar [54,78,79]. In *M. maripaludis*, the archaellin N-glycan is a tetrasaccharide Sug-4-β-ManNAc3NAmA6Thr-4-β-GlcNAc3NAcA-3-β-GalNAc, where Sug is a diglycoside of an aldulose exclusively found in this species. Glycosyltransferases responsible for the 2nd, 3rd and 4th sugars have been identified as AglO, AglA and AglL, respectively [55]. A number of genes encoding enzymes involved in the biosynthesis of the individual sugar components have also been identified [80–82] (Y. Ding, S. Siu and K. Jarrell, unpublished data).

Figure 3. An assembly model for the archaellum of *M. maripaludis*. The N-linked glycan, synthesized by a series of Agl proteins and assembled by Agl glycosyltransferases on an unknown lipid carrier is flipped across the cytoplasmic membrane and attached to the archaellins by the oligosaccharyltransferase AglB. The archaellins are also processed by FlaK which removes the signal peptide and the posttranslationally modified subunits are added to the growing structure by incorporation of new subunits at the base through the activities of the ATPase FlaI in conjunction with the conserved membrane protein FlaJ.

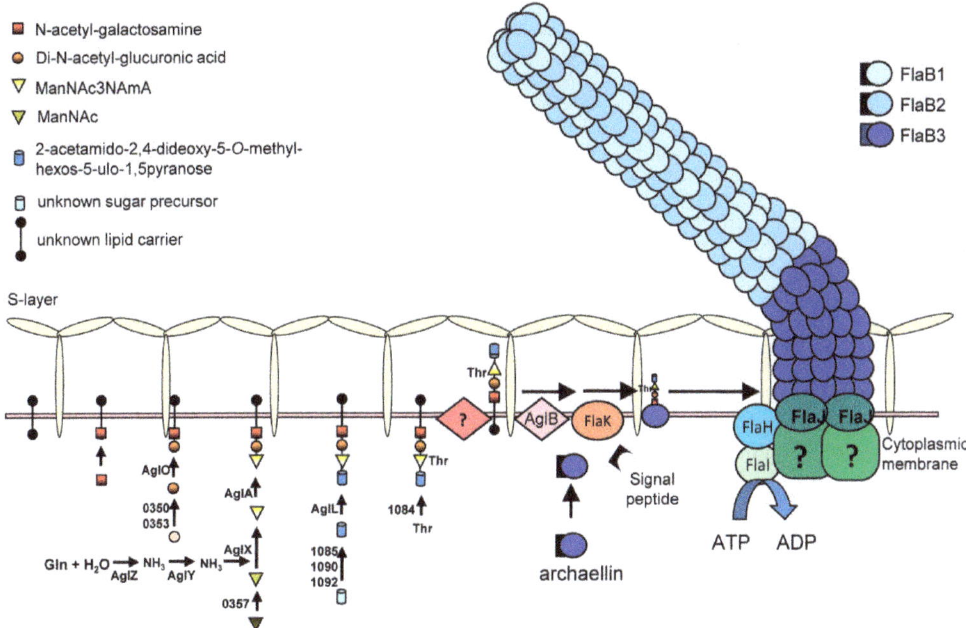

Study of various mutants carrying deletions in *agl* genes that result in truncated glycan have shown that a minimum length of the N-glycan is essential for archaella assembly. In both *M. voltae* and *M. maripaludis*, no archaella were observed on the cell surface when the archaellins were nonglycosylated or carried a glycan that consisted of only a single sugar [54,55], indicating a minimum 2-sugar glycan was necessary for archaellins to assembly into an archaellar filament. When archaellins were modified with a 2- or 3-sugar glycan, the cells assembled archaella but these cells were less motile in swarm plate assays than the wildtype cells that carry the entire 4-sugar glycan [55].

In *Hfx. volcanii*, archaellins were recently found to be modified with the same N-glycan originally found decorating the S layer protein [30,83–85]. The pentasaccharide is composed of a mannose, a methyl ester of hexuronic acid, two hexuronic acids and a hexose [84,86]. As found in *Methanococcus* species, deletion of the oligosaccharyltransferase gene, *aglB,* in *Hfx. volcanii*, led to non-archaellated cells, while swarm plate assays indicated that attachment of a minimum of a 3-sugar glycan to archaellins is necessary for archaella formation [30].

While N-linked glycosylation was shown to be necessary for archaealla formation in *Methanococcus* and *Haloferax*, the archaellins of each species carry multiple N-glycosylation sites. In the case of *Hfx. volcanii* FlgA1 (the major archaellin) all three sites seems to be important in archaella assembly or function since site directed mutagenesis to individually eliminate each site led to non-motility on semi-solid swarm plates [30]. Interestingly, this does not appear to be the case for *M. maripaludis*. In this organism, a spontaneous mutant carrying a FlaB2 version with an asparagine (N) to aspartic acid (D) mutation in the N-X-S/T motif which eliminates the 2nd N-glycosylation site was discovered. This strain however produces functional archaella (Y. Ding and K. Jarrell, in preparation).

3.4. Archaella Regulation

Archaella synthesis is known to be regulated in both *M. jannaschii* and *M. maripaludis*, depending on the availability of H_2, with archaella synthesis induced under H_2 limitation conditions [67,87]. Further study using quantitative proteomics of nutrient-limited *M. maripaludis* indicated that the expression of archaellins was affected by multiple nutritional factors: decreased under nitrogen limitation but increased under phosphate limitation [88]. It was suggested that *M. maripaludis* may respond to nitrogen limitation conditions by shutting down other energy intensive processes like motility when forced to switch to the energy-consuming nitrogen fixation pathway. In *S. solfataricus*, transcription of the archaellin gene is highly increased in the stationary phase and when the cells encounter nitrogen starvation growth conditions [17].

In the *fla* operon of *S. acidocaldarius*, two differentially regulated promoters have been identified [17,39], one lying upstream of the gene encoding the major structural protein FlaB and a second promoter upstream of *flaX* that regulates transcription of the downstream genes *flaX-J*. Transcriptional readthrough also occurs for the *flaB* gene likely due to a weak termination signal. Under tryptone limiting conditions, the expression of both *flaB* and *flaX* was shown by qRT-PCR to dramatically increase showing that the regulation of archaella synthesis can be observed at the transcriptional level. Promoter studies further revealed that in an Aap pilus minus background, the *flaB* promoter activity was dramatically increased under starvation stress, while no difference was observed with *flaX* promoter. To date, no data has been presented on mRNA stability or protein half-lives but nonetheless the results presented by the Albers group so far suggest that the expression of the archaellar accessory and core components is under the constitutive *flaX* promoter to preserve the core complex for quick archaella assembly, which depends then only on the availability of the major structural protein archaellin. Meanwhile, expression of the energy-consuming archaellin subunits depends on the inducible *flaB* promoter in response to environmental cues, such as starvation [17].

In the methanogens where regulation of archaella has been observed, no transcriptional regulators have been reported. In *S. acidocaldarius*, no activators have been found responsible for the induction of *flaB*, but two repressors of archaellation, ArnA and ArnB, have been identified [39]. ArnA is a forkhead-associated (FHA) domain-containing protein. Typically,

FHA-domain containing proteins have phosphopeptide-binding activities. ArnB contains a von Willebrand (vWA) domain and proteins carrying such a domain typically form multi-protein complexes. Indeed, both homologously and heterologously expressed ArnB can be co-purified with a His-tagged version of ArnA and *vice versa,* indicating a strong *in vivo* interaction between these two proteins. Both ArnA and ArnB can be phosphorylated by specific eukaryotic-like protein kinases and dephosphorylated by Ser/Thr phosphatase PPP from the same species *in vitro*. Under tryptone starvation conditions, cells of *ΔarnA* and *ΔarnB* mutants are hypermotile via hyper-archaellation, implying that ArnA and ArnB are repressors for the *fla* operon [39]. While it was expected that ArnA would bind directly to the *flaX* promoter as observed in *S. tokodaii* [89], further studies showed that in *ΔarnA* and *ΔarnB* mutants, the activity of both the *flaB* promoter and *flaX* promoter was not as dramatically upregulated as expected, suggesting that ArnA and ArnB are not acting primarily at the transcriptional level in *S. acidocaldarius* but rather on a protein-protein interaction level. Since Arn homologues are not found in euryarchaeotes, regulation of archaellation in this archaeal group must occur by a different mechanism. Further layers of archaella control are suspected in *S. acidocaldarius,* including a likely positive regulator for the system [39] as well as a role for anti-sense RNAs [90].

In *S. acidocaldarius,* two observations point to a regulatory interplay between the archaella and Aap pili systems that might coordinate the expression of different surface appendages depending on different environmental signals. Firstly, expression of FlaB was dramatically induced in the Aap pilus minus mutant [17]. Secondly, overexpression of the archaella repressor ArnA led to hyper-piliation [39].

Gene deletions that lead to nonarchaellation of *M. maripaludis* have been reported, including ones in the *fla* operon but also ones affecting glycosylation of the archaellins (*agl* genes) [55]. In such mutants, archaellin structural proteins are not detected after subsequent transfers suggesting that secondary mutations have occurred that have resulted in cessation of transcription of the *fla* operon, presumably resulting in energetic savings under conditions where archaella cannot be assembled. These mutations are not located in the promoter region and may be in genes encoding activators for the *fla* operon (G. Jones and K. Jarrell, unpublished observation).

3.5. Archaella Structure

The archaellum has been described as "a bacterial propeller with a pilus-like structure" [40]. Knowledge about archaella structure is very limited, mainly from *H. salinarum* and *Sulfolobus shibatae,* a euryarchaeote and a crenarcheote living in very different extreme environments [21,91,92]. Despite the phylogenetic distance between the two organisms, three dimensional reconstructions of the archaella from the two species are similar in structure and provide a basic symmetry for archaellar filaments that is distinct from bacterial flagella filaments. The outer domain forms a 3-start helix wound around a solid inner core domain, which lacks an internal channel. The inner core domain is conserved both in size and shape, with a diameter of 5 nm in both archaea, and thought to be constructed as alpha-helices by the hydrophobic N-terminal segment of archaellins. The N-terminal sequences of archaellins are highly conserved and homologous to those of type IV pilins [41], where they are known to be involved in subunit-subunit interactions in that bacterial appendage [21]. Considering that mature archaellins have a highly conserved and hydrophobic 30-40 amino acids at the N-terminus [45,71,93], this structure might apply to the whole archaeal domain. Compared with the conserved inner core, the size of the outer domain varies, and is responsible for the differences in the diameter of archaellar filaments in *S. shibatae* (14 nm) and in *H. salinarum* (10 nm). These variable regions might reflect adaptations of the different archaella to optimize their performance to specific harsh environments inhabited by a variety of archaea [21,91].

Scanning 10 amino acid deletion analysis of the archaellin FlaB2 in *M. maripaludis* was recently conducted (Y. Ding and K. Jarrell, in preparation). Complementation of a *flaB2* deletion strain with any of the 10 amino acid deletion versions of *flaB2*, including deletions in the variable region of the protein, did not restore archaellation. Complementation with a *flaB2* variant carrying only a three amino acid deletion in the variable region led to only poor restoration of archaellation, suggesting perhaps that absolute length of the archaellin is important for proper archaella assembly.

3.6. Archaella Function

The archaellum is a motility apparatus widespread throughout the domain Archaea, and helps archaeal cells swim in liquid medium or swarm through semi-solid medium [17,52,53,94]. The archaellum rotation, like that of bacterial flagella, can be clockwise or counterclockwise, at least in *Halobacterium*, indicating the presence of a switch [95]. In bacteria, the rotation of flagella is under the control of the chemotaxis system and binding of phospho-CheY to the switch protein FliM has been documented [96]. In Archaea, a bacterial-like chemotaxis system including *cheY* homologues has been identified in euryarchaeota [97,98] but not in crenarchaeota [99]. Nevertheless, homologues to FliM have not been reported and where the interaction between the archaella and the chemotaxis system occurs has not been identified [100]. As mentioned, three newly identified proteins encoded by genes close to *fla* operon were found to be the mediator between the chemotaxis proteins CheY, CheD and CheC2 and FlaCE/D in *H. salinarum* [66].

Until recently, information on swimming speeds of different Archaea was extremely limited. Cells of *H. salinarum* were reported to swim at speeds of 2-3 μm per second [95] and it was unknown whether such slow speeds were typical of Archaea in general. Recently, using a "thermo-microscope", the swimming speed of selected species of hyperthermophilic Archaea was measured [101]. If speed is measured in bodies per second (bps), the two hyperthermophilic methanogens *M. jannaschii* and *Methanocaldococcus villosus* are the fastest swimmers so far reported, with speeds of close to 400 and 500 bps (absolute velocity of 468 and 589 μm/s, respectively; compare to *E. coli* speed of 20 bps). Swimming speeds were also measured for numerous other Archaea, including ones where genetic studies on archaellation have been done. These include *M. voltae* (128 μm/s, 64 bps), the weakly motile *M. maripaludis* (45 μm/s, 30 bps) and the thermoacidophile, *S. acidocaldarius* (60 μm/s, 40 bps).

Besides swimming, the archaellum plays critical roles in other functions for Archaea such as surface adhesion and cell-cell contact and, perhaps, even intercellular communication. In *Pyrococcus furiosus*, where the archaellum is the only surface appendage, it is responsible for cellular adhesion to different kinds of abiotic surfaces. Bundles of archaella also form cable-like structures mediating cell-cell contact [102]. Interestingly, P. *furiosus* uses archaella to adhere onto cells of another archaeon, *Methanopyrus kandleri,* to form a bi-species biofilm [37]. Cable-like structures composed of archaella were also shown to mediate cell-cell contact and abiotic surface adhesion in *M. villosus* [103] and *M. maripaludis* [31] (Figure 4).

Figure 4. Role of archaella in attachment of Archaea to surfaces and other cells. Scanning electron micrograph of *M. maripaludis* attached to silicon wafer via thick cables of archaella (thick arrows) which can unwind to individual archaellar filaments (thin arrows). Bar = 100 nm (**B**). Connection of *M. maripaludis* cells to each other and underlying nickel EM grid via archaellar bundles. (**A**) and (**B**) reprinted from [31]. Bar = 100 nm (**C**). Scanning electron micrograph showing attachment of *Mcc. villosus* cells to a surface and to other cells via bundles of archaella. Bar = 1 μm. Courtesy of Gerhard Wanner, University of Munich, Germany.

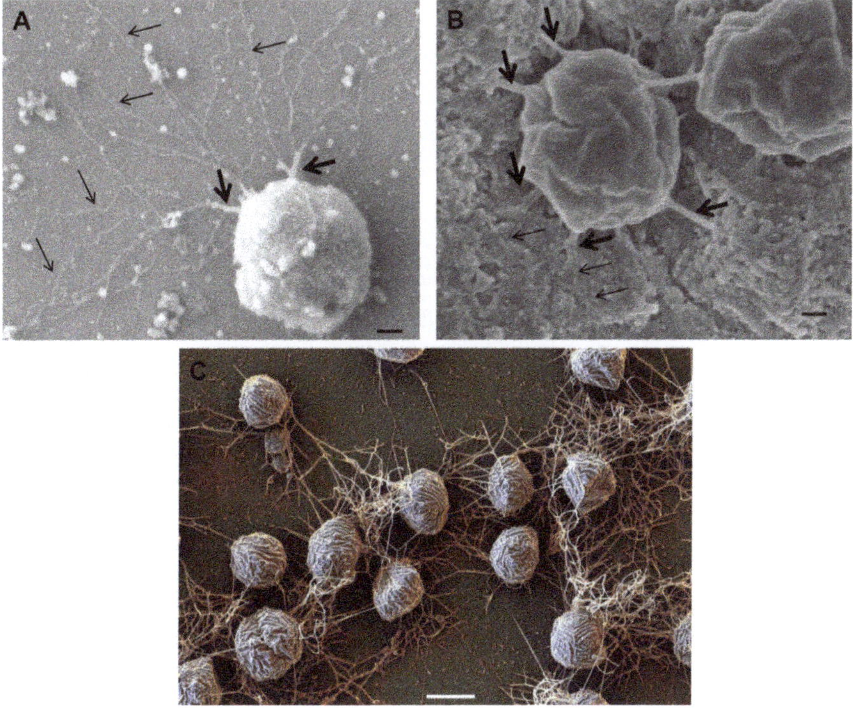

When *P. furiosus* and *M. villosus* were cocultured, the growth of both archaea increased, and cells of these two species were connected via archaella and formed "flocks" in liquid medium, even without the presence of a solid surface. Although the function played by archaella under this situation is not clear, it was speculated that archaella may mediate signalling and interaction between both partners [38]. Such a precedent was reported in the syntrophic relationship between the methanogen *Methanothermobacter thermoautotrophicus* and the bacterial syntroph *Pelotomaculum thermopropionicum*. Here, the archaeon perceives the flagellar cap protein FliD of the bacterium and up-regulates a number of genes involved in methanogenesis, ATP synthesis and hydrogen utilization, thus preparing itself for the onset of the syntrophic interaction and indicating that cell surface appendages can have critical roles in intercellular communication [104].

In some Archaea, the function of the archaella in adhesion and biofilm formation may be intertwined with other surface structures such as pili. In *M. maripaludis* and *S. solfataricus*, both archaella and pili are necessary for attachment to abiotic surfaces [31,105]. In *M. maripaludis*, bundles of archaella are clearly observed in the persistence stage of adherence and seem to be strong enough to maintain the attachment, which lead to the speculation that pili may play an important role in the initial stages of attachment [31]. However, in *S. acidocaldarius*, although archaella, the UV-induced pili (Ups pili) and the adhesive pili (Aap pili) all play roles in surface adhesion and biofilm formation, archaella appear to have only minor effects compared with the other two appendages, but they are speculated to play a role in cell release from the biofilm [32]. In *S. solfataricus*, the expression of archaellin FlaB was significantly reduced in adherent cells, implying that archaella might play important roles in the initial attachment but not the persistence [105]. In Archaea in which archaella play an important role in persistence of adhesion to surfaces (*P. furiosus, M. villosus* and *M. maripaludis)*, bundles of archaella are commonly observed. In contrast, in *S. acidocaldarius*, archaellar bundles have not been reported, suggesting that archaella in this species are used for motility and attachment initiation, but not for the persistence of attachment [32].

4. Pili

Structures believed to be pili were observed in electron microscopic studies of various Archaea decades ago [106,107] but it is only recently that studies specifically focused on the structure, assembly, genetics and function of these appendages have been reported. Most of the recent pili studies have focused on the genetically tractable species within the genera *Sulfolobus* and *Methanococcus.* Bacterial pili are involved in many different functions, such as adhesion, twitching motility, DNA uptake, and biofilm formation [108,109]. In Archaea, various pili functions have also been reported including adhesion, cell aggregation, biofilm formation, and DNA exchange [34,36].

All archaeal pili studied thus far are type IV pili–like with one exception, *i.e.*, the Mth60 fimbriae of *M. thermoautotrophicus* [110]. The Mth60 pili, 5nm in diameter, are composed of a 16kDa glycoprotein with a predicted length of the mature processed pilin of 143 amino acids. While the nature of the glycan attached has not been reported, there are potential N-glycosylation sites present in the protein. There is an interesting regulation in the biosynthesis of the pili since they are barely observed on the surfaces of planktonic cells but found in high numbers when cells were grown on surfaces (Figure 5). The Mth60 pili were the first archaeal pili shown to play a role as adhesins both to abiotic surfaces and to other cells. Interestingly, these pili are a rare case of an archaeal surface structure that can be stained effectively with succinimidyl esters of fluorescent dyes (Alexa dyes) [111]. Recently, another *Methanothermobacter* isolate, *M. tenebrarum,* was reported to possess bundles of polar pili [112].

4.1. Type IV Pili-Like Loci in Archaea

Genetic and structural work on type IV-like pili has been reported in both the crenarchaeota (*Sulfolobus* species) and euryarchaeota (*M. maripaludis*). All *Sulfolobus* species studied (*S. acidocaldarius, S. solfataricus* and *S. tokodaii*) possess UV-inducible pili encoded by the *ups* operon while only *S. acidocaldarius* shows, in addition, the presence of a second type IV pili system termed adhesive (Aap) pili. A comparison of the genetic loci encoding these appendages in different Archaea is presented in Figure 6.

Figure 5. Scanning electron micrograph of *M. thermautotrophicus* grown on gold EM grids and expressing many Mth60 fimbriae. Bar = 1μm. Courtesy of Gerhard Wanner, University of Munich, Germany.

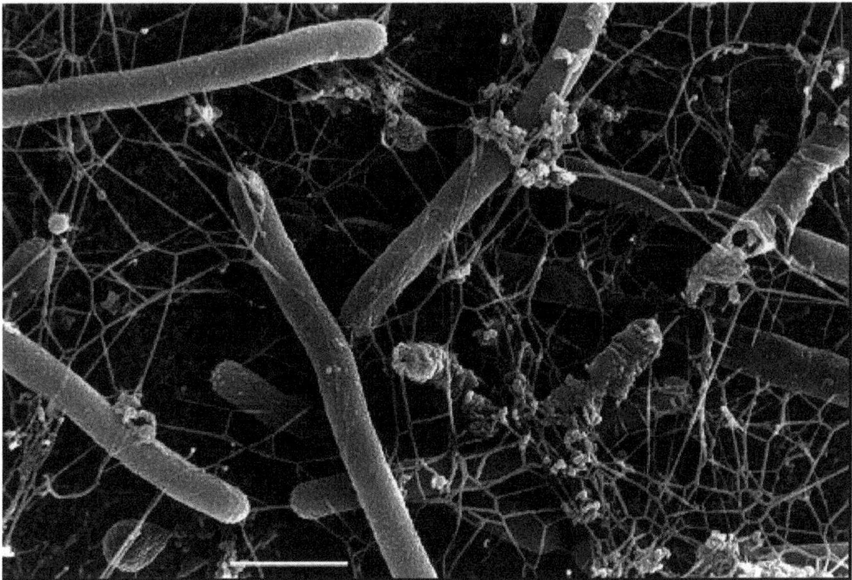

4.1.1. Adhesive (Aap) Pili of *S. acidocaldarius*

The Aap pili locus of *S. acidocaldarius* consists of five genes [9] (Figure 6). A *pilB* ATPase homologue (*aapE*) and a *pilC* integral membrane protein homologue (*aapF*) have been identified in the locus together with *aapX*, a gene encoding an iron-sulfur oxidoreductase. Flanking these genes, but transcribed in the opposite direction are two pilin genes (*aapA* and *aapB*). Mutants containing deletions in genes that prevented both archaellum and Ups pili formation still had pili on their surface. Mass spectrometry of pili, purified from such mutant cells, revealed the major pilin to be encoded by *aapB* [9]. These pilins subunits, like all other preproteins with class III signal peptides in *S. acidocaldarius*, were processed by PibD. Deletion mutants were created in each of the *aap* genes and analyzed for the presence of pili [9]; all five genes were essential for pili biosynthesis.

Figure 6. Type IV pili-like loci found in three different Archaea. Functionally similar genes are given identical colors. (**A**) The adhesive pili locus of *S. acidocaldarius* showing the pilin-like genes *aapA/B*, the iron sulfur oxidoreductase (*aapX*), the ATPase (*aapE*) and the inner membrane protein (*aapF*). (**B**) The UV-inducible pili locus of *S. solfataricus* contains genes for two pilin (*upsA* and *upsB*), an ATPase (*upsE*) and a conserved membrane protein (*upsF*) as well as a protein of unknown function *(upsX)* and (**C**) The adhesive pili locus found in *M. maripaludis* contains genes for three minor pilins (*epdA,B,C*), a prepilin peptidase (*eppA*) and a number of proteins of unknown function. Located outside this locus are genes for the ATPase (MMP0040), inner membrane proteins (MMP0038 and MMP0039) and the major pilin gene (MMP1685).

A. Aap pili locus of *S. acidocaldarius*

B. Ups pili locus of *S. solfataricus*

C. Adhesive pili locus of *M. maripaludis*

4.1.2. UV-Inducible Pili of *Sulfolobus* Species

UV-inducible pili were first reported in *S. solfataricus* [36] where an operon of five genes with a similarity to type IV pili genes was found (Figure 6). The UpsA and UpsB proteins had a class III signal sequence similar to that seen in type IV pilins, and UpsB was shown to be processed by PibD. The signal peptide present on UpsA is too small (see Table 1) to allow separation of the processed from the nonprocessed form in SDS-PAGE in the standard processing assay, although it is likely that UpsA is also a substrate for PibD, since PibD is the only prepilin-like peptidase in this organism. An ATPase (UpsE) and an integral membrane protein (UpsF) which showed similarities to the corresponding proteins, TadA and TadC respectively, in the type IV pili system (Tad pili) of *Aggregatibacter* [36,113] were also identified. The final protein encoded in this locus was a predicted highly hydrophilic protein UpsX that was unique to *Sulfolobales* and its role in Ups pili formation is currently unknown [36]. Deletion analysis demonstrated that the ATPase of the operon was necessary for formation of the pili after UV treatment, linking this operon with the Ups pili structures. Whether UpsA or UpsB is the major structural protein of the pili is unknown.

The UV-induced pili operon is well conserved in other species of *Sulfolobus*, such as *S. acidocaldarius* and *S. tokodaii* [24].

4.1.3. Pili Locus in *M. maripaludis*

A type IV pili locus, containing 11 potential genes (*mmp0231-mmp0241*) was predicted using bioinformatics in the genome of *M. maripaludis* [24] (Figure 6). This locus is unlike the type IV pili loci in the order *Sulfolobales*, where there are only five genes. The 11 genes form a single transcriptional unit that consists of three pilin-like genes (*epdA, epdB* and *epdC*), a prepilin peptidase (*eppA*) and a number of other genes which have no homologues in either bacterial type IV pili systems or the Aap or Ups pili of *Sulfolobus* species. Some are even restricted to *Methanococcus* species (D. Nair and K. Jarrell, unpublished data). Unlike the *Sulfolobus* pili systems, the genes encoding the major pilin (MMP1685), as well as the ATPase and the conserved pilus membrane protein were not found in the locus. The gene encoding the ATPase necessary for assembly of the *M. maripaludis* pili (*mmp0040*) was identified in a small gene cluster, that also included two genes (*mmp0038* and *mmp0039*) with homologies to the conserved membrane proteins of type IV pili systems (D. Nair and K. Jarrell, in preparation). Also unlike in *Sulfolobus* where all substrates with class III signal peptides are processed by PibD, in *M. maripaludis*, there is a devoted prepilin peptidase, EppA, which only cleaves the prepilins while FlaK processes archaellins [22,23]. EppA is a larger protein with additional four transmembrane segments compared to FlaK [24]. The substrates of EppA have a negatively charged amino acid at the +5 position, similar to most of the bacterial type IV pili and missing in the non-EppA substrates (*i.e.*, archaellins) [5]. By creating a genetic hybrid, where the four amino acids (KGAS) around the cleavage site of FlaB2 (archaellin) were substituted for the cleavage site (RGQI) of EpdA (pilin), the importance of a +1 glutamine for cleavage was shown [24].

Deletion analysis of all 11 genes in the original pili locus as well as other type IV pilus gene homologues found outside the locus was conducted. Deletion of most of the genes resulted in a phenotype with no piliation. The three pilin-like genes (*epdA*, *epdB* and *epdC*) [28], the prepilin peptidase (*eppA*), the major pilin protein (*mmp1685*), the ATPase (*mmp0040*), and both conserved membrane proteins (*mmp0038* and *mmp0039*) were all found to be essential for normal piliation (D. Nair and K. Jarrell, in preparation). In addition, some of the other genes in the locus, annotated with unknown function, such as *mmp0234*, *mmp0239*, *mmp0240* and *mmp0241* were also found to be essential for piliation. Repeated efforts to delete *mmp0231* were unsuccessful. The two genes in the cluster that were found not to be essential for pili formation were *mmp0235* and *mmp0238*. *M. maripaludis* pili are known to have a role in attachment on solid surfaces, although only in the presence of archaella [31]. Whether the pili assembled in the absence of MMP0235 or MMP0238 can still function in attachment is currently under investigation. Interestingly, both of these proteins have predicted signal peptides and it is possible that one of them may function as a tip adhesion.

While studies directed at the biogenesis of archaeal type IV-like pili have not been presented, it seems likely to generally follow the process of bacterial type IV pili. However, the presence of essential genes in both *Sulfolobus* and *Methanococcus,* that have no homologues in bacterial systems, suggests that there will be aspects unique to the archaeal domain.

4.2. Pili Regulation

The regulation of archaeal surface structures is in its infancy and information on the regulation of pili is limited to several recent observations in *Sulfolobus* species. Ups pili were first identified when a putative pilus gene locus was strongly upregulated after cells were UV irradiated. This upregulation was apparently dependent upon DNA double strand breaks as other DNA damaging agents, like bleomycin, had similar effects [36]. Understanding of the molecular mechanism behind the induction is currently not known but very recently it was shown to involve the transcriptional regulator Sa-Lrp in *S. acidocaldarius* [114]. Examination of a strain deleted for Sa-lrp demonstrated that it was defective in the UV-induced aggregation mediated by Ups pili and qRT-PCR confirmed that, in the mutant, *upsA* transcript levels were eight-fold lower than that of wildtype cells following UV induction. It was suggested that Sa-Lrp likely acts in conjunction with other regulators at the transcriptional level [114].

Expression of Aap pili in *S. acidocaldarius* is known to be growth phase dependent. Lower numbers of Aap pili are found on cells in stationary phase and qRT-PCR showed reduced levels of transcript for *aap* genes in the stationary phase compared to the exponential phase [9]. Several observations point to an interplay in the regulation of Aap pili and archaella (see Section 3.4.). Archaella are up-regulated in the stationary phase when Aap numbers are reduced. In addition, archaella expression was increased in all *aap* gene deletion strains, especially in the *aapF* deletion strain, which was hyper-archaellated [9,17] and finally overexpression of ArnA, a repressor of archaellation leads to increased Aap pili production under tryptone starvation conditions [39]. Additional layers of regulation are likely as numerous antisense RNAs are predicted to lie with *aapF* [90]. However, their actual effect on Aap pili formation has not yet been reported.

4.3. Pili Structure

The first structure of an archaeal pilus was that from *M. maripaludis* [11]. The 6nm diameter pili in this organism are found in small numbers (5–10 per cell) and located peritrichously. Despite the similarities of the *M. maripaludis* pilins to bacterial type IV pilins, cryo-electron microscopy of *M. maripaludis* pili showed a structure that was different from that of any bacterial pili (type IV or otherwise) or archaella. This study showed that two different helical symmetries existed and that they coexisted within the same archaeal pilus [11]. A hollow lumen with a diameter of 20Å was observed, a feature missing in archaella structures [21].

Mass spectrometry of purified *M. maripaludis* pili samples identified the major structural protein to be MMP1685, a small glycoprotein of only 74 amino acids including a 12 amino acid class III signal peptide [28]. Interestingly, the glycan N-linked to this pilin was not identical to that previously characterized on archaellins but instead had an additional hexose attached to the linking sugar, GalNAc. The purpose of the added sugar and the identity of the glycosyltransferase responsible for the hexose addition are not known. As found in Aap pili of *S. acidocaldarius,* as well as type IV pili systems of bacteria in general, the structure in *M. maripaludis* is composed of a single major pilin although there is genetic evidence for the role of at least three other proteins as minor pilins [28].

Ups pili, with a diameter of 10 nm but of variable lengths, are peritrichously located on the surface of different *Sulfolobus* species [34,36]. The structure of *S. solfataricus* Ups pili showed straight fibers consisting of three evenly spaced helices with a pitch of 15.5 nm.

Very recently, the structure of another archaeal pilus, the extremely stable *S. acidocaldarius* Aap pili, with a diameter of 11 nm, was published [9]. Remarkably, the Aap pilus structure was also different from known bacterial pili or the *M. maripaludis* pili or indeed any other archaeal type IV pili-like structure including archaella and Iho670 fibers. The Aap pilus displayed a rotation per subunit of 138° and a rise per subunit of 5.7 °, very different from those of studied bacterial type IV pili. These studies indicate that although the building blocks of the many archaeal appendages are similar, the small sequence changes can lead to very different quaternary structures [9,11].

4.4. Pili Function

Study of archaeal pili has revealed a variety of functions, mainly related to attachment and biofilm formation, sometimes in conjunction with archaella.

Ups pili appearance is correlated with cellular aggregation which enhances the exchange of chromosomal DNA, likely aiding in the population overcoming the DNA damage which leads to Ups pili biosynthesis [36,115]. Support for this comes from studies on mutants unable to make Ups pili. These strains were unable to exchange DNA when UV-induced and this resulted in decreased survival of the cells. Formation of Ups pili by at least one partner is essential for the exchange of DNA. Even though *S. acidocaldarius, S. solfataricus* and *S. tokodaii* are all capable of synthesising UV inducible pili that lead to cell aggregation, the aggregation only occurs between cells of the same species [34] suggesting that there is a specific recognition of the cell surface by the pili. Perhaps different N-linked glycosylation structures on the pilin subunits are involved in this species specific recognition. In addition to their role in cell aggregation and DNA exchange, Ups pili can also play a role in attachment to surfaces but the importance of Ups pili in surface attachment varies among different *Sulfolobus* species. In *S. acidocaldarius*, where both Aap and Ups pili are made, the Ups pili have little effect on surface attachment [32]. However in *S. solfataricus*, Ups pili, in collaboration with archaella, are essential for adherence, since mutants lacking the ability to form either one of the surface structures were unable to attach to a variety of tested surfaces [105]. Analysis of adherent cells by qRT-PCR showed an upregulation of the Ups pilin genes (*upsA* and *upsB*) with a significant decrease in the transcription of the archaellin gene *flaB*, suggesting the role of archaella may be in the initial attachment but not in persistence after attachment. Ups pili in *S. solfataricus* may have a role in biofilm maturation [33,105] while in *S. acidocaldarius,* mutants defective in Ups pili formation have changes in the structure and development of biofilms [32].

The major function of Aap pili is attachment of *S. acidocaldarius* cells to surfaces. Their exceptional stability reflects the need for these structures to function in the extreme thermoacidophilic environments inhabited by *S. acidocaldarius*. The nature of biofilm formation by *S. acidocaldarius* was also strongly influenced by Aap pili. Biofilms of strains unable to make Aap pili were flat and denser than those of wildtype cells and they lacked the tower structures observed in the biofilms of wildtype cells [9,32].

Studies aimed at elucidating a function for the pili of *M. maripaludis* demonstrated a role in attachment to various abiotic surfaces [31] but this role was co-dependent on the presence of archaella (see Section 3.6.). Since attachment of cells to surfaces and other cells was via bundles of archaella, it may be that the prolonged attachment is mediated by archaella and the pili role may be more in the initial stages, which is exactly the opposite to the roles suggested for Ups pili and archaella in *S. solfataricus* [105].

In *Hfx. volcanii*, nonarchaellated mutants adhere as effectively as wildtype cells to glass cover slips indicating that archaella in this strain are not strictly necessary for attachment, at least under laboratory conditions [26]. However, a *pibD* deletion mutant did not adhere, indicating that another appendage, likely pili, composed of subunits processed by PibD, was responsible for attachment [26]. In addition, while Ups pili are known to lead to cell aggregation and subsequent DNA transfer in *Sulfolobus* species, type IV-like pilus structures are not involved in the conjugative transfer of DNA observed in *Hfx. volcanii* since the rate of conjugation is not affected when *pibD* is deleted [26]. However, formal identification of pili structures in *Hfx. volcanii* has not been reported, although a variety of filaments can be observed on the surface of *Hfx. volcanii* [26].

No evidence has been presented that suggests type IV-like pili in any archaeon are involved in surface motility (twitching), a common function of their bacterial counterpart. Twitching motility involves the extension and retraction of type IV pili through the activity of two separate ATPases, one to add subunits to the base of the structure (extension) and one to remove subunits (retraction) [108]. In archaeal type IV-like pili systems only a single ATPase has been identified, suggesting that retraction may not be possible.

5. Other Unusual Archaeal Surface Structures

5.1. Iho670 Fibers

A novel structure found in *Ignicoccus hospitalis* is the adhesive filament "Iho670 fiber", so-called since they are comprised mainly of the protein encoded by the Igni_0670 gene [35]. These are extremely brittle and gentle handling of the cells must be employed in order to observe the appendages still attached to cells (Figure 7). The Iho670 protein was detected in appreciable amounts in a proteomic study regardless of whether *I. hospitalis* was grown in single culture or in co-culture with *Nanoarchaeum equitans,* an organism with which it can form an "intimate association" or biocoenosis [116,117]. Other than the hydrophobic amino terminus typical of all archaeal type IV pilin-like proteins, the Iho670 protein shows no primary sequence similarity to archaellins, the Mth60 fimbrin of *M. thermoautotrophicus*, the hamus protein of SM1 euryarchaeon or the three cannulae proteins of *Pyrodictium.* It does have a type IV pilin-like signal peptide (Table 1), confirmed by N-terminal sequencing of the mature protein [35].This signal peptide removal, detected by an *in vitro* processing assay, is likely carried out by the single prepilin peptidase (Igni_1405) detected in the completely sequenced genome. Iho670 fibers have a diameter of 14 nm and can be up to 20 μm in length; SDS-PAGE indicated the structure is composed of a single protein of 33 kDa [35]. The Iho670 protein has the potential to be glycosylated as *I. hospitalis* does have a putative oligosaccharyltransferase (Igni_0016; [118]) and the Iho670 protein does possess potential N-linked glycosylation sites. However, Iho670 proteins did not test positively for glycosylation when stained with the PAS reagent, although false negative results have been reported using this method. Structural analysis has revealed that single α-helices form the core of the Iho670 filament. The overall helical symmetry is similar to that of the archaellar filaments of *H. salinarum* and *S. shibatae,* however the quaternary structure of Iho670 fibers is, again, unique [10]. The fibers were also observed to transition between rigid and curved segments down the length of the filament, however the mechanism of switching and supercoiling has not yet been studied.

5.2. Cannulae

A distinctive feature of species of the hyperthermophilic genus *Pyrodictium* is that individual cells grow in a network of extracellular tubules called cannulae (Figure 8) [8]. Cannulae have an outer diameter of 25 nm and appear empty when thin-sectioned or cross-fractured. The cannula structure itself is composed of at least three related glycoproteins, termed CanA, CanB and CanC, with molecular masses in the 20–24 kDa range. The N-terminal 25 amino acids of each protein are identical [8] and Can proteins were reported to contain signal peptides [119]. The genes encoding the three cannulae protein subunits (*canA, canB* and *canC*) were identified [119] and later published in a patent application [120]. The overall cannula structure is remarkably heat resistant and insensitive to denaturing conditions; no loss of structure was observed even after incubation at 140 °C for 60 min or at 100 °C in 2% SDS for 10 min [8].

Figure 7. Electron micrograph of three *I. hospitalis* cells showing numerous Iho670 fibers on the carbon support film. Bar = 2 μm. Courtesy of Carolin Meyer and Reinhard Rachel, University of Regensburg, Germany.

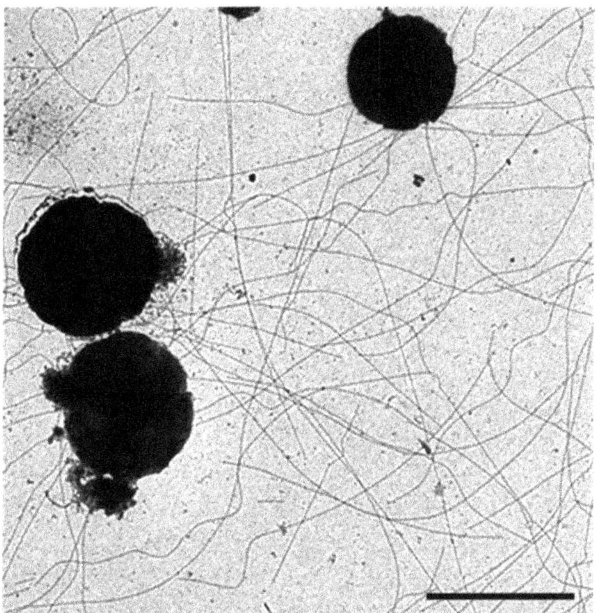

Figure 8. Scanning electron micrograph of a network of *Pyrodictium* cells and cannulae. Bar = 2 µm. Courtesy of Gertraud Rieger & Reinhard Rachel, University of Regensburg, Germany, and René Herrmann, ETH Zürich, Switzerland.

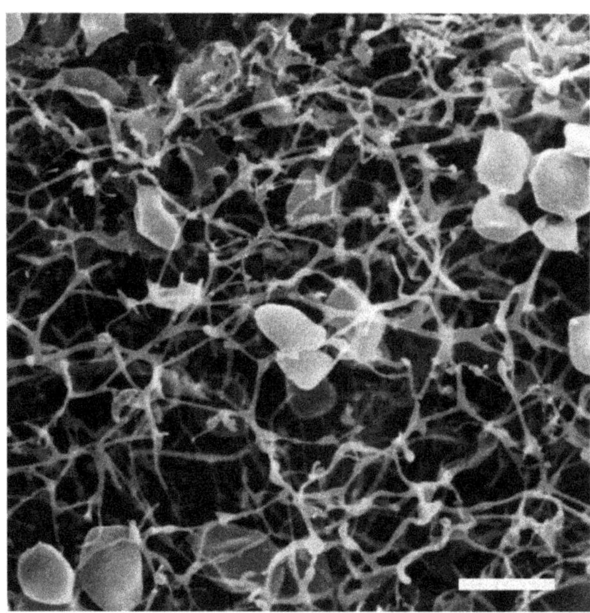

Three dimensional reconstructions of cannula-cell interactions provided the first evidence that cannulae enter the "periplasmic space" but not the cytoplasm of connected *Pyrodictium* cells [119]. Though it has been observed that cannulae elongate, remain attached as *Pyrodictium* cells and undergo binary fission [121], it remains unclear if any material, genetic or nutritive, is actually exchanged between cells through the cannula-cell connections. Cannulae may be necessary for the growth of *Pyrodictium* cells as spontaneous cannulae-free mutants have not been observed in laboratory cultures [119]. The functions of cannulae are unknown although a role in adhesion would not be unexpected [4]. Investigations into possible functions as well as the mechanism of assembly are hampered by the lack of a genetic system in *Pyrodictium* and an inability thus far to obtain strains unable to synthesize the tubules. Nevertheless, the massive amounts of material devoted to the structures point to an important role for these appendages.

5.3. Hami

Studies of Archaea in non-geothermal environments revealed the existence of the SM1 euryarchaeon whose filamentous "hami" represent another cell surface appendage unique to Archaea. The term "hamus" (Latin for hook, barb) is a direct reference to this surface appendage's characteristic structure as each pilus-like fiber ends in a three-pronged tip [7]. The archaeal cells exhibiting hami grow in cold (~10 °C) sulphidic springs as members of archaeal/bacterial communities with mainly *Thiothrix* or IMB1 ε-proteobacterium as the bacterial member [121,122].

These archaeal/bacterial communities resemble a string-of-pearls that are macroscopically visible, as the whitish pearls can reach a diameter of up to 3 mm. The SM1 organisms are small cocci, with approximately 100 filamentous hami emanating peritrichously from each cell (Figure 9A). Hami are 7 to 8 nm in diameter and 1 to 3 μm long; the filament structure is helical with no evidence of a central channel. Three prickles (4 nm in diameter) radiate from the filament every 46 nm, giving the filament the appearance of barbwire. Distally, the filament ends with a tripartite hook (diameter 60 nm); with its thicker ends (diameter 5 nm), this hook region is likened to grappling hooks and anchors (Figure 9B) [7]. The hamus structure remains stable across a wide temperature and pH range (0–70 °C; pH 0.5–11.5), and is formed by a 120 kDa protein. While potential glycosylation of the protein was investigated, it did not stain with the PAS reagent or digest with PNGaseF (Peptide N-glycosidase F). However, neither test rules out conclusively the presence of attached glycan. Hami facilitate a strong adhesion of individual SM1 euryarcheon cells to chemically diverse surfaces as well as to their bacterial partners in sulphidic springs [7,123].

Though the SM1 euryarcheon has withstood attempts at laboratory cultivation, biofilms predominantly consisting of SM1 euryarchaea were harvested from a sulphidic spring near Regensburg, Germany. The opaque, white droplets of SM1 biofilm collected on polyethylene nets were decidedly different from previously characterized SM1/bacterial string-of-pearls communities; the harvested SM1 biofilms predominantly consisted of archaeal cells (>95%) whereas the string-of-pearls communities had archaea to bacteria in ratios closer to 1:1. Examined under confocal laser scanning microscopy, the SM1 cells in the biofilm were observed to be approximately 4 μm apart, surrounded by an extracellular polymeric substance (EPS) composed of proteins and polysaccharides. The regular separation of single cells is thought to be the result of contact between the hami of neighboring cells, whose average length of 2 μm corresponds to the cell-cell distance of 4 μm. Hami also contribute to the EPS as its main protein component; as hami filaments entangle, they create a web between cells which contributes to the overall biofilm structure. The function of hami as mediators of cell-surface attachment and biofilm initialization has been proposed [124] as a variation of the role often played by pili and/or flagella in the formation of many bacterial biofilms.

5.4. Bindosome

The bindosome is a surface structure of *S. solfataricus* comprised of sugar binding proteins which act with ABC transporters to facilitate sugar uptake [125,126]. The sugar binding proteins GlcS and AraS have type IV pilin-like signal peptides which are processed by PibD, a type IV prepilin-like peptidase of *S. solfataricus* [25] . The expression of GlcS and AraS in the cell surface is directed by the bindosome assembly system (Bas), yet another type IV pilus like assembly system is composed of three pilin proteins BasABC, BasE, a PilT-like ATPase, and BasF, a PilC-like integral membrane protein [43]. Deletion of either of these gene groups resulted in severe defects in the ability of the cells to grow on sugars transported by these sugar binding proteins.

Figure 9. The hami of euryarchaeon SM1. (**A**) Electron micrograph of SM1 cells with numerous hami on the surface. Bar = 500nm. (**B**) The hook and prickle region of a hamus filament. Bar = 50 nm. Courtesy of Christine Moissl-Eichinger, Institute for Microbiology and Archaea Center, Regensburg.

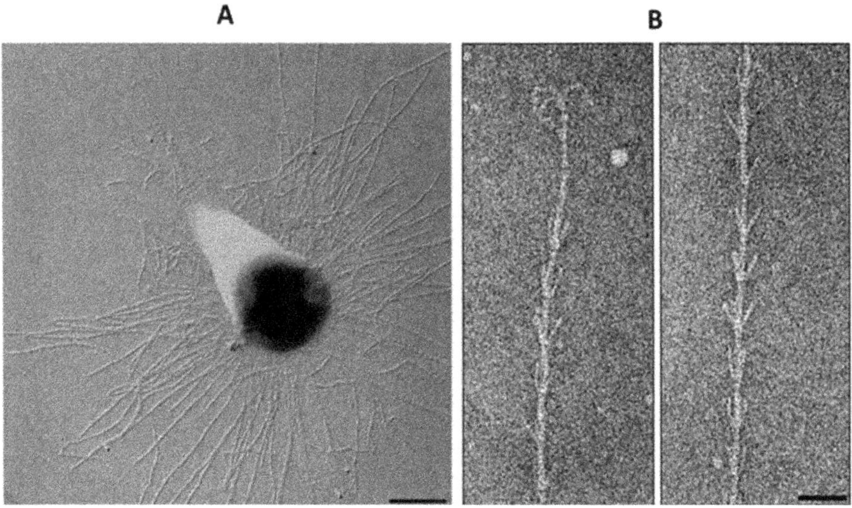

Since the Bas system appears similar to the bacterial type IV pilus assembly system, it was suspected to result in a pilus-like structure. While such a specific bindosome appendage has not been observed on cell surfaces, it is believed that a sugar binding structure embedded into the cell wall is integral to the *S. solfataricus* envelope [5]. It was recently reported that the ATPase activity of BasE is required for glucose uptake, and that the sugar binding proteins occur in complexes of high molecular mass. SlaA, the S layer glycoprotein, was also identified in these complexes, suggesting that the sugar binding proteins in *S. solfataricus* associate with the S-layer. On a cellular level, the deletion of *basEF* resulted in abnormal morphology and S-layer architecture; this further suggests that the sugar binding proteins are a functional part of the S-layer in *S. solfataricus* [127].

While studies are so far limited to *S. solfataricus*, bindosomes may actually be widely distributed in Archaea since many substrate binding proteins with type IV pilin-like signal peptides can be identified in both crenarchaeotes and euryarchaeotes [24].

6. Concluding Remarks

Archaea possess a wide variety of surface structures that contribute to their survival in what are often extremely harsh environmental niches. This requires that these structures be extremely stable to such stresses as high salt, extremely high temperatures and low pH. The mechanisms by which this occurs are currently unknown but may be partially due to the often encountered N-glycosylation of the major subunits of the structures. The Archaea have adapted the type IV pilus model for use in assembly of a diverse repertoire of appendages performing a variety of functions including swimming and adhesion. Some of the most unusual and abundant of the surface structures, like hami and cannulae, are currently only found on Archaea that lack a developed genetic system. Thus how these complex structures are formed, let alone the full listing of their functions, remains a mystery. Continued study of archaeal surface structures will undoubtedly provide a much needed insight into how archaeal cells thrive in their unusual habitats and how they interact with their neighbors and abiotic surfaces. Elucidation of their assembly mechanisms will lead to new discoveries about regulation in archaeal systems. This has recently begun in studies showing the involvement of regulators in the biosynthesis of archaella and pili in *Sulfolobus*. Studies on the roles of posttranslational modifications of surface structure components may shed light on protein stability enhancement via glycosylation and lead to biotechnological advances. As more Archaea become genetically tractable, new discoveries will be made in currently less well-studied, but no less interesting, members of the domain.

We recommend that readers also consult a recent review by the Sonja Albers group [128] which was published during the final stages of preparation of this review.

Acknowledgements

Original research conducted in the Jarrell laboratory is funded by a Discovery Grant from the Natural Sciences and Engineering Research Council of Canada (NSERC) (to KFJ). Y.D. is sponsored by China Scholarship Council (2010622028). The authors are indebted to fellow researchers who generously supplied unpublished figures for this review.

References

1. Jarrell, K.F.; Walters, A.D.; Bochiwal, C.; Borgia, J.M.; Dickinson, T.; Chong, J.P.J. Major players on the microbial stage:why archaea are important. *Microbiology* **2011**, *157*, 919–936.
2. Cavicchioli, R. Archaea-timeline of the third domain. *Nat. Rev. Microbiol.* **2011**, *9*, 51–61.
3. Ng, S.Y.M.; Zolghadr, B.; Driessen, A.J.M.; Albers, S.V.; Jarrell, K.F. Cell surface structures of Archaea. *J. Bacteriol.* **2008**, *190*, 6039–6047.
4. Albers, S.V.; Meyer, B.H. The archaeal cell envelope. *Nat. Rev. Microbiol.* **2011**, *9*, 414–426.
5. Pohlschroder, M.; Ghosh, A.; Tripepi, M.; Albers, S.V. Archaeal type IV pilus-like structures-evolutionarily conserved prokaryotic surface organelles. *Curr. Opin. Microbiol.* **2011**, *14*, 1–7.
6. Albers, S.V.; Pohlschroder, M. Diversity of archaeal type IV pilin-like structures. *Extremophiles.* **2009**, *13*, 403–410.
7. Moissl, C.; Rachel, R.; Briegel, A.; Engelhardt, H.; Huber, R. The unique structure of archaeal 'hami', highly complex cell appendages with nano-grappling hooks. *Mol. Microbiol.* **2005**, *56*, 361–370.
8. Rieger, G.; Rachel, R.; Hermann, R.; Stetter, K.O. Ultrastructure of the hyperthermophilic archaeon *Pyrodictium. abyssi. J. Struct. Biol.* **1995**, *115*, 78–87.
9. Henche, A.L.; Ghosh, A.; Yu, X.; Jeske, T.; Egelman, E.; Albers, S.V. Structure and function of the adhesive type IV pilus of *Sulfolobus. acidocaldarius. Environ. Microbiol.* **2012**, *14*, 3188–3202.

10. Yu, X.; Goforth, C.; Meyer, C.; Rachel, R.; Schröder, G.F.; Egelman, E.H. Filaments from Ignicoccus hospitalis show diversity of packing in proteins containing N-terminal type IV pilin helices. *J. Mol. Biol.* **2012**, *422*, 274–281.
11. Wang, Y.A.; Yu, X.; Ng, S.Y.M.; Jarrell, K.F.; Egelman, E.H. The structure of an archaeal pilus. *J. Mol. Biol.* **2008**, *381*, 456–466.
12. Jarrell, K.F.; Albers, S.V. The archaellum: an old motility structure with a new name. *Trends Microbiol.* **2012**, *20*, 307–312.
13. Jarrell, K.F.; VanDyke, D.J.; Wu, J. Archaeal flagella and pili In *Current Research and Future Trends Pili and Flagella;* Jarrell, K.F., Ed.; Caister Academic Press: Norfolk, UK, 2009; pp 215–234.
14. Jarrell, K.F.; McBride, M.J. The surprisingly diverse ways that prokaryotes move. *Nat. Rev. Microbiol.* **2008**, *6*, 466–476.
15. Ng, S.Y.; Chaban, B.; Jarrell, K.F. Archaeal flagella, bacterial flagella and type IV pili: A comparison of genes and posttranslational modifications. *J. Mol. Microbiol. Biotechnol.* **2006**, *11*, 167–191.
16. Ghosh, A.; Albers, S.V. Assembly and function of the archaeal flagellum. *Biochem. Soc. Trans.* **2011**, *39*, 64–69.
17. Lassak, K.; Neiner, T.; Ghosh, A.; Klingl, A.; Wirth, R.; Albers, S. Molecular analysis of the crenarchaeal flagellum. *Mol. Microbiol.* **2012**, *83*, 110–124.
18. Eichler, J. Response to Jarrell and Albers: the name says it all. *Trends Microbiol.* **2012**, *20*, 512–513.
19. Wirth, R. Response to Jarrell and Albers: seven letters less does not say more. *Trends Microbiol.* **2012**, *20*, 511–512.
20. Leigh, J.A.; Albers, S.V.; Atomi, H.; Allers, T. Model organisms for genetics in the domain archaea: methanogens, halophiles, Thermococcales and Sulfolobales. *FEMS Microbiol. Rev.* **2011**, *35*, 577–608.
21. Cohen-Krausz, S.; Trachtenberg, S. The flagellar filament structure of the extreme acidothermophile *Sulfolobus. shibatae* B12 suggests that archaeabacterial flagella have a unique and common symmetry and design. *J. Mol. Biol.* **2008**, *375*, 1113–1124.
22. Bardy, S.L.; Jarrell, K.F. FlaK of the archaeon *Methanococcus. maripaludis* possesses preflagellin peptidase activity. *FEMS Microbiol. Lett.* **2002**, *208*, 53–59.
23. Bardy, S.L.; Jarrell, K.F. Cleavage of preflagellins by an aspartic acid signal peptidase is essential for flagellation in the archaeon *Methanococcus. voltae*. *Mol. Microbiol.* **2003**, *50*, 1339–1347.
24. Szabo, Z.; Stahl, A.O.; Albers, S.V.; Kissinger, J.C.; Driessen, A.J.M.; Pohlschroder, M. Identification of diverse archaeal proteins with class III signal peptides cleaved by distinct archaeal prepilin peptidases. *J. Bacteriol.* **2007**, *189*, 772–778.
25. Albers, S.V.; Szabo, Z.; Driessen, A.J. M. Archaeal homolog of bacterial type IV prepilin signal peptidases with broad substrate specificity. *J. Bacteriol.* **2003**, *185*, 3918–3925.
26. Tripepi, M.; Imam, S.; Pohlschroder, M. *Haloferax. volcanii* flagella are required for motility but are not involved in PibD-dependent surface adhesion. *J. Bacteriol.* **2010**, *192*, 3093–3102.

27. Jarrell, K.F.; Jones, G.M.; Kandiba, L.; Nair, D.B.; Eichler, J. S-layer glycoproteins and flagellins: reporters of archaeal posttranslational modifications. *Archaea.* **2010**, doi:10.1155/2010/612948.
28. Ng, S.Y.M.; Wu, J.; Nair, D.B.; Logan, S.M.; Robotham, A.; Tessier, L.; Kelly, J.F.; Uchida, K.; Aizawa, S.; Jarrell, K.F. Genetic and mass spectrometry analysis of the unusual type IV-like pili of the archaeon *Methanococcus. maripaludis.*. *J. Bacteriol.* **2011**, *193*, 804–814.
29. Kelly, J.; Logan, S.M.; Jarrell, K.F.; Vandyke, D.J.; Vinogradov, E. A novel N-linked flagellar glycan from *Methanococcus. maripaludis*. *Carbohydr. Res.* **2009**, *344*, 648–653.
30. Tripepi, M.; You, J.; Temel, S.; Önder, Ö.; Brisson, D.; Pohlschröder, M. N-glycosylation of *Haloferax. volcanii* flagellins requires known Agl proteins and is essential for biosynthesis of stable flagella. *J. Bacteriol.* **2012**, *194*, 4876–4887.
31. Jarrell, K.F.; Stark, M.; Nair, D.B.; Chong, J.P.J. Flagella and pili are both necessary for efficient attachment of *Methanococcus. maripaludis* to surfaces. *FEMS Microbiol. Lett.* **2011**, *319*, 44–50.
32. Henche, A.L.; Koerdt, A.; Ghosh, A.; Albers, S.V. Influence of cell surface structures on crenarchaeal biofilm formation using a thermostable green fluorescent protein. *Environ. Microbiol.* **2012**, *14*, 779–793.
33. Koerdt, A.; Gödeke, J.; Berger, J.; Thormann, K.M.; Albers, S.V. Crenarchaeal biofilm formation under extreme conditions. *PloS One* **2010**, doi:10.1371/journal.pone.0014104.
34. Ajon, M.; Fröls, S.; van Wolferen, M.; Stoecker, K.; Teichmann, D.; Driessen, A.J.; Grogan, D.W.; Albers, S.V.; Schleper, C. UV-inducible DNA exchange in hyperthermophilic archaea mediated by type IV pili. *Mol. Microbiol.* **2011**, *82*, 807–817.
35. Muller, D.W.; Meyer, C.; Gurster, S.; Kuper, U.; Huber, H.; Rachel, R.; Wanner, G.; Wirth, R.; Bellack, A. The Iho670 fibers of *Ignicoccus. hospitalis*: A new type of archaeal cell surface appendage. *J. Bacteriol.* **2009**, *191*, 6465–6468.
36. Frols, S.; Ajon, M.; Wagner, M.; Teichmann, D.; Zolghadr, B.; Folea, M.; Boekema, E.J.; Driessen, A.J.; Schleper, C.; Albers, S.V. UV-inducible cellular aggregation of the hyperthermophilic archaeon *Sulfolobus. solfataricus* is mediated by pili formation. *Mol. Microbiol.* **2008**, *70*, 938–952.
37. Schopf, S.; Wanner, G.; Rachel, R.; Wirth, R. An archaeal bi-species biofilm formed by *Pyrococcus. furiosus* and *Methanopyrus. kandleri*. *Arch. Microbiol.* **2008**, *190*, 371–377.
38. Weiner, A.; Schopf, S.; Wanner, G.; Probst, A.; Wirth, R. Positive, neutral and negative interactions in cocultures between *Pyrococcus. furiosus* and different methanogenic Archaea. *Microb. Insights* **2012**, *5*, 1–10.
39. Reimann, J.; Lassak, K.; Khadouma, S.; Ettema, T.J.; Yang, N.; Driessen, A.J.; Klingl, A.; Albers, S.V. Regulation of archaella expression by the FHA and von Willebrand domain-containing proteins ArnA and ArnB in *Sulfolobus. acidocaldarius*. *Mol. Microbiol.* **2012**, *86*, 24–36.
40. Trachtenberg, S.; Cohen-Krausz, S. The archaeabacterial flagellar filament: a bacterial propeller with a pilus-like structure. *J. Mol. Microbiol. Biotechnol.* **2006**, *11*, 208–220.

41. Faguy, D.M.; Jarrell, K.F.; Kuzio, J.; Kalmokoff, M.L. Molecular analysis of archael flagellins: similarity to the type IV pilin-transport superfamily widespread in bacteria. *Can. J. Microbiol.* **1994**, *40*, 67–71.
42. Jarrell, K.F.; Bayley, D.P.; Kostyukova, A.S. The archaeal flagellum: a unique motility structure. *J. Bacteriol.* **1996**, *178*, 5057–5064.
43. Zolghadr, B.; Weber, S.; Szabo, Z.; Driessen, A.J.M.; Albers, S.V. Identification of a system required for the functional surface localization of sugar binding proteins with class III signal peptides in *Sulfolobus. solfataricus*. *Mol. Microbiol.* **2007**, *64*, 795–806.
44. Strom, M.S.; Nunn, D.N.; Lory, S. A single bifunctional enzyme, PilD, catalyzes cleavage and N-methylation of proteins belonging to the type IV pilin family. *Proc. Natl. Acad. Sci. USA* **1993**, *90*, 2404–2408.
45. Bardy, S.L.; Eichler, J.; Jarrell, K.F. Archaeal signal peptides—A comparative survey at the genome level. *Protein Sci.* **2003**, *12*, 1833–1843.
46. Bayley, D.P.; Jarrell, K.F. Further evidence to suggest that archaeal flagella are related to bacterial type IV pili. *J. Mol. Evol.* **1998**, *46*, 370–373.
47. Peabody, C.R.; Chung, Y.J.; Yen, M.R.; Vidal-Ingigliardi, D.; Pugsley, A.P.; Saier, M.H., Jr. Type II protein secretion and its relationship to bacterial type IV pili and archaeal flagella. *Microbiology* **2003**, *149*, 3051–3072.
48. Faguy, D.M.; Koval, S.F.; Jarrell, K.F. Physical characterization of the flagella and flagellins from *Methanospirillum. hungatei*. *J. Bacteriol.* **1994**, *176*, 7491–7498.
49. Thomas, N.A.; Bardy, S.L.; Jarrell, K.F. The archaeal flagellum: A different kind of prokaryotic motility structure. *FEMS Microbiol. Rev.* **2001**, *25*, 147–174.
50. Pyatibratov, M.G.; Beznosov, S.N.; Rachel, R.; Tiktopulo, E.I.; Surin, A.K.; Syutkin, A.S.; Fedorov, O.V. Alternative flagellar filament types in the haloarchaeon *Haloarcula. marismortui*. *Can. J. Microbiol.* **2008**, *54*, 835–844.
51. Bardy, S.L.; Ng, S.Y.; Jarrell, K.F. Recent advances in the structure and assembly of the archaeal flagellum. *J. Mol. Microbiol. Biotechnol.* **2004**, *7*, 41–51.
52. Chaban, B.; Ng, S.Y.; Kanbe, M.; Saltzman, I.; Nimmo, G.; Aizawa, S.I.; Jarrell, K.F. Systematic deletion analyses of the *fla* genes in the flagella operon identify several genes essential for proper assembly and function of flagella in the archaeon, *Methanococcus. maripaludis*. *Mol. Microbiol.* **2007**, *66*, 596–609.
53. Patenge, N.; Berendes, A.; Engelhardt, H.; Schuster, S.C.; Oesterhelt, D. The *fla* gene cluster is involved in the biogenesis of flagella in *Halobacterium. salinarum*. *Mol. Microbiol.* **2001**, *41*, 653–663.
54. Chaban, B.; Voisin, S.; Kelly, J.; Logan, S.M.; Jarrell, K.F. Identification of genes involved in the biosynthesis and attachment of *Methanococcus. voltae* N-linked glycans: Insight into N-linked glycosylation pathways in Archaea. *Mol. Microbiol.* **2006**, *61*, 259–268.
55. Vandyke, D.J.; Wu, J.; Logan, S.M.; Kelly, J.F.; Mizuno, S.; Aizawa, S.I.; Jarrell, K.F. Identification of genes involved in the assembly and attachment of a novel flagellin N-linked tetrasaccharide important for motility in the archaeon *Methanococcus. maripaludis*. *Mol. Microbiol.* **2009**, *72*, 633–644.

56. Jarrell, K.F.; Jones, G.M.; Nair, D.B. Role of N-linked glycosylation in cell surface structures of Archaea with a focus on flagella and S layers. *Int. J. Microbiol.* **2010**, doi: 10.1155/2010/470138.
57. Gerl, L.; Deutzmann, R.; Sumper, M. Halobacterial flagellins are encoded by a multigene family. Identification of all five gene products. *FEBS Lett.* **1989**, *244*, 137–140.
58. Gerl, L.; Sumper, M. Halobacterial flagellins are encoded by a multigene family. Characterization of five flagellin genes. *J. Biol. Chem.* **1988**, *263*, 13246–13251.
59. Tarasov, V.Y.; Pyatibratov, M.G.; Tang, S.L.; Dyall-Smith, M.; Fedorov, O.V. Role of flagellins from A and B loci in flagella formation of *Halobacterium. salinarum*. *Mol. Microbiol.* **2000**, *35*, 69–78.
60. Kalmokoff, M.L.; Jarrell, K.F.; Koval, S.F. Isolation of flagella from the archaebacterium *Methanococcus. voltae* by phase separation with Triton X-114. *J. Bacteriol.* **1988**, *170*, 1752–1758.
61. Bardy, S.L.; Mori, T.; Komoriya, K.; Aizawa, S.; Jarrell, K.F. Identification and localization of flagellins FlaA and FlaB3 within flagella of *Methanococcus. voltae*. *J. Bacteriol.* **2002**, *184*, 5223–5233.
62. Macnab, R.M. How bacteria assemble flagella. *Annu. Rev. Microbiol.* **2003**, *57*, 77–100.
63. Aizawa, S.I. Flagellar assembly in *Salmonella typhimurium*. *Mol. Microbiol.* **1996**, *19*, 1–5.
64. Beznosov, S.N.; Pyatibratov, M.G.; Fedorov, O.V. On the multicomponent nature of *Halobacterium. salinarum* flagella. *Microbiology Russ.* **2007**, *76*, 435–441.
65. Syutkin, A.S.; Pyatibratov, M.G.; Beznosov, S.N.; Fedorov, O.V. Various mechanisms of flagella helicity formation in Halobacteria. *Microbiology Russ.* **2012**, *81*, 573–581.
66. Schlesner, M.; Miller, A.; Streif, S.; Staudinger, W.F.; Muller, J.; Scheffer, B.; Siedler, F.; Oesterhelt, D. Identification of Archaea-specific chemotaxis proteins which interact with the flagellar apparatus. *BMC Microbiol.* **2009**, *9*, 56.
67. Mukhopadhyay, B.; Johnson, E.F.; Wolfe, R.S. A novel pH_2 control on the expression of flagella in the hyperthermophilic strictly hydrogenotrophic methanarchaeaon *Methanococcus. jannaschii*. *Proc. Natl. Acad. Sci. USA* **2000**, *97*, 11522–11527.
68. Thomas, N.A.; Jarrell, K.F. Characterization of flagellum gene families of methanogenic archaea and localization of novel flagellum accessory proteins. *J. Bacteriol.* **2001**, *183*, 7154–7164.
69. Banerjee, A.; Ghosh, A.; Mills, D.J.; Kahnt, J.; Vonck, J.; Albers, S.V. FlaX, a unique component of the crenarchaeal archaellum, forms oligomeric ring-shaped structures and interacts with the motor ATPase FlaI. *J. Biol. Chem.* **2012**, *287*, 43322–43330.
70. Ghosh, A.; Hartung, S.; van der Does, C.; Tainer, J.A.; Albers, S.V. Archaeal flagellar ATPase motor shows ATP-dependent hexameric assembly and activity stimulation by specific lipid binding. *Biochem. J.* **2011**, *437*, 43–52.
71. Kalmokoff, M.L.; Jarrell, K.F. Cloning and sequencing of a multigene family encoding the flagellins of *Methanococcus. voltae*. *J. Bacteriol.* **1991**, *173*, 7113–7125.

72. Szabo, Z.; Albers, S.V.; Driessen, A.J.M. Active-site residues in the type IV prepilin peptidase homologue PibD from the archaeon *Sulfolobus. solfataricus*. *J. Bacteriol.* **2006**, *188*, 1437–1443.
73. Hu, J.; Xue, Y.; Lee, S.; Ha, Y. The crystal structure of GxGD membrane protease FlaK. *Nature* **2011**, *475*, 528–531.
74. Thomas, N.A.; Chao, E.D.; Jarrell, K.F. Identification of amino acids in the leader peptide of *Methanococcus. voltae* preflagellin that are important in posttranslational processing. *Arch. Microbiol.* **2001**, *175*, 263–269.
75. Ng, S.Y.; VanDyke, D.J.; Chaban, B.; Wu, J.; Nosaka, Y.; Aizawa, S.; Jarrell, K.F. Different minimal signal peptide lengths recognized by the archaeal prepilin-like peptidases FlaK and PibD. *J. Bacteriol.* **2009**, *191*, 6732–6740.
76. Sumper, M. Halobacterial glycoprotein biosynthesis. *Biochim. Biophys. Acta.* **1987**, *906*, 69–79.
77. Voisin, S.; Houliston, R.S.; Kelly, J.; Brisson, J.R.; Watson, D.; Bardy, S.L.; Jarrell, K.F.; Logan, S.M. Identification and characterization of the unique N-linked glycan common to the flagellins and S-layer glycoprotein of *Methanococcus. voltae*. *J. Biol. Chem.* **2005**, *280*, 16586–16593.
78. Chaban, B.; Logan, S.M.; Kelly, J.F.; Jarrell, K.F. AglC and AglK are involved in biosynthesis and attachment of diacetylated glucuronic acid to the N-glycan in *Methanococcus. voltae*. *J. Bacteriol.* **2009**, *191*, 187–195.
79. Shams-Eldin, H.; Chaban, B.; Niehus, S.; Schwarz, R.T.; Jarrell, K.F. Identification of the archaeal *alg*7 gene homolog encoding N-acetylglucosamine-1-phosphate transferase of the N-linked glycosylation system by cross-domain complementation in *Saccharomyces. cerevisiae*. *J. Bacteriol.* **2008**, *190*, 2217–2220.
80. Namboori, S.C.; Graham, D.E. Acetamido sugar biosynthesis in the Euryarchaea. *J. Bacteriol.* **2008**, *190*, 2987–2996.
81. Jones, G.M.; Wu, J.; Ding, Y.; Uchida, K.; Aizawa, S.; Robotham, A.; Logan, S.M.; Kelly, J.; Jarrell, K.F. Identification of genes involved in the acetamidino group modification of the flagellin N-linked glycan of *Methanococcus. maripaludis*. *J. Bacteriol.* **2012**, *194*, 2693–2702.
82. VanDyke, D.J.; Wu, J.; Ng, S.Y.; Kanbe, M.; Chaban, B.; Aizawa, S.I.; Jarrell, K.F. Identification of putative acetyltransferase gene, MMP0350, which affects proper assembly of both flagella and pili in the archaeon *Methanococcus. maripaludis*. *J. Bacteriol.* **2008**, *190*, 5300–5307.
83. Calo, D.; Guan, Z.; Eichler, J. Glyco-engineering in Archaea: differential N-glycosylation of the S-layer glycoprotein in a transformed *Haloferax. volcanii* strain. *Microb. Biotechnol.* **2011**, *4*, 461–470.
84. Calo, D.; Kaminski, L.; Eichler, J. Protein glycosylation in Archaea: Sweet and Extreme. *Glycobiology.* **2010**, *20*, 1065–1076.
85. Eichler, J.; Maupin-Furlow, J. Post-translation modification in Archaea: Lessons from *Haloferax. volcanii* and other haloarchaea. *FEMS Microbiol. Rev.* **2012**, doi: 10.1111/1574–6976.12012.

86. Abu-Qarn, M.; Yurist-Doutsch, S.; Giordano, A.; Trauner, A.; Morris, H.R.; Hitchen, P.; Medalia, O.; Dell, A.; Eichler, J. *Haloferax. volcanii* AglB and AglD are involved in N-glycosylation of the S-layer glycoprotein and proper assembly of the surface layer. *J. Mol. Biol.* **2007**, *374*, 1224–1236.
87. Hendrickson, E.L.; Liu, Y.; Rosas-Sandoval, G.; Porat, I.; Soll, D.; Whitman, W.B.; Leigh, J.A. Global responses of *Methanococcus. maripaludis* to specific nutrient limitations and growth rate. *J. Bacteriol.* **2008**, *190*, 2198–2205.
88. Xia, Q.; Wang, T.; Hendrickson, E.L.; Lie, T.J.; Hackett, M.; Leigh, J.A. Quantitative proteomics of nutrient limitation in the hydrogenotrophic methanogen *Methanococcus. maripaludis*. *BMC Microbiol.* **2009**, *9*, 149.
89. Duan, X.; He, Z.G. Characterization of the specific interaction between archael FHA domain-containing protein and the promoter of a flagella-like gene-cluster and its regulation by phosphorylation. *Biochem. Biophys. Res. Commun.* **2011**, *407*, 242–247.
90. Wurtzel, O.; Sapra, R.; Chen, F.; Zhu, Y.; Simmons, B.A.; Sorek, R. A single-base resolution map of an archaeal transcriptome. *Genome Res.* **2010**, *20*, 133–141.
91. Cohen-Krausz, S.; Trachtenberg, S. The structure of the archeabacterial flagellar filament of the extreme halophile *Halobacterium. salinarum* R1M1 and its relation to eubacterial flagellar filaments and type IV pili. *J. Mol. Biol.* **2002**, *321*, 383–395.
92. Trachtenberg, S.; Galkin, V.E.; Egelman, E.H. Refining the structure of the *Halobacterium. salinarum* flagellar filament using the iterative helical real space reconstruction method: Insights into polymorphism. *J. Mol. Biol.* **2005**, *346*, 665–676.
93. Kalmokoff, M.L.; Karnauchow, T.M.; Jarrell, K.F. Conserved N-terminal sequences in the flagellins of archaebacteria. *Biochem. Biophys. Res. Commun.* **1990**, *167*, 154–160.
94. Jarrell, K.F.; Bayley, D.P.; Florian, V.; Klein, A. Isolation and characterization of insertional mutations in flagellin genes in the archaeon *Methanococcus. voltae*. *Mol. Microbiol.* **1996**, *20*, 657–666.
95. Marwan, W.; Alam, M.; Oesterhelt, D. Rotation and switching of the flagellar motor assembly in *Halobacterium. halobium*. *J. Bacteriol.* **1991**, *173*, 1971–1977.
96. Welch, M.; Oosawa, K.; Aizawa, S.; Eisenbach, M. Phosphorylation-dependent binding of a signal molecule to the flagellar switch of bacteria. *PNAS* **1993**, *90*, 8787–8791.
97. Rudolph, J.; Nordmann, B.; Storch, K.F.; Gruenberg, H.; Rodewald, K.; Oesterhelt, D. A family of halobacterial transducer proteins. *FEMS Microbiol. Lett.* **1996**, *139*, 161–168.
98. Rudolph, J.; Oesterhelt, D. Deletion analysis of the *che* operon in the archaeon *Halobacterium. salinarium*. *J. Mol. Biol.* **1996**, *258*, 548–554.
99. Jarrell, K.F.; Ng, S.Y.; Chaban, B. Flagellation and chemotaxis. **In** Cavicchioli, R., Ed.; Archaea: molecular and cellular biology; ASM Press: Washington, DC, USA, 2007; pp 385–410.
100. del Rosario, R.C.; Diener, F.; Diener, M.; Oesterhelt, D. The steady-state phase distribution of the motor switch complex model of *Halobacterium. salinarum*. *Math. Biosci.* **2009**, *222*, 117–126.

101. Herzog, B.; Wirth, R. Swimming behavior of selected species of Archaea. *Appl. Environ. Microbiol.* **2012**, *78*, 1670–1674.
102. Nather, D.J.; Rachel, R.; Wanner, G.; Wirth, R. Flagella of *Pyrococcus. furiosus*: Multifunctional organelles, made for swimming, adhesion to various surfaces, and cell-cell contacts. *J. Bacteriol.* **2006**, *188*, 6915–6923.
103. Bellack, A.; Huber, H.; Rachel, R.; Wanner, G.; Wirth, R. *Methanocaldococcus. villosus* sp. nov., a heavily flagellated archaeon adhering to surfaces and forming cell-cell contacts. *Int. J. Syst. Evol. Microbiol.* **2011**, *61*, 1239–1245.
104. Shimoyama, T.; Kato, S.; Ishii, S.; Watanabe, K. Flagellum mediates symbiosis. *Science* **2009**, *323*, 1574.
105. Zolghadr, B.; Klingl, A.; Koerdt, A.; Driessen, A.J.; Rachel, R.; Albers, S.V. Appendage-mediated surface adherence of *Sulfolobus. solfataricus*. *J. Bacteriol.* **2010**, *192*, 104–110.
106. Weiss, R.L. Attachment of bacteria to sulfur in extreme environments. *J. Gen. Microbiol.* **1973**, *77*, 501–507.
107. Doddema, H.J.; Derksen, J.W.M.; Vogels, G.D. Fimbriae and flagella of methanogenic bacteria. *FEMS Microbiol. Lett.* **1979**, *5*, 135–138.
108. Burrows, L.L. *Pseudomonas aeruginosa* twitching motility: type IV pili in action. *Annu. Rev. Microbiol.* **2012**, *66*, 493–520.
109. Pelicic, V. Type IV pili: e pluribus unum? *Mol. Microbiol.* **2008**, *68*, 827–837.
110. Thoma, C.; Frank, M.; Rachel, R.; Schmid, S.; Nather, D.; Wanner, G.; Wirth, R. The Mth60-fimbriae of *Methanothermobacter. thermoautotrophicus* are functional adhesins. *Environ. Microbiol.* **2008**, *10*, 2785–2795.
111. Wirth, R.; Bellack, A.; Bertl, M.; Bilek, Y.; Heimerl, T.; Herzog, B.; Leisner, M.; Probst, A.; Rachel, R.; Sarbu, C.; Schopf, S.; Wanner, G. The mode of cell wall growth in selected archaea is similar to the general mode of cell wall growth in bacteria as revealed by fluorescent dye analysis. *Appl. Environ. Microbiol.* **2011**, *77*, 1556–1562.
112. Nakamura, K.; Takahashi, A.; Mori, C.; Tamaki, H.; Mochimaru, H.; Nakamura, K.; Takamizawa, K.; Kamagata, Y. *Methanothermobacter. tenebrarum* sp. nov., a hydrogenotrophic thermophilic methanogen isolated from gas-associated formation water of a natural gas field in Japan. *Int. J. Syst. Evol. Microbiol.* **2012**, doi:10.1099/ijs.0.041681–0.
113. Kachlany, S.C.; Planet, P.J.; DeSalle, R.; Fine, D.H.; Figurski, D.H. Genes for tight adherence of *Actinobacillus. actinomycetemcomitans*: from plaque to plague to pond scum. *Trends Microbiol.* **2001**, *9*, 429–437.
114. Vassart, A.; van Wolferen, M.; Orell, A.; Hong, Y.; Peeters, E.; Albers, S.V.; Charlier, D. Sa-Lrp from *Sulfolobus. acidocaldarius* is a versatile, glutamine-responsive, and architectural transcriptional regulator. *Microbiology Open* **2012**, doi: 10.1002/mbo3.58.
115. Frols, S.; Gordon, P.M.; Panlilio, M.A.; Duggin, I.G.; Bell, S.D.; Sensen, C.W.; Schleper, C. Response of the hyperthermophilic archaeon *Sulfolobus. solfataricus* to UV damage. *J. Bacteriol.* **2007**, *189*, 8708–8718.

116. Huber, H.; Küper, U.; Daxer, S.; Rachel, R. The unusual cell biology of the hyperthermophilic Crenarchaeon *Ignicoccus. hospitalis*. *Antonie. van Leeuwenhoek* **2012**, *102*, 203–219.
117. Giannone, R.J.; Huber, H.; Karpinets, T.; Heimerl, T.; Küper, U.; Rachel, R.; Keller, M.; Hettich, R.L.; Podar, M. Proteomic characterization of cellular and molecular processes that enable the *Nanoarchaeum. equitans—Ignicoccus. hospitalis* relationship. *PloS One* **2011**, doi:10.1371/journal.pone.0022942.
118. Magidovich, H.; Eichler, J. Glycosyltransferases and oligosaccharyltransferases in Archaea: putative components of the N-glycosylation pathway in the third domain of life. *FEMS Microbiol. Lett.* **2009**, *300*, 122–130.
119. Nickell, S.; Hegerl, R.; Baumeister, W.; Rachel, R. *Pyrodictium.* cannulae enter the periplasmic space but do not enter the cytoplasm, as revealed by cryo-electron tomography. *J. Struct. Biol.* **2003**, *141*, 34–42.
120. Barton, N.R.; O'Donoghue, E.; Short, R.; Frey, G.; Weiner, D.; Robertson, D.E.; Briggs, S.; Zorner, P. Chimeric cannulae proteins, nucleic acids encoding them and methods for making and using them. International Patent Applic. WO 2005/094543 A2, 2005.
121. Horn, C.; Paulmann, B.; Kerlen, G.; Junker, N.; Huber, H. In vivo observation of cell division of anaerobic hyperthermophiles by using a high-intensity dark-field microscope. *J. Bacteriol.* **1999**, *181*, 5114–5118.
122. Moissl, C.; Rudolph, C.; Huber, R. Natural communities of novel archaea and bacteria with a string-of-pearls-like morphology: molecular analysis of the bacterial partners. *Appl. Environ. Microbiol.* **2002**, *68*, 933–937.
123. Moissl-Eichinger , C.; Huber, H. Archaeal symbionts and parasites. *Curr. Opin. Microbiol.* **2011**, *14*, 364–370.
124. Henneberger, R.; Moissl, C.; Amann, T.; Rudolph, C.; Huber, R. New insights into the lifestyle of the cold-loving SM1 euryarchaeon: natural growth as a monospecies biofilm in the subsurface. *Appl. Environ. Microbiol.* **2006**, *72*, 192–199.
125. Albers, S.V.; Elferink, M.G.; Charlebois, R.L.; Sensen, C.W.; Driessen, A.J.M.; Konings, W.N. Glucose transport in the extremely thermoacidophilic *Sulfolobus. solfataricus* involves a high-affinity membrane-integrated binding protein. *J. Bacteriol.* **1999**, *181*, 4285–4291.
126. Elferink, M.G.; Albers, S.V.; Konings, W.N.; Driessen, A.J. Sugar transport in *Sulfolobus. solfataricus* is mediated by two families of binding protein-dependent ABC transporters. *Mol. Microbiol.* **2001**, *39*, 1494–1503.
127. Zolghadr, B.; Klingl, A.; Rachel, R.; Driessen, A.J.; Albers, S.V. The bindosome is a structural component of the *Sulfolobus. solfataricus* cell envelope. *Extremophiles.* **2011**, *15*, 235–244.
128. Lassak, K.; Ghosh, A.; Albers, S.V. Diversity, assembly and regulation of archaeal type IV pili-like and non-type-IV pili-like surface structures. *Res. Microbiol.* **2012**, *163*, 630–644.

A Survey of Protein Structures from Archaeal Viruses

Nikki Dellas, C. Martin Lawrence and Mark J. Young

Abstract: Viruses that infect the third domain of life, Archaea, are a newly emerging field of interest. To date, all characterized archaeal viruses infect archaea that thrive in extreme conditions, such as halophilic, hyperthermophilic, and methanogenic environments. Viruses in general, especially those replicating in extreme environments, contain highly mosaic genomes with open reading frames (ORFs) whose sequences are often dissimilar to all other known ORFs. It has been estimated that approximately 85% of virally encoded ORFs do not match known sequences in the nucleic acid databases, and this percentage is even higher for archaeal viruses (typically 90%–100%). This statistic suggests that either virus genomes represent a larger segment of sequence space and/or that viruses encode genes of novel fold and/or function. Because the overall three-dimensional fold of a protein evolves more slowly than its sequence, efforts have been geared toward structural characterization of proteins encoded by archaeal viruses in order to gain insight into their potential functions. In this short review, we provide multiple examples where structural characterization of archaeal viral proteins has indeed provided significant functional and evolutionary insight.

Reprinted from *Life*. Cite as: Dellas, N.; Lawrence, C.M.; Young, M.J. A Survey of Protein Structures from Archaeal Viruses. *Life* **2013**, *3*, 118-130.

1. Introduction

Archaeal viruses infect the third domain of life, Archaea. Over 5000 bacterial viruses and eukaryotic viruses have been identified, as compared to the approximately 100 archaeal viruses that have been characterized to date. All identified archaeal viruses infect extremophilic hosts, including acidophiles and hyperthermophiles found in terrestrial hot springs and deep sea vents, alkaliphiles and halophiles that thrive in alkaline and hypersaline environments, and methanogens that survive in anaerobic environments [1,2]. Further investigations will likely uncover archaeal viruses that replicate in mesophilic archaeal hosts [3].

It is well documented that on an evolutionary timescale, the three-dimensional fold of a protein persists longer than its sequence [4]. Thus, in the case of archaeal viral proteins, where there is little sequence similarity at the amino acid level to proteins with known function, efforts have been focused on the utilization of structural homology as a tool to identify distant evolutionary relationships [5]. For example, in many of the cases discussed below, structural characterization has uncovered relationships between archaeal viral proteins and other proteins with known function, some of which belong to different viral families or domains of life. The evolutionary history of archaeal viruses appears very complex, and there is an accumulation of structural and genomic evidence that supports the idea that many genes of a given archaeal viral genome are of different or unknown origin [6–8].

Archaeal viruses that have been identified infect (1) thermoacidophilic or hyperthermophilic members of the phylum Crenarchaeota or (2) halophilic or methanogenic members of the phylum Euryarchaeota. These viruses have been grouped into viral families based on their morphologic and genetic structure [9] (Figure 1). To date, 32 proteins have been structurally characterized from viruses infecting Crenarchaeota belonging to the viral families *Fuselloviridae*, *Lipothrixviridae*, *Rudiviridae*, *Globuloviridae*, *Bicaudaviridae*, and an unclassified viral family to which the viruses *Sulfolobus* turreted icosahedral virus (STIV) and STIV2 belong (Table 1). In comparison, only three cryo-electron microscopy (cryoEM) structures of viruses from Euryarchaeota have been solved, including the icosahedral virus *Haloarcula hispanica* virus (SH1) [10], and two head-tail viruses: (*Haloarcula vallismortis* tailed virus 1 (HVTV-1) and *Halorubrum sodomense* tailed virus 2 (HSTV-2) [11].

Table 1. Overview of viruses, viral families, and potential hosts.

Virus	Virus Family	Host
Haloarcula hispanica virus (SH1)	unclassified	*Haloarcula hispanica, Halorubrum, Haloferax* [12,13]
Sulfolobus turreted icosahedral virus (STIV)	unclassified	*Sulfolobus* [14]
Acidianus filamentous virus 1 (AFV1)	*Lipothrixviridae*	*Acidianus* [15]
Sulfolobus islandicus rod-shaped virus 1 (SIRV1)	*Rudiviridae*	*Sulfolobus* [16]
Sulfolobus spindle-shaped virus 1 (SSV1)	*Fuselloviridae*	*Sulfolobus* [17]

Figure 1. Examples of archaeal virus morphologies. (**a**) *Sulfolobus* turreted icosahedral virus (STIV), belonging to an unclassified Crenarchaeal virus family. (**b**) two viruses are represented: *Sulfolobus islandicus* rod-shaped virus (SIRV) from the family *Rudiviridae*, which are the smaller particles shaped as straight rods, and *Sulfolobus islandicus* filamentous virus (SIFV) belonging to the family *Lipothrixviridae*, which are larger, filamentous particles. (**c**) *Sulfolobus* spindle-shaped virus 1 (SSV1) belonging to the family *Fuselloviridae*. All micrographs are the courtesy of S. Brumfield.

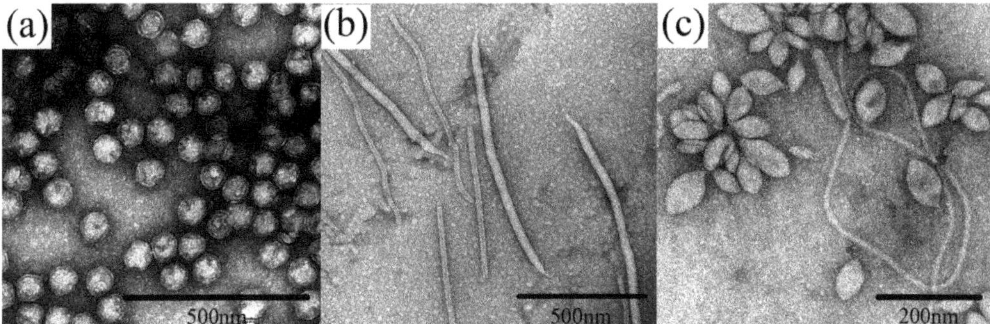

2. Discussion

2.1. Evolutionary Links Revealed Through Structures of Archaeal Virus Major Capsid Proteins

The coat protein of a virus represents a component of the virion that provides physiochemical stability and protection to the enclosed viral genome. Enveloped viruses contain a host-derived lipid membrane that surrounds the coat protein (as observed in archaeal viruses of the filamentous *Lipothrixviridae*), while other viruses contain an internal lipid membrane. STIV is one such inner lipid membrane-containing virus that belongs to an unclassified viral family within Crenarchaeota. The crystal structure of the major coat protein (MCP) from STIV (B345) reveals a double β-barrel fold, termed the "double jelly roll fold" (Figure 2) [18,19]. Interestingly, the double β-barrel fold is observed in other viruses that infect Bacteria (for example, PRD1 [20]) and Eukarya (for example, PBCV-1 [21]) (Figure 2). In addition to a homologous coat protein fold, these viruses also share other features, such as an internal lipid membrane and conserved genes (for example, a packaging ATPase) [14,22,23]. The cryo-EM structure of SH1, which infects hosts in Euryarchaeota, reveals that its coat protein adopts a single β-barrel fold [10]. It is therefore speculated that SH1 represents an ancient version of the coat protein fold that was then duplicated to generate the double β-barrel fold observed in STIV, PRD1, PBCV-1 and other viruses thought to be part of this lineage [10]. STIV and STIV2 are the only characterized viruses from this lineage that infect thermoacidophilic hosts. There are a variety of features of thermostability observed within the coat protein of STIV, such as its fold compactness and lack of cavities. However, other viruses from this lineage also contain such features and it is not known which of these specifically contribute to STIV's ability to maintain structural integrity within a thermophilic environment [19].

Figure 2. The double β-barrel fold is conserved in a viral lineage found within all three domains of life. The conserved coat protein fold of STIV B345 (PDB ID 2BBD [19]), PRD1 P3 (PDB ID 1GW7 [24]), and PBCV-1 Vp54 (PDB ID 1J5Q [21]) from viruses that infect hosts in Archaea, Bacteria, and Eukarya, respectively. Each jelly roll fold is shaded differently within the three structures.

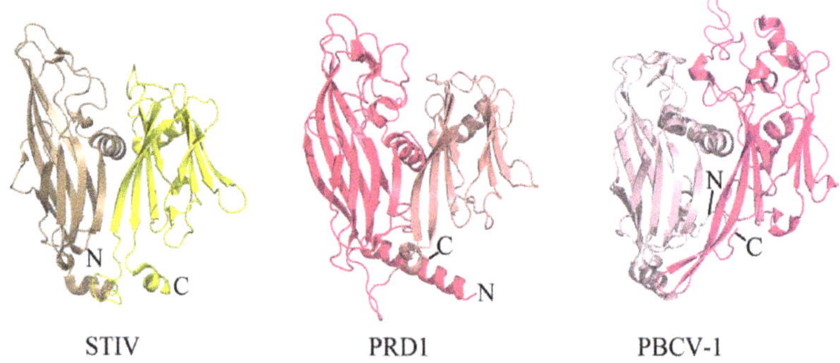

Certain viruses belonging to *Lipothrixviridae* and *Rudiviridae* have also demonstrated a conserved coat protein fold. In fact, a viral order (*Ligamenvirales*) encompassing both of these families has been proposed on the basis of genomic similarities between members of these two families that extend beyond conservation of the gene encoding the coat protein [9]. Viral members of both families are linear in morphology; however, those from *Lipothrixviridae* are enveloped and exist as (400–1,950) × 24–38 nm flexible, filamentous particles, while those from *Rudiviridae* are (610–900) × 23 nm non-enveloped, stiff, and rod-shaped in morphology [9]. *Acidianus* filamentous virus 1 (AFV1) belongs to *Lipothrixviridae* and encodes two structural proteins (orf132 and orf140), both of which contain an anti-parallel four-helix bundle fold that is structurally homologous to the C-terminal domain of the single MCP from *Sulfolobus islandicus* rod-shaped virus from Yellowstone National Park (SIRV-YNP) (Figure 3) [25]. Interestingly, tobacco mosaic virus, a thermostable, eukaryotic positive strand RNA virus, utilizes a heavily decorated 4-helix bundle to assemble a rod-like helical structure [26]. In contrast to tobacco mosaic virus, it is not yet known how these archaeal proteins are assembled into their presumably helical rod- or filamentous forms. However, a model has been proposed for coat protein assembly within AFV1: double-stranded DNA is thought to wrap around the positively charged protein AFV1 orf132, while AFV1 orf140 is proposed to interact with exterior of the DNA-protein bundle through its N-terminus and with the lipid membrane through its hydrophobic C-terminal helix [25].

2.2. Common DNA-Binding Motifs Observed in Proteins from Archaeal Viruses

Many of the published structures of archaeal viral proteins reveal DNA-binding motifs that are found in other organisms, such as the winged helix-turn-helix (wHTH) and the ribbon helix-helix (RHH) folds. The core of the wHTH fold is composed of a right handed three helix bundle followed by two β-strands, forming a 2 stranded antiparallel β-sheet known as the wing (Figure 4). The third helix of the three-helix bundle is the recognition helix, which inserts into the major groove of DNA [27], while the wing is often involved in nonspecific interactions with the ribose phosphate backbone. The wHTH motif is present in all three domains of life and often functions as a DNA recognition component of various transcription factors. The fold has also been found less frequently in proteins within RNA metabolism and those that are involved in protein-protein interactions [27]. Many archaeal virus proteins that adopt the wHTH motif are thus suggested to play roles in transcriptional regulation, although for many of these, specific binding sites within the viral or host genomes have yet to be characterized. Examples include F93 from STIV (unclassified viral family) and F93 from *Sulfolobus* spindle-shaped virus 1 (SSV1) (*Fuselloviridae*) (Figure 4). These dimeric wHTH proteins are structurally homologous to the prokaryotic MarR/SlyA protein family of transcription regulators [28,29] and are therefore expected to recognize (pseudo-) palindromic DNA targets.

The RHH fold is well named, beginning with a β-strand that precedes two α-helices. This β-strand interacts with the β-strand of another subunit, forming an anti-parallel two-stranded β-sheet that inserts into the major groove of DNA (Figure 5) [31]. While the wHTH motif is found in all domains of life, the RHH motif is found strictly in prokaryotes. Archaeal viral proteins that adopt the RHH fold (or elaborated versions of this) include E73 from *Sulfolobus* spindle-shaped virus from Ragged Hills (SSV-RH, *Fuselloviridae*) [32] and SvtR from SIRV1 (*Rudiviridae*) (Figure 5) [33]. SvtR is currently the best-characterized example of an archaeal virus protein containing the RHH motif. It is structurally homologous to bacterial RHH proteins and was determined to bind four target sequences from the SIRV1 genome. Target sequences include those preceding the coding region for its own gene as well as that for the coat protein; transcription of both the coat protein gene and its own gene were blocked by SvtR in an *in vitro* transcription assay [33].

Figure 3. A conserved 4-helix bundle arrangement found in archaeal viruses from *Lipothrixviridae* and *Rudiviridae*. The two structural proteins from AFV1, orf132 (PDB ID 3FBL [25]) and orf140 (PDB ID 3FBZ [25]) adopt the same 4 α-helix bundle arrangements as the C-terminal domain of the single coat protein from SIRV-YNP (PDB ID 3F2E [30]). It is noteworthy that the first 50 amino acids from AFV1 orf132 are absent from the x-ray data. The N-terminal and C-terminal ends of each protein structure are labeled with "N" and "C", respectively. The α-helices are labeled for SIRV-YNP CP and follow the same arrangement in the other two protein structures.

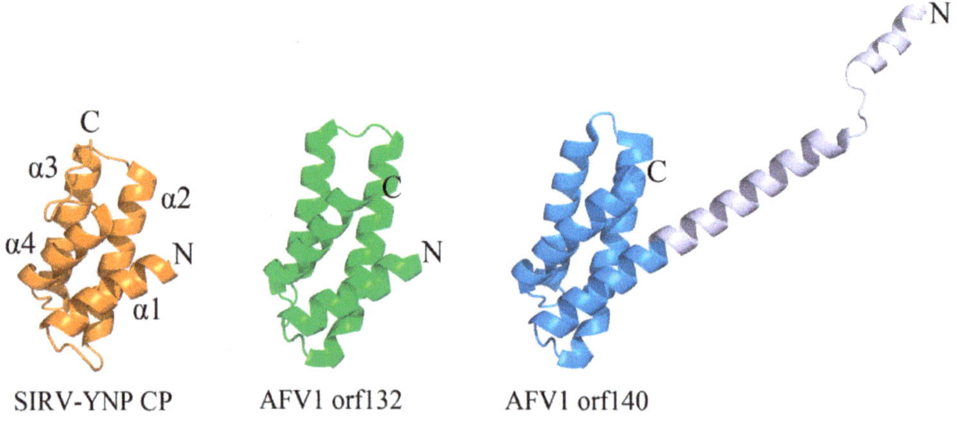

Figure 4. wHTH motif found in two structural homologs from STIV and SSV1. The overall folds of STIV F93 (left, PDB ID 2CO5 [28]) and SSV1 F93 (right, PDB ID 1TBX [29]) highlight the components of the wHTH motif, including (in order of N- to C-terminal) α1 (red), α2 (orange), α3 (the recognition helix, yellow), and a β-sheet (pink, comprising β2, β3 for STIV F93 and β1, β2 for SSV1 F93).

Figure 5. RHH motifs in two unrelated proteins encoded by viruses from *Fuselloviridae* and *Rudiviridae*. The overall folds of SIRV1 SvtR (left, PDB ID 2KEL [33]) and SSV-RH E73 (right, PDB ID 4AAI [32]) are shown. The components of the RHH fold are highlighted for each dimer, including (in order of N- to C-terminal) β1 (turquoise), α1 (blue), and α2 (purple). SSV-RH E73 has an elaboration of the RHH fold, containing an additional α-helix (α3, colored gray) that may enhance the stability of its dimer and/or contribute to an additional ligand binding site.

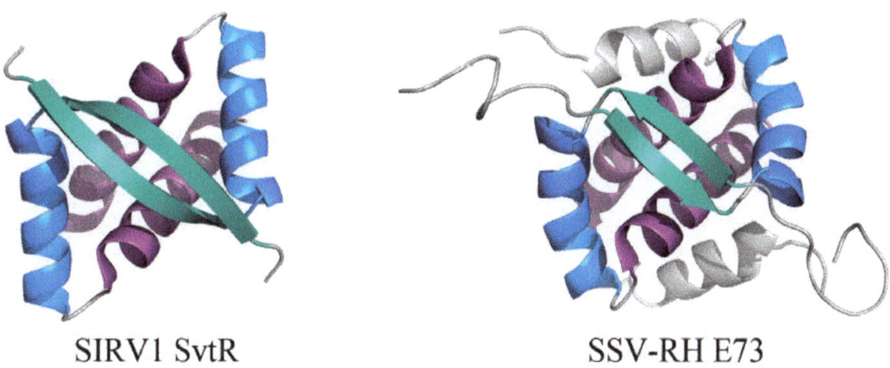

2.3. Structurally Characterized Enzymes of Archaeal Viruses

To date, there have been five structurally characterized proteins that are thought to serve an enzymatic role in the context of a viral infection. Four out of five structurally characterized enzymes from archaeal viruses are involved in nucleic acid metabolism. These include crystal structures of two unrelated nucleases [34,35] as well as the SSV1 viral integrase (SSV1Int) [36,37] and a Rep protein (Figure 6) [38]. In addition, structural studies have also identified a putative gylcosyltransferase (Figure 7) [39]. Each of these enzymes is suggested to have a different evolutionary origin, highlighting the putative complexity and mosaicity of archaeal viral genomes.

Putative and characterized nucleases include SSV-RH D212 (*Fuselloviridae*) and AFV1 orf157 (*Lipothrixviridae*), respectively. SSV-RH D212 adopts the PD-D/EXK nuclease superfamily fold, however its activity remains uncharacterized and is therefore only a putative nuclease. This protein most closely resembles an archaeal Holliday junction resolvase. The full-length version is not active in the traditional Holliday junction cleavage assay and a clipped version only shows very low levels of metal-dependent nuclease activity, suggesting that the DNA binding surfaces are distinct between SSV-RH D212 and its closest homologs (Figure 6a) [35]. AFV1 orf157 very distantly resembles a trimmed down version of the two-layer sandwich fold of HIV1-integrase from the phosphonucleotidyl superfamily, which successfully guided experiments toward the functional characterization of orf157 as a nuclease (Figure 6b) [34].

Figure 6. Fold Topologies from archaeal viral enzymes involved in nucleic acid metabolism. **(a)** SSV-RH D212 (left, PDB ID 2W8M [35]) with a fold that is similar to those of members belonging to the PD-D/EXK nuclease superfamily. An archaeal member of this family, HJC Holliday junction resolving enzyme from *Sulfolobus solfataricus* (right, PDB ID 1HH1 [40]) aligns quite well with a single D212 subunit (middle), however the dimer interfaces are substantially different for the two. Each subunit within the dimer is colored differently. **(b)** AFV1 orf157 (left, PDB ID 3II2 [34]) very distantly resembles the two-layer sandwich fold of the phosphonucleotidyl superfamily to which HIV1-integrase belongs (right, PDB ID 1ITG [41]), however it only has one layer of α-helices and not two. A catalytic residue, Glu86, is colored magenta. **(c)** SSV1 integrase (left, PDB ID 3VCF [36]) shares a conserved fold and catalytic residues with members belonging to the tyrosine recombinase family (highlighted in magenta), including the catalytic pentad (cluster of amino acids in the center of the protein) and the tyrosine nucleophile (distal to the active site). SIRV1 orf119 (right, PDB ID 2X3G [38]) has been identified as a novel type of Rep protein with the conserved fold of the superfamily II rep protein group. It shares conserved features with members of this group, including a conserved catalytic tyrosine (highlighted in magenta). Each subunit for the Rep protein dimer is colored differently.

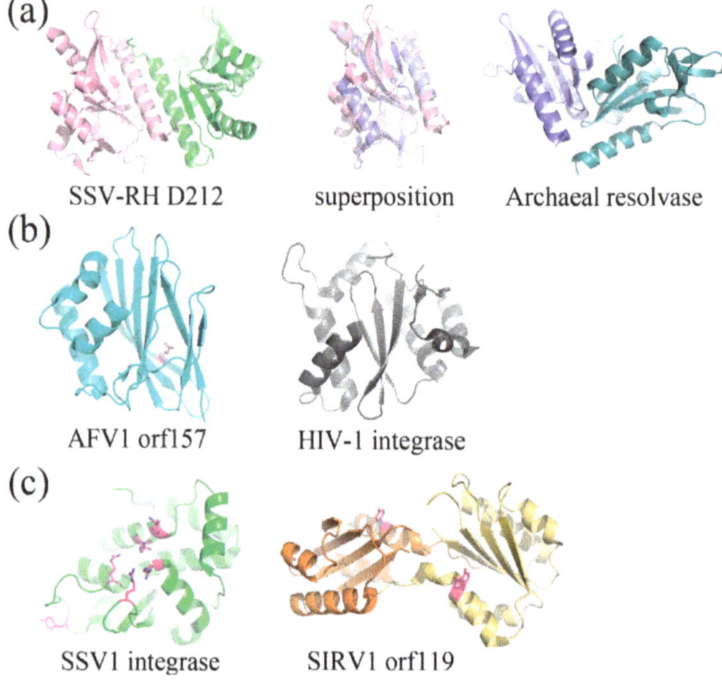

The structurally and functionally characterized viral integrase from SSV1 (SSV1Int, *Fuselloviridae*) is a member of the tyrosine recombinase superfamily (Figure 6c). Interestingly, the biochemical [42] and structural analyses suggests that during strand exchange and cleavage, the enzyme assembles its active site in *trans*, which is consistent with mechanisms utilized by eukaryotic and not bacterial tyrosine recombinases [36,37]. SIRV1 orf119 (*Rudiviridae*) adopts a fold that is similar to superfamily II of Rep proteins, involved in the initiation of genomic replication (Figure 6c) [38]. Orf119 implements a novel "flip-flop" mechanism that is unique to conventional rolling circle replication (RCR) or rolling hairpin replication (RHR). This flip-flop mechanism may be employed by Rep proteins in other viruses that have closed, linear dsDNA genomes, such as members of the eukaryotic viral family, *Poxviridae* and bacteriophage N15 [38].

The crystal structure of STIV A197 reveals a putative glycosyltransferase adopting the GT-A fold whose closest structural homologs mainly consist of eukaryotic glycosyltransferases (Figure 7) [39]. Because the major coat protein of STIV is known to be glycosylated [14] and the virus assembles in the cytosol [43], A197 is a strong candidate for this activity, however it presently remains uncharacterized.

Figure 7. The crystal structure of a putative glycosyltransferase from the archaeal virus, STIV. STIV A197 (PDB ID 2C0N [39]) adopts the glycosyltransferase GT-A family fold and contains many of the conserved catalytic residues seen throughout the family, including the DXD motif (located in a loop, colored magenta), a catalytic base, (Asp151, located in a helix, colored magenta), and a conserved manganese binding site (not depicted).

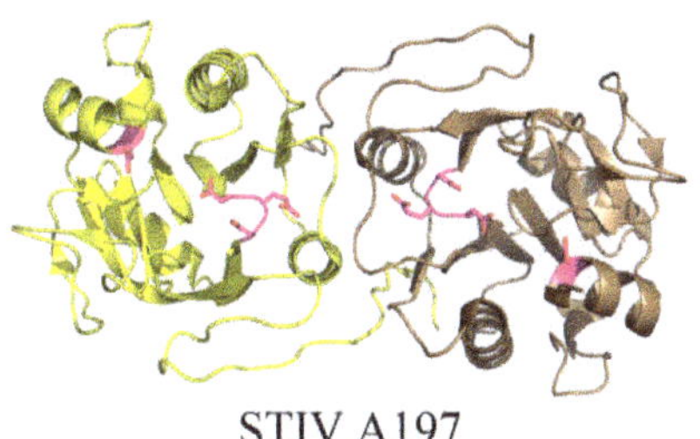

STIV A197

3. Conclusions

Structural characterization of proteins from archaeal viruses has lead to several conclusions. One conclusion is that certain archaeal viruses may be extant versions of ancient viral lineages. For example, the single β-barrel fold from SH1 [10] and double β-barrel fold from STIV [19] share homologs in viruses infecting Bacteria and Eukarya, suggesting that this fold was present in a common ancestor that existed before the domains of life emerged [19,20]. While viruses from this lineage share certain similarities, the majority of the genes are not obviously conserved by sequence homology. This observation supports the idea that archaeal viruses, like most viruses in general, act as mobile genetic elements and can potentially gain, lose, and transfer genetic material quite easily [44]. Collectively, the examples of structurally characterized archaeal viral enzymes also emphasize the fact that the genes comprising an archaeal virus genome are mosaic and may originate from different domains of life [8]. For example, the tyrosine recombinase from SSV1 employs a mechanism that parallels that utilized for the same enzyme in eukaryotes [36,37]). The fact that archaeal viruses (as well as viruses from Bacteria and Eukarya) code for and maintain a high abundance of DNA-binding motifs suggests that it is an indispensable feature that is present among all domains of life [8,27].

The majority of structurally characterized archaeal virus proteins infect hyperthermophilic archaea of the phylum Crenarchaeota. It is therefore not surprising that many structural studies of proteins from archaeal viruses report features of thermostability including compactness of fold [45], absence of cavities [46], a high number of salt bridges [46], a high ratio of charged to uncharged residues [45,47], short loops [45], oligomerization with (in some cases) extensive subunit interfaces [32,33] and the presence of disulfide bonds [28,36,37,46,48–50]. In contrast to the cellular proteins of mesophilic organisms, which do not generally utilize disulfide bonds, there is strong genomic [28,48,51–53] and metagenomic [5] evidence for the use of stabilizing disulfides in hyperthermophiles and their viruses.

Currently, there exist a relatively small number of archaeal viral protein structures compared to those in the bacterial and eukaryotic viral domains. Archaeal viruses therefore remain the largest group of unexplored territory in the realm of protein structural and functional characterization. Given the vast sequence diversity of archaeal viral proteins and lack of identifiable protein homologs, it is tempting to speculate that fold novelty would be commonly observed for structures of these proteins. While there are several structures of archaeal viral proteins with novel folds [47,50,54,55], the majority that have been characterized thus far share structural homology to proteins of known function. However, it is likely that the small subset of archaeal viral protein structures biases our current perspective, and that as more of these proteins are structurally characterized, fold novelty may become more commonly observed from viruses infecting this domain of life. Therefore, in addition to the examples discussed herein of fold conservation masked by sequence diversity, future examples of fold novelty may also explain the higher than usual level of unidentifiable genes in archaeal virus genomes.

Structural studies of archaeal viral proteins have facilitated the investigation of distant evolutionary relationships and the identification of functionally characterized structural homologs. High levels of sequence divergence have hindered the ability to rely on sequence for functional annotation, and as a consequence, structural characterization has become invaluable as a tool for archaeal viral protein characterization. The vast amount of sequence space covered by archaeal viruses may be a function of evolutionary distance and selective pressures (such as high temperature) imposed on the virus to maintain viability in the context of both the hyperthermophilic environment and the host. As proven through structural annotation, such sequences often mask a conserved protein fold, however examples of fold novelty also exist and most likely contribute to the high levels of sequence diversity observed among archaeal viral genes.

Acknowledgments

This work has been supported by National Science Foundation Grants DEB-0936178, EF-080220, and MCB-0920312 as well as National Aeronautics and Space Administration grant NNA-08CN85A.

References

1. Meile, L.; Jenal, U.; Studer, D.; Jordan, M.; Leisinger, T. Characterization of psi-m1, a virulent phage of methanobacterium-thermoautotrophicum marburg. *Arch. Microbiol.* **1989**, *152*, 105–110.
2. Pfister, P.; Wesserfallen, A.; Stettler, R.; Leisinger, T. Molecular analysis of methanobacterium phage psi m2. *Mol. Microbiol.* **1998**, *30*, 233–244.
3. Brochier-Armanet, C.; Boussau, B.; Gribaldo, S.; Forterre, P. Mesophilic crenarchaeota: Proposal for a third archaeal phylum, the thaumarchaeota. *Nat. Rev. Microbiol.* **2008**, *6*, 245–252.
4. Williamson, M. *How Proteins Work*; Garland Publishing Inc: New York, NY, **2011**; p. 464.
5. Lawrence, C.M.; Menon, S.; Eilers, B.J.; Bothner, B.; Khayat, R.; Douglas, T.; Young, M.J. Structural and functional studies of archaeal viruses. *J. Biol. Chem.* **2009**, *284*, 12599–12603.
6. Prangishvili, D. Evolutionary insights from studies on viruses of hyperthermophilic archaea. *Res. Microbiol.* **2003**, *154*, 289–294.
7. Le Romancer, M.; Gaillard, M.; Geslin, C.; Prieur, D. Viruses in extreme environments. *Rev. Environ. Sci. Biotechnol.* **2007**, *6*, 17–31.
8. Prangishvill, D.; Garrett, R.A.; Koonin, E.V. Evolutionary genomics of archaeal viruses: Unique viral genomes in the third domain of life. *Virus Res.* **2006**, *117*, 52–67.
9. Prangishvili, D.; Forterre, P.; Garrett, R.A. Viruses of the archaea: A unifying view. *Nat. Rev. Microbiol.* **2006**, *4*, 837–848.
10. Jaalinoja, H.T.; Roine, E.; Laurinmaki, P.; Kivela, H.M.; Bamford, D.H.; Butcher, S.J. Structure and host-cell interaction of sh1, a membrane-containing, halophilic euryarchaeal virus. *Proc. Natl. Acad. Sci. USA* **2008**, *105*, 8008–8013.
11. Pietila, M.K.; Laurinmaki, P.; Russell, D.A.; Ko, C.C.; Jacobs-Sera, D.; Butcher, S.J.; Bamford, D.H.; Hendrix, R.W. Insights into head-tailed viruses infecting extremely halophilic archaea. *J. Virol.* **2013**.
12. Porter, K.; Kukkaro, P.; Bamford, J.K.; Bath, C.; Kivela, H.M.; Dyall-Smith, M.L.; Bamford, D.H. Sh1: A novel, spherical halovirus isolated from an australian hypersaline lake. *Virology* **2005**, *335*, 22–33.
13. Pina, M.; Bize, A.; Forterre, P.; Prangishvili, D. The archeoviruses. *FEMS Microbiol. Rev.* **2011**, *35*, 1035–1054.
14. Maaty, W.S.; Ortmann, A.C.; Dlakic, M.; Schulstad, K.; Hilmer, J.K.; Liepold, L.; Weidenheft, B.; Khayat, R.; Douglas, T.; Young, M.J.; *et al.* Characterization of the archaeal thermophile sulfolobus turreted icosahedral virus validates an evolutionary link among double-stranded DNA viruses from all domains of life. *J. Virol.* **2006**, *80*, 7625–7635.
15. Bettstetter, M.; Peng, X.; Garrett, R.A.; Prangishvili, D. Afv1, a novel virus infecting hyperthermophilic archaea of the genus acidianus. *Virology* **2003**, *315*, 68–79.
16. Zillig, W.; Kletzin, A.; Schleper, C.; Holz, I.; Janekovic, D.; Hain, J.; Lanzendorfer, M.; Kristjansson, J.K. Screening for sulfolobales, their plasmids and their viruses in icelandic solfataras. *Systematic and Appl. Microbiol.* **1994**, *16*, 609–628.

17. Palm, P.; Schleper, C.; Grampp, B.; Yeats, S.; Mcwilliam, P.; Reiter, W.D.; Zillig, W. Complete nucleotide-sequence of the virus ssv1 of the archaebacterium sulfolobus-shibatae. *Virology* **1991**, *185*, 242–250.
18. Richardson, J.S. The anatomy and taxonomy of protein structure. *Adv. Protein. Chem.* **1981**, *34*, 167–339.
19. Khayat, R.; Tang, L.; Larson, E.T.; Lawrence, C.M.; Young, M.; Johnson, J.E. Structure of an archaeal virus capsid protein reveals a common ancestry to eukaryotic and bacterial viruses. *Proc. Natl. Acad. Sci. USA* **2005**, *102*, 18944–18949.
20. Benson, S.D.; Bamford, J.K.; Bamford, D.H.; Burnett, R.M. Viral evolution revealed by bacteriophage prd1 and human adenovirus coat protein structures. *Cell* **1999**, *98*, 825–833.
21. Nandhagopal, N.; Simpson, A.A.; Gurnon, J.R.; Yan, X.; Baker, T.S.; Graves, M.V.; Van Etten, J.L.; Rossmann, M.G. The structure and evolution of the major capsid protein of a large, lipid-containing DNA virus. *Proc. Natl. Acad. Sci. USA* **2002**, *99*, 14758–14763.
22. Stromsten, N.J.; Bamford, D.H.; Bamford, J.K. In vitro DNA packaging of prd1: A common mechanism for internal-membrane viruses. *J Mol. Biol.* **2005**, *348*, 617–629.
23. Happonen, L.J.; Redder, P.; Peng, X.; Reigstad, L.J.; Prangishvili, D.; Butcher, S.J. Familial relationships in hyperthermo- and acidophilic archaeal viruses. *J. Virol.* **2010**, *84*, 4747–4754.
24. San Martin, C.; Huiskonen, J.T.; Bamford, J.K.; Butcher, S.J.; Fuller, S.D.; Bamford, D.H.; Burnett, R.M. Minor proteins, mobile arms and membrane-capsid interactions in the bacteriophage prd1 capsid. *Nat. Struct. Biol.* **2002**, *9*, 756–763.
25. Goulet, A.; Blangy, S.; Redder, P.; Prangishvili, D.; Felisberto-Rodrigues, C.; Forterre, P.; Campanacci, V.; Cambillau, C. Acidianus filamentous virus 1 coat proteins display a helical fold spanning the filamentous archaeal viruses lineage. *Proc. Natl. Acad. Sci. USA* **2009**, *106*, 21155–21160.
26. Namba, K.; Stubbs, G. Structure of tobacco mosaic virus at 3.6 a resolution: Implications for assembly. *Science* **1986**, *231*, 1401–1406.
27. Aravind, L.; Anantharaman, V.; Balaji, S.; Babu, M.M.; Iyer, L.M. The many faces of the helix-turn-helix domain: Transcription regulation and beyond. *FEMS Microbiol. Rev.* **2005**, *29*, 231–262.
28. Larson, E.T.; Eilers, B.; Menon, S.; Reiter, D.; Ortmann, A.; Young, M.J.; Lawrence, C.M. A winged-helix protein from sulfolobus turreted icosahedral virus points toward stabilizing disulfide bonds in the intracellular proteins of a hyperthermophilic virus. *Virology* **2007**, *368*, 249–261.
29. Kraft, P.; Oeckinghaus, A.; Kummel, D.; Gauss, G.H.; Gilmore, J.; Wiedenheft, B.; Young, M.; Lawrence, C.M. Crystal structure of f-93 from sulfolobus spindle-shaped virus 1, a winged-helix DNA binding protein. *J. Virol.* **2004**, *78*, 11544–11550.
30. Szymczyna, B.R.; Taurog, R.E.; Young, M.J.; Snyder, J.C.; Johnson, J.E.; Williamson, J.R. Synergy of nmr, computation, and x-ray crystallography for structural biology. *Structure* **2009**, *17*, 499–507.
31. Schreiter, E.R.; Drennan, C.L. Ribbon-helix-helix transcription factors: Variations on a theme. *Nat. Rev. Microbiol.* **2007**, *5*, 710–720.

32. Schlenker, C.; Goel, A.; Tripet, B.P.; Menon, S.; Willi, T.; Dlakic, M.; Young, M.J.; Lawrence, C.M.; Copie, V. Structural studies of e73 from a hyperthermophilic archaeal virus identify the "rh3" domain, an elaborated ribbon-helix-helix motif involved in DNA recognition. *Biochemistry* **2012**, *51*, 2899–2910.
33. Guilliere, F.; Peixeiro, N.; Kessler, A.; Raynal, B.; Desnoues, N.; Keller, J.; Delepierre, M.; Prangishvili, D.; Sezonov, G.; Guijarro, J.I. Structure, function, and targets of the transcriptional regulator svtr from the hyperthermophilic archaeal virus sirv1. *J. Biol. Chem.* **2009**, *284*, 22222–22237.
34. Goulet, A.; Pina, M.; Redder, P.; Prangishvili, D.; Vera, L.; Lichiere, J.; Leulliot, N.; van Tilbeurgh, H.; Ortiz-Lombardia, M.; Campanacci, V.; *et al.* Orf157 from the archaeal virus acidianus filamentous virus 1 defines a new class of nuclease. *J. Virol.* **2010**, *84*, 5025–5031.
35. Menon, S.K.; Eilers, B.J.; Young, M.J.; Lawrence, C.M. The crystal structure of d212 from sulfolobus spindle-shaped virus ragged hills reveals a new member of the pd-(d/e)xk nuclease superfamily. *J. Virol.* **2010**, *84*, 5890–5897.
36. Eilers, B.J.; Young, M.J.; Lawrence, C.M. The structure of an archaeal viral integrase reveals an evolutionarily conserved catalytic core yet supports a mechanism of DNA cleavage in trans. *J. Virol.* **2012**, *86*, 8309–8313.
37. Zhan, Z.; Ouyang, S.; Liang, W.; Zhang, Z.; Liu, Z.J.; Huang, L. Structural and functional characterization of the c-terminal catalytic domain of ssv1 integrase. *Acta Crystallogr. D Biol. Crystallogr.* **2012**, *68*, 659–670.
38. Oke, M.; Kerou, M.; Liu, H.T.; Peng, X.; Garrett, R.A.; Prangishvili, D.; Naismith, J.H.; White, M.F. A dimeric rep protein initiates replication of a linear archaeal virus genome: Implications for the rep mechanism and viral replication. *J. Virol.* **2011**, *85*, 925–931.
39. Larson, E.T.; Reiter, D.; Young, M.; Lawrence, C.M. Structure of a197 from sulfolobus turreted icosahedral virus: A crenarchaeal viral glycosyltransferase exhibiting the gt-a fold. *J. Virol.* **2006**, *80*, 7636–7644.
40. Bond, C.S.; Kvaratskhelia, M.; Richard, D.; White, M.F.; Hunter, W.N. Structure of hjc, a holliday junction resolvase, from sulfolobus solfataricus. *Proc. Natl. Acad. Sci. USA* **2001**, *98*, 5509–5514.
41. Dyda, F.; Hickman, A.B.; Jenkins, T.M.; Engelman, A.; Craigie, R.; Davies, D.R. Crystal structure of the catalytic domain of HIV-1 integrase: Similarity to other polynucleotidyl transferases. *Science* **1994**, *266*, 1981–1986.
42. Letzelter, C.; Duguet, M.; Serre, M.C. Mutational analysis of the archaeal tyrosine recombinase ssv1 integrase suggests a mechanism of DNA cleavage in trans. *J. Biol. Chem.* **2004**, *279*, 28936–28944.
43. Fu, C.Y.; Wang, K.; Gan, L.; Lanman, J.; Khayat, R.; Young, M.J.; Jensen, G.J.; Doerschuk, P.C.; Johnson, J.E. *In vivo* assembly of an archaeal virus studied with whole-cell electron cryotomography. *Structure* **2010**, *18*, 1579–1586.

44. Viruses, Plasmids, and Transposable Genetic Elements. In *Molecular Biology of the Cell*, 3rd edition; Alberts B., Johnson, A., Lewis J., Raff,.M., Roberts, K., Walter, P., Eds.; Garland Science: New York, NY, USA, 1994;. Available online: http://www.ncbi.nlm.nih.gov/books/NBK28286/ (accessed on 24 November 2012).
45. Goulet, A.; Spinelli, S.; Blangy, S.; van Tilbeurgh, H.; Leulliot, N.; Basta, T.; Prangishvili, D.; Cambillau, C.; Campanacci, V. The thermo- and acido-stable orf-99 from the archaeal virus afv1. *Protein Sci.* **2009**, *18*, 1316–1320.
46. Felisberto-Rodrigues, C.; Blangy, S.; Goulet, A.; Vestergaard, G.; Cambillau, C.; Garrett, R.A.; Ortiz-Lombardia, M. Crystal structure of atv (orf273), a new fold for a thermo- and acido-stable protein from the acidianus two-tailed virus. *Plos One* **2012**, doi:10.1371/journal.pone.0045847.
47. Goulet, A.; Spinelli, S.; Blangy, S.; van Tilbeurgh, H.; Leulliot, N.; Basta, T.; Prangishvili, D.; Cambillau, C.; Campanacci, V. The crystal structure of orf14 from sulfolobus islandicus filamentous virus. *Proteins* **2009**, *76*, 1020–1022.
48. Menon, S.K.; Maaty, W.S.; Corn, G.J.; Kwok, S.C.; Eilers, B.J.; Kraft, P.; Gillitzer, E.; Young, M.J.; Bothner, B.; Lawrence, C.M. Cysteine usage in sulfolobus spindle-shaped virus 1 and extension to hyperthermophilic viruses in general. *Virology* **2008**, *376*, 270–278.
49. Keller, J.; Leulliot, N.; Collinet, B.; Campanacci, V.; Cambillau, C.; Pranghisvilli, D.; van Tilbeurgh, H. Crystal structure of afv1–102, a protein from the acidianus filamentous virus 1. *Protein Sci.* **2009**, *18*, 845–849.
50. Larson, E.T.; Eilers, B.J.; Reiter, D.; Ortmann, A.C.; Young, M.J.; Lawrence, C.M. A new DNA binding protein highly conserved in diverse crenarchaeal viruses. *Virology* **2007**, *363*, 387–396.
51. Beeby, M.; O'Connor, B.D.; Ryttersgaard, C.; Boutz, D.R.; Perry, L.J.; Yeates, T.O. The genomics of disulfide bonding and protein stabilization in thermophiles. *PLoS Biol* **2005**, *3*, doi:10.1371/journal.pbio.0030309.
52. Mallick, P.; Boutz, D.R.; Eisenberg, D.; Yeates, T.O. Genomic evidence that the intracellular proteins of archaeal microbes contain disulfide bonds. *Proc. Natl Acad Sci USA* **2002**, *99*, 9679–9684.
53. Jorda, J.; Yeates, T.O. Widespread disulfide bonding in proteins from thermophilic archaea. *Archaea* **2011**, *2011*, 409156.
54. Keller, J.; Leulliot, N.; Cambillau, C.; Campanacci, V.; Porciero, S.; Prangishvili, D.; Forterre, P.; Cortez, D.; Quevillon-Cheruel, S.; van Tilbeurgh, H. Crystal structure of afv3–109, a highly conserved protein from crenarchaeal viruses. *Virol. J.* **2007**, *4*, doi:10.1186/1743-422X-4-12.
55. Oke, M.; Carter, L.G.; Johnson, K.A.; Liu, H.; McMahon, S.A.; Yan, X.; Kerou, M.; Weikart, N.D.; Kadi, N.; Sheikh, M.A.; *et al.* The scottish structural proteomics facility: Targets, methods and outputs. *J. Struct. Funct .Genomics* **2010**, *11*, 167–180.

Domain Structures and Inter-Domain Interactions Defining the Holoenzyme Architecture of Archaeal D-Family DNA Polymerase

Ikuo Matsui, Eriko Matsui, Kazuhiko Yamasaki and Hideshi Yokoyama

Abstract: Archaea-specific D-family DNA polymerase (PolD) forms a dimeric heterodimer consisting of two large polymerase subunits and two small exonuclease subunits. According to the protein-protein interactions identified among the domains of large and small subunits of PolD, a symmetrical model for the domain topology of the PolD holoenzyme is proposed. The experimental evidence supports various aspects of the model. The conserved amphipathic nature of the *N*-terminal putative α-helix of the large subunit plays a key role in the homodimeric assembly and the self-cyclization of the large subunit and is deeply involved in the archaeal PolD stability and activity. We also discuss the evolutional transformation from archaeal D-family to eukaryotic B-family polymerase on the basis of the structural information.

Reprinted from *Life*. Cite as: Matsui, I.; Matsui, E.; Yamasaki, K.; Yokoyama, H. Domain Structures and Inter-Domain Interactions Defining the Holoenzyme Architecture of Archaeal D-Family DNA Polymerase. *Life* **2013**, *3*, 375-385.

1. Introduction

Replicative DNA polymerases (DNA Pols) are divided into archaeal-eukaryotic and bacterial types that appear not to be homologous to each other [1,2]. All archaea and eukaryotes, as well as viruses, encode B-family polymerases that are responsible for genome replication [3,4], while bacterial replication is performed by C-family polymerases that are not found in archaea or eukaryotes. All eukaryotes possess four paralogous B-family polymerases, Pols α, δ, ε, and ξ, involved in DNA replication and repair [5,6]. In addition, Euryarchaeota, a subdomain of archaea, have a distinct type of polymerase, D-family DNA polymerases (PolDs) unrelated to B- or C-family polymerases [7–9]. Euryarchaeal DNA-replication is not understood in as much detail as bacteria or eukaryotic replication, whereas all methanogens, key players in greenhouse-gas and bio-fuel production, probably including formation of deep-sea methane hydrate, belong to Euryarchaeota [10].

PolDs, originally discovered from Euryarchaeota [7], were also identified in the archaeal phyla diverged early from the major archaeal phyla Crenearchaeota and Euryarchaeota [11–13]. From recent analysis of the evolution of DNA replication apparatus, it is likely that the last common ancestor of archaea had two DNA polymerases of the B-family and one of the D-family [14]. Recent data demonstrated that PolD is an essential DNA polymerase [15]. Indeed, PolD from *Pyrococcus abyssi* (PabPolD) plays an important role in chromosomal replication, together with the B-family DNA polymerase, and is capable of RNA primer elongation [9,16]. We also confirmed that PolD from *P. horikoshii* (PhoPolD) uses RNA primer for DNA synthesis even to a lesser extent than that of the DNA primer, whereas PolB uses only the DNA primer. This strongly suggested that PolD is a key enzyme responsible for lagging strand synthesis. By analogy with the eukaryotic mechanism, it was proposed that PolD completes Okazaki fragments and lagging strand synthesis [9].

In contrast to the biochemical properties of PolD, knowledge about the molecular structure of PolD is still scarce. Therefore, in the present article, we review the molecular structure and domain topology of both of the subunits. Especially, we point out that the amphipathic nature of the *N*-terminal ~50 residues of the large subunit is conserved completely in Euryarchaeota and Korarchaeota, and that it possesses a possible function to modify the efficiency of the large-subunit folding and the assembly of the PolD holoenzyme.

2. Background

2.1. Functional Information for PolD

PhoPolD was proposed to be a dimeric heterodimer (molecular weight: 420 kDa) consisting of two small subunits (DP1s) (PH0123, NCBI accession number NP_142131; 622 amino acids), and two large subunits (DP2s) (PH0121, NCBI accession number NP_142130; 1434 amino acids) [8]. DP2 is the catalytic subunit of DNA polymerase [8], while DP1 is the catalytic subunit of 3'–5' exonuclease. The *C*-terminal domain of DP1 contains five Mre11-like nuclease motifs [17]. Mre11 is a nuclease involved in double-stranded DNA break repair and belongs to the calcineurin-like phosphoesterase superfamily [18]. The second subunits of eukaryotic B-family DNA polymerases (Pols α, β, ε, and ξ) also show similarities to the Mre11-like exonuclease region, although the catalytic residues of the second subunits are replaced by non-catalytic residues [18–20]. The second subunits of the eukaryotic B-family DNA polymerases play a specific role in regulating the first catalytic subunits. Interestingly, it was reported that PolD demonstrated strong DNA polymerase and 3'–5' exonuclease activities when the two subunits were mixed or co-expressed, although each individual subunit demonstrated only a weak activity [17,21]. The domain containing the *N*-terminal 300 residues of DP2 [abbreviated as DP2(1-300); similar descriptions for other fragments will be used in the present manuscript] was reported to be essential for the folding of PolD and is probably the oligomerization domain [22].

The domain topology essential for complex formation of PolD was characterized with the yeast two-hybrid (Y2H) and surface plasmon resonance (SPR) assays [23]. DP2(1-100) interacts with another region in the same subunit, DP2(792-1163), containing the catalytic residues for DNA polymerization, Asp1122 and Asp1124, to form a ring-shaped structure. A putative third acidic-residue involved in the catalytic reaction remains unidentified. Catalytic DP2(792-1163) also interacts with the inter-subunit domain, DP1(1-200). It is noticeable that the polypeptide DP2(792-1163) was expressed as an insoluble form in *E. coli* probably due to its hydrophobicity [23]. As the molecular mechanisms of the protein folding for DP2 were unknown, refolding of the hydrophobic domain DP2(792-1163) harbouring catalytic Asp1122 and Asp1124 residues was investigated by mixing with an equimolar amount of the *N*-terminal domain of DP2 in 3 M urea, and successive stepwise dialysis to remove urea [24]. Before and after dialysis, sampling was carried out and refolding efficiency was examined by SDS-PAGE after removing the precipitate. These results suggest that DP2(1-50) is a minimum and essential element for the refolding of DP2(792-1163) to maintain the complex in soluble form. Furthermore, the complexes DP2(1-100)DP2(792-1163) and DP2(1-300)DP2(792-1163) were recovered completely without any precipitate and were purified completely by gel filtration, indicating that the *N*-terminal 100 amino-acid region, DP2(1-100), is sufficient to refold the catalytic DP2(792-1163) domain as a soluble complex form. According to each band intensity and molecular weight, the molar ratio of DP2(1-100) to DP2(792-1163) and DP2(1-300) to DP2(792-1163) was estimated to be 1 to 1. The molecular weight and the structural uniformity of the purified complex, DP2(1-100)DP2(792-1163) were confirmed by gel filtration [24]. The protein was eluted as a sharp peak and its molecular mass was estimated to be 72 kDa, indicating a uniform dimer, $[DP2(1\text{-}100)DP2(792\text{-}1163)]_2$. Then, the thermostability of the purified dimer, $[DP2(1\text{-}100)DP2(792\text{-}1163)]_2$, was analyzed using a circular dichroism (CD) spectrometer and a fluorometric method between 20 °C and 85 °C [24]. These results indicate that the dimer $[DP2(1\text{-}100)DP2(792\text{-}1163)]_2$ is stable at 85 °C with no dissociation to monomers or drastic conformational changes. The ~50 *N*-terminal residues play essential roles in the dimeric assembly and the self-cyclization of the DP2 subunit.

Using surface plasmon resonance (SPR) assay, we measured the dissociation constant (K_D) of the DP2(1-300) domain against DNA [24]. The K_D value of the DP2(1-300) domain against 3'-recess DNA is moderate ($K_D = 1.3 \times 10^{-6}$ (M)). Since the K_D value of the whole PolD molecule to 3'-recess DNA was determined to be 2.7×10^{-10} (M), the DP2(1-300) domain seems to play a supplementary role in the DNA-binding mechanism of the dimeric heterodimer PolD. The details of how the domain interacts with DNA need further investigation.

The *C*-terminal domain of DP2, DP2(1164-1434), contains cysteine-cluster (^{1289}**C**VK**C**NTKFR RPPLDGK**C**PI**C**^{1308}; cysteine residues that are candidates for zinc-binding residues are shown in bold), associates with an inter-subunit domain, DP1(1-200) [23].

2.2. Structural Information for Isolated Domains and Fragments

DP2(1-300) has been reported to be essential for the folding of PolD and is probably the oligomerization domain, because the deletion of this part from the PolD holoenzyme caused complete loss of the specific bands on SDS-PAGE analysis of the recombinant-cell extract after heating at 85 °C for 30 min as reported previously [22]. Since the molecular mechanisms of the protein folding and biochemical function of the DP2(1-300) domain were unknown, the crystal structure of DP2(1-300) was determined at 2.2 Å resolution according to the multiwavelength anomalous dispersion (MAD) method [24]. The refined model contains the 48–291 region, although ~50 *N*-terminal residues are disordered. Hereafter, the crystal structure is designated as DP2(48-291). DP2(48-291) has an ellipsoidal shape, and its dimensions are approximately $45 \times 30 \times 30$ Å^3. DP2(48-291) mainly constitutes an α-helical structure containing nine α helices and three β strands. Three β strands (β1, β2, and β3) form a twisted β-sheet at the center of one face, and the β sheets are surrounded by α4, α5, and α8 helices. The β3 strand and α8 are connected by a 22-residue-long kinked loop, which is located on the surface and is in the vicinity of the β-sheet and α6 helices. The relatively long α8 helix is kinked at Ala-260.

The coordinates of DP2(48-291) were submitted to the web server SSM (Secondary Structure Matching program) to search for other proteins with a similar folding pattern in the PDB. Due to its low structural similarity with other proteins, except for archaeal DP2 (highest Z score = 1.4), the folding of DP2(48-291) was considered to be novel.

Furthermore, the NMR analysis revealed that region DP1(1-72) contains a folded structure, although the succeeding DP1(73-200) is unfolded [25]. DP1(1-72) part of the domain has only 72 aa. The structure of DP1(1-72) was determined by multi-dimensional NMR methods (Figure 1). The revealed globular structure contains four α-helices and a very short two-stranded parallel β-sheet which was identified in a region connecting the α-helices. Searching the Protein Data Bank (PDB) by the DALI program [26] identified structures similar to DP1(1-72). The similar structures identified with high reliability scores include the *N*-terminal domains of the second subunits of the eukaryotic B-family DNA polymerase, *i.e.*, human Pols α and ε (Z-scores 7.2 and 7.1, respectively) [27] (Figure 1). The similar structures also include the δ subunit of the clamp loader γ complex of *E. coli* DNA polymerase III [28], and the domain II of *Thermotoga maritima* RuvB protein [29]. These are classified into AAA+ ATPases, which are chaperonine-like ATPases associated with a variety of cellular activities including DNA replication, recombination, proteolysis, and membrane fusion [30], in which the *C*-domain is similar to DP1(1-72).

The oligomeric state of DP1(1-72) in solution was elucidated by an analytical ultracentrifugation method [25]. Sedimentation equilibrium data in the concentration of 11–350 μM as a monomeric protein showed a slightly curved distribution in a radius^2 *vs.* ln(A_{280}) plot, yielding an average molecular weight of 12.6 kDa indicating an equilibrium between the monomeric and dimeric states (8.5 and 17.1 kDa, respectively). Dimerization is likely to be achieved by hydrophobic interactions as well as electrostatic attractions, although the dimerization mode is not necessarily fixed but is probably rather dynamic. This expands knowledge regarding the domain topology of the holoenzyme. Although this region is connected to the *C*-terminal unstructured portion, a long ~130 amino-acid region, its position in the holoenzyme is probably fixed after association with the remaining part of DP1 and/or DP2. Since DP1(1–200) interacts with PCNA [23], it is possible that the dimer-monomer equilibrium may be influenced by such accessory components, and that the domain presumably possesses a function like a sensor.

The X-ray structure of the complex of the regulatory second subunit (p50) of human DNA Pol δ with the *N*-terminal domain ($p66_N$) of the regulatory third subunit was reported (PDB ID 3E0J) [30]. It was also suggested that the second subunit (p50) of human Pol δ lacks the region equivalent to the *N*-terminal domains of the second subunits of human Pol α and ε, and DP1(1-72), and that, instead, the third subunit (p66) supplies an equivalent domain, with a weak sequence similarity [31]. From the structure similarity in the *N*-terminal region and the prominent sequence similarity in the *C*-terminal exonuclease-like region [18,32,33], it is now evident that the DP1 subunit of archaeal PolDs and the second subunit of eukaryotic Pol α and ε are evolutionary related, although that of Pol δ is rather distant. It was recently reported that the first subunit of the eukaryotic Pol ε was derived from chimeric origins between archaeal B-family and D-family polymerases, where only the *C*-terminal zinc finger-like region of the DP2 subunit of PolD was incorporated into the polymerase subunit of Pol ε [14]. We reported that *P. horikoshii* DP1(1–200) interacts with a synthetic peptide corresponding to this zinc finger-like region, DP2(1290–1310) [23]. We suggested, therefore, that, during the evolutionary transformation from archaeal D-family to eukaryotic B-family polymerases the interacting pair of DP1 and the *C*-terminal region of DP2 is conserved, probably in order to maintain the holoenzyme structure [25].

Figure 1. NMR structures of DP1(1-72) of *P. horikoshii* PolD (left) and the *N*-terminal domains of the second subunits of human Pol α and ε (middle, right).

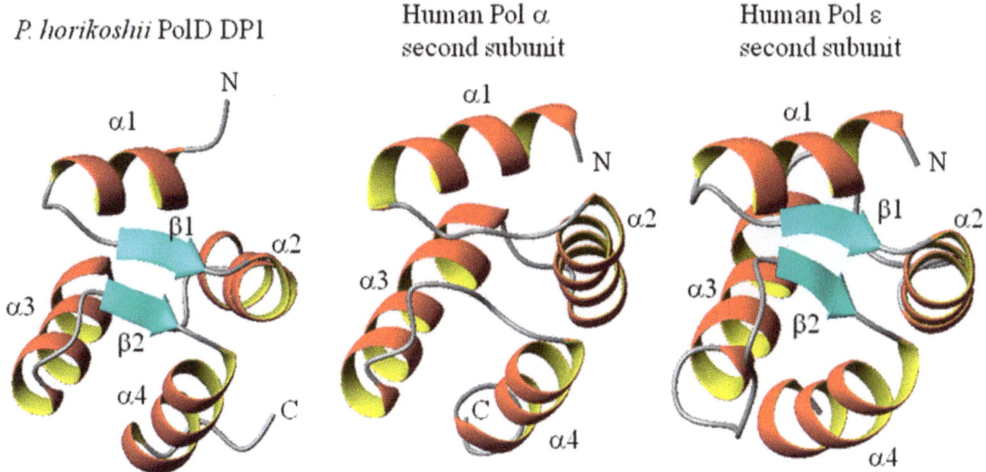

3. A Model for the PolD Holoenzyme

We reported the domain topology essential for complex formation and interaction with other proteins, which was characterized with the yeast two-hybrid (Y2H) and surface plasmon resonance (SPR) assays [23]. Refolding of the catalytic domain DP2(792-1163) was investigated by mixing with an equimolar amount of the *N*-terminal domain of DP2. The dimer, $[DP2(1\text{-}100)DP2(792\text{-}1163)]_2$ was reported to be heat stable and a uniform molecule [24]. The results of the refolding experiments suggest that the dimer, $[DP2(1\text{-}100)DP2(792\text{-}1163)]_2$ forms a central core of the dimeric large subunit, $(DP2)_2$, in which DP2(1-50) is an essential element for the refolding of DP2(792-1163) to maintain the complex in soluble form. On the basis of the results, here we propose a revised domain topology of the PolD holoenzyme, probably associated in a symmetric manner as shown in Figure 2. In the symmetric model, two small subunits (DP1) are associated with a dimeric large subunit, $(DP2)_2$ by multiple protein-protein interactions shown with arrows in orange (Figure 2b). It is noticeable that the two small subunits (DP1) are also placed in the anti-parallel direction from the central two-fold axis in Figure 2b.

Figure 2. Domain structure and topology of the PhoPolD holoenzyme. (**a**) Schematic maps of the domains for the small and large subunits. The mini-intein insertion site and the catalytic center for polymerization in the large subunit (L) are shown with arrows and vertical lines, respectively; (**b**) The domain topology of dimeric heterodimer. Each domain is colored in the same manner as (a). The ~50 *N*-terminal residues of the large subunit are depicted with dotted red circles. The red letters "N" in the dotted circles indicate the *N*-terminus of the large subunit. The interactions between each domain are shown with arrows in orange [23]. The replication factor, proliferating cell nuclear antigen is abbreviated as PCNA.

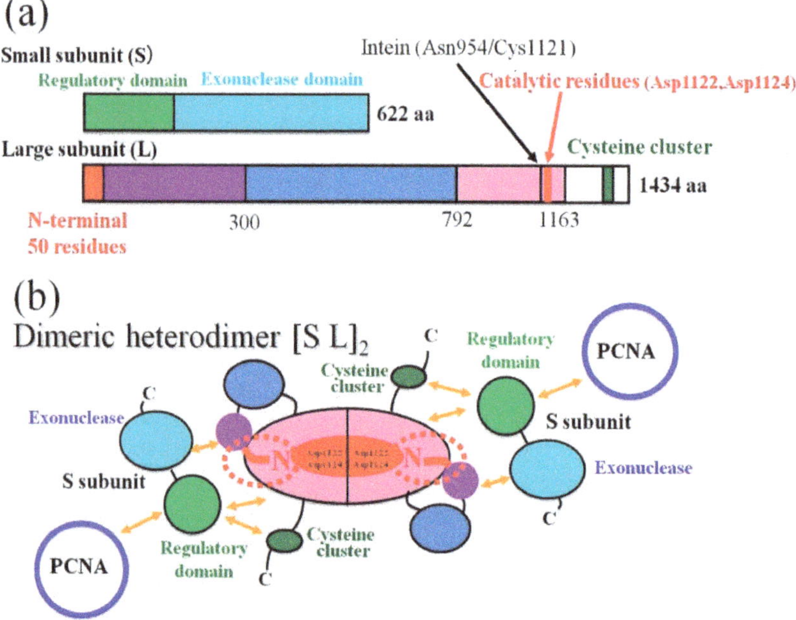

4. The Amphipathic Nature of the ~50 *N*-Terminal Residues Conserved Completely in PolD of Archaea

According to secondary-structure prediction using the amino acid sequence and the PSIPRED program [34], the region between residues 8 and 33 of DP2 is likely to form an α-helix. In order to estimate how the disordered ~50 *N*-terminal residues of DP2 form the dimeric assembly, we manually built a model of residues 1 to 47 in the crystal packing of DP2(48-291). The modeled α-helices were fitted well in the remaining space of the crystal packing for DP2(48-291) as shown in a stereo view of Figure 3, suggesting one possible conformation of the ~50 *N*-terminal residues after trapping in the crystal lattice. One DP2(48-291) molecule and the tentative model of DP2(1-47) are shown in red and blue, and symmetry-related molecules are shown in pink and cyan, respectively.

Figure 3. Crystal packing of DP2(48-291) with modeled α-helices DP2(1-47) is shown in stereo view. One DP2(48-291) molecule and the tentative model of DP2(1-47) are shown in red and blue, and symmetry-related molecules are shown in pink and cyan, respectively.

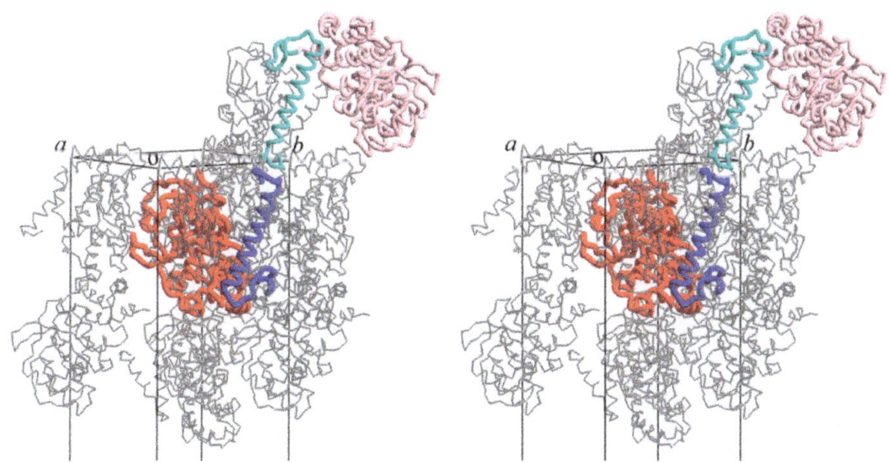

Since the disordered ~50 *N*-terminal residues was likely to form an α-helix with complex formation according to secondary-structure prediction and molecular modeling, helical wheel analysis of DP2(8-33) was carried out with the DNASIS-Mac ver 2.0 software. As shown in Figure 4A, half of the wheel is hydrophobic and covered with 11 hydrophobic residues (Met10, Tyr13, Phe14, Met16, Leu17, Ile21, Ala24, Tyr25, Ile27, Ala28, Ala31), and the other side is hydrophilic [24]. Since these hydrophobic residues are conserved well in all euryarchaeal DP2s as shown in Figure 4B, the α-helix DP2(8-33) should associate with hydrophobic domain DP2(792-1163) on the hydrophobic interface. We have already found three hydrophobic regions in the *C*-terminus of DP2(792-1163) [23]. One or more hydrophobic regions forming the catalytic center might associate with the amphipathic α-helix (8-33). Interestingly, it was recently reported that the mutant of (G)-PYF box motif located at one of the three hydrophobic regions forming the catalytic center of PabPolD from *P. abyssi* were rapidly degraded, suggesting the (G)-PYF box inside the hydrophobic domain DP2(792-1163) has a major role in PolD stability and in polymerase activity [35]. We have previously reported that the deletion of DP2(1-300) from the PolD holoenzyme made it unable to detect the PolD proteins from the recombinant-cell extract after heating, indicating the major role of DP2(1-300) in PolD stability [22]. We also reported that its ~50 *N*-terminal residues play a key role in PolD folding and stability. The (G)-PYF box motif might be a counterpart to assemble with the putative α-helix, DP2(8-33), although further work would be necessary to confirm the interaction. Since PolDs are essential in DNA replication and repair [15] and the amphipathic nature of the ~50 *N*-terminal residues is conserved completely in Euryarchaeota and Korarchaeota as shown in Figure 4B, mutations to modify the amphipathic nature might change the phenotypes of these archaea, especially in PolD stability and in polymerase activity. The *N*-terminus of DP2 seems to be a major target for the DNA replication control.

Figure 4. The amphipathic nature of the *N*-terminal extremity conserved well in the large subunits DP2 of archaea. (**a**) Helical wheel projection of DP2(8-33). Hydrophobic amino-acid residues are colored in red. A hydrophobic interface is indicated with a red vertical line; (**b**) Sequence alignment and the hydrophobic profile conserved in the ~50 *N*-terminal residues of archaeal DP2. The hydrophobic profile is emphasized with gray shading in the alignment. The eight hydrophobic residues in the conserved region between residues 8 and 33 of PhoDP2 are depicted on the upper side of the alignment. The starting point of the secondary element (α1 helix) of PhoDP2 is also indicated with an arrow. The figure was produced with EMBL-EBI tool ClustalW. The asterisks indicate identical residues, and the period and colon indicate similar residues among species. The sequences are from five archaeal species, *P. horikoshii* (*P. hori*, NCBI accession number: NP_142130), *Methanococcus jannaschii* (MC. jann, accession number: U67603-3), *Halobacterium* sp. NRC-1 (*Hal.* sp., accession number: AE005116-6), *Thermoplasma acidophilum* (T. acid, accession number: AL445063-36), and *Korarchaeum cryptofilum* (Kor. cryp, NCBI accession number: NC_010482).

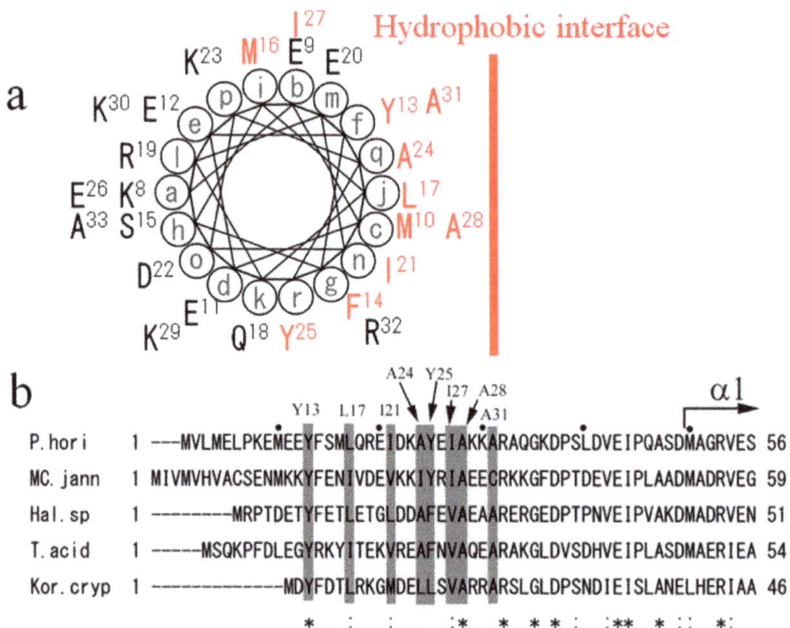

5. Conclusions

Archaea-specific D-family DNA polymerase (PolD) forms a tetramer consisting of two large polymerase subunits (DP2) and two small exonuclease subunits (DP1). PolDs, originally discovered from Euryarchaeota, were also identified in the putative phyla Nanoarchaeota, Thaumarchaeota (formally mesophillic Crenarchaeota), and Korarchaeota, which may have diverged early from the major archaeal phyla Crenearchaeota and Euryarchaeota.

Interestingly, the *C*-terminal part of DP1, eukaryotic Mre11-like nuclease domain, shows low but significant homology to the non-catalytic second subunit of eukaryotic B-family DNA polymerases (Pols α, δ, and ε). We reported that the *N*-terminal domain of *Pyrococcus horikoshii* DP1 interacts with a synthetic peptide corresponding to the *C*-terminal zinc finger-like region of the DP2 [23]. The NMR structure analysis of the *N*-terminal domain of DP1 suggested that the interacting pair of DP1 and the *C*-terminal zinc finger-like region of the DP2 are conserved during the evolutionary transformation from archaeal D-family to eukaryotic B-family polymerases, probably in order to maintain the holoenzyme structure [25].

We reported the *N*-terminal (1–300) domain structure determined by X-ray crystallography, although ~50 *N*-terminal residues were disordered [24]. The determined structure consists of nine α helices and three β strands. We also identified the DNA-binding ability of the domain by SPR measurement, suggesting that the structure shows a novel DNA-associating fold. Refolding of the catalytic domain for the large subunit (DP2) by mixing with its *N*-terminal domain of DP2 suggested that the disordered *N*-terminus (~50 residues) play a key role in self-cyclization and homodimeric assembly of DP2. According to the molecular structure of the *N*-terminal region of DP2 and the symmetrical topology model of the PolD holoenzyme, the amphipathic nature of the *N*-terminus interacting with the active center of DP2 might be a potent tool to control the archaeal PolD stability and polymerase activity.

Acknowledgments

We thank K. Hiramoto for technical support. This work was supported in part by a Grant-in-Aid for Science Research (18608005 to I. M.) from the Japan Society for the Promotion of Science.

Conflict of Interest

The authors declare no conflict of interest.

References

1. Leipe, D.D.; Aravind, L.; Koonin, E.V. Did DNA replication evolve twice independently? *Nucleic Acids Res.* **1999**, *27*, 3389–3401.
2. Bailey, S.; Wing, R.A.; Steitz, T.A. The structure of *T. aquaticus* DNA polymerase III is distinct from eukaryotic replicative DNA polymerases. *Cell* **2006**, *126*, 893–904.
3. Burgers, P.M.; Koonin, E.V.; Bruford, E.; Blanco, L.; Burtis, K.C.; Christman, M.F.; Copeland, W.C.; Friedberg, E.C.; Hanaoka, F.; Hinkle, D.C.; *et al.* Eukaryotic DNA polymerases: Proposal for a revised nomenclature. *J. Biol. Chem.* **2001**, *276*, 43487–43490.
4. Grabowski, B.; Kelman, Z. Archaeal DNA replication: Eukaryal proteins in a bacterial context. *Annu. Rev. Microbiol.* **2003**, *57*, 487–516.
5. Pavlov, Y.I.; Shcherbakova, P.V.; Rogozin, I.B. Roles of DNA polymerase in replication, repair, and recombination in eukaryotes. *Int. Rev. Cytol.* **2006**, *255*, 41–132.

6. Hubscher, U.; Maga, G.; Spadari, S. Eukaryotic DNA polymerases. *Annu. Rev. Biochem.* **2002**, *71*, 133–163.
7. Ishino, Y.; Komori, K.; Cann, I.K.; Koga, Y. A novel DNA polymerase family found in Archaea. *J. Bacteriol.* **1998**, *180*, 2232–2236.
8. Shen, Y.; Musti, K.; Hiramoto, M.; Kikuchi, H.; Kawarabayasi, Y.; Matsui, I. Invariant Asp-1122 and Asp-1124 are essential residues for polymerization catalysis of family D DNA polymerase from *Pyrococcus horikoshii*. *J. Biol. Chem.* **2001**, *276*, 27376–27383.
9. Henneke, G.; Flament, D.; Hübscher, U.; Querellou, J.; Raffin, J.P. The hyperthermophilic euryarchaeota *Pyrococcus abyssi* likely requires the two DNA polymerases D and B for DNA replication. *J. Mol. Biol.* **2005**, *350*, 53–64.
10. Kurr, M.; Huber, R.; Konig, H.; Jannasch, H.W.; Fricke, H.; Trincone, A.; Kristjansson, J.K.; Stetter, K.O. *Methanopyrus kandleri*, gen. and sp. nov. represents a novel group of hyperthermophilic methanogen, growing at 110 °C. *Arch. Microbiol.* **1991**, *156*, 239–247.
11. Waters, E.; Hohn, M.J.; Ahel, I.; Graham, D.E.; Adams, M.D.; Barnstead, M.; Beeson, K.Y.; Bibbs, L.; Bolanos, R.; Keller, M.; *et al.* The genome of *Nanoarchaeum equitans*: Insights into early archaeal evolution and derived parasitism. *Proc. Natl. Acad. Sci. USA* **2003**, *100*, 12984–12988.
12. Brochier-Armanet, C.; Boussau, B.; Gribaldo, S.; Forterre, P. Mesophillic Crenarchaeota: Proposal for a third archaeal phylum, the Thaumarchaeota. *Nat. Rev. Microbiol.* **2008**, *6*, 245–252.
13. Elkins, J.G.; Podar, M.; Graham, D.E.; Makarova, K.S.; Wolf, Y.; Randau, L.; Hedlund, B.P.; Brochier-Armanet, C.; Kunin, V.; Anderson, I.; *et al.* A korarchaeal genome reveals insights into the evolution of the Archaea. *Proc. Natl. Acad. Sci. USA* **2008**, *105*, 8102–8107.
14. Tahirov, T.H.; Makarova, K.S.; Rogozin, I.B.; Pavlov, Y.I.; Koonin, E.V. Evolution of DNA polymerases: An inactivated polymerase-exonuclease module in Pol ε and a chimeric origin of eukaryotic polymerases from two classes of Archaeal ancestors. *Biol. Direct* **2009**, *4*, 11–15.
15. Berquist, B.R.; DasSarma, P.; DasSarma, S. Essential and non-essential DNA replication genes in the model halophilic archaeon, *Halobacterium* sp. NRC-1. *BMC Genet.* **2007**, *8*, e31.
16. Rouillon, C.; Henneke, G.; Flament, D.; Querellou, J.; Raffin, J.P. DNA polymerase switching on homotrimeric PCNA at the replication fork of the Euryarchaea *Pyrococcus abyssi*. *J. Mol. Biol.* **2007**, *350*, 53–64.
17. Shen, Y.; Tang, X.-F.; Yokoyama, H.; Matsui, E.; Matsui, I. A 21-amino acid peptide from the cysteine cluster II of the family D DNA polymerase from *Pyrococcus horikoshii* stimulate its nuclease activity which is Mre-11-like and prefers manganese ion as the cofactor. *Nucleic Acids Res.* **2004**, *32*, 158–168.
18. Aravind, L.; Koonin, E.V. Phosphoesterase domains associated with DNA polymerases of diverse origins. *Nucleic Acids Res.* **1998**, *26*, 3746–3752.
19. Gueguen, Y.; Rolland, J.; Lecompte, O.; Azam, P.; Romancer, G.L.; Flament, D.; Raffin, J.P.; Dietrich, J. Characterization of two DNA polymerases from the hyperthermophilic euryarchaeon *Pyrococcus abyssi*. *Eur. J. Biochem.* **2001**, *268*, 5961–5969.

20. Hopfner, K.; Karcher, A.; Craig, L.; Woo, T.T.; Carney, J.P.; Tainer, J.A. Structural biochemistry and interaction architecture of the DNA double-strand break repair Mre11 nuclease and Rad50-ATPase. *Cell* **2001**, *105*, 473–485.
21. Uemori, T.; Sato, Y.; Kato, I.; Doi, H.; Ishino, Y. A novel DNA polymerase in the hyperthermophilic archaeon, *Pyrococcus furiosus*: Gene cloning, expression, and characterization. *Genes Cells* **1997**, *2*, 499–512.
22. Shen, Y.; Tang, X.-F.; Matsui, I. Subunit interaction and regulation of activity through terminal domains of the family D DNA polymerase from *Pyrococcus horikoshii. J. Biol. Chem.* **2003**, *278*, 21247–21257.
23. Tang, X.-F.; Shen, Y.; Matsui, E.; Matsui, I. Domain topology of the DNA polymerase D complex from hyperthermophilic archaeon *Pyrococcus horikoshii. Biochemistry* **2004**, *43*, 11818–11827.
24. Matsui, I.; Urushibata, Y.; Shen, Y.; Matsui, E.; Yokoyama, H. Novel structure of an *N*-terminal domain that is crucial for the dimeric assembly and DNA-binding of an archaeal DNA polymerase D large subunit from *Pyrococcus horikoshi. FEBS Lett.* **2011**, *585*, 452–458.
25. Yamasaki, K.; Urushibata, Y.; Yamasaki, T.; Arisaka, F.; Matsui, I. Solution structure of the *N*-terminal domain of the archaeal D-family DNA polymerase small subunit reveals evolutionary relationship to eukaryotic B-family polymerase. *FEBS Lett.* **2010**, *584*, 3370–3375.
26. Holm, L.; Sander, C. Protein structure comparison by alignment of distance matrices. *J. Mol. Biol.* **1993**, *233*, 123–138.
27. Nuutinen, T.; Tossavainen, H.; Fredriksson, K.; Pirilä, P.; Permi, P.; Pospiech, H.; Syvaoja, J.E. The solution structure of the amino-terminal domain of human DNA polymerase ε subunit B is homologous to *C*-domains of AAA+ proteins. *Nucleic Acids Res.* **2008**, *36*, 5102–5110.
28. Jeruzalmi, D.; Yurieva, O.; Zhao, Y.; Young, M.; Stewart, J.; Hingorani, M.; O'Donnel, M.; Kuriyan, J. Mechanism of processivity clamp opening by the delta subunit wrench of the clamp loader complex of *E. coli* DNA polymerase III. *Cell* **2001**, *106*, 417–428.
29. Putnam, C.D.; Clancy, S.B.; Tsuruta, H.; Gonzalez, S.; Wetmur, J.G.; Tainer, J.A. Structure and mechanism of the RuvB Holliday junction branch migration motor. *J. Mol. Biol.* **2001**, *311*, 297–310.
30. Neuwald, A.F.; Aravind, L.; Spouge, J.L.; Koonin, E.V. AAA$^+$: A class of chaperonine-like ATPases associated with the assemply, operation, and disassembly of protein complexes. *Genome Res.* **1999**, *9*, 27–43.
31. Baranovskiy, A.G.; Babayeva, N.D.; Liston, V.G.; Rogozin, I.B.; Koonin, E.V.; Pavlov, Y.I.; Vassylyev, D.G.; Tahirov, T.H. X-ray structure of the complex of regulatory subunits of human DNA polymerase delta. *Cell Cycle* **2008**, *7*, 3026–3036.
32. Cann, I.K.; Komori, K.; Toh, H.; Kanai, S.; Ishino, Y. A heterodimeric DNA polymerase: Evidence that members of Euryarchaeota possess a distinct DNA polymerase. *Proc. Natl. Acad. Sci. USA* **1998**, *95*, 14250–14255.
33. Mäkiniemi, M.; Pospiech, H.; Kilpeläinen, S.; Jokela, M.; Vihinen, M.; Syväoja, J.E. A novel family of DNA-polymerase-associated B subunits. *Trends Biochem. Sci.* **1999**, *24*, 14–16.

34. Bryson, K.; McGuffin, L.J.; Marsden, R.L.; Ward, J.J.; Sodhi, J.S.; Jones, D.T. Protein structure prediction servers at University College London. *Nucleic Acids Res.* **2005**, *33*, W36–W38.
35. Castrec, B.; Laurent, S.; Henneke, G.; Flament, D.; Raffin, J.P. The glycine-rich motif of *Pyrococcus abyssi* DNA polymerase is critical for protein stability. *J. Mol. Biol.* **2010**, *396*, 840–848.

The Function of Gas Vesicles in Halophilic Archaea and Bacteria: Theories and Experimental Evidence

Aharon Oren

Abstract: A few extremely halophilic Archaea (*Halobacterium salinarum*, *Haloquadratum walsbyi*, *Haloferax mediterranei*, *Halorubrum vacuolatum*, *Halogeometricum borinquense*, *Haloplanus* spp.) possess gas vesicles that bestow buoyancy on the cells. Gas vesicles are also produced by the anaerobic endospore-forming halophilic Bacteria *Sporohalobacter lortetii* and *Orenia sivashensis*. We have extensive information on the properties of gas vesicles in *Hbt. salinarum* and *Hfx. mediterranei* and the regulation of their formation. Different functions were suggested for gas vesicle synthesis: buoying cells towards oxygen-rich surface layers in hypersaline water bodies to prevent oxygen limitation, reaching higher light intensities for the light-driven proton pump bacteriorhodopsin, positioning the cells optimally for light absorption, light shielding, reducing the cytoplasmic volume leading to a higher surface-area-to-volume ratio (for the Archaea) and dispersal of endospores (for the anaerobic spore-forming Bacteria). Except for *Hqr. walsbyi* which abounds in saltern crystallizer brines, gas-vacuolate halophiles are not among the dominant life forms in hypersaline environments. There only has been little research on gas vesicles in natural communities of halophilic microorganisms, and the few existing studies failed to provide clear evidence for their possible function. This paper summarizes the current status of the different theories why gas vesicles may provide a selective advantage to some halophilic microorganisms.

Reprinted from *Life*. Cite as: Oren, A. The Function of Gas Vesicles in Halophilic Archaea and Bacteria: Theories and Experimental Evidence. *Life* **2013**, *3*, 1-20.

1. Introduction

A small number of species of extremely halophilic Archaea of the family *Halobacteriaceae* (8 out of the 137 species with names with standing in the nomenclature as of August 2012) [1] are able to produce gas vesicles. These include two organisms that have been investigated in-depth in the past decades as models for the study of gas vesicle production: *Halobacterium salinarum* and *Haloferax mediterranei*. These also include the intriguing flat, square or rectangular *Haloquadratum walsbyi*, which is abundantly found in hypersaline brines e.g. in saltern crystallizer ponds worldwide. Such cells are buoyant: liquid cultures of *Hbt. salinarum* left standing form a pellicle at the surface.

Gas vesicle production is by no means restricted to the *Halobacteriaceae*. Many representatives of the bacteria belonging to different classes possess them. The ability to form gas vesicles is especially widespread among the cyanobacteria. Gas vesicles often occur in clusters ('gas vacuoles') visible as bright refractile bodies in the phase contrast microscope. Masses of gas-vacuolate cyanobacteria are often found floating on the surface of freshwater lakes, buoyed up due to the high content of gas vesicles that bestow buoyancy on the cells. There are also a few methanogenic Archaea that produce gas vesicles.

Presence of refractile 'gas vacuoles' in 'Bacterium halobium' isolated from salted herring, a strain of the species now known as *Hbt. salinarum*, was first reported by Helena Petter in the early 1930s [2,3]. She grew red colonies from salted herring and dried cod: three isolates formed transparent colonies; four had opaque colonies with gas vesicles. She also published the first drawings of *Halobacterium* cells with gas vacuoles. Already then she suggested that the presence of gas vesicles and the buoyancy these vesicles bestow on the cells can be of considerable ecological advantage: the vesicles may buoy the cells to the surface of brine pools and salt lakes, where they would benefit from higher concentrations of oxygen. Oxygen and other gases are much less soluble in saturated salt solutions than in freshwater or in seawater. For example, distilled water at 20 °C in equilibrium with air contains 9.10 mg/L (284 μM) O_2, while at 260 ppt (parts per thousand; g/kg of solution) salinity there is only 1.67 mg/L (52 μM). At 35 °C the values are reduced to 6.92 mg/L (216 μM) and 1.51 mg/L (47 μM), respectively [4]. Therefore oxygen might become a limiting factor at high salt. This interpretation was quickly adopted by others. Trijntje Hof described in 1935 a similar gas-vacuolate strain [5]. From those days onward, *Halobacterium* became a popular object for the study of gas vesicles, as shown by the admirable electron micrographs—even including early stereopictures of surprisingly high quality—published by Houwink in 1956 [6] and by the early physiological studies by Helge Larsen and his coworkers [7].

Since those early times many possible functions have been suggested for gas vesicle synthesis in different members of the *Halobacteriaceae*: buoying the cells towards more oxygen-rich surface layers in hypersaline water bodies to prevent oxygen limitation, reaching higher light intensities for the light-driven proton pump bacteriorhodopsin, positioning the cells in an optimal orientation for light absorption, light shielding, and reducing the cytoplasmic volume leading to a higher surface-area-to-volume ratio. These conclusions were mainly based on laboratory studies, and very few investigations on gas vesicles were performed in the field in natural communities of halophilic microorganisms. In this paper I will attempt to evaluate the relative merit of the different theories proposed to explain why gas vesicles may provide a selective advantage to some halophilic microorganisms.

2. What Gas Vesicles Are

Gas vesicles are hollow cylindrical or spindle-shaped structures, built of protein subunits. In the *Halobacteriaceae* the size of the vesicles varies between 0.2–1.5 μM in length and they are ~0.2 μM in diameter. A cell of *Hbt. salinarum* can contain up to 70 spindle-shaped gas vesicles when grown aerobically at 40 °C [8]. The wall of the vesicles consists of a single layer of the 7–8 kDa GvpA protein, which forms 4.6 nm wide 'ribs' that run nearly perpendicular to the long axis of the vesicle. A second structural component, the 31–42 kDa GvpC, is a protein that contains internal repeats. It strengthens the vesicle by attaching to its outer surface [8,9]. The majority of the gas vesicles in *Hbt. salinarum* have the form of a wide, rounded bicone ('lemon-shaped' or 'spindle-shaped'). The same organism also produces some longer, narrower and cylindrical gas vesicles [8].

The GvpA protein has a very high content of hydrophobic amino acids, and is highly conserved in all prokaryotes that produce gas vesicles, Archaea as well as Bacteria [9]. The GvpA protein is one of the few proteins in halophilic Archaea that do not require salt for stabilization. The GvpC protein strengthens the structure of the vesicles, assists in their assembly, and to a large extent determines the shape of the gas vesicles.

Similar to the gas vesicles of other prokaryotes, the vesicles found in the members of the *Halobacteriaceae* are sensitive to pressure. They are typically weaker than the gas vesicles of cyanobacteria which generally withstand pressures of up to 0.2–0.3 MPa (~2–3 atmospheres): a pressure of 0.09 MPa causes collapse of half of the gas vesicles in *Hbt. salinarum*, while the weakest gas vesicles within the cells are already destroyed by a pressure of 0.05 MPa [9,10]. These critical collapse pressures set a limit to the depth in a water column in which gas-vacuolate cells can occur. A pressure of 0.09 MPa corresponds to a water depth of ~9 m in fresh water or ~7.3 m in salt-saturated brine. Part of the variation in critical pressure can be explained by the variation in cylinder radius of the gas vesicles. The density of gas vesicles in different prokaryotes was estimated to vary from 60 kg m^{-3} for the widest vesicles to 210 kg m^{-3} for the narrowest ones, present in the marine cyanobacterium *Trichodesmium* [9].

In-depth studies on the genes involved in gas vesicle production and their regulation were thus far performed only with *Hbt. salinarum* and *Hfx. mediterranei*. In both species gas vesicle production increases in the stationary phase, and the ability to produce gas vesicles is easily lost by mutation [8,11,12]. Formation of gas vesicles requires 8–14 different proteins, including the two structural proteins GvpA and GvpC [13–15]. The product of the gene *gvpD* is negative regulator. Transformation of *Hfx. volcanii* with an *mc-vac* construct containing a *gvpD* deletion leads to cells with high numbers of gas vesicles [16]. Further information about the (putative) functions of the different genes involved can be found in a number of papers [17–19], including a recent review article [8]. How environmental factors such as oxygen and salt concentration that affect gas vesicle biosynthesis are transduced to the regulators and influence transcription is largely unknown.

3. How Common Are Gas Vesicles Among the Species of *Halobacteriaceae*?

The ability to produce gas vesicles is not widely distributed among the halophilic Archaea. Out of the 40 genera and 137 species of *Halobacteriaceae* (as of August 2012) with names with standing in the nomenclature [1], no more than eight species belonging to six genera were reported to possess gas vesicles (Table 1).

Gas-vacuolate *Halobacterium* strains have been isolated many times, and these include the isolates from salted fish used in the early studies by Petter, Houwink, Larsen, and others [1–3,5–7]. All known gas-vacuolate strains of *Halobacterium* can be assigned to the species *Hbt. salinarum*, including the widely studied strain NRC-1 [27]. The other species of the genus, *Hbt. jilantaiense* retrieved from a salt lake in Inner Mongolia, China, *Hbt. noricense* from a salt mine and '*Hbt. piscisalsi*' from fermented fish in Thailand (later described as a junior synonym of *Hbt. salinarum*) lack the property.

Out of the 11 currently recognized species within the genus *Haloferax*, *Hfx. mediterranei* is the only one with gas vesicles. It was isolated from an enrichment culture for extreme halophiles able to grow on single carbon sources, using brine from a Spanish saltern pond as inoculum [20].

Table 1. Gas vesicle producing halophilic Archaea (family *Halobacteriaceae*).

Genus	Species	Source of isolation	Flagellar motility	Bacteriorhodopsin / halorhodopsin	References
Halobacterium	*Hbt. salinarum*	Salted fish	+	+	[1]
Haloferax	*Hfx. mediterranei* (basonym: *Halobacterium mediterranei*)	Saltern, Spain	weak	- [a]	[20]
Halogeometricum	*Hgm. borinquense*	Saltern, Puerto Rico	-	- [a]	[21]
Haloplanus	*Hpl. natans*	Experimental outdoor pond, Dead Sea, Israel	-	NR	[22]
	Hpl. vescus	Saltern, China	+	NR	[23]
	Hpl. aerogenes	Saltern, China	+	NR	[24]
Haloquadratum	*Hqr. walsbyi*	Salterns, Australia and Spain	-	+	[25]
Halorubrum	*Hrr. vacuolatum* (basonym: *Natronobacterium vacuolatum* corrig.)	Lake Magadi, Kenya	-	NR	[26]

[a] as deduced from the genome sequence; NR = not reported.

The genus *Halogeometricum* contains the gas-vacuolate *Hgm. borinquense* from a saltern pond in Puerto Rico [21]. No further studies on its gas vesicles were reported. The description of the second species of the genus, *Hgm. rufum*, did not mention the presence of gas vesicles.

All three described species of the genus *Haloplanus* are gas-vacuolate: *Hpl. aerogenes*, *Hpl. natans* (Figure 1), and *Hpl. vescus*. *Hpl. aerogenes* and *Hpl. vescus* were isolated from solar salterns [23,24]; *Hpl. natans* was obtained from outdoor simulation ponds in which mixtures of Dead Sea and Red Sea waters were incubated [22].

The genus *Haloquadratum* consists of a single species, *Hqr. walsbyi*. This is the flat, square-rectangular archaeon first observed by Walsby in a brine pool on the Sinai peninsula, Egypt [28], and only isolated more than two decades later from saltern crystallizer ponds [25,29–31]. *Hqr. walsbyi* is probably the only gas-vacuolate halophilic archaeon that is abundantly found in hypersaline brines worldwide. Figure 2 shows a picture of such square gas-vacuolate cells in the crystallizer brine of a saltern in Israel. Generally the gas vesicles are not distributed evenly throughout the cells, but they are concentrated near the edges of the squares. This can be seen in the left panel of Figure 2 and in many published micrographs and electron micrographs [30,31–33].

Figure 1. Phase-contrast micrograph of gas-vacuolate cells of *Haloplanus natans* strain RE-101^T (**a**) and cells after centrifugation, causing collapse of the gas vesicles (**b**) Bars, 10 µM. Note the concentrations of gas vesicles at the periphery of the cells in panel (**a**). From Elevi Bardavid *et al.*, 1997 [22], reproduced with permission from the Society for General Microbiology, Reading, UK.

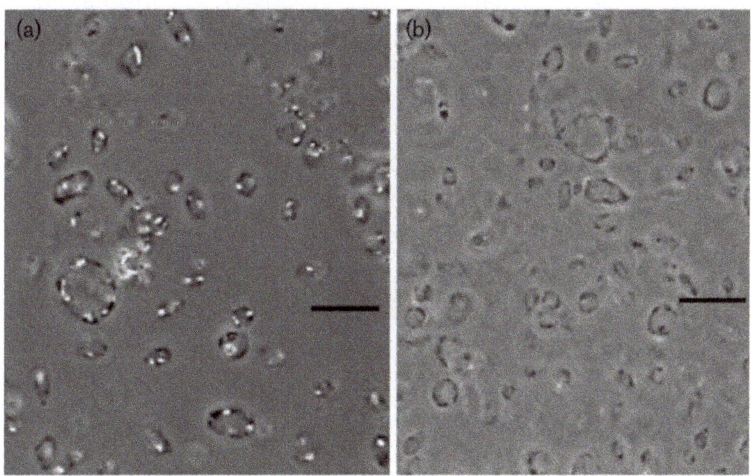

Halorubrum vacuolatum from Lake Magadi, Kenya [26], an alkaliphilic species of small cells originally described as *Natronobacterium vacuolatum* (*vacuolata*), is the only of the 25 species of the genus *Halorubrum* that carries gas vesicles. Its cells are very small, ~1–1.5 µM in the stationary phase. Little information is available on the properties of its gas vesicles and on the regulation of their production. It lacks gas vesicles when grown in salt concentrations below 15% [18,34].

Figure 2. Flat, square to rectangular, *Haloquadratum walsbyi*-type cells from the saltern crystallizer ponds of Eilat, Israel. Note the division plane in the left panel. Photograph: O. Shapiro, N.Siboni and N.Pri-El.

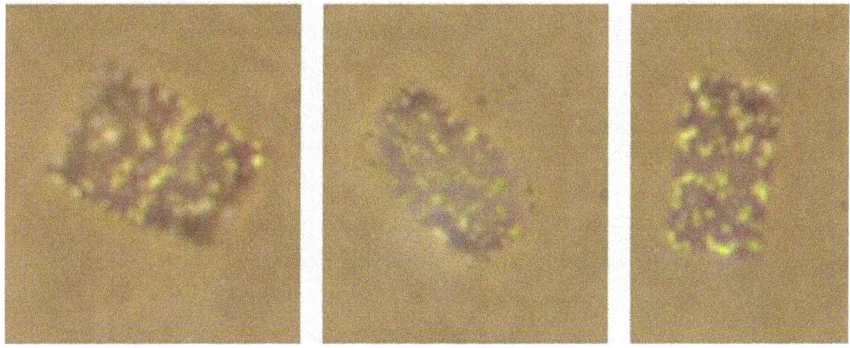

Some gas-vacuolate species are actively motile by means of flagella, so that their mode of positioning in the water column may be determined both by their buoyancy due to the gas vesicles and by flagella-driven motility. Calculations have shown that, at least in slow growing cells, upward movement mediated by gas vesicles is energetically less expensive than by rotating flagella [9,35]. This does not imply that passive movement mediated by gas vesicles is not energetically costly. To produce sufficient gas vesicles so that 10% of the cell volume is occupied by gas, an equivalent of ~7.2% of the total protein synthesized or ~4.3% of the dry cell mass must be gas vesicle protein (values determined for the cyanobacterium *Anabaena*) [9]. To justify this cost there must be some selective advantage to the production of gas vesicles.

4. The Possible Advantages of Gas Vesicles to Halophilic Archaea: Laboratory Studies

To elucidate the possible selective advantages of the production of gas vesicles by members of the *Halobacteriaceae*, laboratory studies have been performed mostly with gas-vacuolate *Halobacterium* strains, with *Hfx. mediterranei*, or with transformants of *Hfx. volcanii* (a non-vacuolate species) carrying genes for gas vesicle production derived from *Hfx. mediterranei*.

4.1. Competition for Limiting Oxygen

The idea that gas vesicles may buoy cells to the surface of brine pools and salt lakes where they would benefit from higher concentrations of oxygen was already proposed by Petter in the 1930s [2,3]. The ability to float to the brine surface may be advantageous for an aerobic halophilic microorganism as the solubility of oxygen and other gases in salt-saturated brines is low. To what extent oxygen is indeed a limiting factor in those environments inhabited by gas-vacuolate members of the *Halobacteriaceae* will be discussed in Section 5.2.

To experimentally test whether under oxygen limitation there may be a selective advantage to gas-vacuolate types, competition experiments were set up between a gas-vacuolate strain of *Hbt. salinarum* (strain PHH1) and a mutant affected in gas vesicle synthesis [36]. In shaken cultures both strains grew equally well, but in deep static cultures, where, due to the community respiration, steep vertical oxygen concentration gradients were established, cells of the wild type floated and became dominant, probably due to their successful competition for oxygen which was in short supply. In shallow static cultures, however, the gas-vesicle deficient mutant won the competition. A possible explanation is that under the conditions employed the wild type wasted much energy to produce unnecessary gas vesicles, while the mutant had a lower protein burden.

4.2. Can Gas Vesicles Function as Intracellular Oxygen Reservoirs?

The O_2 carrying capacity of water at 37 °C is ~9.7 mol m^{-3} MPa^{-1}. For gas vesicles the value is 388 mol m^{-3} MPa^{-1}, *i.e.*, 40-times as large. Therefore the use of isolated gas vesicles (from the cyanobacterium *Anabaena flos-aquae*) was explored as oxygen carriers to increase the oxygen supply in mammalian cell culture systems [37]. Addition of 1% volume of gas vesicles thus leads to a 39% increase in oxygen carrying capacity, a 10% gas vesicle volume results in a 390% increase. The question can therefore be asked whether gas vesicles inside halophilic Archaea may serve as intracellular oxygen reservoirs. The answer probably must be negative. The gas vesicle wall and the cell membrane are both highly permeable to gases, so gases will rapidly equilibrate between the cell and the surrounding brine. Therefore the cell cannot "store" oxygen to keep it for its own use, and the oxygen content of the gas vesicles will rapidly decrease when oxygen is consumed by community respiration. Even in dense communities of halophilic Archaea the total volume occupied by the gas vesicles is still small. For example, in a hypothetical case in which a brine is inhabited by 10^8 *Hqr. walsbyi* cells that are 3 × 3 × 0.2 μM in size and have 20% of their intracellular volume occupied by gas vesicles, the total O_2 carrying capacity of the brine will be increased only by ~0.7% due to the presence of the gas vesicles.

Diffusion of oxygen in air is 5,700 times (at 20 °C) to 20,000 times (40 °C) more rapid than in water [38]. This may mean that oxygen present in the gas vesicles may be made accessible much more rapidly than oxygen dissolved in the cytoplasm or in the surrounding brine. Whether this is indeed of physiological importance for these small cells, in which the relevant diffusion distances are less than 1 μM, may be doubted.

4.3. Is Gas Vesicle Biosynthesis Induced by Anaerobiosis?

Although basically an organism with an aerobic life style, *Hbt. salinarum* possesses different modes of anaerobic growth. These include anaerobic respiration using dimethylsulfoxide (DMSO) or trimethylamine *N*-oxide (TMAO) as the electron acceptor [39] and fermentative growth on arginine [40,41]. Arginine fermentation is not commonly found among the *Halobacteriaceae*, and appears to be a special feature of the genus *Halobacterium*. Therefore a specific enrichment procedure could be developed for this genus, based on its ability to grow anaerobically in arginine-containing media [42].

In contrast to what might be expected, anaerobiosis represses the formation of gas vesicles. When grown or incubated under anoxic conditions, three strains of *Hbt. salinarum*, *Hfx. mediterranei* and *Hfx. volcanii* transformants lacked gas vesicles altogether, or had a much lower content than aerobically grown cells. Even the gas vesicle-overproducing Δ*gvpD* transformants did not produce gas vesicles under anaerobic conditions, demonstrating that the repressing protein GvpD was not involved. Presence of large amounts of GvpA in the cells' cytoplasm implied that the assembly of gas vesicles was inhibited [43]. *Hbt. salinarum* cells grown by arginine fermentation contain only a few groups of tiny gas vesicles [43], whereas cells that grow using TMAO respiration contain few larger vesicles [44]. This may explain why in the anaerobic enrichment cultures for *Halobacterium* using arginine as energy source, gas vesicles were only seldom observed [42]; whether the enriched strains may have possessed the potential for gas vesicle production was not ascertained.

In oligonucleotide microarray studies with *Halobacterium* NRC-1, slight upregulation of *gvpA* and *gvpC* was reported following anaerobic incubation with arginine, while anaerobic cultures with TMAO showed apparent stronger (2–5-fold) upregulation of *gvpA*, *gvpC*, *gvpN*, and *gvpO*; these cells showed an abundance of gas vesicles [44,45]. This was probably due to the presence of citrate in the growth medium and not to the growth condition employed: citrate stimulates gas vesicle production. In a different medium lacking citrate, anaerobiosis strongly inhibited gas vesicle formation. Therefore only the gas vesicles already produced under oxic conditions might help the cells to avoid the anoxic zones of the brine [43].

4.4. Induction of Gas Vesicle Formation at High Salinity

Hfx. mediterranei is a versatile organism, able to grow over a wide range of salt concentrations. It produces gas vesicles only when the salt concentration in the medium exceeds 170 g/L [16,17–19,46,47]. The relative abundance of *mc-vac* mRNA in cells grown at 250 g/L salt was sevenfold higher than in cells grown in 150 g/L [16].

How the cells sense the salt concentration and transduce the information to regulate the transcription of gas vesicle genes is still unknown. The reason why high-salt-grown cells need more gas vesicles than low-salt-grown cells is also not clear, but it could be related to the lowered solubility of oxygen and other gases in concentrated brines. *Hbt. salinarum* has a much more restricted salinity range for growth, and therefore similar experiments with *Halobacterium* have not been reported. Effects of medium salinity on gas vesicle production were also examined in *Hrr. vacuolatum* [34].

4.5. Induction of Gas Vesicle Formation at Low Temperature

Different strains of *Hbt. salinarum* showed increased contents of gas vesicles when grown at a low temperature (15 °C). Growth is slow, and the cells formed are tightly filled with gas vesicles [48]. Oligonucleotide microarray experiments with *Halobacterium* NRC-1 showed a 1.8-fold to 8-fold increase in *gvp*ACNO expression and a 3-fold increase in expression of *gvp*DE in the cold [49].

As oxygen solubility in aqueous solutions is increased at lowered temperatures, the effect cannot be understood as an adaptation to an increased oxygen requirement. Growth and cell metabolism are also very slow at 15 °C, resulting in a low oxygen demand.

4.6. Are Gas Vesicles Formed to Increase Light Availability?

Some members of the *Halobacteriaceae* contain the membrane-bound light-driven proton pump bacteriorhodopsin and/or the light-driven chloride pump halorhodopsin. Such cells can directly convert light energy to a proton gradient (to drive generation of ATP, ion transport processes, *etc.*) or to pump chloride into the cells against the thermodynamic gradient. In some cases photoheterotrophic growth is even possible based on the energy of photons absorbed by bacteriorhodopsin [40,41,50]. Light and low oxygen concentrations are among the factors that trigger bacteriorhodopsin production with the formation of 'purple membrane' in *Hbt. salinarum.*

Some, but not all halophilic Archaea that produce gas vesicles also possess bacteriorhodopsin and halorhodopsin. Functional bacteriorhodopsin proton pumps are active in *Halobacterium* and in *Haloquadratum* (Table 1; see also [51]), but genes for bacteriorhodopsin or halorhodopsin were not detected in the genomes of *Hfx. mediterranei* and *Hgm. borinquense*; presence of the purple retinal pigments was never ascertained in *Haloplanus* spp. and in *Hrr. vacuolatum.*

Although upward flotation toward the brine surface may be beneficial for bacteriorhodopsin-containing cells to be able to harvest more light, a direct correlation between light intensity, presence or absence of bacteriorhodopsin, and gas vesicle production was never documented. These advantages could also be tested in competition experiments. This suggestion, made by Kessel *et al.* in 1985 [52], has not yet been followed up.

4.7. Are Gas Vesicles Formed as a Means of Protection Against Excess Light?

Another possible function was suggested for the gas vesicles, namely their use as light-shielding organelles to protect the organism exposed to high light intensities in its environment against harmful ultraviolet radiation. Light is scattered due to the large difference in refractive index between the gas vesicles and the cell's cytoplasm or the surrounding medium. This hypothesis was not supported by controlled laboratory experiments. No significant differences in sensitivity to UV radiation were observed between a gas-vacuolate *Halobacterium* strain and a gas-vesicle-deficient mutant. Therefore the gas vesicles are not effective as light-shielding organelles [53].

5. The Possible Advantages of Gas Vesicles to Halophilic Archaea: Field Studies

5.1. How Successful Are Gas-Vacuolate Species of Halobacteriaceae in Colonizing Hypersaline Environments?

One of the most compelling arguments against a great ecological advantage of the production of gas vesicles is the observation that hypersaline environments (natural salt lakes, saltern evaporation and crystallizer ponds, *etc.*) are seldom dominated by gas-vacuolate types of Archaea. The only exception appears to be *Haloquadratum*, which is often found in very high numbers, as discussed below. It must, however, be realized that culture-dependent approaches in which colonies developing on agar plates are examined are problematic due to the low recovery percentage of colonies as compared to the microscopically observed numbers. Microscopical examination of brines after concentration of the cells by centrifugation is also not effective, as high-speed centrifugation causes collapse of the vesicles due to the high pressure applied to them. It is therefore possible that gas-vacuolate types are more abundant that generally realized, but data are lacking. It would be interesting to examine the abundance at which *gvp* genes turn up in the metagenomes of different hypersaline environments.

Hbt. salinarum and *Hfx. mediterranei*, the two species used in most experiments on archaeal gas vesicles, are not at all abundant in aquatic hypersaline environments. Gas-vacuolate *Halobacterium* strains were generally isolated not from salt lakes but from salted fish and salted hides [1–3,7,54], and the genus *Halobacterium* contributes very little to the prokaryote community in saltern ponds and salt lakes. *Halobacterium* strains can be specifically enriched and isolated from salterns based on their ability to grow anaerobically on arginine (see above), but their numbers appear to be small [36]. *Halobacterium* 16S rRNA gene sequences seldom turn up in metagenomes and in 16S rRNA gene libraries prepared from such environments. Moreover, the *Halobacterium*-specific glycolipids (sulfated triglycosyl and tetraglycosyl diphytanyl diether lipids) form only a very small fraction, if they are detectable at all, of the total glycolipids in the community [55].

Hfx. mediterranei was never shown to be present in high numbers in any environment, in spite of its extremely high versatility: it can grow in a wide range of salt concentrations, use many more substrates for growth than most other halophilic Archaea, can digest a range of polymeric compounds, has a very high growth rate, and excretes halocins—protein antibiotics that inhibit growth of many other members of the *Halobacteriaceae* [1,20].

Haloplanus strains were only occasionally isolated, and it is not clear how widespread this genus might be. *Haloplanus*-related 16S rRNA gene sequences were retrieved from an oilfield and from a gypsum crust in a solar saltern [22]. They were also found, albeit at a low frequency, in the metagenome of the Dead Sea in 2007, at a time community densities in the lake were very low [56]. At the time the Dead Sea was subject to deep mixing, so that presence of intact gas vesicles could not be expected. We also know little about the worldwide abundance of gas-vacuolate *Halogeometricum*.

The only extreme halophile with gas vesicles that appears to contribute significantly to the prokaryote community in hypersaline lakes, both natural and artificial, is the square *Hqr. walsbyi*. Its characteristic flat square cells are abundantly found in saltern crystallizer ponds worldwide. First found to be present in numbers as high as 7×10^7 cells per mL of brine in a coastal salt pool on the Sinai peninsula, Egypt, it was since encountered in salterns in Spain, Israel, Mexico, USA, Australia, and elsewhere, sometimes contributing more than half of the prokaryote numbers [57,58]; they exist in certain natural salt lakes as well [59].

5.2. Are Natural Communities of Halophilic Archaea Ever Oxygen-Limited?

From the time of the early studies by Petter it was often assumed that the main function of gas vesicles in the *Halobacteriaceae* may be to obtain access to oxygen, potentially a limiting factor in natural brines. Indeed, the community densities of halophilic Archaea and other microorganisms in hypersaline brines in nature are often high (densities of 10^7–10^8 cells/mL are not exceptional), and the solubility of oxygen is reduced as salinity increases.

There have, however, only been very few measurements of actual oxygen concentrations of natural brines and of community respiration rates in such brines. Using a method based on the Winkler titration, Rodriguez-Valera *et al.* [60] measured 0.3–0.8 mg/L (9–25 μM) oxygen in salt-saturated saltern ponds in Spain. Similar values (~20–27 μM) were estimated for crystallizer brine in the Eilat, Israel salterns [61]. Such concentrations are probably not limiting to the community of aerobic halophilic microorganisms in any way. When the Eilat brine, containing 3.9×10^7 prokaryote cells per mL, most of which were of the *Haloquadratum* type, was incubated in the dark at 30 °C in a closed containers, it took as much as 50 h before oxygen was completely depleted [61]; later experiments in which the oxygen uptake in Eilat crystallizer brine (3×10^7 prokaryote cells per mL) was monitored at 35 °C by an oxygen optode showed oxygen depletion after 32–38 h (R. Pinhassi, E. Maimon, R. Horwitz and A. Oren, unpublished results). In the natural system oxygen is continuously supplied by diffusion from the air and mixing by waves, as well as by photosynthesis by the unicellular halophilic alga *Dunaliella salina* during daytime. Experiments in which the time course of oxygen depletion by cultures of *Hbt. salinarum*, *Hqr. walsbyi* and by Eilat crystallizer brine was followed by means of a Clark-type oxygen microelectrode showed a linear decrease in O_2 concentrations down to values below 1 μM (M. Krause, G. Panasia, N. Meyer, and A. Oren, unpublished results). This shows that the affinity of halophilic Archaea for oxygen is sufficiently high so that oxygen cannot be expected ever to become limiting in the shallow crystallizer ponds in spite of the presence of the dense microbial communities. Therefore the cells under these situations do not need buoyancy by gas vesicles to obtain oxygen.

5.3. Studies on Haloquadratum in Coastal Brine Pools, Sinai Peninsula

The flat, square *Haloquadratum*-type microorganisms were first recognized by Walsby in brine pools on the Red Sea coast at the southern end of the Sinai Peninsula, Egypt [28,35]. Until the organism was brought in culture in 2004 [25,29–31], most studies on these fascinating organisms were based on samples collected from these pools. The early reports mentioned positive buoyancy of the square cells, which often were present in pairs, groups of four, eight, sixteen [28,62]; in one case even an assembly of sixty-four cells was observed "like postage stamps in a sheet" [63]. Cells were reported to have accumulated in large numbers on the surface of the brine pool, buoyed up by their gas vesicles [63]. For further studies, gas-vacuolate cells were concentrated by leaving brine samples standing for a few days, collecting cells from the surface, followed by further concentration by low speed centrifugation (accelerated flotation), taking care not to exceed the critical pressure at which gas vesicles start collapsing [28,63,64]. In other microscopy and electron microscopy studies high-speed centrifugation was employed to concentrate the cells, leading to collapse of the gas vesicles [62]. A definite reduction in gas vesicle numbers per cell was seen after application of 0.15 MPa, most of the vesicles had disappeared by 0.25 MPa, and none remained at pressures beyond 0.3 MPa [28].

5.4. Studies on Haloquadratum in Saxkoye Lake, Ukraine

Flat square gas-vacuolate cells were collected from the surface waters of the hypersaline Saxkoye Lake, Ukraine [59]. In this study, performed shortly after Walsby had discovered this type of cells in the Sinai brine pool, the nature of the structures found was misinterpreted as square microcolonies, the gas vesicles erroneously considered to be the cells. Although details are lacking, the published information makes it understood that the cells were collected from the surface water film by floating electron microscope grids on top of the brine. Whether indeed these square structures were present at a higher density at the brine surface than in the deeper waters was not reported.

5.5. Studies on Haloquadratum in the Crystallizer Ponds of the Eilat Salterns

In the saltern crystallizer ponds of Eilat, square gas-vacuolate cells of the *Hqr. walsbyi* type consistently make up a high percentage of the microbial community; at least 70%–80% of all cells show this characteristic morphology (see also Figure 2). This fact enabled us to obtain information on the polar lipid composition of *Haloquadratum* long before the organism had been brought into culture [58].

The presence of these dense populations in saltern crystallizer ponds presented a unique opportunity to study the possible role of the gas vesicles in the life of *Haloquadratum*. No indications for positive buoyancy were obtained [65]:

(1) When samples of brine were placed in a Petroff-Hauser counting chamber and left to stand for up to 4 hours, cells had not accumulated near the cover slip, but remained distributed evenly within the 20 μM thick space between slide and cover slip. However, when the brine sample was first subjected to pressurization causing collapse of all gas vesicles, most cells sank to the bottom.

(2) When the brine was incubated for up to 5 days at room temperature in diffuse daylight in 1 liter glass cylinders, little change in the vertical distribution of the cells was demonstrated, except for a tendency for a small decrease of cell numbers in the upper layer toward the end of the incubation period. To prevent convection currents to cause mixing of the brine, the experiment was repeated by establishing a salinity gradient from 100% brine below to 90% brine – 10% distilled water on top. The result was similar.

(3) In "accelerated flotation" experiments in which brine samples were centrifuged at speeds (26 x *g*) for periods of up to 12 h, the cells were still homogeneously distributed throughout the tubes. A similar result was obtained following 60 h centrifugation in a swing-out rotor at 39.1 x *g*. In all these cases the calculated maximum pressure exerted on the cells did not exceed 0.014 MPa, a pressure insufficient to cause collapse of even the weakest gas vesicles. When the brine had earlier been subjected to pressure above 0.2 MPa and all gas vesicles had been collapsed, cell densities at the lower end of the centrifuge tube were significantly higher than at near the surface after 60 h of low speed centrifugation.

These observations suggest that the gas vesicles present in the square halophilic Archaea in the saltern ponds of Eilat provide negligible floating/sinking velocity to the cells. The gas vesicles were sufficient to provide neutral buoyancy, but not positive buoyancy that may allow the cells to float under optimal conditions. In this respect the properties of the *Haloquadratum* community in the Eilat salterns may have differed from that in the Sinai brine pool. If the gas-vacuolate flat square cells do not float in a test tube in the laboratory, they cannot be expected to buoy up in the natural environment where wind, waves, and water currents will tend to disperse them equally at all depths [65].

6. Cell Size and Colony Size as Critical Parameters Affecting Buoyancy of Gas-Vacuolate Prokaryotes

The rate at which a particle rises or sinks in a liquid medium depends not only on the difference in density between the particle and the solution and on the viscosity of the solution (which is about twice as high in saturated brines than in freshwater), but also on the size of the particle. When estimating the rate of sinking or flotation of a prokaryotic cell in a water column, the first approximation is that determined by Stokes's equation for a spherical particle [66]:

$$V = \frac{2 \cdot g \cdot r^2 \cdot (\rho_1 - \rho_2)}{9\eta}$$

where V is the velocity of fall, g is the acceleration of gravity, r is the 'equivalent' radius of the particle, ρ_1 is the density of particle, ρ_2 is the density of medium, and η is the dynamic viscosity of medium. For a non-spherical particle such as a flat *Haloquadratum* cell certain corrections must be made, as the hydrodynamic behavior of the square cells cannot be simulated by a sphere. A flat particle will encounter increased drag forces compared to a spherical particle of the same volume when moving in the water column with the direction of motion parallel to the short axis of that particle. For the flat square cells, the shape increases the effective Stokes radius approximately two-fold [65,67].

Based on different assumptions to estimate the density of *Haloquadratum* cells, flotation rates calculated from Stokes's equation were in the order of a few millimeters per day only. Thus, even when the water column is not subjected to any mixing by waves and currents, the square Archaea in the salterns will not float toward the surface of the brine at a significant speed. Other gas-vacuolate members of the *Halobacteriaceae* such as *Hrr. vacuolatum* and *Hfx. mediterranei* also have cells smaller than 2–3 mM [1,20,26], and therefore they cannot be expected to float at higher speeds.

Gas vesicle-containing cyanobacteria such as *Microcystis*, *Anabaena*, and *Trichodesmium* can form surface blooms because their cells are largeand their content of gas vesicles can be very high, causing a large difference in density between the cells and their medium. Furthermore, they generally grow in colonies, filaments or bundles of filaments, thus increasing their effective radius in Stokes's equation, and they live in a medium of half the viscosity of that of the halophilic Archaea. Increasing the effective radius, resulting in higher flotation rates, may be an option for *Haloquadratum*, which is known to form small sheets when cells do not separate after division. The material from the Sinai brine pool showed an abundance of sheets of four to eight cells [28,52,63,64], and the observation of a sheet of 64 cells [63] suggests that formation of such 'colonies' may increase the efficiency of flotation, provided their density is indeed lower than that of the brine.

7. Do the Gas Vesicles of *Haloquadratum* Serve to Optimize Light Absorption?

Another hypothesis to explain the possible function of gas vesicles in *Haloquadratum*, first proposed by Bolhuis *et al.* [33], is based on the observation that the gas vesicles are often mainly located close to the cell periphery. Photographs of cells with most or all gas vesicles close to the edges are found in many publications [28,31,64]. Sheets are so thin that they bulge slightly with gas vesicles along their edges. Light absorption by thin sheets can be highly efficient, and especially by sheets oriented normal to the incoming light. It was postulated that the arrangement of the gas vesicles may aid the cells to position themselves parallel to the water surface. Such a horizontal positioning would aid the cells in collecting as many photons as possible to be absorbed by the bacteriorhodopsin proton pump, present in *Haloquadratum*, to generate ATP. The fact that *Haloquadratum* lacks flagella may be important here: flagellar movement would have caused the cells to rotate [35]. A similar location of peripheral gas vesicles in larger cells was found in *Hpn. natans* (Figure 1, left panel), an organism that also does not show active motility.

One possible problem with this hypothesis is the fact that the salt lakes and shallow pools where organisms such as *Haloquadratum* thrive are generally exposed to very high light intensities, so that light is not likely to become a limiting factor. The opposite may be true: the cells have different mechanisms to protect themselves against damage by high levels of radiation. Another question to be asked is whether the geometry of the cells and the spatial arrangement of the gas vesicles indeed allow the cells to position themselves parallel to the water surface and so maximize light absorption. The small cells, with their accordingly low Reynolds number [66,68] experience their medium as extremely viscous, and will not easily rotate; Brownian motion will further tend to randomize the cells' orientation. Moreover, in a macroscopic model in which much larger square "cells" (with an accordingly higher Reynolds number) with near-neutral buoyancy provided by peripheral "gas vesicles", the cells remained oriented randomly in the water rather than positioning themselves horizontally (Figure 3).

Positioning of the square cells to maximize light exposure can probably be expected only if the cells are sufficiently flexible so that the margins with the gas vesicles will bend upward and the cells become somewhat cup-shaped. Indeed the flat cells show a degree of flexibility as shown by photographs of exceptionally large "folded" cells [29,30]. It remains to be ascertained whether 'standard' small square cells with a diameter of 2–3 μM are also flexible enough to become deformed by the buoyancy of the gas vesicles.

Figure 3. Random orientation of model *Haloquadratum*-type "Archaea" with peripheral "gas vesicles" suspended in water. The model "cells" were crafted from thin glass plates (10 cm diameter) glass with bubble wrap at both sides along the edges to provide near-neutral buoyancy. With thanks to the staff of the Max Planck Institute for Marine Microbiology, Bremen, for the facilities provided.

8. Do Gas Vesicles Serve to Increase the Cell Surface/Cytoplasmic Volume Ratio?

An altogether different idea why some halophilic Archaea may possess gas vesicles, a hypothesis that has nothing to do with cell buoyancy, oxygen limitation and light harvesting, was proposed already in the 1950s by Houwink [6] in his electron microscope study of *Hbt. salinarum*: "Since the gas vacuoles fill part of the space within the cell wall, the ratio of volume of the cytoplasm/surface area of the cell wall is smaller than it would be if the gas vacuoles were lacking." And he added: "It is quite another question whether a large surface area with respect to the volume of the cytoplasm is profitable to the organism. There was found no indication that this was the case with *H. halobium*" (now: *Hbt. salinarum*).

This hypothesis was revived in several recent studies. It was argued that a larger surface-to volume-ratio, resulting in shorter diffusion times, may be important for especially for organisms growing at low temperatures [18]. Gas vesicle production leads to a large surface area of the cell for nutrient acquisition but a relatively small volume of the cytoplasm [8,18,43]. A low growth temperature is one of the factors inducing gas vesicle synthesis in *Halobacterium* [47,69]. This may explain why the gas-vacuolate *Halobacterium* strains NRC-1 and PHH1 grow faster in the cold compared to the gas vesicle-negative strain PHH4 [48].

9. The Function of Gas Vesicles in the Life of Endospore-Forming Anaerobic Bacteria in the Sediments of Hypersaline Lakes

Even more enigmatic than the function of gas vesicles in the aerobic halophilic Archaea is the finding of gas vesicles in two obligate anaerobic halophilic representatives of the domain Bacteria (*Firmicutes*, Order *Halanaerobiales*): *Sporohalobacter lortetii* (basonym: *Clostridium lortetii*) [70,71] and *Orenia sivashensis* [72]. These were isolated from Dead Sea sediment and from sediment of Lake Sivash, Crimea, respectively.

Vegetative cells of *S. lortetii* do not carry gas vesicles. These structures are synthesized concomitant with the production of endospores, and they remain attached to the mature spores. Also in *O. sivashensis* gas vesicles were found in the mature spores, located between the inner membrane of the exosporium and the spore coat, but also in different parts of the vegetative cells. Similar anaerobes that produce endospores with attached gas vesicles are known from soils and other non-hypersaline environments. It was postulated that such spores, which are oxygen-resistant, may rise to the water surface and become dispersed by water currents until reaching a new anaerobic environment suitable for germination [73]. However, to what extent this mechanism may indeed function is unknown. Endospores of *S. lortetii* and *O. sivashensis* were never shown to float. Moreover, *S. lortetii* was recovered from Dead Sea sediment at a depth of 60 m, where the hydrostatic pressure far exceeds the critical collapse pressure of the gas vesicles, so that functional gas vesicles cannot exist at the site.

10. Epilogue

Helge Larsen, who in 1967 wrote his pioneering paper on the nature of the gas vesicles of *Halobacterium* [7], delivered in 1972 a lecture entitled "The Halobacteria's Confusion to Biology" [74], a title paraphrasing Kluyver and van Niel's "The Microbe's Contribution to Biology". Now, forty years later, considerable confusion still exists about the possible function of the gas vesicles in the life of *Halobacterium* and its gas-vacuolate relatives among the *Halobacteriaceae*. Whatever the ecological advantage of the production of gas vesicles may be, offsetting the cost of their production is as yet unclear.

We have learned very much about the structure of gas vesicles, the genes involved in their formation, and the regulation of their production in two model species: strains of *Hbt. salinarum* and *Hfx. mediterranei*. All those studies, however, have not yet led to an unequivocal answer why some members of the family need gas vesicles to remain competitive in their natural environment. A *Haloquadratum* culture left standing on the bench does not contain cells at the bottom; rather, the cells will stay in the water column, but will not float to the surface. Thriving at the surface of lakes and ponds would be very harsh: exposure to UV light, even more dryness, and if rainfall comes the cells will disrupt and die. Also, most isolates of *Hbt. salinarum* are unable to float; only the laboratory strains PHH1 and NRC-1 float at the surface when the culture is left standing on the bench. The *c*-vac expressing strains SB3, GN101, and PHH4 will remain distributed in the water column. This may show that the 'overproduction' of gas vesicles by PHH1 and NRC-1 might be a laboratory artifact. Still, the fact that the property is maintained in nature, in spite of the obvious cost to the cell that produces the vesicles, proves that there must be some benefits involved. In conclusion, the ecological advantage of gas vesicle production is not yet understood.

Kessel *et al.*, in a 1985 paper on the square gas-vacuolate microorganisms from the Sinai brine pools [52], wrote: "Although information is accumulating on the physiology and chemistry of gas vesicles in halobacteria, we still need experimental confirmation of their ecological significance in these organisms", and: "Laboratory experiments… will at best only test whether it is *feasible* for the gas vacuole to provide benefits through buoyancy. The actual benefits can only be assessed by making observations and performing experiments in the field". These words remain very true indeed, more than a quarter of century later.

Acknowledgments

I thank three anonymous reviewers for their helpful comments.

References

1. Oren, A. Family *Halobacteriaceae*. In *The Prokaryotes. A Handbook on the Biology of Bacteria: Ecophysiology and Biochemistry*, 4th ed.; Rosenberg, E., DeLong, E.F., Thompson, F., Lory, S., Stackebrandt, E., Eds.; Springer: New York, NY, USA, 2013; in press.
2. Petter, H.F.M. On bacteria of salted fish. *Proc. Kon. Akad. Wetensch. Ser. B* **1931**, *34*, 1417–1423.
3. Petter, H.F.M. Over Roode en Andere Bacteriën van Gezouten Visch (in Dutch). Ph.D. thesis, University of Utrecht, Utrecht, The Netherlands, 1932.
4. Sherwood, J.E.; Stagnitti, F.; Kokkinn, M.J.; Williams, W.D. A standard table for predicting equilibrium dissolved oxygen concentrations in salt lakes dominated by sodium chloride. *Int. J. Salt Lake Res.* **1992**, *1*, 1–6.

5. Hof, T. Investigations concerning bacterial life in strong brines. *Rec. Trav. Bot. Neerl.* **1935**, *32*, 92–173.
6. Houwink, A.L. Flagella, gas vacuoles and cell-wall structure in *Halobacterium. halobium*: An electron microscope study. *J. Gen. Microbiol.* **1956**, *15*, 146–150.
7. Larsen, H.; Omang, S.; Steensland, H. On the gas vacuoles of the halobacteria. *Arch. Mikrobiol.* **1967**, *59*, 197–203.
8. Pfeifer, F. Distribution, formation and regulation of gas vesicles. *Nature Rev. Microbiol.* **2012**, *10*, 705–715.
9. Walsby, A.E. Gas vesicles. *Microbiol. Rev.* **1994**, *58*, 94–144.
10. Walsby, A.E. The pressure relationships of gas vacuoles. *Proc. R. Soc. London B* **1971**, *178*, 301–326.
11. DasSarma, S.; Damerval, T.; Jones, J.G.; Tandeau de Marsac, N. A plasmid-encoded gas vesicle protein gene in a halophilic archaebacterium. *Mol. Microbiol.* **1987**, *1*, 365–370.
12. Pfeifer, F.; Weidinger, G.; Goebel, W. Genetic variability in *Halobacterium halobium*. *J. Bacteriol.* **1981**, *145*, 371–381.
13. Horne, M.; Englert, C.; Wimmer, C.; Pfeifer, F. A DNA region of 9 kbp contains all genes necessary for gas vesicle synthesis in halophilic archaebacteria. *Mol. Microbiol.* **1991**, *5*, 1159–1174.
14. Englert, C.; Krüger, K.; Offner, S.; Pfeifer, F. Three different but related gene clusters encoding gas vesicles in halophilic archaea. *J. Mol. Biol.* **1992**, *227*, 586–592.
15. Offner, S.; Hofacker, A.; Wanner, G.; Pfeifer, F. Eight of fourteen *gvp* genes are sufficient for formation of gas vesicles in halophilic archaea. *J. Bacteriol.* **2000**, *182*, 4328–4336.
16. Englert, C.; Wanner, G.; Pfeifer, F. Functional analysis of the gas vesicle gene cluster of the halophilic Archaea *Haloferax mediterranei* defines the vac-region boundary and suggests a regulatory role for the *gvpD* gene or its product. *Mol. Microbiol.* **1992**, *6*, 3543–3550.
17. Offner, S.; Ziese, U.; Wanner, G.; Typke, D.; Pfeifer, F. Structural characteristics of halobacterial gas vesicles. *Microbiology UK* **1998**, *144*, 1331–1342.
18. Pfeifer, F.; Krüger, K.; Röder, R.; Mayr, A.; Ziesche, S.; Offner, S. Gas vesicle formation in halophilic Archaea. *Arch. Microbiol.* **1997**, *167*, 259–268.
19. Pfeifer, F.; Gregor, D.; Hofacker, A.; Ploßer, P.; Zimmermann, P. Regulation of gas vesicle formation in halophilic archaea. *J. Mol. Microbiol. Biotechnol.* **2002**, *4*, 175–181.
20. Rodriguez-Valera, F.; Juez, G.; Kushner, D.J. *Halobacterium mediterranei* spec. nov., a new carbohydrate-utilizing extreme halophile. *System. Appl. Microbiol.* **1983**, *4*, 369–381.
21. Montalvo-Rodríguez, R.; Vreeland, R.H.; Oren, A.; Kessel, M.; Betancourt, C.; López-Garriga, J. *Halogeometricum borinquense* gen. nov., sp. nov., a novel halophilic archaeon from Puerto Rico. *Int. J. Syst. Bacteriol.* **1998**, *48*, 1305–1312.
22. Elevi Bardavid, R.; Mana, L.; Oren, A. *Haloplanus natans* gen. nov., sp. nov., an extremely halophilic gas-vacuolate archaeon from Dead Sea–Red Sea water mixtures in experimental mesocosms. *Int. J. Syst. Evol. Microbiol.* **2007**, *57*, 780–783.

23. Cui, H.-L.; Gao, X.; Li, X.-Y.; Xu, X.-W.; Zhou, Y.-G.; Liu, H.-C.; Zhou, P.-J. *Haloplanus vescus* sp. nov., an extremely *halophilic Archaea* from a marine solar saltern, and emended description of the genus *Haloplanus*. *Int. J. Syst. Evol. Microbiol.* **2010**, *60*, 1824–1827.
24. Cui, H.-L.; Gao, X.; Yang, X.; Xu, X.-W. *Haloplanus aerogenes* sp. nov., an extremely halophilic archaeon from a marine solar saltern. *Int. J. Syst. Evol. Microbiol.* **2011**, *61*, 965–968.
25. Burns, D.G.; Janssen, P.H.; Itoh, T.; Kamekura, M.; Li, Z.; Jensen, G.; Rodríguez-Valera, F.; Bolhuis, H.; Dyall-Smith, M.L. *Haloquadratum walsbyi* gen. nov., sp. nov., the square haloarchaeon of Walsby, isolated from saltern crystallizers in Australia and Spain. *Int. J. Syst. Evol. Microbiol.* **2007**, *57*, 387–392.
26. Mwatha, W.E.; Grant, W.D. *Natronobacterium vacuolata*, a haloalkaliphilic archaeon isolated from Lake Magadi, Kenya. *Int. J. Syst. Bacteriol.* **1993**, *43*, 401–404.
27. Gruber, C.; Legat, A.; Pfaffenhuemer, M.; Radax, C.; Weidler, G.; Busse, H.-J.; Stan-Lotter, H. *Halobacterium noricense* sp. nov., an archaeal isolate from a bore core of an alpine Permian salt deposit, classification of *Halobacterium* sp. NRC-1 as a strain of *H. salinarum* and emended description of *H. salinarum*. *Extremophiles* **2004**, *8*, 431–439.
28. Walsby, A.E. A square bacterium. *Nature* **1980**, *283*, 69–71.
29. Bolhuis, H. Walsby's square archaeon. It's hip to be square, but even more hip to be culturable. In *Adaptation to Life at High Salt Concentrations in Archaea, Bacteria, and Eukarya*; Gunde-Cimerman, N., Oren, A., Plemenitaš, A., Eds.; Springer: Dordrecht, The Netherlands, 2005; pp. 187–199.
30. Bolhuis, H.; te Poele, E.M.; Rodríguez-Valera, F. Isolation and cultivation of Walsby's square archaeon. *Environ. Microbiol.* **2004**, *6*, 1287–1291.
31. Burns, D.G.; Camakaris, H.M.; Janssen, P.H.; Dyall-Smith, M.L. Cultivation of Walsby's square haloarchaeon. *FEMS Microbiol. Lett.* **2004**, *238*, 469–473.
32. Sublimi-Saponetti, M.; Bobba, F.; Salerno, G.; Scarfato, A.; Corcelli, A.; Cucolo, A.M. Morphological and structural aspects of the extremely halophilic archaeon *Haloquadratum walsbyi*. *PLoS One* **2011**, doi:10.1371/journal.pone.0018653.
33. Bolhuis, H.; Palm, P.; Wende, A.; Falb, M.; Rampp, M.; Rodriguez-Valera, F.; Pfeiffer, F.; Oesterhelt, D. The genome of the square archaeon *Haloquadratum walsbyi*: life at the limits of water activity. *BMC Genomics* **2006**, *7*, 169.
34. Mayr, A.; Pfeifer, F. The characterization of the nv-*gpvACNOFGH* gene cluster involved in gas vesicle formation in *Natronobacterium vacuolatum*. *Arch. Microbiol.* **1997**, *168*, 24–32.
35. Walsby, A.E. Archaea with square cells. *Trends Microbiol.* **2005**, *13*, 193–195.
36. Beard, S.J.; Hayes, P.K.; Walsby, A.E. Growth competition between *Halobacterium salinarum* strain PHH1 and mutants affected in gas vesicle synthesis. *Microbiology UK* **1997**, *143*, 467–473.
37. Sundararajan, A.; Ju, L.-K. Use of cyanobacterial gas vesicles as oxygen carriers. *Cytotechnology* **2006**, *52*, 139–149.
38. Richard, T. Calculating the oxygen diffusion coefficient in water. Available online: http://compost.css.cornell.edu/oxygen/oxygen.diff.water.html (accessed on 27 November 2012).

39. Oren, A.; Trüper, H.G. Anaerobic growth of halophilic archaeobacteria by reduction of dimethylsulfoxide and trimethylamine *N*-oxide. *FEMS Microbiol. Lett.* **1990**, *70*, 33–36.
40. Hartmann, R.; Sickinger, H.-D.; Oesterhelt, D. Anaerobic growth of halobacteria. *Proc. Natl. Acad. Sci. USA* **1980**, *77*, 3821–3825.
41. Oesterhelt, D. Anaerobic growth of halobacteria. *Meth. Enzymol.* **1982**, *88*, 417–420.
42. Oren, A.; Litchfield, C.D. A procedure for the enrichment and isolation of *Halobacterium* species. *FEMS Microbiol. Lett.* **1999**, *173*, 353–358.
43. Hechler, T.; Pfeifer, F. Anaerobiosis inhibits gas vesicle formation in halophilic Archaea. *Mol. Microbiol.* **2009**, *71*, 132–145.
44. DasSarma, P.; Zamora, R.C.; Müller, J.A.; DasSarma, S. Genome-wide responses of the model archaeon *Halobacterium* sp. strain NRC-1 to oxygen limitation. *J. Bacteriol.* **2012**, *194*, 5530–5537.
45. Müller, J.A.; DasSarma, S. Genomic analysis of anaerobic respiration in the archaeon *Halobacterium* sp. strain NRC-1: Dimethyl sulfoxide and trimethylamine *N*-oxide as terminal electron acceptors. *J. Bacteriol.* **2005**, *187*, 1659–1667.
46. Englert, C.; Horne, M.; Pfeifer, F. Expression of the major gas vesicle protein in the halophilic archaebacterium *Haloferax mediterranei* is modulated by salt. *Mol. Gen. Genet.* **1990**, *222*, 225–232.
47. Röder, R.; Pfeifer, F. Influence of salt on the transcription of the gas-vesicle gene of *Haloferax mediterranei* and identification of the endogeneous transcriptional activator. *Microbiology UK* **1996**, *142*, 1715–1723.
48. Bleiholder, A.; Frommherz, R.; Teufel, K.; Pfeifer, F. Expression of multiple *tfb* genes in different *Halobacterium salinarum* strains and interaction of TFB with transcriptional activator GvpE. *Arch. Microbiol.* **2012**, *194*, 269–279.
49. Coker, J.; DasSarma, P.; Kumar, J.; Müller, J.; DasSarma, S. Transcriptional profiling of the model archaeon *Halobacterium* sp. NRC-1: Responses to changes in salinity and temperature. *Saline Syst.* **2007**, *3*, 6.
50. Bickel-Sandkötter, S.; Gärtner, W.; Dane, M. Conversion of energy in halobacteria: ATP synthesis and phototaxis. *Arch. Microbiol.* **1996**, *166*, 1–11.
51. Lobasso, S.; Lopalco, P.; Vitale, R.; Sublimi Saponetti, M.; Capitanio, G.; Mangini, V.; Milano, F.; Trotta, M.; Corcelli, A. The light-activated proton pump BopI of the archaeon *Haloquadratum walsbyi*. *Photochem. Photobiol.* **2012**, *88*, 690–700.
52. Kessel, M.; Cohen, Y.; Walsby, A.E. Structure and physiology of square-shaped and other halophilic bacteria from the Gavish Sabkha. In *Hypersaline Ecosystems. The Gavish Sabkha*; Friedman, G.M., Krumbein, W.E., Eds.; Springer-Verlag: Berlin, Germany, 1985; pp. 267–287.
53. Simon, R.D. Interactions between light and gas vacuoles in *Halobacterium salinarium* strain 5: effect of ultraviolet light. *Appl. Environ. Microbiol.* **1980**, *40*, 984–987.
54. Oren, A. *Halophilic Microorganisms and Their Environments*; Kluwer Scientific Publishers: Dordrecht, The Netherlands, 2002.

55. Lopalco, P.; Lobasso, S.; Baronio, M.; Angelini, R.; Corcelli, A. Chapter 6, In *Halophiles and Hypersaline Environments*; Ventosa, A., Ed.; Springer-Verlag: Berlin, Germany, 2011; pp. 123–135.
56. Bodaker, I.; Sharon, I.; Suzuki, M.T.; Reingersch, R.; Shmoish, M.; Andreishcheva, E.; Sogin, M.L.; Rosenberg, M.; Belkin, S.; Oren, A.; Béjà, O. Comparative community genomics in the Dead Sea: an increasingly extreme environment. *ISME J.* **2010**, *4*, 399–407.
57. Antón, J.; Llobet-Brossa, E.; Rodríguez-Valera, F.; Amann, R. Fluorescence *in situ* hybridization analysis of the prokaryotic community inhabiting crystallizer ponds. *Environ. Microbiol.* **1999**, *1*, 517–523.
58. Oren, A.; Duker, S.; Ritter, S. The polar lipid composition of Walsby's square bacterium. *FEMS Microbiol. Lett.* **1996**, *138*, 135–140.
59. Romanenko, V.I. Square microcolonies in the surface water film of the Saxkoye lake (in Russian). *Mikrobiologiya (USSR)* **1981**, *50*, 571–574.
60. Rodriguez-Valera, F.; Ventosa, A.; Juez, G.; Imhoff, J.F. Variation of environmental features and microbial populations with salt concentrations in a multi-pond saltern. *Microb. Ecol.* **1985**, *11*, 107–115.
61. Warkentin, M.; Schumann, R.; Oren, A. Community respiration studies in saltern crystallizer ponds. *Aquat. Microb. Ecol.* **2009**, *56*, 255–261.
62. Stoeckenius, W. Walsby's square bacterium: Fine structure of an orthogonal prokaryote. *J. Bacteriol.* **1981**, *148*, 352–360.
63. Kessel, M.; Cohen, Y. Ultrastructure of square bacteria from a brine pool in southern Sinai. *J. Bacteriol.* **1982**, *150*, 851–860.
64. Parkes, K.; Walsby, A.E. Ultrastructure of a gas-vacuolate square bacterium. *J. Gen. Microbiol.* **1981**, *126*, 503–506.
65. Oren, A.; Priel, N.; Shapiro, O.; Siboni, N. Buoyancy studies in natural communities of square gas-vacuolate archaea in saltern crystallizer ponds. *Saline Syst.* **2006**, *2*, 4.
66. Denny, M.W. *Air and Water: The Biology and Physics of Life's Media*; Princeton University Press: Princeton, New Jersey, USA, 1993.
67. McNown, J.S.; Malaika, J. Effect of particle shape on settling velocity at low Reynolds numbers. *Trans. Amer. Geophys. Union* **1950**, *31*, 74–82.
68. Purcell, E.M. Life at low Reynolds number. *Amer. J. Phys.* **1977**, *45*, 3–11.
69. Coker, J.A.; DasSarma, P.; Kumar, J.; Müller, J.A.; DasSarma, S. Transcriptional profiling of the model archaeon *Halobacterium* sp. NRC-1: Responses to changes in salinity and temperature. *Saline Syst.* **2007**, *3*, 6.
70. Oren, A. *Clostridium lortetii* sp. nov., a halophilic obligately anaerobic bacterium producing endospores with attached gas vacuoles. *Arch. Microbiol.* **1983**, *136*, 42–48.
71. Oren, A.; Pohla, H.; Stackebrandt, E. Transfer of *Clostridium lortetii* to a new genus *Sporohalobacter* gen. nov. as *Sporohalobacter lortetii* comb. nov., and description of *Sporohalobacter marismortui* sp. nov. *System. Appl. Microbiol.* **1987**, *9*, 239–246.

72. Zhilina, T.N.; Tourova, T.P.; Kuznetsov, B.B.; Kostrikina, N.A.; Lysenko, A.M. *Orenia sivashensis* sp. nov., a new moderately halophilic anaerobic bacterium from lake Sivash lagoons. *Microbiology* (Russia) **1999**, *68*, 452–459.
73. Duda, V.I.; Makar'eva, E.D. Morphogenesis and function of gas caps on spores of anaerobic bacteria of the genus (in Russian). *Mikrobiologiya* **1978**, *70*, 689–694.
74. Larsen, H. The halobacteria's confusion to biology. *Antonie van Leeuwenhoek* **1973**, *39*, 383–396.

Pivotal Enzyme in Glutamate Metabolism of Poly-γ-Glutamate-Producing Microbes

Makoto Ashiuchi, Takashi Yamamoto and Tohru Kamei

Abstract: The extremely halophilic archaeon *Natrialba aegyptiaca* secretes the L-homo type of poly-γ-glutamate (PGA) as an extremolyte. We examined the enzymes involved in glutamate metabolism and verified the presence of L-glutamate dehydrogenases, L-aspartate aminotransferase, and L-glutamate synthase. However, neither glutamate racemase nor D-amino acid aminotransferase activity was detected, suggesting the absence of sources of D-glutamate. In contrast, D-glutamate-rich PGA producers mostly possess such intracellular sources of D-glutamate. The results of our present study indicate that the D-glutamate-anabolic enzyme "glutamate racemase" is pivotal in the biosynthesis of PGA.

Reprinted from *Life*. Cite as: Ashiuchi, M.; Yamamoto, T.; Kamei, T. Pivotal Enzyme in Glutamate Metabolism of Poly-γ-Glutamate-Producing Microbes. *Life* **2013**, *3*, 181-188.

1. Introduction

"No homochirality, no life" [1]. This saying may be accepted as true from the study of biologically essential polymers. Most proteins and polysaccharides are indeed comprised of L-amino acids and D-dominant sugars, respectively, and are considered homochiral molecules. Furthermore, D-ribose is required for *in vivo* synthesis of nucleic acids. Although a large body of evidence indicates that life originated in an aqueous environment and was accompanied by the generation of such chiral molecules of different sizes (Figure 1a), there are some contradictions from the perspective of polymer chemistry. Generally, the chemical polymerization of substrate monomers is performed under water-limited (or anhydrous) conditions, because this process depends on condensation reactions involving dehydration (removal of a H_2O molecule) between two functional groups of interest. Biopolymers in water will therefore undergo hydrolysis (for degradation) rather than elongation (for synthesis). This stance does not contradict knowledge that *in vivo* syntheses of biopolymers involves strictly water-limited spaces, *e.g.*, the cell membranes or hydrophobic clefts of supramolecular machineries. Also to be considered is that the homochirality of substrate monomers facilitates cleavage catalyzed by water [2,3]. Maybe, water was indispensable for the final changes in the chemical transformations that made life possible (Figure 1b). It is well known that a crystal is formed by the systematic assembly of molecules possessing the same conformation. Thus, if researchers had proved that the primitive, spontaneous polymerization reactions occurred in the absence of water, specifically during crystallization, they may not be perplexed by the origin of homochirality in biomass. In contrast, the search for the occurrence of heterochirality in the biological world is becoming a rather attractive pursuit for researchers.

Figure 1. A speculative chemical process in which water made life possible. (a) A general scheme in which water continuously participated in all steps in the creation of life (*solid arrows*). (b) An alternative view indicating that water participated intermittently in some limited reactions (*dotted arrows*). When various materials containing homochiral polymers are assembled to generate supramolecular machineries with biological functions, several physicochemical forces mediated by water molecules, *e.g.*, the formation of hydrophobic interaction and hydrogen bonds, are indispensable. Essential water molecules may participate in a later stage of the process.

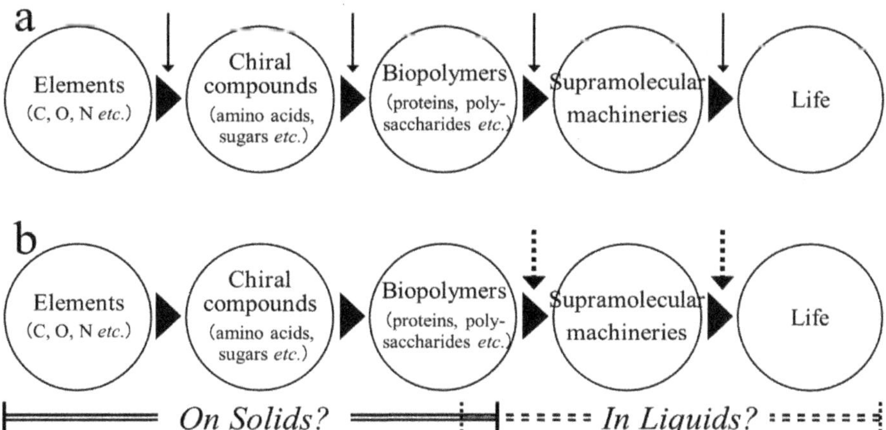

The extremely halophilic archaeon *Natrialba aegyptiaca* synthesizes the extracellular polyamide poly-γ-glutamate (L-PGA), which consists of more than 10,000 molecules of L-glutamate polymerized *via* γ-amide linkages and encompasses a chiral center in every glutamyl residue. This archaeal polymer therefore exhibits protein-like homochirality. However, well-characterized PGAs generally contain D-glutamate [4] and endogenous D-glutamate is likely to be a substrate for bacterial PGA synthetases [5]. In the present study, we focus on glutamate metabolism by *N. aegyptiaca* and also discuss a pivotal enzyme involved in the synthesis of microbial PGAs.

2. Results and Discussion

2.1. Determination of Sources of L-glutamate in the Production of L-PGA by N. aegyptiaca

L-PGA is known to possess extremolyte-like applicability [6]. Extremolytes are exclusively isolated from extremophiles and protect biological macromolecules and cells from damage by external stress [7]. As L-glutamate serves as the best substrate for *in vivo* synthesis of microbial PGAs [4], we first assayed for the activities of enzymes involved in L-glutamate synthesis in *N. aegyptiaca*. Figure 2 revealed the presence of NAD(P)H-dependent L-glutamate dehydrogenases (GDH; *a* and *b*), pyridoxal 5'-phosphate-dependent L-aspartate aminotransferase (GOT; *c*), and NADPH-dependent L-glutamate synthase (GS; *d*). We were unable to detect the activity of $NAD(P)^+$-dependent dehydrogenases using other amino acids (e.g., L-alanine, L-aspartate, L-lysine, L-phenylalanine, and L-serine) as a substrate. L-Glutaminase (GLS; *e*) was also inactive.

N. aegyptiaca hence possesses multiple enzymatic sources of L-glutamate, similarly to well-characterized PGA producers from bacilli [4]. Moreover, the absence of glutamate oxidases indicated that glutamate anabolism predominates compared with its catabolism.

Figure 2. Intracellular activities for L-glutamate sources of *N. aegyptiaca* showing higher L-PGA productivity (means ± standard deviation; n = 5). One unit (U) was defined as the amount of enzyme that catalyzed the formation of 1 μmol of L-glutamate per min. Enzymes *a*, GDH (NADH); *b*, GDH (NADPH); *c*, GOT; *d*, GS; and *e*, GLS.

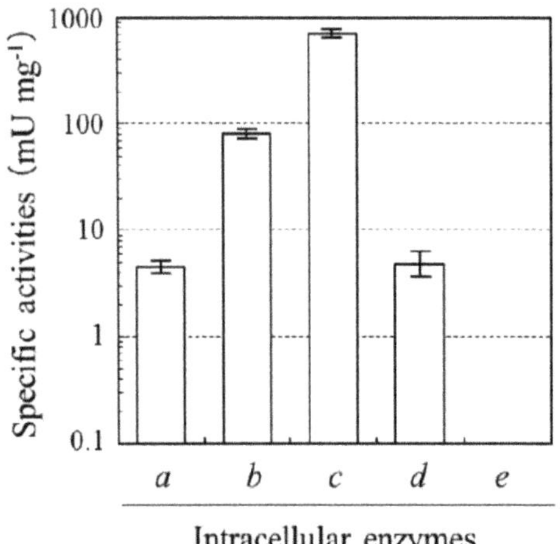

(*a*) α-Ketoglutarate + NH_3 + NADH
⇄ L-Glutamate + H_2O + NAD^+

(*b*) α-Ketoglutarate + NH_3 + NADPH
⇄ L-Glutamate + H_2O + $NADP^+$

(*c*) α-Ketoglutarate + L-Aspartate
⇄ L-Glutamate + Oxaloacetate

(*d*) α-Ketoglutarate + L-Glutamine + NADPH
⇄ 2×L-Glutamate + $NADP^+$

(*e*) L-Glutamine ⇄ L-Glutamate + NH_3

2.2. Absence of D-glutamate Suppliers in N. aegyptiaca

Although the pathways involved in D-glutamate metabolism have not, to our knowledge, been identified in *N. aegyptiaca*, the use of well-characterized PGA producers allowed us to conduct an in-depth investigation of the biosynthesis of D-glutamate [4]. Two distinct types of the suppliers have been confirmed [4] as follows: glutamate racemase (1) and D-amino acid aminotransferase (2).

$$\text{L-Glutamate} \rightleftarrows \underline{\text{D-glutamate}} \quad (1)$$

$$\text{D-Alanine} + \alpha\text{-Ketoglutarate} \rightleftarrows \text{Pyruvate} + \underline{\text{D-Glutamate}} \quad (2)$$

In addition to these suppliers, we also examined the activity of an enzyme, NAD(P)H-dependent D-glutamate dehydrogenases (3).

$$\alpha\text{-Ketoglutarate} + NH_3 + \text{NAD(P)H} \rightleftarrows \underline{\text{D-Glutamate}} + H_2O + \text{NAD(P)}^+ \quad (3)$$

Under the conditions tested, neither glutamate racemase nor D-amino acid aminotransferase activity was detected. Moreover, our attempts to discover NAD(P)H-dependent D-glutamate dehydrogenases were unsuccessful, suggesting that the normal metabolism of *N. aegyptiaca* prevents the synthesis of D-glutamate and can eventually produce the L-homo type of PGA. This conclusion does not conflict with findings that PGA with different stereochemical conformations can be synthesized by the co-expression of a D-glutamate supplier, glutamate racemase, in an *Escherichia coli* strain genetically-engineered to produce PGA [8]. Besides, other amino acid racemases were absent in *N. aegyptiaca* as well, implying that this archaeon essentially has no potential for the production of D-amino acids.

Figure 3. Time course of "L to D" (*open circles*) and "D to L" (*closed circles*) conversions in glutamate racemization of *B. subtilis*. If the D concentration is exceedingly higher than the L concentration in the coexistence of both enantiomers, the intracellular racemase may suffer a substrate inhibition due to its peculiar conformational change [9,10].

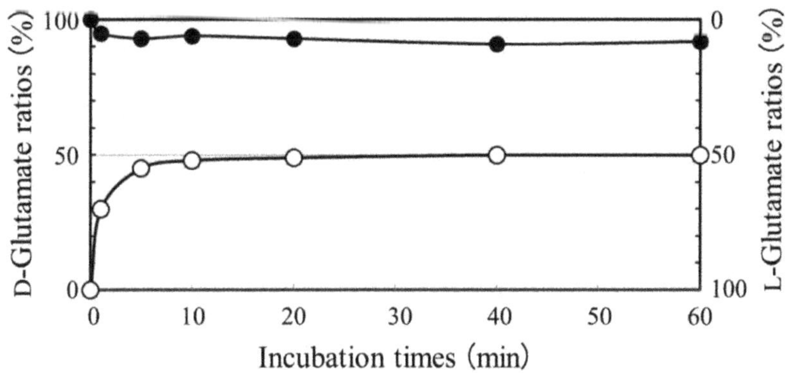

2.3. Analysis of a Pivotal D-glutamate Supplier in B. Subtilis Producing DL-PGA

Bacillus subtilis and *B. anthracis* produce DL- and D-PGA, respectively. The chirality of *B. subtilis* PGA is thus heterogeneous, whereas *B. anthracis* PGA possesses homochirality, which is symmetric to the PGA produced by *N. aegyptiaca*. In the present study, we focused our attention on amino acid racemase, a pivotal enzyme that can connect the L-amino acid world with its mirror world. In fact, only this racemase utilizes both enantiomers of amino acid substrates among known enzymes involved in amino acid biosynthesis. We thus examined the role of glutamate racemase in glutamate metabolism using the cytosolic fraction of *B. subtilis*. Figure 3 shows kinetics of the conversion (racemization) of glutamate from "L to D" and "D to L". The data indicate that *B. subtilis* prefers the former conversion. Therefore, the glutamate racemase probably serves physiologically in the anabolism of PGA (to supply D-glutamate) rather than to catabolize PGA (degradation of the resulting D-glutamate) in a manner that contradicts published speculations [11,12].

3. Experimental Section

3.1. Microbes and Culture Conditions

N. aegyptiaca was cultured at 37 °C for 4 days in 200 mL of an S medium (pH 7.2) containing 25% NaCl, 0.2% KCl, 1% trisodium citrate, 0.75% casamino acids, and 1% yeast extract, and the cells were harvested by centrifugation. To enhance PGA productivity, harvested cells (0.4 g) were then cultured on a 2% agar plate (145 mm diameter) containing S medium (50 ml) at 37 °C for 2 weeks. Under these conditions, the productivity reached ~1.2 g/g of cells.

B. subtilis was cultured in 200 ml of Luria-Bertani medium [13] at 37 °C for 16 h and harvested by centrifugation. To induce PGA synthesis [14], harvested cells (0.4 g) were cultured at 37 °C for 4 days on a 1.5% agar plate (145 mm diameter) containing a GS medium (50 ml, pH 7.0) containing 2% L-glutamate, 5% sucrose, 1% $(NH_4)_2SO_4$, 0.5% $MgSO_4$•$7H_2O$, 0.27% KH_2PO_4, 0.42% Na_2HPO_4, 0.05% NaCl, and 1% yeast extract. Under these conditions, PGA productivity reached ~0.8 g/g of cells.

3.2. Preparation of Cytosolic Fractions

The harvested cells (1 g) were suspended in 1 mL of a standard buffer (0.1 M Mops-NaOH (pH 7.0), 0.2 M KCl, and 5 mM dithiothreitol), sonicated (Branson, CT, USA), centrifuged at 12,000 × G for 10 min, and further centrifuged at 39,000 × G for 30 min. After dialysis at 4°C overnight against a 1,000-fold volume of the same buffer, all particulates were removed by filtration. The resulting supernatant (cytosolic fraction) does not show the activity of extracellular γ-glutamyltransferase, which liberates glutamate monomers from the N-terminal end of PGA [12], and was analyzed as described below.

3.3. Enzyme Assays

3.3.1. Sources of L-Glutamate

Dehydrogenase: The reaction mixture contained 0.1 M Tris-HCl (pH 8.0), 5 mM α-ketoglutarate, 0.4 mM NADH (or NADPH), and the cytosolic fraction (1 mg of proteins). Reactions were initiated by the addition of 80 mM NH_4Cl and assayed by following NAD(P)H oxidation (decrease in absorbance at 340 nm or 360 nm) at 37 °C.

Aminotransferase: The reaction mixture contained 0.1 M Tris-HCl (pH 8.0), 50 mM L-aspartate, 0.1 mM pyridoxal 5'-phosphate, 0.4 mM NADH, commercially available malate dehydrogenase (MDH; 2.5 U), and the cytosolic fraction. Reactions were initiated by the addition of 10 mM α-ketoglutarate and assayed by coupling the reaction to MDH and following NADH oxidation (decrease in absorbance at 340 nm) at 37 °C.

Synthase: The reaction mixture contained 0.1 M Tris-HCl (pH 8.0), 5 mM α-ketoglutarate, 0.4 mM NADPH, and the cytosolic fraction. Reactions were initiated by the addition of 80 mM L-glutamine, and assayed by following NADPH oxidation (decrease in absorbance at 360 nm) at 30 °C.

Glutaminase (Amidohydrolase): The reaction mixture contained 0.1 M Tris-HCl (pH 7.5), 5 mM NAD^+, commercially available GDH (0.5 U), and the cytosolic fraction. Reactions were initiated by the addition of 50 mM L-glutamine and assayed by coupling the reaction to GDH and following NAD^+ reduction (increase in absorbance at 340 nm) at 30 °C.

3.3.2. D-Glutamate Suppliers

Reaction mixtures (0.2 mL) were incubated at 37 °C for indicated times, inactivated with 8 μL of 12 M HCl, neutralized with 16 μL of 6 M NaOH, and then diluted 5-fold with 2 mM $CuSO_4$. A 5-μL sample was withdrawn and loaded onto a CHIRALPAK MA(+) column (4.6 × 50 mm; DAISEL, Tokyo, Japan) using an LC-10 HPLC system (Shimadzu, Kyoto, Japan). The column was eluted with 2 mM $CuSO_4$ solution at the flow rate of 1 ml min^{-1}, and the absorbance of the eluate was monitored at 235 nm. The elution volumes of D- and L-glutamate were approximately 10 and 15 ml, respectively. Yields (fmol) were estimated using the following equations: $Y_D = 2.97\times$ and $Y_L = 2.91\times$ (where × represents each peak area on the HPLC profiles) [15].

Racemase: The reaction mixture contained 0.1 M Tris-HCl (pH 8.0), 10 mM L-glutamate (or D-glutamate), and the cytosolic fraction.

Aminotransferase: The reaction mixture contained 0.1 M Tris-HCl (pH 8.0), 50 mM D-alanine, 10 mM α-ketoglutarate, 0.1 mM pyridoxal 5'-phosphate, and the cytosolic fraction.

Dehydrogenase: The reaction mixture consisted of the same components as the mixtures for the L-glutamate dehydrogenase assays described above.

3.3.3. Other Enzymes

Glutamate Oxidases catalyze the following reaction: Glutamate + O_2 + H_2O → α-Ketoglutarate + NH_3 + H_2O_2. These are two distinct types of glutamate catabolism catalyzed by L- or D-glutamate oxidases. The reaction mixture contained 0.1 M Tris-HCl (pH 8.0), 10 mM L- or D-glutamate, 0.03% 2,4-dinitrophenylhydrozine, and the cytosolic fraction. The formation of 2,4-dinitrophenylhydrozone of α-ketoglutarate after incubation at 37°C was followed by measuring the increase in absorbance at 550 mm.

Other amino acid racemases catalyze the following reaction: L-amino acid(s) ⇄ D-amino acid(s). The reaction mixtures contained 0.1 M Tris-HCl (pH 8.0), 10 mM L-amino acid other than L-glutamate (e.g., L-alanine, L-aspartate, L-lysine, L-phenylalanine, and L-serine), 0.1 mM pyridoxal 5'-phosphate, and the cytosolic fraction of *N. aegyptiaca*. After incubation at 37 °C, they were assayed by the HPLC methods [16].

3.4. Protein Assay

Protein concentrations were determined using a protein assay kit (Bio-Rad, CA, USA) using bovine serum albumin as a standard.

4. Concluding Remarks

From the perspective of polymer chemistry, a question arises whether homochiral biopolymers that possess highly complex structures that contribute to the selection and persistence of chirality were generated spontaneously in water, a solvent that randomizes the chirality of chemical compounds. We believe that it is necessary to discuss in detail the possibility that water could have participated in the chemical reactions that made life possible (Figure 1), and that the disruption of molecular chirality that characterizes biological processes in life may not have occurred in the environment before the birth of life. Archaea essentially inhabit an L-amino acid world owing to the absence of amino acid racemases, while eubacteria, to survive from adverse biological impacts, such as attack by proteases, envelop cell surfaces with peptidoglycans containing certain D-amino acids.

Mammals were once believed to reside in an L-amino acid world along with archaea; however, D-serine has been identified in humans and other animals as a mediator of memory, synthesized by the enzyme "serine racemase" [17]. Although D-PGA is not toxic to mammals, it nullifies the immunity of hosts and can contribute to the severity of diseases "anthrax" [4]. These may suggest a novel story based on the view that "no chirality–breaking, no evolution (or no adaptation)" for life based on homochirality. If so, when did the first glutamate racemase appear in the L-amino acid world? Can a mirror form of glutamate racemase, created only from D-amino acids, yet catalyze the racemization of glutamate? It may be the case that the glutamate racemase will be considered as the pivotal enzyme responsible for modifying the present L-dominant glutamate-metabolic pathways in nature (Figure 4).

Figure 4. L-Glutamate-dominant world, its mirror world, and a pivotal enzyme connecting both worlds. Enzymes *a*, GDHs; *b*, L-amino acid aminotransferases including GOT; *c*, GS; *d*, glutamate racemase; and *e*, D-amino acid aminotransferase.

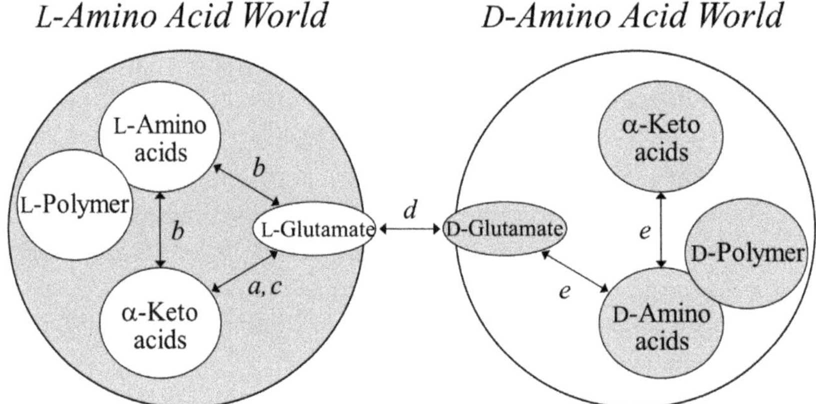

References

1. Shimada, A.; Ozaki, H. Flexible enantioselectivity of tryptophanase attributable to benzene ring in heterocyclic moiety of D-tryptophan. *Life* **2012**, *2*, 215–228.
2. Cohen, B.A.; Chyba, C.F. Racemization of meteoritic amino acids. *Icarus* **2000**, *145*, 272–281.
3. Bada, J.L.; Miller, S.L. Racemization and the origin of optically active organic compounds in living organisms. *Biosystems* **1987**, *20*, 21–26.
4. Ashiuchi, M.; Misono, H. Poly-γ-glutamic acid, In *Biopolymers*; Fahnestock, S.R., Steinbüchel, A., Eds.; Wiley-VCH: Weinheim, Germany, 2002; Volume 7, Chapter 6, pp. 123–174.

5. Ashiuchi, M. Occurrence and biosynthetic mechanism of poly-γ-glutamic acid, In *Microbiol. Monogr.*: *Amino-Acid Homopolymers Occurring in Nature*; Hamano, Y., Ed.; Springer-Verlag: Heidelberg, Germany, 2010; Volume 15, pp. 77–94.
6. Yamasaki, D.; Minouchi, Y.; Ashiuchi, M. Extremolyte-like applicability of an archaeal exo-polymer, poly-γ-L-glutamate. *Environ. Technol.* **2010**, *31*, 1129–1134.
7. Lentzen, G.; Schwarz, T. Extremolytes: natural compounds from extremophiles for versatile applications. *Appl. Microbiol. Biotechnol.* **2006**, *72*, 623–634.
8. Ashiuchi, M.; Soda, K.; Misono, H. A poly-γ-glutamate synthetic system of *Bacillus subtilis* IFO 3336: Gene cloning and biochemical analysis of poly-γ-glutamate produced by *Escherichia coli* clone cells. *Biochem. Biophys. Res. Commun.* **1999**, *263*, 6–12.
9. Ruzheinikov, S.N.; Taal, M.A.; Sedelnikova, S.E.; Baker, P.J.; Rice, D.W. Substrate-induced conformational changes in *Bacillus subtilis* glutamate racemase and their implications for drug discovery. *Structure* **2005**, *13*, 1707–1713.
10. Lundqvist, T.; Fisher, S.L.; Kern, G.; Folmer, R.H.A.; Xue, Y.; Newton, D.T.; Keating, T.A.; Alm, R.A.; De Jonge, B.L.M. Exploitation of structural and regulatory diversity in glutamate racemases. *Nature* **2007**, *447*, 817–822.
11. Kimura, K.; Tran, L.-S.P.; Itoh, Y. Roles and regulation of the glutamate racemase isogenes, *racE* and *yrpC*, in *Bacillus subtilis. Microbiology* **2004**, *150*, 2911–2920.
12. Kimura, K.; Tran, L.-S.P.; Uchida, I.; Itoh, Y. Characterization of *Bacillus subtilis* γ-glutamyltrans-ferase and its involvement in the degradation of capsule poly-γ-glutamate. *Microbiology* **2004**, *150*, 4115–4123.
13. Sambrook, J.; Fritsch, E.F.; Maniatis, T. *Molecular Cloning: A Laboratory Manual*, 2nd ed.; Cold Spring Harbor Laboratory: Cold Spring Harbor, NY, USA, 1989.
14. Ashiuchi, M.; Kamei, T.; Baek, D.-H.; Shin, S.-Y.; Sung, M.-H.; Soda, K.; Yagi, T.; Misono, H. Isolation of *Bacillus subtilis* (*chungkookjang*), a poly-γ-glutamate producer with high genetic competence. *Appl. Microbiol. Biotechnol.* **2001**, *57*, 764–769.
15. Ashiuchi, M.; Shimanouchi, K.; Nakamura, H.; Kamei, T.; Soda, K.; Park, C.; Sung, M.-H.; Misono, H. Enzymatic synthesis of high-molecular-mass poly-γ-glutamate and regulation of its stereochemistry. *Appl. Environ. Microbiol.* **2004**, *70*, 4249–4255.
16. Ashiuchi, M. Analytical approaches to poly-γ-glutamate: Rapid quantification, molecular size determination, and stereochemistry investigation. *J. Chromatogr. B* **2011**, *879*, 3096–3101.
17. Wolosker, H.; Blackshaw, S.; Snyder, S.H. Serine racemase: A glial enzyme synthesizing D-serine to regulate glutamate-*N*-methyl-D-aspartate neurotransmission. *Proc. Natl. Acad. Sci. USA* **1999**, *96*, 13409–13414.

Halophilic Bacteria as a Source of Novel Hydrolytic Enzymes

María de Lourdes Moreno, Dolores Pérez, María Teresa García and Encarnación Mellado

Abstract: Hydrolases constitute a class of enzymes widely distributed in nature from bacteria to higher eukaryotes. The halotolerance of many enzymes derived from halophilic bacteria can be exploited wherever enzymatic transformations are required to function under physical and chemical conditions, such as in the presence of organic solvents and extremes in temperature and salt content. In recent years, different screening programs have been performed in saline habitats in order to isolate and characterize novel enzymatic activities with different properties to those of conventional enzymes. Several halophilic hydrolases have been described, including amylases, lipases and proteases, and then used for biotechnological applications. Moreover, the discovery of biopolymer-degrading enzymes offers a new solution for the treatment of oilfield waste, where high temperature and salinity are typically found, while providing valuable information about heterotrophic processes in saline environments. In this work, we describe the results obtained in different screening programs specially focused on the diversity of halophiles showing hydrolytic activities in saline and hypersaline habitats, including the description of enzymes with special biochemical properties. The intracellular lipolytic enzyme LipBL, produced by the moderately halophilic bacterium *Marinobacter lipolyticus,* showed advantages over other lipases, being an enzyme active over a wide range of pH values and temperatures. The immobilized LipBL derivatives obtained and tested in regio- and enantioselective reactions, showed an excellent behavior in the production of free polyunsaturated fatty acids (PUFAs). On the other hand, the extremely halophilic bacterium, *Salicola marasensis* sp. IC10 showing lipase and protease activities, was studied for its ability to produce promising enzymes in terms of its resistance to temperature and salinity.

Reprinted from *Life*. Cite as: de Lourdes Moreno, M.; Pérez, D.; García, M.T.; Mellado, E. Halophilic Bacteria as a Source of Novel Hydrolytic Enzymes. *Life* **2013**, *3*, 38-51.

1. Introduction

Microbial life can be found over a wide range of extreme conditions (salinity, pH, temperature, pressure, light intensity, oxygen and nutrient conditions). Hypersaline environments constitute typical examples of environments with extreme conditions due to their high salinity, exposure to high and low temperatures, low oxygen conditions and in some cases, high pH values. Bacteria and Archaea are the most widely distributed organisms in these environments [1].

The classification of Kushner and Kamekura [2] defines different categories of halophilic microorganisms based on the optimal salt concentration wherein they show optimal growth, and it includes four categories: non-halophilic organisms are defined as those requiring less than 1% NaCl, whereas if they can tolerate high salt concentrations are considered as halotolerant microorganisms. With respect to halophilic microorganisms, the classification distinguishes among slight halophiles (marine bacteria), which grow best in media with 1% to 3% NaCl, moderate halophiles, growing best in media with 3% to 15% NaCl, and extreme halophiles, which show optimal growth in media containing 15% to 30% NaCl.

Halophiles have developed two different adaptive strategies to cope with the osmotic pressure induced by the high NaCl concentration of the normal environments they inhabit [3,4]. The halobacteria and some extremely halophilic bacteria accumulate inorganic ions in the cytoplasm (K^+, Na^+, Cl^-) to balance the osmotic pressure of the medium, and they have developed specific proteins that are stable and active in the presence of salts. In contrast, moderate halophiles accumulate in the cytoplasm high amounts of specific organic osmolytes, which function as osmoprotectants, providing osmotic balance without interfering with the normal metabolism of the cell [5].

In recent years, halophilic microorganisms have been explored for their biotechnological potential in different fields [6]. The applications range from the use of different products, such as the compatible solutes, biopolymers or carotenoids in a variety of industries or the use of these microorganisms in environmental bioremediation processes. Besides being intrinsically stable and active at high salt concentrations, halophilic enzymes offer important opportunities in biotechnological applications, such as food processing, environmental bioremediation and biosynthetic processes. In this sense, the finding of novel enzymes showing optimal activities at various ranges of salt concentrations, temperatures and pH values is of great importance [7]. It is important to highlight that the use of enzymes from halophiles in industrial applications are not limited to their stability at high salt concentrations, since these extremozymes usually are also tolerant to high temperatures and they are stable in presence of organic solvents [8].

In general, low water activity (a_w) produces conformation changes in the enzymes affecting the catalytic activity due to the reduced hydration. However, halophilic enzymes are active and stable in media with low water activity because even at low a_w enough water is present to retain suitable charge distribution at the active site maintaining the conformation of the enzyme [9]. Organic solvents with a log P lower than 2, generally are considered to cause enzyme denaturation, producing the distortion of the water-biocatalyst interactions. However, several studies have been reported on hydrolases produced by extremophiles isolated from saline environments that are very stable in solutions containing organic solvents [10].

This review will be focused on the diversity and biotechnological applications of the novel described enzymes from two groups of bacteria: the moderate and the extreme halophiles. The adaptation to live in hypersaline environments give rise to these extremophiles advantages to be exploited from a biotechnological point of view.

2. Diversity of Halophilic Bacteria Showing Hydrolytic Activities

Most research studies performed on hypersaline environments have been focused on the microbial diversity and ecology of these environments and progress in understanding the systematic, cellular function and metabolic activities of halophiles have been achieved. However, environmental studies based on the diversity of halophiles showing hydrolytic activities in saline habitats remain largely unexplored; however, the hydrolysis of high-molecular-weight biopolymers constitutes an initial step in the metabolism of organic compounds in the different ecosystems, playing an important role in the geochemical cycling of nutrients in salterns. In this sense, these studies must be considered of great interest due to the great biotechnological potential exhibited by these enzymes.

In the last few years, different screening programs have been carried out to study the diversity of microorganisms producing hydrolytic enzymes throughout direct plating on agar media supplemented with specific substrates for the enzymes of interest. A limited number of halophilic bacteria showing hydrolytic activities have been isolated and characterized from different hypersaline habitats, such as solar salterns, salt lakes, saline deserts and saline deposits (Table 1).

2.1. Screening on Solar Salterns

Sánchez-Porro *et al.* [11] showed the abundance of five hydrolases including amylase, protease, lipase, DNase and pullulanase in a community of moderate halophiles isolated from water and sediment (superficial layer) in Spanish salterns, describing amylase producers as the most abundant isolates. Most environmental isolates able to produce hydrolytic enzymes belonged to the Gram-negative genera *Salinivibrio* and *Halomonas*, two genera widely distributed in hypersaline environments [12,13]. Among the Gram-positive, representatives of the genera *Bacillus* and *Salibacillus* were predominant. Isolates producing lipases were very diverse from a phylogenetic point of view. However, the pullulanase producers were limited to representatives of the genera *Salinivibrio*, *Halomonas* and *Bacillus-Salibacillus*. Four strains presented the five hydrolytic activities tested and multiple hydrolytic activities were detected in a few strains. A moderately halophilic bacterium (strain SM19^T) displaying lipolytic activity was isolated and characterized. Strain SM19^T is a Gram-negative rod that grows optimally in culture media containing 7.5% NaCl under aerobic conditions and was classified in the genus *Marinobacter* as a novel species *Marinobacter lipolyticus* sp. nov. [14].

Moreno *et al.* [15] studied the diversity of extreme halophiles producing lipase, protease, amylase and nuclease in crystallizer ponds at two solar salterns in South Spain, concluding that 70% of total of the hydrolytic isolates were also amylase producers and no DNAse producers were detected among the screened population. Multiple hydrolytic activities were also found in this study. A clear dominance of Archaea was found, although a population of Bacteria was also present, accounting for around 7% of the total hydrolytic community isolated. Only three isolated strains were characterized as extremely halophilic bacteria (genera *Salicola*, *Salinibacter* and *Pseudomonas*). The genera *Salinibacter* and *Salicola* have been previously reported to compete with halophilic members of the Archaea in crystallizer ponds of different solar salterns [16,17]. The extremely halophilic strain, *Salicola* strain IC10 was selected for further studies for two main reasons, the potential biotechnological applications due to its dual hydrolytic activity (lipase and protease) and its use as a model for fundamental studies aimed to unravel the adaptations of halophilic enzymes, which allow them to be stable and active at high salt concentrations.

2.2. Screening on Salt Lakes

In a hypersaline lake in Iran, Rohban *et al.* [18] investigated the ability of halophilic strains to produce different extracellular hydrolases (lipase, amylase, protease, xylanase, DNase, inulinase, pectinase, cellulase and pulullanase) and a wider distribution of hydrolytic activity was observed among Gram-positive bacteria. Extreme halophiles, less represented in comparison with moderate halophiles, showed higher interest as producers of amylases, lipases, cellulases and pectinases. This study reported *Salicola* as the predominant genus among Gram-negative isolates, in contrast with the study of Sanchez-Porro *et al.* [11], according to which most of Gram-negative isolates belonged to the genus *Halomonas*. Among the Gram-positive hydrolase-producing isolates, representatives of the genera *Virgibacillus* and *Thalassobacillus* were predominant. On the other hand, strains of the genus *Salinivibrio* were not found, probably due to the higher concentration of salt in Howz Soltan sea site in comparison with those saline habitats studied in Spain.

Table 1. Microorganisms able to produce hydrolytic enzymes isolated from different hypersaline environments.

Isolation Site	Hydrolytic Activity Assayed	Most Abundant Hydrolytic Activity	Isolate Affiliation	References
Salterns in Almeria, Cadiz and Huelva (Spain)	amylase protease lipase DNase pullulanase	amylase	*Salinivibrio* *Halomonas* *Chromohalobacter* *Bacillus-Salibacillus* *Salinicoccus* *Marinococcus*	[11]
Saltern in Huelva (Spain)	lipase protease amylase nuclease	amylase	*Halorubrum* *Haloarcula* *Halobacterium* *Salicola* *Salinibacter* *Pseudomonas*	[15]
Howz Soltan Lake (Iran)	lipase amylase protease xylanase DNase inulinase pectinase cellulase pulullanase	lipase	*Salicola* *Halovibrio* *Halomonas* *Oceanobacillus* *Thalassobacillus* *Halobacillus* *Virgibacillus* *Gracilibacillus* *Salinicoccus* *Piscibacillus*	[18]
Maharlu Salt Lake (Iran)	protease lipase	ND	*Bacillus* *Paenibacillus* *Halobacterium* *Aeromonas* *Staphylococcus*	[19,20]

Table 1. *Cont.*

Isolation Site	Hydrolytic Activity Assayed	Most Abundant Hydrolytic Activity	Isolate Affiliation	References
Deep-sea sediments of the Southern Okinawa Trough (China)	amylase protease lipase DNase	amylase	*Alcanivorax* *Bacillus* *Cobetia* *Halomonas* *Methylarcula* *Micrococcus* *Myroides* *Paracoccus* *Planococcus* *Pseudomonas* *Psychrobacter* *Sporosarcina* *Sufflavibacter* *Wangia*	[25]
Slanic Prahova salt mine (Romania)	amylase gelatinase lipase protease cellulase xylanase	lipase protease	ND	[27]
Atacama Desert (Chile)	amylase protease lipase DNase xylanase pullulanase	DNase	*Bacillus* *Halobacillus* *Pseudomonas* *Halomonas* *Staphylococcus*	[28]
Saline desert "Indian Wild Ass Sanctuary" (India)	amylase	ND	*Bacillus*	[32]

ND: Not determined

Ghasemi and coworkers [19] performed a study focused to identify moderate halophiles from a hypersaline lake in the southern area of Iran. They isolated 16 strains exhibiting proteolytic activity and in comparison to Gram-negative bacteria, the Gram-positive rods displayed higher proteolytic activities. These data are in agreement with other findings previously reported that revealed that Gram-positive bacteria are the dominant proteolytic isolates in the saline environments. Representatives of the genera *Bacillus*, *Paenibacillus, Halobacterium* and *Aeromonas* were described. One species of *Bacillus* was found as the highest protease producer and it was further studied. Other study, concerning lipolytic activity, investigated the presence of halophilic bacteria producing lipases from a Maharlu salt lake in Iran [20]. All strains obtained in this study were moderate halophiles assigned to the genera *Bacillus* and *Staphylococcus*.

2.3. Screening on Saline Deposits

Deep-sea sediments have also been considered interesting habitats as source of novel enzymes, due to their extreme conditions; however, they have been practically unexplored on earth [21–24]. For this reason, Dang *et al.* [25] screened deep-sea sediments of the Southern Okinawa Trough in order to show the diversity and abundance of bacterial isolates secreting enzymes (amylases, proteases, lipases, DNases and chitinase). The isolates on this study were quite diverse and a total of 14 different genera were identified, including *Alcanivorax*, *Bacillus*, *Cobetia*, *Halomonas*, *Methylarcula*, *Micrococcus*, *Myroides*, *Paracoccus*, *Planococcus*, *Pseudomonas*, *Psychrobacter*, *Sporosarcina*, *Sufflavibacter* and *Wangia*. Bacteria of the *γ-Proteobacteria* lineage, especially those from the *Halomonas* and *Psychrobacter* groups, dominated in the culturable bacteria assemblage. This predominance could indicate that this population is ubiquitous in marine sediments, at least in the west Pacific and the distinct distribution of the other bacterial groups might indicate that their distributions could be restricted by certain environmental conditions.

According to the studies of Sánchez-Porro *et al.* [11] and Moreno *et al.* [15], the strains producing amylolytic enzymes were the most diverse and abundant physiological group among the hydrolytic producers. Some isolated strains even harbored all the extracellular hydrolytic activities screened, except for the chitinase activity. The lack of chitinase activity on the bacterial isolates could indicate that the terrestrial export of the particulate organic matters may be the major source of the biopolymers buried in the studied deep sea sediments. Dell' Anno and Danovaro [26] suggested that the microbial degradation of extracellular DNA in deep-sea ecosystem might provide another suitable C and N source for sediment prokaryote metabolism.

In a similar study, Cojoc and colleagues [27] elucidated the extracellular hydrolytic activities of halophilic bacteria collected from a salt deposit having 45.5 to 499 m deep. The lipolytic and proteolytic activities were predominant among the isolated strains and one of the strains hydrolyzed six different substrates. The relatively low number of microorganisms in the investigated environment can be correlated with the low temperature exhibited by the investigated area.

2.4. Screening on Saline Deserts

Recently, in heavy-metal-contaminated soils with extreme salt conditions from the Atacama Desert, Moreno *et al.* [28] carried out a screening to isolate hydrolytic enzyme producers. The most frequent hydrolytic activity detected in the study was DNase, followed by amylase and lipase in contrast with the studies of Sánchez-Porro *et al.* [11] and Moreno *et al.* [15]. In the analyzed community, pullulanase and protease producers were also detected, being xylanases the least represented among the hydrolases tested. It is interesting to emphasize that multiple hydrolytic activity was frequently detected in the isolates reported in this study supporting previous studies in other hypersaline habitats [11,15]. As reported in the study of Rohban *et al.* [18], most environmental isolates able to produce hydrolytic enzymes were Gram-positive bacteria, although the isolates were assigned to the family *Bacillaceae*, comprising species of the genera *Bacillus, Halobacillus* and *Thalassobacillus*. Only two isolates were related to the Gram-negative bacteria *Pseudomonas halophila* [29] and *Halomonas organivorans* [30] and the other characterized isolates were related to *Salinicoccus roseus* [31].

Khunt *et al.* [32] isolated from an Indian saline desert environment, moderate halophiles able to produce extracellular amylases and they reported that isolates were Gram-positive, non-capsulated bacteria assigned to the genus *Bacillus*.

In summary, we can conclude that there is a wide taxonomic diversity of microorganisms showing hydrolytic activity in saline environments and in most cases multiple activities are present. Aerial distribution of the dormant spores probably explains the occurrence of *Bacillus* in most habitats. This genus is well known as an extracellular enzyme producer and many industrial processes use species of this genus for commercial production of enzymes [33].

3. Biotechnological Potential of Bacterial Halophilic Hydrolases

Although halophilic enzymes are considered a novel alternative for use as biocatalysts in different industries, there are relatively few studies on halophilic enzymes, with some being based on their isolation and others on their characterization. However, it is important to thoroughly study these enzymes in order to use them in biotechnological processes [8,34]. Only a few industrial applications of halophilic enzymes, mainly in the manufacturing of solar salt from seawater, fermented food, textile, pharmaceutical and leather industries, have been reported. Highlighted halophilic intracellular or extracellular hydrolases mentioned in this review are summarized in the Table 2.

3.1. Hydrolases Produced by Moderately Halophilic Bacteria

3.1.1. Bacterial Lipolytic Enzymes

Bacterial lipolytic enzymes are valuable biocatalysts due to their broad substrate specificity and high chemo-, regio- and stereoselectivity [35–40]. Thus, these enzymes are currently used as detergent additives, in the food and paper industries, and as enantioselective biocatalysts for the production of fine chemicals [37,41–46]. However, industrial applications of lipases are often hampered by their low stability in the processes, including low thermostability and loss of activity in presence of the organic solvents, where most of reactions are performed. In this sense, the lipases isolated from extreme microorganisms constitute an excellent alternative in the industrial processes [47].

Table 2. Selected enzymes from extremely and moderately halophilic bacteria with potential biotechnological applications.

Source	Bacteria	Enzyme	Localization	References
Extremely halophilic bacteria	*Salicola marasensis* sp. IC10	Lipase LipL	Extracellular	[15]
		Protease SaliPro	Extracellular	[15]
Moderately halophilic bacteria	*Marinobacter lipolyticus*	Lipase LipBL	Intracellular	[48–50]
	Pseudoalteromonas ruthenica	Haloprotease CP1	Extracellular	[51,54]
	Halobacillus karajensis	Protease	Extracellular	[55]
	Nesterenkonia sp. strain F	α-amylase	Extracellular	[57]
	Thalassobacillus sp. LY18	α-amylase	Extracellular	[58]

An interesting intracellular lipolytic enzyme produced by the moderately halophilic bacterium *Marinobacter lipolyticus* SM19 was isolated and characterized [48]. This enzyme, designated LipBL, was assigned to the family VIII of lipolytic enzymes and it was expressed in *Escherichia coli*. LipBL is a protein of 404 amino acids with a molecular mass of 45.3 kDa and high identity to class C β-lactamases. LipBL was purified and biochemically characterized. The temperature for its maximal activity was 80 °C and the pH optimum determined at 25 °C was pH 7.0, showing optimal activity without sodium chloride, while maintaining 20% activity in a wide range of NaCl concentrations. This enzyme exhibited optimal activity against short-medium length acyl chain substrates, although it also hydrolyzes olive oil and fish oil. The enzyme is also active towards different chiral and prochiral esters. Exposure of LipBL to buffer-solvent mixtures showed that the enzyme had remarkable activity and stability in all organic solvents tested. For improving the stability and for use in industrial processes LipBL was immobilized in different supports [48].

The fish oil hydrolysis using LipBL results in an enrichment of free eicosapentaenoic acid (EPA), but not docosahexaenoic acid (DHA), relative to its levels present in fish oil. The fish oil used in the experiment contained 18.6% of EPA, increasing to 27% after the treatment with LipBL-CNBr immobilized derivative, representing an increase of 45.2% of the total concentration, however, the content in DHA only increase from 12% to 12.5%, representing an increase of 4.2%. These results indicated that LipBL showed more selectivity towards the hydrolysis of DHA increasing EPA, being a good candidate for use at an industrial scale for the production of fish oils enriched in polyunsaturated fatty acids (PUFAs) [48,49]. Due to its biotechnological interest, LipBL mutants were obtained in order to determine the influence of different residues in the functionality of LipBL. Mutants were constructed by site directed mutagenesis obtaining LipBL variants with different properties (specificity to different substrates, optimal pH values and thermostability) [50].

3.1.2. Bacterial Proteases

Proteases constitute one of the most important groups of industrial enzymes, comprising currently the majority of worldwide enzyme sales. They have been widely used in industry for a long time, especially in washing detergent, baking, brewing, cheese industry and tanning industry [51,52]. Due to the stability and properties of halophilic proteases, these enzymes are good candidates for use in industrial processes. An interesting extracellular protease, designated haloprotease CP1 has been isolated from the moderately halophilic bacterium *Pseudoalteromonas ruthenica* [51]. The maximal production of the protease CP1 by *P. ruthenica* CP76 was detected at the end of the exponential growth phase at 37 °C, in media containing 7.5% salt and supplemented with sucrose (50 mM). Protease CP1 was purified from *P. ruthenica* CP76 using ion exchange and gel filtration chromatography. The purified enzyme was biochemically characterized, showing optimal activity at 55 °C, pH 8.5 and high tolerance to a wide range of NaCl concentrations (0 to 4 M NaCl). It has a molecular mass of 38 kDa and the activity was strongly inhibited by ethylenediamunetetraacetic acid (EDTA), phenylmethylsulfonyl fluoride (PMSF) and Pefabloc [53,54]. Another enzyme characterized was the protease produced by the moderately halophilic bacterium *Halobacillus karajensis* strain MA-2. Effect of various temperatures, initial pH, salt and different nutrient sources on protease production revealed that the maximum secretion of the enzyme occurred at 34 °C, pH 8.0–8.5, and in the presence of gelatin. The maximum enzyme activity was obtained at pH values ranged from 8.0 to 10.0, with 55% and 50% activity remaining at pH 6 and 11, respectively. Moreover, the enzyme activity was strongly inhibited by PMSF, Pefabloc SC and EDTA; indicating that this enzyme probably belongs to the subclass of serine metalloproteases. These findings suggested that the protease secreted by *Halobacillus karajensis* presents a great potential for biotechnological applications when alkaline conditions are required [55].

3.1.3. Bacterial Amylases

Amylases constitute a group of interesting enzymes from the biotechnological point of view [56]. Most interesting applications are in the clinical and analytical chemistries, as well as their widespread applications in starch saccharification and in the textile, food, brewing and distilling industries [56], where halophilic amylases highlighted due to their stability and versatility. In the last years, several extracellular halophilic α-amylases have been purified from moderate halophiles. One of them was purified from *Nesterenkonia* sp. strain F [57]. This enzyme showed an apparent molecular weight of 110 kDa by SDS-PAGE and it exhibited maximal activity at pH 7–7.5, being relatively stable at pH 6.5–7.5. Optimal temperature for the amylase activity and stability was 45 °C. The purified enzyme was highly active in a broad range of NaCl concentrations (0–4 M) with optimal activity at 0.25 M NaCl. This amylase was highly stable in the presence of 3-4 M NaCl and the activity was not influenced by Ca^{2+}, Rb^{+}, Li^{+}, Cs^{+}, Mg^{2+} and Hg^{2+}, whereas Fe^{2+}, Cu^{2+}, Zn^{2+} and Al^{2+} strongly inhibited the enzyme. Moreover, this α-amylase was inhibited by EDTA, but was not inhibited by PMSF and β-mercaptoethanol [57].

An extracellular halophilic α-amylase was purified from *Thalassobacillus* sp. LY18 [58]. This enzyme showed a molecular mass of 31 kDa and its optimal enzyme activity was found to be at 70 °C, pH 9.0, and 10% NaCl. The α-amylase was highly stable over a broad range of temperatures (30–90 °C), pH (6.0–12.0), and NaCl concentrations (0%–20%), showing excellent thermostable, alkalistable, and halotolerant nature. The enzyme was stimulated by Ca^{2+}, but greatly inhibited by EDTA, indicating it was a metalloenzyme. Complete inhibition by diethyl pyrocarbonate and β-mercaptoethanol revealed that histidine residue and disulfide bond were essential for enzyme catalysis. Furthermore, it showed high activity and stability in the presence of water-insoluble organic solvents with log P (ow) ≥ 2.13 [58].

3.2. Hydrolases Produced by Extremely Halophilic Bacteria. Bacterial Lipolytic and Proteolytic Enzymes

On the other hand, there are some hydrolases characterized from extremely halophilic bacteria. The extremely halophilic bacterium *Salicola marasensis* IC10 produces an extracellular protease, designated Salipro, and at least an intracellular lipase, which was named LipL [15]. The lipolytic activity of the *S. marasensis* IC10 is mainly located in the cytoplasmatic fraction. This enzyme is active in presence of different compounds as substrates: *p*-nitrophenyl butyrate, *p*-nitrophenylvalerate, *p*-nitrophenylcaprilate and *p*-nitrophenyldecanoate as well as 4-methylumbelliferone and the enzyme production is maximal at the end of the exponential phase. The effect of different factors on the lipolytic activity was analyzed and the maximum values were obtained in the presence of 1 M NaCl, although the tolerance was found in the range from 0 to 4 M NaCl, with 6 mM of betaine. The lipolytic activity increased in the presence of organic solvents as 1-buthanol, 2-buthanol and acetone (5% and 10%, v/v), EDTA 1% (v/v) and metal ions as Ni^{2+} and Ca^{2+}.

To identify the optimal conditions for protease production, the characterization of the intracellular fraction of *Salicola* sp. IC10 during growth was performed, finding the optimal conditions at pH 8.0, 40 °C and a medium with 15%–20% (w/v) NaCl. Thus, there is a correlation between the optimal conditions for cultivation of the strain and the maximum production of the proteolytic enzyme. This protease showed the capability to effectively catalyze the hydrolysis of various proteins. The most specific substrate to the enzyme was egg albumin, followed by gelatine (97% relative activity) [15].

Therefore, we can conclude that enzymes produced by halophilic bacteria show interesting properties for use in different industries.

References

1. Ventosa, A. Unusual microorganisms from unusual habitats: hypersaline environments. In *Prokaryotic Diversity-Mechanism and Significance*; Logan, N.A., Lppin-Scott, H.M., Oyston, P.C.F., Eds.; Cambridge University Press: Cambridge, UK, 2006; pp. 223–253.
2. Kushner, D.J.; Kamekura, M. Physiology of halophilic eubacteria. In *Halophilic Bacteria*; Rodríguez-Varela, F., Ed.; CRC Press: Boca Raton, FL, USA, 1988; pp. 109–138.

3. Madigan, M.T.; Oren, A. Thermophilic and halophilic extremophiles. *Curr. Opin. Microbiol.* **1999**, *2*, 265–269.
4. Oren, A. Diversity of halophilic microorganisms: Environments, phylogeny, physiology, and applications. *J. Ind. Microbiol. Biotechnol.* **2002**, *28*, 58–63.
5. Nieto, J.J.; Vargas, C. Synthesis of osmoprotectants by moderately halophilic bacteria: Genetic and applied aspects. *Recent. Res. Devel. Microbiol.* **2002**, *6*, 403–418.
6. Mellado, E.; Ventosa, A. Biotechnological potential of moderately and extremely halophilic microorganisms. In *Microorganisms for Health Care, Food and Enzyme Production*; Barredo, J.L., Ed.; Research Signpost: Kerala, India, 2003; pp. 233–256.
7. Gómez, J.; Steiner, W. The biocatalytic potential of extremophiles and extremozymes. *Food Technol. Biotechnol.* **2004**, *2*, 223–235.
8. Oren, A. Industrial and environmental applications of halophilic microorganisms. *Environ. Technol.* **2010**, *31*, 825–834.
9. Zaccai, G. The effect of water on protein dynamics. *Philos. Trans. R. Soc. Lond. B. Biol. Sci.* **2004**, *359*, 1269–1275.
10. Salameh, M.; Wiegel, J. Lipases from extremophiles and potential for industrial applications. *Adv. Appl. Microbiol.* **2007**, *61*, 253–283.
11. Sánchez-Porro, C.; Martín, S.; Mellado, E.; Ventosa, A. Diversity of moderately halophilic bacteria producing extracellular hydrolytic enzymes. *J. Appl. Microbiol.* **2003**, *94*, 295–300.
12. Ventosa, A. Taxonomy of moderately halophilic heterotrophic eubacteria. In Halophilic bacteria. Rodríguez-Valera, F., Ed.; CRC Press: Boca Raton, FL, USA, 1988; pp. 71–84.
13. Ventosa, A.; Nieto, J.J.; Oren, A. Biology of moderately halophilic aerobic bacteria. *Microbiol. Mol. Biol. Rev.* **1998**, *62*, 504–544.
14. Martín, S.; Márquez, M.C.; Sánchez-Porro, C.; Mellado, E.; Arahal, D.R.; Ventosa, A. *Marinobacter lipolyticus* sp. nov., a novel moderate halophile with lipolytic activity. *Int. J. Syst. Evol. Microbiol.* **2003**, *53*, 1383–1387.
15. Moreno, M.L.; García, M.T.; Ventosa, A.; Mellado, E. Characterization of *Salicola* sp. IC10, a lipase- and protease-producing extreme halophile. *FEMS Microbiol. Ecol.* **2009**, *68*, 59–71.
16. Ovreas, L.; Bourne, D.; Sandaa, R.A.; Casamayor, E.O.; Benlloch, S.; Goddard, V. Response of bacterial and viral communities to nutrient manipulations in sea water mesocosms. *Aquat. Microbiol. Ecol.* **2003**, *31*, 109–121.
17. Maturrano, L.; Valens-Vadell, M.; Roselló-Mora, R.; Antón, J. *Salicola marasensis* gen. nov., sp. nov., an extremely halophilic bacterium isolated from the Maras solar salterns in Perú. *Int. J. Syst. Evol. Microbiol.* **2006**, *56*, 1685–1691.
18. Rohban, R.; Amoozegar, M.A.; Ventosa, A. Screening and isolation of halophilic bacteria producing extracellular hydrolyses from Howz Soltan Lake, Iran. *J. Ind. Microbiol. Biotechnol.* **2009**, *36*, 333–340.
19. Ghasemi, Y.; Rasoul-Amini, S.; Ebrahiminezhad, A.; Kazemi, A.; Shahbazia, M.; Talebniaa, N. Screening and Isolation of Extracellular Protease Producing Bacteria from the Maharloo Salt Lake. *Iran. J. Pharm. Sci.* **2011**, *7*, 175–180.

20. Ghasemi, Y.; Rasoul-Amini, S.; Kazemi, A.; Zarrini, G.; Morowvat, M.T.; Kargar, M. Isolation and Characterization of Some Moderately Halophilic Bacteria with Lipase Activity. *Microbiology* **2011**, *80*, 483–487.
21. Whitman, W.B; Coleman, D.C; Wiebe, W.J. Prokaryotes: The unseen majority. *Proc. Natl. Acad. Sci. USA* **1998**, *95*, 6578–6583.
22. D'Hondt, S.; Rutherford, S.; Spivack, A.J. Metabolic activity of subsurface life in deep-sea sediments. *Science* **2012**, *295*, 2067–2070.
23. Parkes, R.J.; Webster, G.; Cragg, B.A.; Weightman, A.J.; Newberry, C.J.; Ferdelman, T.G.; Kallmeyer, J.; Jorgensen, B.B.; Aiello, I.W.; Fry, J.C. Deep sub-seafloor prokaryotes stimulated at interfaces over geological time. *Nature* **2005**, *436*, 390–394.
24. Schippers, A.; Neretin, L.N.; Kallmeyer, J.; Ferdelman, T.G.; Cragg, B.A.; Parkes, R.J.; Jorgensen, B.B. Prokaryotic cells of the deep sub-seafloor biosphere identified as living bacteria. *Nature* **2005**, *433*, 861–864.
25. Dang, H.; Zhu, H.; Wang, J.; Li, T. Extracellular hydrolytic enzyme screening of culturable heterotrophic bacteria from deep-sea sediments of the Southern Okinawa Trough. *World J. Microbiol. Biotechnol.* **2009**, *25*, 71–79.
26. Dell'Anno, A.; Danovaro, R. Extracellular DNA plays a key role in deep-sea ecosystem functioning. *Science* **2005**, *309*, 2179.
27. Cojoc, R.; Merciu, S.; Popescu, G.; Dumitru, L.; Kamekura, M.; Enache, M. Extracellular hydrolytic enzymes of halophilic bacteria isolated from a subterranean rock salt crystal. *Rom. Biotechnol. Lett.* **2009**, *14*, 4658–4664.
28. Moreno, M.L.; Piubeli, F.; Bonfá, M.R.; García, M.T.; Durrant, L.R.; Mellado, E. Analysis and characterization of cultivable extremophilic hydrolytic bacterial community in heavy-metal-contaminated soils from the Atacama Desert and their biotechnological potentials. *J. Appl. Microbiol.* **2012**, *113*, 550–559.
29. Sorokin, D.Y.; Tindall, B.J. The status of the genus name *Halovibrio.* Fendrich 1988 and the identity of the strains *Pseudomonas halophila* DSM 3050 and *Halomonas. variabilis* DSM 3051. Request for an opinion. *Int. J. Syst. Evol. Microbiol.* **2006**, *56*, 487–489.
30. García, M.T.; Mellado, E.; Ostos, J.C.; Ventosa, A. *Halomonas. organivorans* sp. nov., a moderate halophile able to degrade aromatic compounds. *Int. J. Syst. Evol. Microbiol.* **2004**, *54*, 1723–1728.
31. Ventosa, A.; Marquez, M.C.; Ruiz-Berraquero, F.; Kocur, M. *Salinicoccus. roseus* gen. nov., a new moderately halophilic Gram-positive coccus. *Syst. Appl. Microbiol.* **1990**, *13*, 29–33.
32. Khunt, M.; Pandhi, N.; Rana, A. Amylase from moderate halophiles isolated from wild ass excreta. *Int. J. Pharm. Bio. Sci.* **2011**, *1*, 586–592.
33. Schallmey, M.; Singh, A.; Ward, O.P. Developments in the use of *Bacillus* species for industrial production. *Can. J. Microbiol.* **2004**, *50*, 1–17.
34. Delgado-García, M.; Valdivia-Urdiales, B.; Aguilar-González, C.N.; Contreras-Esquivel, J.C.; Rodríguez-Herrera, R. Halophilic hydrolases as a new tool for the biotechnological industries. *J. Sci. Food. Agric.* **2012**, *92*, 2575–2580.

35. Chahinian, H.; Ali, Y.B.; Abousalham, A.; Petry, S.; Mandrich, L.; Manco, G.; Canaan, S.; Sarda, L. Substrate specificity and kinetic properties of enzymes belonging to the hormone-sensitive lipase family: Comparison with non-lipolytic and lipolytic carboxyl esterases. *Biochim. Biophys. Acta.* **2005**, *1738*, 29–36.
36. Houde, A.; Kademi, A.; Leblanc, D. Lipases and their industrial applications: An overview. *Appl. Biochem. Biotechnol.* **2004**, *46*, 155–170.
37. Jaeger, K.E.; Eggert, T. Lipases for biotechnology. *Curr. Opin. Biotechnol.* **2002**, *13*, 390–397.
38. Park, J.H.; Ha, H.J.; Lee, W.K.; Généreux-Vincent, T.; Kazlauskas, R.J. Molecular basis for the stereoselective ammoniolysis of N-alkyl aziridine-2-carboxylates catalyzed by *Candida antarctica* lipase B. *Chembiochem* **2009**, *10*, 2213–2222.
39. Rodriguez, J.A.; Mendoza, L.D.; Pezzotti, F.; Vanthuyne, N.; Leclaire, J.; Verger, R.; Buono, G.; Carriere, F.; Fotiadu, F. Novel chromatographic resolution of chiral diacylglycerols and analysis of the stereoselective hydrolysis of triacylglycerols by lipases. *Anal. Biochem.* **2008**, *375*, 196–208.
40. Snellman, E.A.; Colwell, R.R. *Acinetobacter* lipases: molecular biology, biochemical properties and biotechnological potential. *J. Ind. Microbiol. Biotechnol.* **2004**, *31*, 391–400.
41. Breuer, M.; Ditrich, K.; Habicher, T.; Hauer, B.; Kesseler, M.; Stürmer, R.; Zelinski, T. Industrial methods for the production of optically active intermediates. *Angew. Chem. Int. Ed. Engl.* **2004**, *43*, 788–824.
42. Hasan, F.; Shah, A.A.; Hameed, A. Industrial applications of microbial lipases. *Enzym. Microbiol. Technol.* **2005**, *39*, 235–251.
43. Jaeger, K.E.; Holliger, P. Chemical biotechnology a marriage of convenience and necessity. *Curr. Opin. Biotechnol.* **2010**, *21*,711–712.
44. Jaeger, K.E.; Reetz, M.T. Microbial lipases form versatile tools for biotechnology. *Trends Biotechnol.* **1998**, *16*, 396–403.
45. Snellman, E.A.; Sullivan, E.R.; Colwell, R.R. Purification and properties of the extracellular lipase, LipA of *Acinetobacter.* sp. RAG-1. *FEBS. J.* **2002**, *269*, 5771–5779.
46. Schmid, A.; Dordick, J.S.; Hauer, B.; Kiener, A.; Wubbolts, M.; Witholt, B. Industrial biocatalysis today and tomorrow. *Nature* **2001**, *409*, 258–268.
47. Pikuta, E.V.; Hoover, R.B.; Tang, J. Microbial extremophiles at the limits of life. *Crit. Rev. Microbiol.* **2007**, *33*, 183–209.
48. Pérez, D.; Martín, S.; Fernández-Lorente, G.; Filice, M.; Guisán, J.M.; Ventosa, A.; García, M.T.; Mellado, E. A novel halophilic lipase, LipBL, with applications in synthesis of Eicosapentaenoic acid (EPA). *PlosOne*. **2011**, doi:10.1371/journal.pone.0023325.
49. Pérez, D.; Ventosa, A.; Mellado, E.; Guisán, J.M.; Fernández-Lorente, G.; Filice, M. Lipasa LipBL y sus aplicaciones. Spanish Patent P201031636, 8 November 2010.
50. Pérez, D.; Kovacic, F.; Wilhelm, S.; Jaeger, K.E.; García, M.T.; Ventosa, A.; Mellado, E. Identification of amino acids involved in the hydrolytic activity of lipase LipBL from *Marinobacter lipolyticus*. *Microbiology* **2012**, *158*, 2192–2203.

51. Chand, S.; Mishra, P. Research and Application of Microbial Enzymes. India's. Contribution. *Adv. Biochem. Eng. Biotechnol.* **2003**, *85*, 95–124.
52. Li, A.N.; Li, D.C. Cloning, expression and characterization of the serine protease gene from *Chaetomium thermophilum*. *J. Appl. Microbiol.* **2009**, *106*, 369–380.
53. Sánchez-Porro, C.; Mellado, E.; Bertoldo, C.; Antranikian, G.; Ventosa, A. Screening and characterization of the protease CP1 produced by the moderately halophilic bacterium *Pseudoalteromonas ruthenica.* sp. strain CP76. *Extremophiles*. **2003**, *7*, 221–228.
54. Sánchez-Porromona, C.; Mellado, E.; Martín, S.; Ventosa, A. Proteasa producida por una bacteria halófila moderada: modo de producción de la enzima. Spanish Patent P200300745, 26 March 2003.
55. Karbalaei-Heidari, H.R.; Amoozegar, M.A.; Hajighasemi, M.; Ziaee, A.A.; Ventosa, A. Production, optimization and purification of a novel extracellular protease from the moderately halophilic bacterium *Halobacillus karajensis*. *J. Ind. Microbiol. Biotechnol.* **2009**, *36*, 21–27.
56. Pandey, A.; Nigam, P.; Soccol, C.R.; Soccol, V.T.; Singh, D.; Mohan, R. Advances in microbial amylases. *Biotechnol. Appl. Biochem.* **2000**, *31*, 135–152.
57. Shafiei, M.; Ziaee, A.A.; Amoozegar, M.A. Purification and characterization of a halophilic α-amylase with increased activity in the presence of organic solvents from the moderately halophilic *Nesterenkonia* sp. strain F. *Extremophiles* **2012**, *16*, 627–635.
58. Li, X.; Yu, H.Y. Characterization of an organic solvent-tolerant α-amylase from a halophilic isolate, *Thalassobacillus* sp. LY18. *Folia Microbiol.* **2012**, *57*, 447–453.

Periplasmic Binding Proteins in Thermophiles: Characterization and Potential Application of an Arginine-Binding Protein from *Thermotoga maritima*: A Brief Thermo-Story

Alessio Ausili, Maria Staiano, Jonathan Dattelbaum, Antonio Varriale, Alessandro Capo and Sabato D'Auria

Abstract: Arginine-binding protein from the extremophile *Thermotoga maritima* is a 27.7 kDa protein possessing the typical two-domain structure of the periplasmic binding proteins family. The protein is characterized by a very high specificity and affinity to bind to arginine, also at high temperatures. Due to its features, this protein could be taken into account as a potential candidate for the design of a biosensor for arginine. It is important to investigate the stability of proteins when they are used for biotechnological applications. In this article, we review the structural and functional features of an arginine-binding protein from the extremophile *Thermotoga maritima* with a particular eye on its potential biotechnological applications.

Reprinted from *Life*. Cite as: Ausili, A.; Staiano, M.; Dattelbaum, J.; Varriale, A.; Capo, A.; D'Auria, S. Periplasmic Binding Proteins in Thermophiles: Characterization and Potential Application of an Arginine-Binding Protein from *Thermotoga maritima*: A Brief Thermo-Story. *Life* **2013**, *3*, 149-160.

1. Introduction

1.1. ABC Transporter System and Periplasmic Binding Proteins

ATP-binding cassette (ABC) transporters are a ubiquitous distributed transmembrane protein superfamily involved in the active translocation through the membrane of a big variety of substrates, including ions, sugars, vitamins, drugs, metabolic products, lipids, amino acids and polypeptides. ABC transporters are involved in a wide number of cellular processes and they utilize the energy obtained by the ATP hydrolysis to carry out their functions, while their dysfunction underlies a number of genetic diseases [1–3]. ABC proteins are formed by four domains that may be organized in several modes: two highly conserved cytoplasmic nucleotide-binding domains (NBDs) which bind and hydrolyze ATP driving the translocation of the bound ligand and two variable transmembrane domains (TMDs) consisting of α-helices that allow the translocation of solutes across the membranes and provide specificity [4,5]. ABC transporters can be divided into two main functional groups that differ significantly both in their functioning mechanism and in their structural organization, mainly of TMD region: importers that are involved in substrate up-taking and exporters that are responsible for the secretion of various molecules [6–8]. Importers, that are present only in prokaryotes, are often accompanied by additional high-affinity binding proteins that are localized in the periplasmic space of gram-negative bacteria. These proteins named periplasmic binding proteins (PBPs) interact and associate with extreme specificity and affinity with different substrates that are distributed to the appropriate ABC transporter anchored to the inner membrane. The mechanism of ligand uptake by the synergic action of importers and PBPs is well functionally characterized for the maltose uptake system of *E. coli* and the structure of the entire complex (MBP-MalFGK$_2$) has also been described in atomic details [9]. PBPs can recruit a broad number of nutrients and they are also present in other bacteria without periplasm [10] and in archaea [11] constituting one of the largest and widely distributed protein family in bacteria and archaea. To this day, many structures of different wild-type PBPs have been solved at very high resolution, and in all cases, they show some similar peculiar characteristics. Typically, these binding proteins are monomers consisting of two well-defined globular domains joined by a hinge forming a groove between them that constitutes the ligand-binding site. This structure allows the proteins to assume two different conformations depending on the presence or the absence of the ligand (closed and open form respectively), indeed the two domains act like a clamp and move closer each other as a consequence of the ligand binding [12]. Despite of their different sizes that vary between 20 and 60 kDa and a relatively little sequence homology [13], all PBPs share not only the previously described three dimensional conformation but also a similar secondary structure. Each domain consists of a β-sheet core formed by five or six β-strands surrounded by helices and depending on number and order of the strands, two structural classes of PBP superfamily can be identified: the class I possess six β-strands in the order $\beta_2\beta_1\beta_3\beta_4\beta_5\beta_6$ and the two domains are linked by three peptide hinges, while the class II is formed by five strands that take the form of $\beta_2\beta_1\beta_3\beta_5\beta_4$ with two connecting segments between the two domains [12,14,15]. Recently, a third class of PBPs has been identified which do not fall into any of the previously described structural groups and it is characterized by a single long helix hinge [16,17].

1.2. Arginine-Binding Protein from Thermotoga Maritima and Its Potential Applications

Arginine-binding protein (ArgBP) is a member of PBPs superfamily belonging to the class II and it is capable to bind to arginine with very high affinity [18] and a ligand-binding specificity only for arginine [19] or for arginine, ornithine and/or lysine [20,21]. Besides the extremely important role in regulation of intracellular arginine concentration, ArgBP as many other PBPs can be employed in a wide range of biotechnological applications [22], among them ArgBP could be used as sensitive element in biosensor systems for detection of arginine in urine, serum and blood. Indeed, high levels of arginine in bodily fluids is an important signal of enzyme arginase malfunctioning in urea cycle, a disorder that can lead to hyper-argininemia, an autosomal recessive disease caused by high concentration of arginine and ammonia in the blood with dramatic consequences to human health, especially in neonatal age [23,24]. Moreover, arginine is also the precursor for the synthesis of endothelium-derived nitric oxide by the enzyme nitric oxide synthase [25]. Nitric oxide is an important factor for the regulation of blood flow and pressure, therefore anomalous levels of arginine can be the cause of disorders related to nitric oxide synthesis that can lead to vascular diseases like atherosclerosis and hypercholesterolemia [26,27]. Hence, the possibility to get a real-time and continuous detection and monitoring of arginine levels could be fundamental for diagnosis and treatment of these disorders. A reagentless optical biosensor based on ArgBP fluorescence could be a cheap, easy and powerful tool with a broad range of applications. Indeed, several PBPs have already been studied and applied in fluorescence-based biosensors for the detection of other molecules such as glutamine and glucose. In these cases glutamine-binding protein and glucose/galactose-binding protein were proposed as biosensors for optical assay of glutamine and glucose, respectively, by exploiting the intrinsic characteristics of the wild-type proteins, or by introducing residues in specific locations in order to label the proteins with fluorescent dyes and/or to decrease the ligand affinity of the binding protein to physiologically relevant ranges [28–30]. In general, the peculiar characteristic of these binding proteins of assuming open or closed form depending on the ligand-binding with the consequent large conformational change may be exploited to monitor possible changes in fluorescence emission of the aromatic residues or the fluorescent probes expressly linked to the protein. For this reason and for the advantage of avoiding the secondary components production and the modification of the sensor itself, PBPs could be employed as good scaffolds in fluorescence biosensing applications. High affinity and specificity of binding to the ligands combined to high sensitivity of the fluorescent probe is a very important starting point for the development of optical biosensors. However, this is a necessary but not sufficient property for practical applications of the biosensors. High stability under a wide range of different environmental conditions is also required, and an effective partial solution for this problem is given by extremophiles. Indeed, the use of proteins from extremophilic organisms is often preferred to that of their mesophilic homologues, since this provides the essential robustness to the biosensor. Here, a thermostable arginine-binding protein from the extremophile *Thermotoga maritima* (TmArgBP) is described with a view to its potential application as a scaffold for the creation of a solid fluorescent biosensor. *T. maritima* is a hyper-thermophilic gram-negative eubacterium originally isolated from different geothermal heated marine sediments in Italy, Iceland and the Azores, and it can perfectly grow in a range of temperature from 55 up to 90 °C with an optimum of 80 °C [31]. Though a eubacterium, *T. maritima* shows a large number of archaeal gene homologues that reveals an occurred lateral gene transfer during the evolution, demonstrated by a high gene sequence similarity [32,33]. Some thermostable sugar-binding proteins from this microorganism have been already partially characterized and potentially they could also be employed in biosensing [34–36]. The genomic sequence of *T. maritima* contains several open reading frames (ORFs) annotated as putative periplasmic binding proteins [32], among which *tm0593* encodes a protein with similar sequence to several polar amino acid-binding protein although from this homology it could not be possible to predict the specific cognate ligand for the protein. The protein was determined to be an arginine-binding protein with a very high affinity and specificity for arginine as cognate ligand [37]. Innovative experiments of nano-flow electrospray ionization mass spectrometry (nano-ESI-MS) showed that this PBP had a strong preferential selectivity for binding to arginine, this technique allowed to detect the formation of the protein-ligand complex in the presence of arginine as ligand, on the contrary the presence of other amino acids such as histidine or asparagine did not lead to the formation of any complex, in fact in these cases only the free protein was observed [37]. Hence, in order to measure the binding affinity to the arginine, surface plasmon resonance (SPR) experiments were performed and a dissociation constant (K_d) of 20 μM was determined. This K_d value was also compared to that obtained using glutamine instead of arginine, that was 160 μM, since the high sequence homology between this arginine-binding protein and a glutamine-binding protein from *E. coli* [38], confirming that the protein had high affinity and specificity only for the amino acid arginine [37].

2. TmArgBP Structure

Having established that the protein encoded by *tm0593* was a specific arginine-binding protein (TmArgBP) with known dissociation constant, its molecular organization, conformational dynamics and structural stability were studied by means of different methods, including low-resolution biophysical techniques, preliminary X-ray crystallographic analysis, differential scanning calorimetry and molecular dynamics simulation. TmArgBP has a monomeric molecular weight of 27.7 kDa, unlike the most of PBPs, TmArgBP can exist as monomer, homodimer and homotrimer even under strongly denaturing condition [37]. The existence of oligomeric conformations was previously observed in other PBPs from *T. maritima* and also from other organisms and it seems to be related to the ligand transport and the protein membrane anchorage even if the precise specific function is still not well understood [39–41]. Since the three-dimensional structure of the TmArgBP monomer has not been already solved at high resolution, 3D protein models with and without the ligand were created by homology modeling with Modeller 9.5 (Figure 1 B and A, respectively) [42] using the structure of the ligand-bound Arg-, Lys-, His-binding protein from *G. stearothermophilus* (RMSD = 0.807 Å measured by InsightII – superimpose structure module) [43] and ligand-free Gln-binding protein from *E. coli* as templates (RMSD = 0.788 Å) [44] and as predictable, the TmArgBP models showed the characteristic PBP motif. In fact, TmArgBP monomer has the typical structure of PBPs family members, with a single polypeptide chain that folds into two domains connected by a hinge region. This conformation is strictly related to the ligand-binding function for all the PBPs, which is why even with a moderate level of sequence similarity they conserve this tertiary structure. In TmArgBP models, the single domains in open and closed forms were almost totally superimposable with a perfect coincidence of one of the two (the one containing both C- and N-terminal) with a RMSD of 0.66 Å, while the other domain showed a rotation of 50 °C around the hinge from the open to the closed conformation. While the binding to arginine induced the hinge-bending motion between the two domains, it had not any effect on the secondary structure. In both cases, the two domains conserved the typical β-sheet hydrophobic core surrounded by helices with no differences in secondary structures content, which was also confirmed by FTIR and CD experiments [42,45].

A deeper molecular insight of the binding cleft of TmArgBP with bound arginine was provided by a visual analysis of the closed form model (Figure 2). The ligand-protein docking could involve several residues within 5 Å radius centered on the ligand arginine, but some of them appear to play a crucial role for the ligand docking to the binding site. In particular, E42, S35, S93 and Q142 are at a suitable distance to form H-bonds with the guanidine moiety of the ligand arginine. The residues D183 and G94 can interact with the amine group while T147, T146 and R101 form interactions with the carboxylic acid group. Finally, F38, F76 and T96 lie in a favorable position to create hydrophobic bonds with the arginine aliphatic straight chain.

Figure 1. 3D cartoon models of TmArgBP in the absence (**A**) and in the presence (**B**) of arginine (displayed in spheres). The ligand-binding induces the bending of the two domains that makes the protein assumes the closed form.

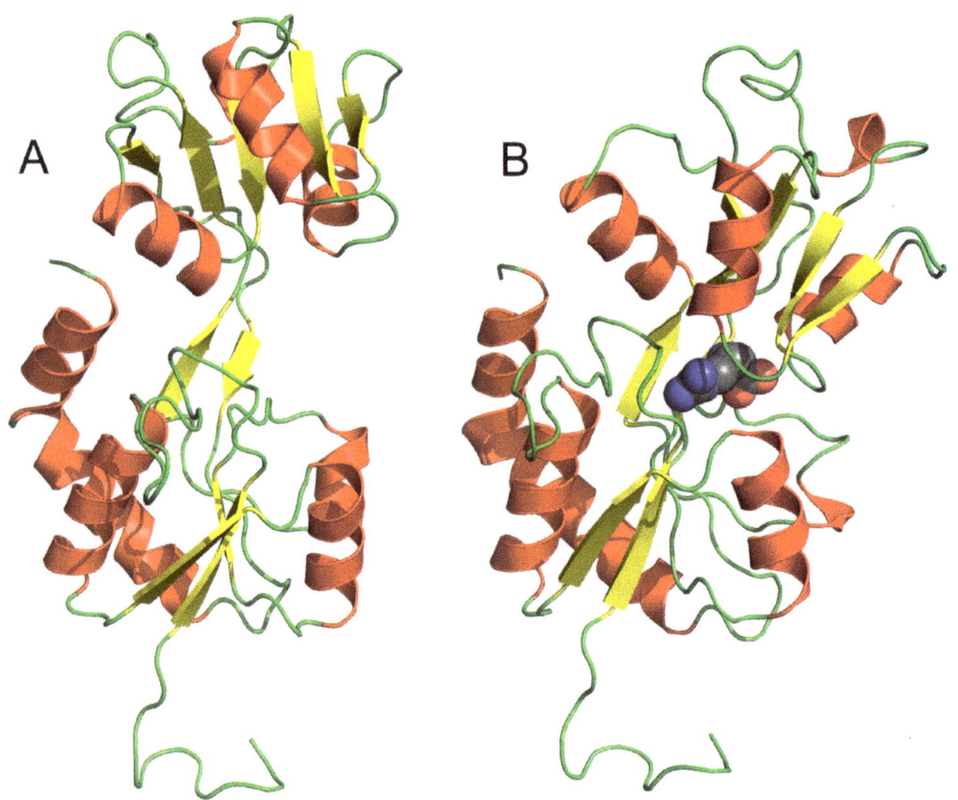

Crystallization trials of both arginine-bound and arginine-free TmArgBP forms were also performed and some preliminary X-ray results were obtained [46]. In the presence of the ligand, ordered crystals of the protein were obtained in one day by using the detergent LDAO. The crystals presented a primitive hexagonal symmetry with an elongated unit-cell (a = 78.2, b = 78.2, c =434.7 Å) and the existence of three or four molecules per asymmetric unit. Otherwise, in the absence of arginine, ordered crystals suitable for X-ray diffraction experiments were obtained in one week by using PEG 3350 as precipitant. These crystals were orthorhombic with unit-cell parameters a = 51.8, b = 91.9, c = 117.9 Å and with the presence of two molecules per asymmetric unit. Arginine-bound and arginine-free TmArgBP crystals were diffracted to 2.7 and 2.25 Å resolution, respectively [46]. Trials to solve the structures of both protein forms at high resolution are in progress.

Figure 2. Close-up view of TmArgBP binding site with bound arginine. All the amino acidic residues lying within a distance of 5 Å from the ligand arginine are shown and represented in stick mode and labeled, while arginine is displayed in spheres. All the possible interactions that can be formed given charge, position and distance of each residue from the arginine are displayed in dashed lines.

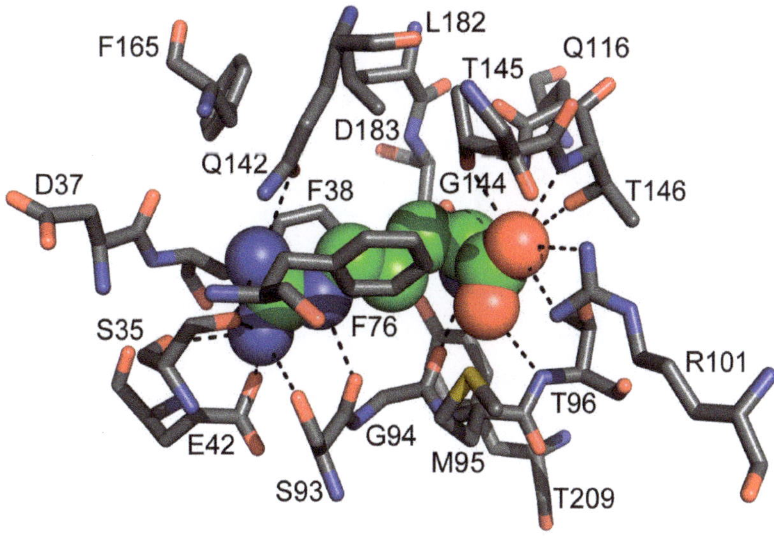

3. Stability of TmArgBP

As expected for proteins from extremophilic organisms, TmArgBP is endowed with an outstanding stability towards thermal and chemical denaturation. The complete protein thermal unfolding occurs around 120 °C while the melting temperature (T_m) has been first estimated by Luchansky *et al.* [37] and then directly observed by Ausili *et al.* [45] at 115 °C.
The denaturation process is irreversible, one-step and cooperative, with a thermodynamic equilibrium model like:

$$N_2 \leftrightarrows 2U$$

where N_2 and 2U are the homo-dimeric native and the unfolded state, respectively. This model suggests that, despite TmArgBP exists as a dimer, and each monomer is formed by two well defined structural domains, only one energetic domain is present during the unfolding process [45]. Moreover, it was observed that the β-structure core is more stable than the more exposed α-helices and in particular, the three highly hydrophobic and completely buried β-strands of the central β-sheet are the most thermostable [42].

This strong protein stability under extremely high temperature conditions was also found in the presence of chemical denaturants such as SDS and guanidine hydrochloride (GdmCl) and at alkaline conditions. At 20 °C high concentrations of SDS (3.5% w/v) induce a partial loss of α-helices and β-sheet elements, however, the protein conserves a well-defined secondary structure that is progressively lost increasing the temperature up its complete unfolding at 95 °C [42]. A similar behavior was observed in the presence of GdmCl. The complete protein chemical denaturation can be achieved at 20 °C only in conditions of GdmCl saturation, while at 95 °C a concentration of at least 2.6 M GdmCl is required to fully denature TmArgBP [45]. At pH 10.5, the protein structure undergoes only little changes at 20 °C, in particular it is noted a partial lost of α-helix structures and a tertiary structure loosening. Moreover, at high pH values, the presence of two β-sheet populations located in the protein hydrophobic core is detectable during the denaturation process [42].

In any case, the cognate ligand binding exerts an important stabilizing effect on the protein structure. Indeed, the presence of arginine increases the T_m from 115 °C to 119 °C without affecting the overall unfolding mechanism, as well as it enhances its resistance to alkaline and chemical denaturant conditions [37,42,45].

4. Tryptophan Microenvironment

In TmArgBP only one tryptophan residue (Trp^{226}) is present and it is localized in the α-helix at C-terminal of the protein at the opposite side of the binding site (Figure 3). Trp^{226} results to be stabilized by electrostatic interactions with near residues. In particular it is possible to notice the presence of stabilizing interactions between the carboxyl group of Asp^{39} facing the positive end of the Trp^{226} indole dipole (N1) (atomic distance = 3.0 Å in the open form) and between the positive charge of Lys^{225} facing the negative end of the Trp^{226} dipole (C4–C5) (atomic distance = 4.2 Å in the open form) (Figure 4A). Trp^{226} environment is highly homogeneous, polar and rather rigid, while the tryptophan indole ring is largely but superficially buried within the protein fold. In fact, the indole ring is not in direct contact with the aqueous phase, but it is very near to it (< 0.5 Å). Moreover, the region of Trp^{226} is not involved in the association of the subunits and also that monomer formation has no significant impact on the protein structure near the indolic residue [45]. Finally, Trp^{226} environment not only is far from the binding site but also its local structure/dynamics is totally disconnected from the binding site region. That was both observed from the analysis of Trp^{226} position and interactions involving the Trp^{226} microenvironment in the predicted structures illustrated in Figure 3 and 4, and inferred from the absolute absence of any change in fluorescence and phosphorescence tryptophan intrinsic emission when the protein assumes the closed form in the presence of the ligand [37,45]. This phenomenon is quite unusual for proteins, which undergo evident conformational changes like the PBPs and in our best knowledge this is the first protein that does not show changes in phosphorescence emission upon ligand binding.

Figure 3. Position of Trp^{226} in TmArgBP. The indole side chain is in evidence, displayed in sticks and labeled. Panel A and B depict the open and closed form, respectively. The residue is far from the binding site and enough buried from the aqueous interface to avoid the direct exposition to the solvent. Notice that the presence of the ligand (displayed in spheres) does not affect the Trp^{226} position.

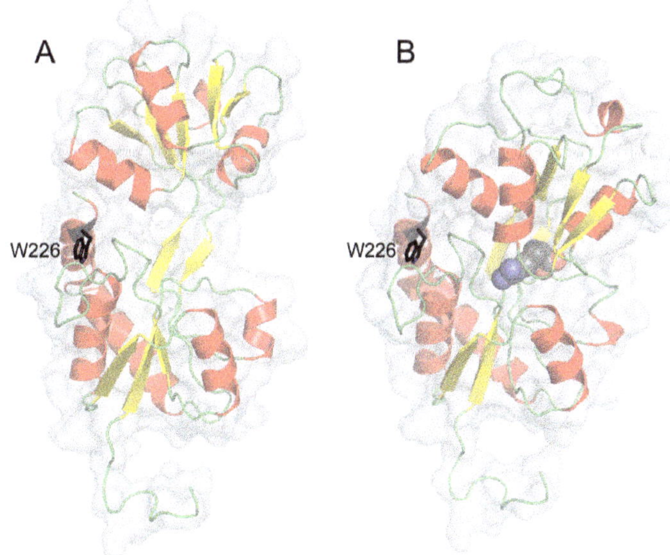

Figure 4. Close-up view of Trp^{226} microenvironment. Trp^{226} and the residues that possibly interact with it are in evidence, displayed in sticks and labeled. The black dashed lines represent the distances between the functional groups involved in the Trp^{226} stabilizing interactions. A and B show the tryptophan microenvironment for open and closed form, respectively. The interaction distances, labeled in the figure, remain roughly unvaried.

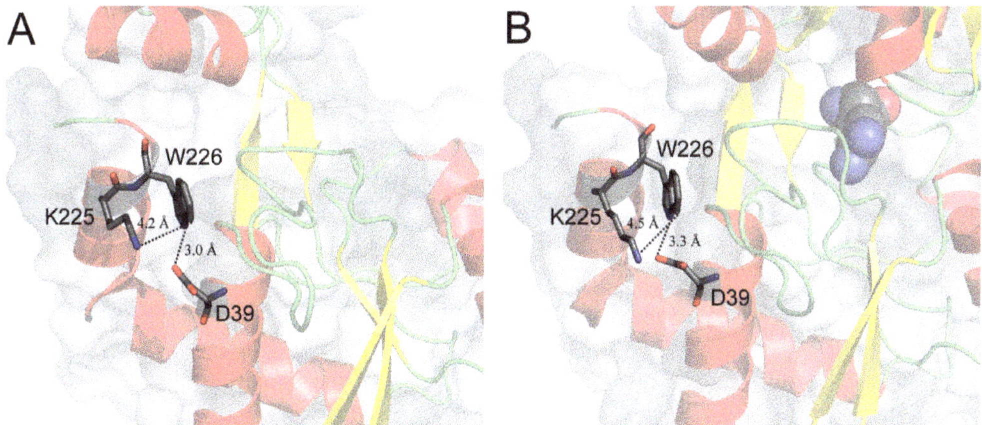

5. Conclusions

TmArgBP could be employed in several biotechnological applications including optical biosensing for arginine detection based on fluorescence. Unfortunately, the possibility of using the intrinsic Trp^{226} emission for this purpose has to be discarded in TmArgBP, but its extremely advantageous characteristics of structural stability and selectivity/sensibility for arginine-binding could make this protein a potential candidate for the design of such a biosensor. Attempts to produce new mutant proteins aimed at providing to TmArgBP an optical sensitivity for conformational changes due to arginine binding are currently in progress with attractive outlooks. The idea is to create double mutants inserting reactive residues in order to attach fluorescent dyes in advantageous positions. The location of the mutations will allow the approach of the dyes when the protein assumes the closed form in the presence of arginine that will be detectable by means of RET or PET techniques. A biosensor based on this methodology would avoid many disadvantages of traditional arginine-detection systems such as high costs, response slowness and laborious sample preparations by qualified operators. For these reasons, we are confident that the design of a fluorescence-based biosensor would be an important advance in arginine detection.

Acknowledgements

The authors wish to thank Anna Marabotti for her kind help with homology modeling analysis.

References

1. Higgins, C.F. ABC transporters: From microorganisms to man. *Annu. Rev. Cell Biol.* **1992**, *8*, 67–113.
2. Jones, P.M.; George, A.M. The ABC transporter structure and mechanism: Perspectives on recent research. *Cell Mol. Life Sci.* **2004**, *61*, 682–699.
3. Linton, K.J. Structure and function of ABC transporters. *Physiology* **2007**, *22*, 122–130.
4. Zolnerciks, J.K.; Andress, E.J.; Nicolaou, M.; Linton, K.J. Structure of ABC transporters. *Essays Biochem.* **2011**, *50*, 43–61.
5. Kos, V.; Ford, R.C. The ATP-binding cassette family: a structural perspective. *Cell Mol. Life Sci.* **2009**, *66*, 3111–3126.
6. Davidson, A.L.; Dassa, E.; Orelle, C.; Chen, J. Structure, function, and evolution of bacterial ATP-binding cassette systems. *Microbiol. Mol. Biol. Rev.* **2008**, *72*, 317–364.
7. Rees, D.C.; Johnson, E.; Lewinson, O. ABC transporters: the power to change. *Nat. Rev. Mol. Cell. Biol.* **2009**, *10*, 218–227.
8. Oldham, M.L.; Davidson, A.L.; Chen, J. Structural insights into ABC transporter mechanism. *Curr. Opin. Struct. Biol.* **2008**, *18*, 726–733.
9. Oldham, M.L.; Khare, D.; Quiocho, F.A.; Davidson, A.L.; Chen, J. Crystal structure of a catalytic intermediate of the maltose transporter. *Nature* **2007**, *450*, 515–521.
10. Gilson, E.; Alloing, G.; Schmidt, T.; Claverys, J.P.; Dudler, R.; Hofnung, M. Evidence for high affinity binding-protein dependent transport systems in gram-positive bacteria and in *Mycoplasma*. *EMBO J.* **1988**, *7*, 3971–3974.

11. Montesinos, M.L.; Herrero, A.; Flores, E. Amino acid transport in taxonomically diverse cyanobacteria and identification of two genes encoding elements of a neutral amino acid permease putatively involved in recapture of leaked hydrophobic amino acids. *J. Bacteriol.* **1997**, *179*, 853–862.
12. Quiocho, F.A.; Ledvina, P.S. Atomic structure and specificity of bacterial periplasmic receptors for active transport and chemotaxis: variation of common themes. *Mol. Microbiol.* **1996**, *20*, 17–25.
13. Tam, R.; Saier, M.H., Jr. Structural, functional, and evolutionary relationships among extracellular solute-binding receptors of bacteria. *Microbiol. Rev.* **1993**, *57*, 320–346.
14. Dwyer, M.A.; Hellinga, H.W. Periplasmic binding proteins: a versatile superfamily for protein engineering. *Curr. Opin. Struct. Biol.* **2004**, *14*, 495–504.
15. Fukami-Kobayashi, K.; Tateno, Y.; Nishikawa, K. Domain dislocation: a change of core structure in periplasmic binding proteins in their evolutionary history. *J. Mol. Biol.* **1999**, *286*, 279–290.
16. Lee, Y.H.; Dorwart, M.R.; Hazlett, K.R.; Deka, R.K.; Norgard, M.V.; Radolf, J.D.; Hasemann, C.A. The crystal structure of Zn(II)-free *Treponema pallidum* TroA, a periplasmic metal-binding protein, reveals a closed conformation. *J. Bacteriol.* **2002**, *184*, 2300–2304.
17. Karpowich, N.K.; Huang, H.H.; Smith, P.C.; Hunt, J.F. Crystal structures of the BtuF periplasmic-binding protein for vitamin B12 suggest a functionally important reduction in protein mobility upon ligand binding. *J. Biol. Chem.* **2003**, *278*, 8429–8434.
18. Wissenbach, U.; Six, S.; Bongaerts, J.; Ternes, D.; Steinwachs, S.; Unden, G. A third periplasmic transport system for L-arginine in *Escherichia coli*: molecular characterization of the artPIQMJ genes, arginine binding and transport. *Mol. Microbiol.* **1995**, *17*, 675–686.
19. Stamp, A.L.; Owen, P.; El Omari, K.; Lockyer, M.; Lamb, H.K.; Charles, I.G.; Hawkins, A.R.; Stammers, D.K. Crystallographic and microcalorimetric analyses reveal the structural basis for high arginine specificity in the *Salmonella enterica* serovar Typhimurium periplasmic binding protein STM4351. *Proteins* **2011**, *79*, 2352–2357.
20. Oh, B.H.; Pandit, J.; Kang, C.H.; Nikaido, K.; Gokcen, S.; Ames, G.F.; Kim, S.H. Three-dimensional structures of the periplasmic lysine/arginine/ornithine-binding protein with and without a ligand. *J. Biol. Chem.* **1993**, *268*, 11348–11355.
21. Celis, R.T.; Leadlay, P.F.; Roy, I.; Hansen, A. Phosphorylation of the periplasmic binding protein in two transport systems for arginine incorporation in *Escherichia coli* K-12 is unrelated to the function of the transport system. *J. Bacteriol.* **1998**, *180*, 4828–4833.
22. De Wolf, F.A.; Brett, G.M. Ligand-binding proteins: their potential for application in systems for controlled delivery and uptake of ligands. *Pharmacol. Rev.* **2000**, *52*, 207–236.
23. Jain-Ghai, S.; Nagamani, S.C.; Blaser, S.; Siriwardena, K.; Feigenbaum, A. Arginase I deficiency: severe infantile presentation with hyperammonemia: more common than reported? *Mol. Genet. Metab.* **2011**, *104*, 107–111.
24. Brusilov, S.W.; Horwich, A.L. Urea cycle enzymes. In *The Metabolic Basis of Inherited Disease*; Scriver, C.R., Beaudet, A.L., Sly, W.S., Valle, D., Eds.; McGraw-Hill: New York, NY, USA, 1989; pp 629–663.

25. Rees, D.D.; Palmer, R.M.; Moncada, S. Role of endothelium-derived nitric oxide in the regulation of blood pressure. *Proc. Natl. Acad. Sci. USA* **1989**, *86*, 3375–3378.
26. Cooke, J.P. Does ADMA cause endothelial dysfunction? *Arterioscler. Thromb. Vasc. Biol.* **2000**, *20*, 2032–2037.
27. Vallance, P.; Collier, J.; Moncada, S. Effects of endothelium-derived nitric oxide on peripheral arteriolar tone in man. *Lancet* **1989**, *2*, 997–1000.
28. Dattelbaum, J.D.; Lakowicz, J.R. Optical determination of glutamine using a genetically engineered protein. *Anal. Biochem.* **2001**, *291*, 89–95.
29. Amiss, T.J.; Sherman, D.B.; Nycz, C.M.; Andaluz, S.A.; Pitner, J.B. Engineering and rapid selection of a low-affinity glucose/galactose-binding protein for a glucose biosensor. *Protein Sci.* **2007**, *16*, 2350–2359.
30. Staiano, M.; Bazzicalupo, P.; Rossi, M.; D'Auria, S. Glucose biosensors as models for the development of advanced protein-based biosensors. *Mol. Biosyst.* **2005**, *1*, 354–362.
31. Huber, R.; Langworthy, T.A.; König, H.; Thomm, M.; Woese, C.R.; Sleytr, U.B.; Stetter, K.O. *Thermotoga maritima* sp. nov. represents a new genus of unique extremely thermophilic eubacteria growing up to 90 °C. *Arch. Microbiol.* **1986**, *144*, 324–333.
32. Nelson, K.E.; Clayton, R.A.; Gill, S.R.; Gwinn, M.L.; Dodson, R.J.; Haft, D.H.; Hickey, E.K.; Peterson, J.D.; Nelson, W.C.; Ketchum, K.A.; *et al.* Evidence for lateral gene transfer between Archaea and bacteria from genome sequence of *Thermotoga maritima*. *Nature* **1999**, *399*, 323–329.
33. Worning, P.; Jensen, L.J.; Nelson, K.E.; Brunak, S.; Ussery, D.W. Structural analysis of DNA sequence: evidence for lateral gene transfer in *Thermotoga maritima*. *Nucleic Acids Res.* **2000**, *28*, 706–709.
34. Tian, Y.; Cuneo, M.J.; Changela, A.; Hocker, B.; Beese, L.S.; Hellinga, H.W. Structure-based design of robust glucose biosensors using a *Thermotoga maritima* periplasmic glucose-binding protein. *Protein Sci.* **2007**, *16*, 2240–2250.
35. Fox, J.D.; Routzahn, K.M.; Bucher, M.H.; Waugh, D.S. Maltodextrin-binding proteins from diverse bacteria and archaea are potent solubility enhancers. *FEBS Lett.* **2003**, *537*, 53–57.
36. Nanavati, D.M.; Thirangoon, K.; Noll, K.M. Several archaeal homologs of putative oligopeptide-binding proteins encoded by *Thermotoga maritima* bind sugars. *Appl Environ. Microbiol.* **2006**, *72*, 1336–1345.
37. Luchansky, M.S.; Der, B.S.; D'Auria, S.; Pocsfalvi, G.; Iozzino, L.; Marasco, D.; Dattelbaum, J.D. Amino acid transport in thermophiles: characterization of an arginine-binding protein in *Thermotoga maritima*. *Mol. Biosyst.* **2010**, *6*, 142–151.
38. Sun, Y.J.; Rose, J.; Wang, B.C.; Hsiao, C.D. The structure of glutamine-binding protein complexed with glutamine at 1.94 A resolution: Comparisons with other amino acid binding proteins. *J. Mol. Biol.* **1998**, *278*, 219–229.
39. Cuneo, M.J.; Changela, A.; Miklos, A.E.; Beese, L.S.; Krueger, J.K.; Hellinga, H.W. Structural analysis of a periplasmic binding protein in the tripartite ATP-independent transporter family reveals a tetrameric assembly that may have a role in ligand transport. *J. Biol. Chem.* **2008**, *283*, 32812–32820.

40. Gonin, S.; Arnoux, P.; Pierru, B.; Lavergne, J.; Alonso, B.; Sabaty, M.; Pignol, D. Crystal structures of an Extracytoplasmic Solute Receptor from a TRAP transporter in its open and closed forms reveal a helix-swapped dimer requiring a cation for α-keto acid binding. *BMC Struct. Biol.* **2007**, doi:10.1186/1472-6807-7-11.
41. Fang, Y.; Kolmakova-Partensky, L.; Miller, C. A bacterial arginine-agmatine exchange transporter involved in extreme acid resistance. *J. Biol. Chem.* **2007**, *282*, 176–182.
42. Scire, A.; Marabotti, A.; Staiano, M.; Iozzino, L.; Luchansky, M.S.; Der, B.S.; Dattelbaum, J.D.; Tanfani, F.; D'Auria, S. Amino acid transport in thermophiles: characterization of an arginine-binding protein in *Thermotoga maritima*. 2. Molecular organization and structural stability. *Mol. Biosyst.* **2010**, *6*, 687–698.
43. Vahedi-Faridi, A.; Eckey, V.; Scheffel, F.; Alings, C.; Landmesser, H.; Schneider, E.; Saenger, W. Crystal structures and mutational analysis of the arginine-, lysine-, histidine-binding protein ArtJ from *Geobacillus stearothermophilus*. Implications for interactions of ArtJ with its cognate ATP-binding cassette transporter, Art(MP)2. *J. Mol. Biol.* **2008**, *375*, 448–459.
44. Hsiao, C.D.; Sun, Y.J.; Rose, J.; Wang, B.C. The crystal structure of glutamine-binding protein from *Escherichia coli*. *J. Mol. Biol.* **1996**, *262*, 225–242.
45. Ausili, A.; Pennacchio, A.; Staiano, M.; Dattelbaum, J.D.; Fessas, D.; Schiraldi, A.; D'Auria, S. Amino acid transport in thermophiles: Characterization of an arginine-binding protein from *Thermotoga maritima*. 3. Conformational dynamics and stability. *J. Photochem. Photobiol. B* **2012**, in press.
46. Ruggiero, A.; Dattelbaum, J.D.; Pennacchio, A.; Iozzino, L.; Staiano, M.; Luchansky, M.S.; Der, B.S.; Berisio, R.; D'Auria, S.; Vitagliano, L. Crystallization and preliminary X-ray crystallographic analysis of ligand-free and arginine-bound forms of *Thermotoga maritima* arginine-binding protein. *Acta Crystallogr. Sect. F Struct. Biol. Cryst. Commun.* **2011**, *67*, 1462–1465.

Biohydrogen Production by the Thermophilic Bacterium *Caldicellulosiruptor saccharolyticus*: Current Status and Perspectives

Abraham A. M. Bielen, Marcel R. A. Verhaart, John van der Oost and Servé W. M. Kengen

Abstract: *Caldicellulosiruptor saccharolyticus* is one of the most thermophilic cellulolytic organisms known to date. This Gram-positive anaerobic bacterium ferments a broad spectrum of mono-, di- and polysaccharides to mainly acetate, CO_2 and hydrogen. With hydrogen yields approaching the theoretical limit for dark fermentation of 4 mol hydrogen per mol hexose, this organism has proven itself to be an excellent candidate for biological hydrogen production. This review provides an overview of the research on *C. saccharolyticus* with respect to the hydrolytic capability, sugar metabolism, hydrogen formation, mechanisms involved in hydrogen inhibition, and the regulation of the redox and carbon metabolism. Analysis of currently available fermentation data reveal decreased hydrogen yields under non-ideal cultivation conditions, which are mainly associated with the accumulation of hydrogen in the liquid phase. Thermodynamic considerations concerning the reactions involved in hydrogen formation are discussed with respect to the dissolved hydrogen concentration. Novel cultivation data demonstrate the sensitivity of *C. saccharolyticus* to increased hydrogen levels regarding substrate load and nitrogen limitation. In addition, special attention is given to the rhamnose metabolism, which represents an unusual type of redox balancing. Finally, several approaches are suggested to improve biohydrogen production by *C. saccharolyticus*.

Reprinted from *Life*. Cite as: Bielen, A.A.M.; Verhaart, M.R.A.; van der Oost, J.; Kengen, S.W.M. Biohydrogen Production by the Thermophilic Bacterium *Caldicellulosiruptor saccharolyticus*: Current Status and Perspectives. *Life* **2013**, *3*, 52-85.

1. Introduction

The use of renewable plant biomass for the production of biofuels, chemicals or other biocommodities can provide a realistic alternative for fossil fuel based processes [1,2]. The implementation of lignocellulosic biomass for biofuel production requires the degradation of recalcitrant substrates like cellulose, hemicellulose or lignin. Lignin is either removed or modified [3], while cellulose and hemicellulose are converted into more readily fermentable mono-, di- and oligo-saccharides. Although this can be achieved by different (thermo)chemical or enzymatic pre-treatments, a more desirable process combines both substrate hydrolysis and fermentation of complex plant biomass. Such a "consolidated bioprocess" (CBP) circumvents the negative environmental impact inherent to (thermo)chemical pre-treatment and might limit overall process costs [1,4].

Hydrogen gas (H_2) is considered an alternative for the non-renewable fossil fuels and can be produced in a carbon neutral process. The controlled biological production of H_2 would allow for capturing the CO_2 released during the process, preventing it to dissipate into the environment. In addition, compared to carbon based (bio)fuel types, H_2 has the advantage that (i) during its oxidation only H_2O is released and (ii) that H_2 fuel cells can be used, which are more energy efficient than the presently used combustion engines [5]. Biohydrogen can be produced from renewable feedstocks in an anaerobic fermentation process, which is often referred to as dark fermentation to distinguish it from photofermentative hydrogen production.

Both plant biomass degradation and biological H_2 formation appear advantageous under thermophilic conditions. Moreover, thermophiles display an extensive glycoside hydrolase inventory aiding in the lignocellulosic biomass breakdown [6–8]. Based on thermodynamic considerations H_2 formation is more feasible at elevated temperatures [9,10]. Correspondingly, H_2 yields are in general higher for (hyper)thermophiles, reaching the theoretical limit of 4 mol H_2 per mol of hexose, compared to the mesophilic hydrogen producers [9,11,12].

Since its isolation in the mid-eighties it has become clear that the thermophilic anaerobic bacterium *Caldicellulosiruptor saccharolyticus* [13] displays both the desirable polysaccharide degrading capabilities (including cellulose) and H_2 producing characteristics, making it an outstanding source for thermostable glycoside hydrolases and an excellent candidate for biohydrogen production from renewable biomass.

This review will discuss the available scientific data on *C. saccharolyticus* regarding its lignocellulolytic capability, substrate specificity, catabolism and H_2 producing capacity, which has made *C. saccharolyticus* to become a model organism for the study of fermentative hydrogen formation at elevated temperatures.

2. Isolation and Initial Characterization

The foreseen commercial value of thermostable cellulolytic enzymes in biotechnological applications triggered the investigation of new sources of these types of enzymes. In the search for novel thermophilic cellulolytic micro-organisms, several anaerobic bacteria have been isolated from natural enrichment sites from the Rotorua-Taupo thermal area in New Zealand [14]. One of the isolated strains, TP8.T 6331 [14] also referred to as TP8 [15] or "*Caldocellum saccharolyticum*" [16], revealed thermostable cellulase activity up to 85 °C [14,15] but also lignocellulolytic biomass decomposition capabilities [16]. Strain TP8.T 6331 was assigned to a new genus *Caldicellulosiruptor* as *Caldicellulosiruptor saccharolyticus* and was characterized as a Gram-positive, asporogenous, extremely thermophilic and strictly anaerobic bacterium capable of sustaining growth at a temperature range of 45–80 °C (T_{opt} =70 °C) and pH range of 5.5–8.0 (pH_{opt} = 7) [13]. Acid production could be detected for a broad substrate range including different pentoses and hexoses, di-saccharides and polysaccharides like cellulose and xylan [13–16]. In particular, the capacity to use cellulose at high temperatures was exceptional.

Ever since, several cellulolytic and weakly cellulolytic *Caldicellulosiruptor* species have been identified, all of which are isolated from terrestrial geothermal regions. [13,14,17–28]. The availability of the fully sequenced genomes of 8 of these *Caldicellulosiruptor* species allows the investigation of the possible differences in their cellulolytic traits and the analysis of other remarkable features of this genus [29–33].

3. Hydrolytic Capacity and Complex Biomass Decomposition

For the decomposition of recalcitrant plant polysaccharides *C. saccharolyticus* does not employ cellulosome-like structures, as described for some *Clostridium* species [34], but wields a variety of free-acting endo- and exo-glycoside hydrolases (GH) capable of hydrolyzing the glycosidic bonds of α- and β-glucans like starch, pullulan and cellulose, but also xylan and hetero-polysaccharides like hemicelluloses and pectin [8,13,33,35,36]. Actually, *Caldicellulosiruptor* species, together with *Thermoanaerobacter* species, are one of the most thermophilic crystalline cellulose-degrading organisms known to date that use free-acting primary cellulases [7]. *C. saccharolyticus* contains 59 open reading frames (ORFs) that include GH catalytic domains [29]. Some of these ORFs code for multifunctional, multi-domain proteins that contain glycoside hydrolase domains, belonging to different GH families, and multiple carbon binding modules [6,29,36]. The catalytic properties and structural organization of some of the glycoside hydrolases from *C. saccharolyticus*, have been extensively investigated.

A β-glucosidase (BglA) [37,38], β-xylosidase [39], β-1,4-xylanase [40] and a type I pullulanase [41] from *C. saccharolyticus* have been cloned into *E. coli* and characterized. The majority of the genes encoding xylan degradation associated enzymes appear to be clustered on the genome (*xynB-xynF*, Csac_2404-2411) [6,36]. Both XynA and XynE exhibit endoxylanase and xylosidase activity [36,42], while XynB only acts as a β-D-xylosidase [43]. XynC does not contain a GH domain and was shown to be an acetyl esterase [44], XynD showed to be active on xylan [36] and although the cloning and expression of intact multi-domain XynF could not be achieved, its *N*- and *C*-terminal parts revealed catalytic activity on arabinoxylan. Hence XynF was proposed to be involved in the degradation of the arabinoxylan component of hemicellulose [36]. A second locus covers several genes coding for multidomain proteins involved in glucan and mannan hydrolysis (*celA-manB*, Csac_1076-1080) [6,36]. *CelA* is coding for a multidomain cellulase [45], the bifunctional cellulase CelB exhibited both endo-β-1,4-glucanase and exo-β-1,4-glucanase activity [36,46,47] and CelC was characterized as an endo-1,4-β-D-glucanase [48]. ManA was characterized as a β-mannanase [49], but *ManB* codes for an inactive mannase, which after correcting for a frame shift in the nucleotide sequence, exhibited β-mannanase activity [48]. Several ORFs of the described *celA-manB* and *xynB-xynF* loci were differentially transcribed on pretreated poplar and switch grass compared to the monosaccharides glucose and xylose, showing their involvement in the decomposition of complex carbohydrates [36].

While some of the GH proteins act intracellularly, others are excreted, allowing the decomposition of non-soluble substrates to smaller oligo- or mono-saccharides. Early findings by Reynolds *et al.* already indicated that a significant percentage of the cellulolytic activity was found to be associated with insoluble substrate [15]. These interactions, between GH and substrate, are facilitated by carbon binding modules (CBM). CBMs allow the positioning of the GH catalytic domains in the vicinity of the substrate, thus increasing the rate of catalysis. Interestingly, most multi-domain GHs identified in *C. saccharolyticus*, containing one or more CBMs, possess a signal peptide, which mark them for excretion [36]. A relative higher amount of GH related proteins could be observed in the substrate bound protein fraction, from *Caldicellulosiruptor* species grown on Avicel, with respect to the whole cell proteome [29]. Additionally, these proteome studies allowed the identification of specific GHs which interact with crystalline cellulose. CelA, a multi-domain GH consisting of two GH domains (GH9 and GH48) and three CBM3 modules, was found to be the most abundant substrate bound protein for strong cellulolytic *Caldicellulosiruptor* species [29]. The sequenced genomes of *Caldicellulosiruptor* species reveal differences in glycoside hydrolytic capacity, which reflects their difference in biomass degrading capabilities [29,35]. The secretome of *C. saccharolyticus* grown on glucose contains several carbohydrate-degrading enzymes including CelA, ManA, CelB and CelC (protein sequence ID A4XIF5/6/7/8 respectively), which indicates that these GH are constitutively expressed even under non-cellulose degrading conditions [50].

In addition to the interactions between substrate and glycoside hydrolases, interactions between whole cells and a substrate have also been observed for *C. saccharolyticus*. These interactions appeared to be substrate specific. For instance, a higher degree of cell-to-substrate attachment was observed for cells grown on switch grass, compared to poplar [36]. Several S-layer homology (SLH) domain containing proteins, which have been identified in *Caldicellulosiruptor* species, are proposed to have a role in such cell substrate interactions. These SLH domain proteins contain both glycoside hydrolases domains and non-catalytic carbohydrate binding domains, which are utilized in lignocellulose degradation by both recruiting and degrading complex biomass via cell substrate interactions [51]. In addition, cell immobilization on a support matrix, like pine wood shavings, supports cell survival and improves the H_2 evolving capacity of *C. saccharolyticus* [52].

4. Sugar Catabolism and Pathway Regulation

4.1. Sugar Uptake and Fermentation

Soluble sugar substrates can enter *C. saccharolyticus* cells either as mono-, di- or oligo-saccharides via several ABC-transporters, which facilitate substrate transport across the membrane at the expense of ATP. In addition, *C. saccharolyticus* contains one fructose specific phosphotransferase system (PTS). During PTS-mediated transport the substrate is both transported and phosphorylated at the expense of phosphoenolpyruvate (PEP). The substrate specificity of the 24 sugar ABC transport systems, identified in *C. saccharolyticus*, has been assigned based on bioinformatic analysis and functional genomics [53]. Most of the identified ABC transporters have a broad predicted substrate specificity and for some transporters the annotated substrate specificity was confirmed by transcriptional data obtained from cells grown on different mono-saccharides [53]. Some substrates can be transported by multiple transporter systems (Figure 1).

The growth of *C. saccharolyticus* on sugar mixtures revealed the co-utilization of hexoses and pentoses, without any signs of carbon catabolite repression (CCR) [33,53]. The absence of CCR is in principle a very advantageous characteristic of *C. saccharolyticus*, as it enables the simultaneous fermentation of hexoses and pentoses [33]. Substrate co-utilization has also been confirmed for biomass derived hydrolysates [54,55]. Despite the absence of CCR, a somewhat higher preference for the pentose sugars (xylose and arabinose) with respect to the hexose sugars (glucose, mannose and galactose) was demonstrated, but the highest preference was observed for the PTS-transported hexose fructose [53].

Once inside the cell, the sugar substrates are converted into a glycolytic intermediate. NMR analysis of the fermentation end-products of *C. saccharolyticus* grown on ^{13}C-labeled glucose showed that the Embden-Meyerhof pathway is the main route for glycolysis [56]. All genes encoding components of the Embden-Meyerhof and non-oxidative pentose phosphate pathway have been identified in the genome. There is no evidence for the presence of the oxidative branch of the pentose phosphate pathway or the Entner-Doudoroff pathway [33] (Figure 1).

Each sugar substrate, with the exception of rhamnose is completely catabolized to glyceraldehyde 3-phosphate (GAP) (rhamnose catabolism will be discussed separately in more detail below). The subsequent conversion of GAP to pyruvate, via the C-3 part of glycolysis, results in the formation of the reduced electron carrier NADH. Pyruvate can be further oxidized to acetyl-CoA by pyruvate: ferredoxin oxidoreductase (POR), which is coupled to the generation of reduced ferredoxin (Fd_{red}). Finally, acetyl-CoA can be converted to the fermentation end-product acetate (Figure 1).

Both types of reduced electron carriers (NADH and Fd_{red}) can be used by hydrogenases for proton reduction, thus forming H_2. The genome of *C. saccharolyticus* contains a gene cluster coding for an NADH-dependent cytosolic hetero-tetrameric Fe-only hydrogenase (*hyd*) and a cluster encoding a membrane bound multimeric [NiFe] hydrogenase (*ech*), which presumably couples the oxidation of Fd_{red} to H_2 production [33]. Under optimal cultivation conditions, when all reductants are used for H_2 formation, the complete oxidation of glucose yields 4 mol of H_2 per mol of glucose consumed (Equation 1).

Figure 1. Overview of the central carbon metabolism of *Caldicellulosiruptor saccharolyticus*.

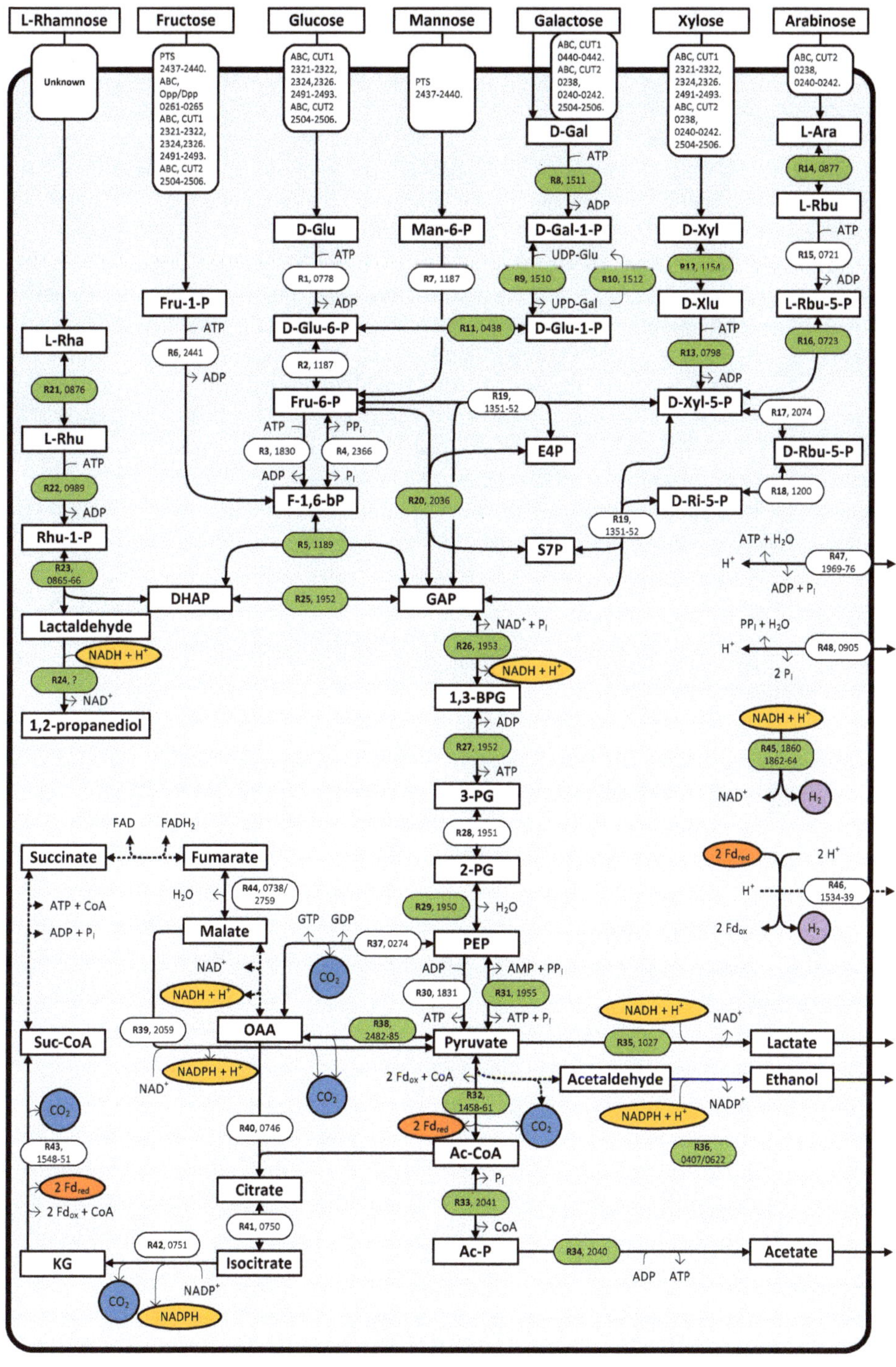

Figure 1. *Cont.*

During sugar catabolism electron carriers (NAD^+ and Fd_{ox}) are reduced. These reduced electron carriers can be re-oxidized by hydrogenases to form H_2 (reaction 45 and 46). For each reaction (numbered) the locus tags of the associated genes are given. When given in green they have been shown to be regulated by a specific mono-saccharide [33,53]. Dotted lines: no candidate genes can be identified. Reactions: **R1**, glucokinase, EC 2.7.1.2; **R2**, phosphoglucose isomerase, EC 5.3.1.9; **R3**, ATP-dependent 6-phosphofructokinase, EC 2.7.1.11; **R4**, PPi-dependent 6-phosphofructokinase, EC 2.7.1.90; **R5**, fructose-bisphosphate aldolase, EC 4.1.2.13; **R6**, 1-phosphofructokinase, EC 2.7.1.56; **R7**, phosphomannose isomerase, EC 5.3.1.8; **R8**, galactokinase, EC 2.7.1.6; **R9**, galactose 1-phosphate uridyl transferase, EC 2.7.7.12; **R10**, UDP-galactose 4-epimerase, EC 5.1.3.2; **R11**, phosphoglucomutase, EC 5.4.2.6; **R12**, D-xylose isomerase, EC 5.3.1.5; **R13**, xylulokinase, EC 2.7.1.17; **R14**, L-arabinose isomerase, EC 5.3.1.4; **R15**, L-ribulokinase, EC 2.7.1.16; **R16**, L-xylulose 5-phosphate 3-epimerase, EC 5.1.3.22; **R17**, ribulose-phosphate 3-epimerase, EC 5.1.3.1; **R18**, ribose-5-phosphate isomerase, EC 5.3.1.6; **R19**, transketolase, EC 2.2.1.1; **R20**, transaldolase, EC 2.2.1.2; **R21**, L-rhamnose isomerase, EC 5.3.1.14; **R22**, L-rhamnulose kinase, EC 2.7.1.5; **R23**, L-rhamnulose-1-phosphate aldolase, EC 4.1.2.19; **R24**, L-1,2-propanediol oxidoreductase, EC 1.1.1.77; **R25**, triose phosphate isomerase, EC 5.3.1.1; **R26**, glyceraldehyde-3-phosphate dehydrogenase, EC 1.2.1.12; **R27**, phosphoglycerate kinase, EC 2.7.2.3; **R28**, phosphoglycerate mutase, EC 5.4.2.1; **R29**, phosphopyruvate hydratase (enolase), EC 4.2.1.11; **R30**, pyruvate kinase, EC 2.7.1.40; **R31**, pyruvate phosphate dikinase, EC 2.7.9.1; **R32**, pyruvate:ferredoxin oxidoreductase, EC 1.2.7.1; **R33**, phosphotransacetylase, EC 2.3.1.8; **R34**, acetate kinase, EC 2.7.2.1; **R35**, L-lactate dehydrogenase, EC 1.1.1.27; **R36**, alcohol dehydrogenase, EC 1.1.1.1; **R37**, phosphoenolpyruvate carboxykinase, 4.1.1.32; **R38**, oxaloacetate decarboxylase (Na^+ Pump), EC 4.1.1.3; **R39**, malic enzyme 1.1.1.40; **R40**, citrate synthase, EC 2.3.3.1; **R41**, aconitase, EC 4.2.1.3; **R42**, isocitrate dehydrogenase, EC 1.1.1.42; **R43**, 2-oxoglutarate:ferredoxin oxidoreductase, EC 1.2.7.3; **R44**, fumarase, EC 4.2.1.2; **R45**, NADH-dependent Fe-only hydrogenase (*hyd*); **R46**, ferredoxin-dependent [NiFe] hydrogenase (*ech*); **R47**, H^+-ATPase, EC 3.6.3.14; **R48**, V-type H^+-translocating pyrophosphatase. Abbreviations: D-Glu, D-Glucose; D-Glu-6-P, D-Glucose-6-phosphate; Fru-6-P, Fructose-6-phosphate; F-1,6-bP, Fructose-1,6-bisphosphate; Fru-1-P, Fructose-1-phosphate; Man-6-P, Mannose-6-phosphate; D-Gal, D-Galactose; D-Gal-1-P, D-Galactose-1-phosphate; D-Glu-1-P, D-Glucose-1-phosphate; D-Xyl, D-Xylose; D-Xlu, D-Xylulose; D-Xyl-5-P, Xylulose 5-phosphate; L-Ara, L-Arabinose; L-Rbu, L-Ribulose; Rbu-5-P, L-Ribulose-5-P; D-Rbu-5-P, D-Ribulose-5-phosphate; D-Ri-5-P, D-Ribose-5-P; E4P, D-Erythrose 4-phosphate; S7P, D-Sedoheptulose-7-phosphate; DHAP, Dihydroxyacetone phosphate; GAP, Glyceraldehyde 3-phosphate; 1,3-BPG, 1,3-Bisphosphoglycerate; 3-PG, 3-Phosphoglycerate; 2-PG, 2-Phosphoglycerate; PEP, Phosphoenolpyruvate; Ac-CoA, Acetyl-CoA; Ac-P, Acetyl phosphate; OAA, Oxaloacetate; KG, 2-Oxoglutarate; Suc-CoA, Succinyl-CoA; H_2O, Water; H^+, Proton; H_2, Hydrogen; CO_2, Carbon dioxide; Fd_{ox} oxidized Ferredoxin; Fd_{red} reduced Ferredoxin; $NADP^+$, Nicotinamide adenine dinucleotide phosphate; NADPH, reduced Nicotinamide adenine dinucleotide phosphate; NAD^+, Nicotinamide adenine dinucleotide; NADH, reduced Nicotinamide adenine dinucleotide; FAD, Flavin adenine dinucleotide; $FADH_2$, reduced Flavin adenine dinucleotide; ATP, Adenosine triphosphate; ADP, Adenosine diphosphate; AMP, Adenosine monophosphate; GTP, Guanosine triphosphate; GDP, Guanosine diphosphate; PP_i, pyrophosphate; P_i, inorganic phosphate; CoA, Coenzyme A; UPD-Glu, UDP-Glucose; UDP-Gal, UPD-Galactose.

$$\text{glucose} + 4\cdot H_2O \rightarrow 2\ \text{acetate}^- + 2\cdot HCO_3^- + 4\cdot H^+ + 4\cdot H_2 \quad (1)$$

$$\text{glucose} + 2\cdot H_2O \rightarrow 2\ \text{ethanol} + 2\cdot HCO_3^- + 2\cdot H^+ \quad (2)$$

$$\text{glucose} \rightarrow 2\ \text{lactate}^- + 2\cdot H^+ \quad (3)$$

However, suboptimal growth conditions lead to a mixed acid fermentation with ethanol (Equation 2) and lactate (Equation 3) as end products in addition to acetate and H_2. Lactate is produced from pyruvate, using NADH as electron donor and catalyzed by lactate dehydrogenase (LDH). The corresponding *ldh* gene could be easily identified in the genome [33,57]. However, the identity of the enzymes and genes involved in ethanol formation, are less clear. Two alcohol dehydrogenase (ADH) genes have been identified in the genome which, based on transcriptional data, can both be involved in ethanol formation from acetaldehyde. The way acetaldehyde is produced, is however, not known. Acetaldehyde can be produced from pyruvate by a pyruvate decarboxylase, as described for yeast, or from acetyl-CoA by an acetaldehyde dehydrogenase, as is commonly seen in fermentative bacteria. In several thermophilic ethanol-producing bacteria, acetaldehyde is produced by a bifunctional acetaldehyde/ethanol dehydrogenase [58,59]. However, no candidate gene could be identified for any of these alternatives [33,60]. A third option might be that acetaldehyde is formed from pyruvate in a CoA-dependent side reaction of the pyruvate:ferredoxin oxidoreductase as described for *Pyrococcus furiosus* [61]. There is no experimental evidence for such a side reaction in *C. saccharolyticus*, but the absence of a dedicated enzymatic acetaldehyde-forming step might explain the low ethanologenic capacity of *C. saccharolyticus*. The difference in the observed lower ethanol to acetate ratio for *C. saccharolyticus* with respect to *Clostridium thermocellum* [16] can be explained by the fact that the *C. saccharolyticus* genome does not have a similar pathway for ethanol formation as identified in *C. thermocellum* [58]. Similar to other high yield ethanol-producing thermophiles, like *Thermoanaerobacter ethanolicus* [62] or *Thermoanaerobacterium saccharolyticum* [63], *Clostridium thermocellum* has a bifunctional acetaldehyde-CoA/alcohol dehydrogenase, which catalyzes ethanol formation from acetyl-coA [58,59]. Such a pathway is absent in low level ethanol-producing thermophiles like *C. saccharolyticus* [33], *Thermoanaerobacter tengcongensis* [64] and *Pyrococcus furiosus* [65]. Although *C. saccharolyticus* is able to produce some ethanol, the flux through the ethanol-forming pathway is apparently limited resulting in the lower ethanol to acetate ratio.

NADPH is assumed to be the preferred substrate for the ethanol-forming ADH reaction [33,57]. A potential source of NAPDH could be the isocitrate dehydrogenase in the oxidative branch of the incomplete TCA cycle (Figure 1). Alternatively, NADPH can be produced from NADH, but so far no candidate genes coding for such transhydrogenase have been identified in *C. saccharolyticus* [33,60]. With respect to H_2 production, it is important to realize that only the production of acetate is coupled to H_2 formation since no net reducing power remains for H_2 formation when ethanol or lactate is produced.

4.2. Rhamnose Fermentation

Compared to glucose, fermentative growth on the deoxy sugar rhamnose is associated with a different carbon and electron metabolism. The proposed pathway for rhamnose degradation (Figure 1) implies that during rhamnose catabolism half of the generated reduced electron-carriers is used for the reduction of lactaldehyde to 1,2-propanediol, while the other half can be recycled through H_2 formation [33]. Indeed, fermentation of rhamnose by *C. saccharolyticus* results in the production of 1,2-propanediol, acetate, H_2 and CO_2 in a 1:1:1:1 ratio ([66], Figure 2a). This ratio suggests that all NADH is used for 1,2-propanediol formation and that all Fd_{red} is used for H_2 formation. However, when *C. saccharolyticus* is grown on rhamnose under a headspace of carbon monoxide (CO), which is an established competitive inhibitor of both NiFe- [67] and Fe-only [68] hydrogenases, H_2 evolution is significantly inhibited ([66], Figure 2b). The CO cultivation condition does not affect 1,2-propanediol formation, but less H_2 is produced and the remaining reduced electron-carriers are now recycled by the formation of lactate and ethanol, consequently leading to a decrease in the acetate level. These findings suggest that, based on the substrate specificity of the lactate dehydrogenase (NADH, [57]) and the ethanol dehydrogenase(s) (NADPH, [57]), electron exchange between the Fd_{red} and $NAD(P)^+$ is required. However, no genes have been identified in *C. saccharolyticus* coding for an enzyme capable of catalyzing such a reaction [33,60]. Transcript levels of the genes from the gene cluster containing both the L-rhamnose isomerase and L-rhamnulose-1-phosphate aldolase are highly upregulated during growth on rhamnose (Figure 3) [33]. Although the function of the other genes in the cluster with respect to rhamnose catabolism remains unclear, the absence of such gene cluster and the rhamnose kinase gene from the genome of other *Caldicellulosiruptor* species, strongly correlates with their inability of rhamnose degradation (Table 1). This difference in rhamnose degrading capability between *Caldicellulosiruptor* species reflects the open nature of the *Caldicellulosiruptor* pan genome [29].

Figure 2. Fermentation profile of *C. saccharolyticus* grown on rhamnose batch cultivation: (**a**) without CO in the headspace (**b**) with a 100% CO headspace. Rhamnose (diamonds), acetate (open triangles), 1,2 propanediol (asterisks), lactate (open squares), ethanol (crosses), H_2 (circles) and OD_{660} (plus sign).

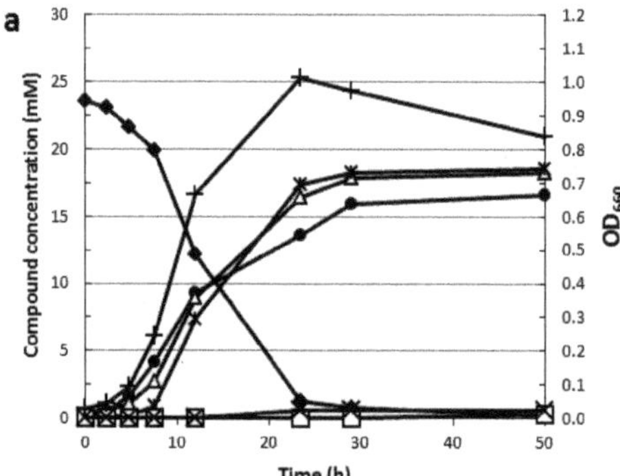

Figure 2. *Cont.*

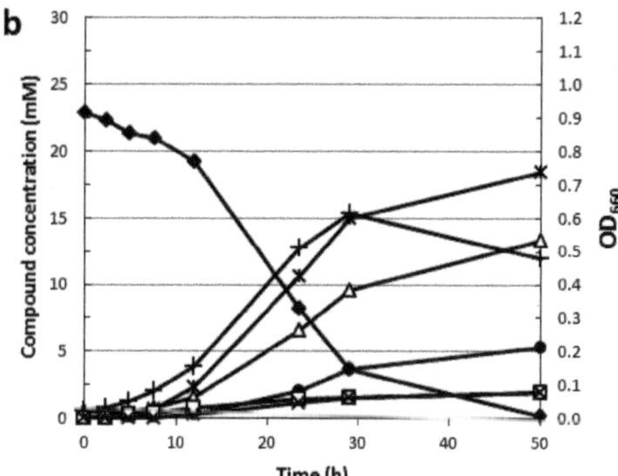

Figure 3. Schematic representation of two rhamnose associated gene clusters from *C. saccharolyticus* (Csac_0865-Csac_0876 and Csac_0989-Csac_0990). For each member of the cluster (grey arrows) the proposed function (text box) and locus tag number (four digit number) is given. Presented log2 values represent the ratio between transcription levels of the specific gene during growth on rhamnose with respect to growth on glucose; (+) upregulated, (−) downregulated on rhamnose versus glucose [33].

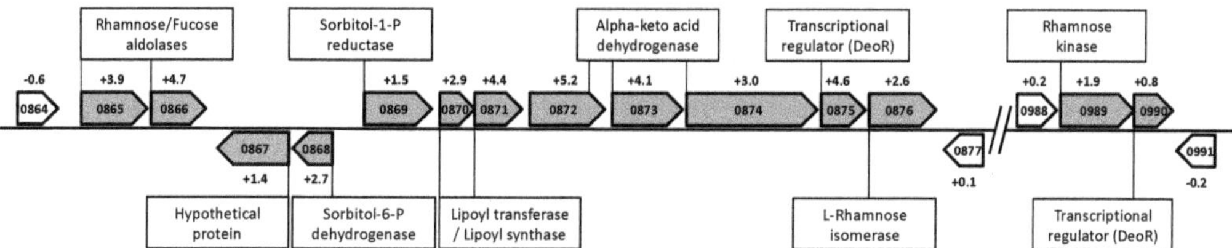

Table 1. Overview of *Caldicellulosiruptor* species which are able to grow on rhamnose. There is a correlation between the ability to grow on rhamnose and the presence of gene clusters orthologous to the rhamnose associated gene clusters identified in *C. saccharolyticus* (Figure 3). nt, not tested for growth on rhamnose. (+), able to grow of rhamnose/gene cluster is present in genome. (−), no growth observed on rhamnose/gene cluster is not present in genome. ns, not sequenced.

Strain	Growth on Rhamnose	Gene cluster Csac_0865-76	Gene cluster Csac_0989-90	Reference
C. saccharolyticus	+	+	+	[13]
C. bescii	+	+	+	[28]
C. owensensis	+	+	+	[21]
C. obsidiansis	nt	+	+	[20]
C. kronotskyensis	nt	+	+	[22]
C. hydrothermalis	nt	+	+	[22]
C. kristjanssonii	−	−	−	[17]
C. lactoaceticus	−	−	−	[23]
C. acetigenus	nt	ns	ns	[26]

4.3. Involvement of Pyrophosphate in the Energy Metabolism

From a bioenergetics point of view, glucose fermentation is optimal when acetate is the only end product, because an additional ATP is generated during the final acetate-forming step (Figure 1). When it is assumed that the ABC-transporter mediated substrate uptake requires 1 ATP, overall ATP yields become 1.5 ATP per acetate versus only 0.5 ATP per lactate or ethanol. However, ATP yields might even be higher when the involvement of pyrophosphate (PP_i) as an energy carrier is considered.

PP_i is a by-product of biosynthesis reactions like DNA and RNA synthesis or is generated when the amino acids are coupled to their tRNAs during protein synthesis [69]. Since these reactions are close to equilibrium accumulation of PP_i is believed to have an inhibitory effect on growth, and only the effective removal of PP_i drives these biosynthetic reactions forward [70]. When PP_i is hydrolysed to P_i by a cytosolic inorganic pyrophosphatase (PPase), the free energy just dissipates as heat. *C. saccharolyticus* does not contain a cytosolic PPase but possesses a membrane bound H^+-translocating PPase [71], which allows the free energy released upon PP_i hydrolysis to be preserved as a proton motive force. The high-energy phosphate bond of PP_i can also be used for the phosphorylation of fructose 6-phosphate, catalyzed by a PP_i-dependent phosphofructokinase (Figure 1).

Furthermore, PP_i is consumed during the catabolic conversion of phosphoenolpyruvate to pyruvate, catalyzed by pyruvate phosphate dikinase (PPDK). Such a catabolic role for PPDK was proposed based on the increase in transcript level of the *ppdk* gene under increased glycolytic fluxes [33,71]. Altogether, the use of PP_i as an energy donor could be a way for the organism to deal with the relative low ATP yields which are usually associated with fermentation [72].

4.4. Mechanism Involved in Mixed Acid Fermentation

H_2 has been reported as a growth inhibitor for *C. saccharolyticus* [16,73] and a critical dissolved H_2 concentration, which leads to the complete inhibition of growth, of 2.2 mmol/L has been determined for controlled batch cultivations [74]. In addition, an elevated P_{H2} has been shown to cause a switch in the fermentation profile, leading to increased formation of ethanol and lactate, in both controlled batch and chemostat cultivations [12,60,74,75]. Willquist *et al.* reported that in controlled batch fermentations the initiation of lactate formation coincided with an increment in both the internal $NADH/NAD^+$ ratio and the P_{H2} of the system [57,75].

An increase of the overall carbon flux through glycolysis results in a higher NADH production rate. To maintain a constant $NADH/NAD^+$ ratio a subsequently higher H_2 formation rate is required. However, when the H_2 formation rate (volumetric H_2 production rate) exceeds the H_2 liquid to gas mass transfer rate, H_2 will accumulate in the liquid phase. In some cases this can even lead to super saturation of the liquid, meaning that the dissolved H_2 concentration exceeds the maximal theoretical H_2 solubility [74,76]. Such high dissolved H_2 levels inhibit H_2 formation and presumably cause an increase in the $NADH/NAD^+$ ratio. An increased $NADH/NAD^+$ ratio has been shown to have an inhibitory effect on GAPDH activity thus limiting the glycolytic flux [75] and consequently leads to a decrease in substrate consumption and growth rate. A switch to lactate formation alleviates the inhibitory effect of an increased $NADH/NAD^+$ ratio by causing a decrease of the $NADH/NAD^+$ ratio through the reduction of pyruvate.

A clear switch to lactate formation caused by an increase in glycolytic flux can be observed for *C. saccharolyticus* grown in a chemostat under high P_{H2} ([77], Figure 4a). When the glucose load is increased from 20 mM to 40 mM, end-product formation is switched from mainly acetate to mainly lactate, respectively. After the switch to 40 mM glucose some adaptation time is required before a new steady state is achieved. During this adaptation period substantially higher amounts of glucose were detected in the culture effluent, probably indicating an inhibition of the glycolytic flux at the level of GAPDH. When a new steady state was achieved, residual glucose was only slightly higher with respect to the 20 mM condition. As a consequence of the increase in glycolytic flux, the lactate concentration dramatically increased while the acetate concentration hardly changed, indicating that during both substrate loads a similar volumetric H_2 productivity was maintained. These data indicate that under these conditions the organism is not capable of dealing with the increased glycolytic flux by increasing its H_2 productivity, but requires a switch to lactate formation. Although oxaloacetate formation was discernible under a 20 mM glucose load, significantly higher and therefore quantifiable levels of oxaloacetate were detected under the 40 mM glucose load condition. This observation together with the increased flux towards lactate suggests that in the newly achieved steady state, a bottleneck exists at the level of pyruvate.

Ethanol formation can also serve as reductant sink in *C. saccharolyticus*. For some chemostat cultivation conditions a decrease in H_2 yield is only associated with ethanol formation and not with lactate formation [56,60,75]. These conditions concern low substrate loads, and the observed increase in ethanol formation was triggered by an increase of the dilution rate [56,75] or a change in mode of gas flushing [60], both potentially generating a moderate increase in the H_2 level of the system.

4.5. Regulation of Reductant Disposal Pathways

Hydrogen, ethanol and lactate formation are the main routes for reductant disposal in *C. saccharolyticus*. The gene expression of the hydrogenases and both alcohol dehydrogenases involved in ethanol formation are proposed to be under the control of the $NADH/NAD^+$ sensitive transcriptional regulator Rex [60] (Figure 5). Cultivation of *C. saccharolyticus* under high P_{H2} conditions was shown to lead to an upregulation of both the hydrogenases and alcohol dehydrogenases [60]. In addition, the *ldh* transcript level is also upregulated under high P_{H2} conditions, but in silico analysis of the *ldh* promoter region did not reveal a likely Rex operator binding sequence [60]. Thus, the exact regulatory mechanism triggering *ldh* transcription remains therefore unclear. Nonetheless, lactate dehydrogenase activity has been shown to be regulated at the enzyme level, with fructose 1,6-bisphosphate and ATP acting as allosteric activators and both NAD^+ and PP_i as competitive inhibitors [57]. Under high P_{H2} cultivation conditions, enzyme activity assays revealed increased LDH activity with respect to low P_{H2} conditions [57,75]. A hampered glycolytic flux at the level of GAPDH, potentially triggered by NADH build-up, might lead to the accumulation of fructose 1,6-bisphosphate. In turn, the accumulated level of fructose

1,6-bisphosphate could stimulate lactate formation (Figure 5), explaining the switch to lactate under high P_{H2} at the enzyme level.

Figure 4. Fermentation profile of *C. saccharolyticus* (**a**) grown in a chemostat under high H_2 partial pressure at two different glucose concentrations (see Section 4.4), (**b**) grown in a chemostat without NH_4^+ in the medium under low and high H_2 partial pressure (see Section 6.4). The switch from 20 mM to 40 mM glucose containing-medium and the switch from low to high H_2 partial pressure cultivation condition is indicated by a vertical line. Acetate (open triangles), lactate (open squares), ethanol (crosses), CO_2 (open circles), oxaloacetate (asterisks), glucose out (open diamonds) and CDW (plus sign). Chemostat cultivation parameters (3 L reactor, 1 L working volume, pH = 7.0 (NaOH), temp 70 °C, D = 0.1 h^{-1}, low P_{H2} [sparging (4 L/h) with N_2 gas and stirring speed = 250 rpm]), high P_{H2} [headspace flushing (4 L/h) with H_2 gas and stirring speed = 50 rpm]).

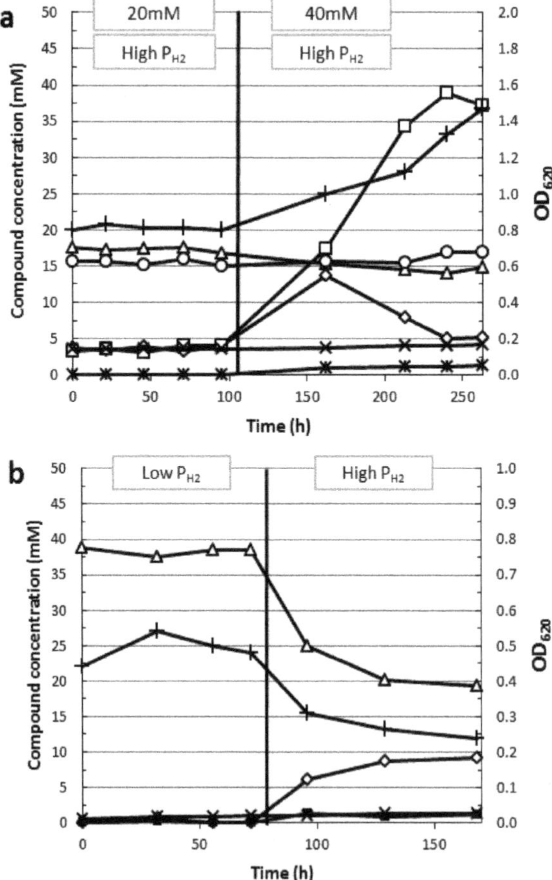

Figure 5. Overview of the regulatory mechanisms involved in the central metabolism of *Caldicellulosiruptor saccharolyticus*. The abbreviations of the compounds and reactions (circled numbers) are given in the legend of Figure 1. For the enzyme reactions given in green (circled numbers, green) the encoding genes are upregulated under increased P_{H2} [60]. Those which are under the control of the REX transcriptional regulator are marked with a green dot. H_2 inhibits its own formation (**R45**, **R46**) and accumulation of NADH inhibits glyceraldehyde-3-phosphate dehydrogenase (**R26**). PP_i, a by-product of biosynthesis, acts as an inhibitor of both pyruvate kinase (**R30**) and lactate dehydrogenase (**R35**) activity. Both ATP and F-1,6-bP are activators of lactate dehydrogenase (**R35**) activity.

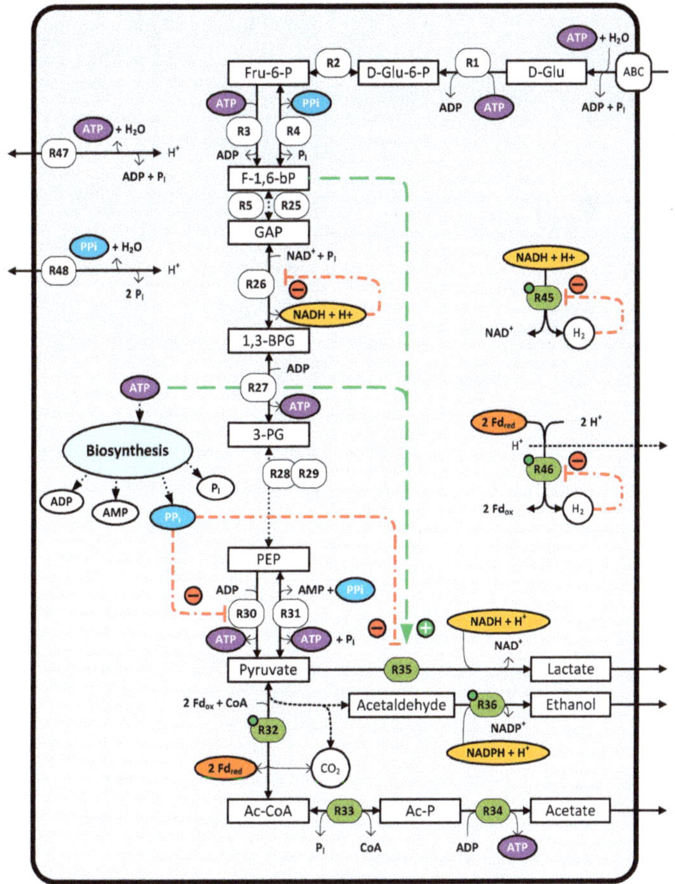

Interestingly, LDH enzyme activity can also be measured under non-lactate producing conditions, which suggest that additional factors control lactate formation [57,75]. Low levels of lactate formation can be observed at the end of growth during the transition to stationary phase for cultures grown in controlled batch systems [55,57,75,78,79]. This initiation of the lactate formation coincided with a relative increase in ATP levels and a relative decrease in PP_i levels [57,71]. These observed changes in ATP and PP_i levels could release the inhibitory effect of PP_i and stimulate LDH activity [57] (Figure 5). For this latter mode of LDH activity regulation, the lactate formation is proposed to be coupled to the energy metabolism. Accordingly, lactate formation is prevented during exponential growth, which is associated with a high anabolic activity and high PP_i levels. Whereas, during the transition to stationary phase, which is associated with low anabolic activity and relative low PP_i levels, lactate formation inhibition is alleviated [12,57].

5. Thermodynamic Considerations of Glucose Conversion and H_2 Formation

For *C. saccharolyticus* an elevated hydrogen concentration has been shown to affect fermentation performance, leading to a mixed acid fermentation [60,73–75]. These observed changes in the fermentation profile as a function of the H_2 concentration can be conceptually explained by considering the thermodynamics of the reactions leading to H_2 formation [9,10]. The Gibbs energy ($\Delta G'$) for a specific reaction can be calculated from the standard Gibbs energy ($\Delta G^{0\prime}$) and the reactant concentrations by the following relation:

$$\Delta G' = \Delta G^{0\prime} + RT\ \ln([C]^c[D]^d/[A]^a[B]^b) \qquad (4)$$

where $\Delta G^{0\prime}$ is the standard Gibbs energy (J/mol, 1 mol concentration of all reactants, at a neutral pH and at a specific temperature), R is the gas constant (J/mol*K); T is the temperature (K), A and B are the substrate concentrations with respective stoichiometric reaction coefficients a, b; and C and D are the reaction products with respective stoichiometric reaction coefficients c, d.

When H_2 formation occurs in the aqueous phase H_2 supersaturation can occur. For instance, over-saturation of 12 to 34 times the equilibrium concentration has been reported for *C. saccharolyticus* cultivations [74]. This indicates that estimating the dissolved H_2 concentration from the measured H_2 partial pressure by using the equilibrium constant does not always give an accurate representation of the state of the system. So H_2(aq), instead of the H_2 partial pressure (P_{H2}), should be used as a parameter when investigating the effect of H_2 on the metabolism. Therefore, all Gibbs energy calculations discussed herein were performed using the $\Delta_f G^0$ of the dissolved H_2 concentration (H_2(aq)).

Under standard conditions (25 °C), acetate formation (Equation 1, $\Delta G^{0\prime} = -142.6$ kJ/reaction) is energetically less favorable then lactate (Equation 3, $\Delta G^{0\prime} = -195.0$ kJ/reaction) and ethanol (Equation 2, $\Delta G^{0\prime} = -231.8$ kJ/reaction) formation. The Gibbs energy ($\Delta G'$) for the complete conversion of glucose to acetate depends, however, on the H_2 concentration (Equation 1), whereas the $\Delta G'$ for both ethanol and lactate formation from glucose is independent of the H_2 concentration (Equations 2 and 3, respectively). This means that the complete oxidation of glucose to acetate becomes energetically more favorable when the dissolved H_2 concentration is lowered. For example, acetate formation becomes energetically favorable compared to lactate or ethanol formation when the dissolved H_2 concentration drops below 5.0 or 0.12 mM, respectively (Figure 6a, supplementary data S1).

Figure 6. Effect of the H_2 concentration on the Gibbs energy change of reactions involved in H_2 formation: (**a**) ΔG' of the complete oxidation of 1 mol of glucose to acetate (closed diamonds), lactate (open squares) and ethanol (closed circles) at 25 °C (solid lines) and 70 °C (dashed lines); (**b**) ΔG' of H_2 formation from NADH (dotted line), reduced ferredoxin (dashed line) and via the bifurcating system (50% NADH and 50% Fd_{red}) (solid line) at 25 °C and 70 °C. Values were calculated from data presented in ([80–84]).

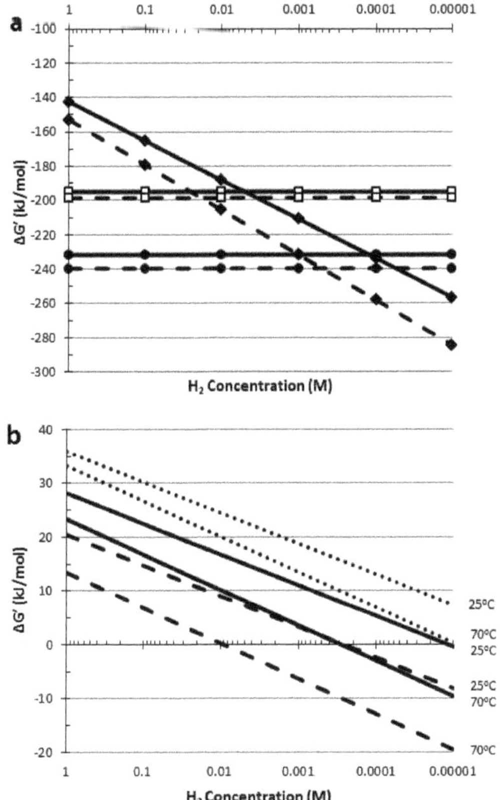

However, considering the Gibbs energies of the overall conversions can be misleading, as the thermodynamics of the involved partial reactions can be less favorable. Acetate formation is inevitably linked to H_2 formation and the $\Delta G^{0\prime}$ of the redox reaction coupling the reoxidation of the reduced electron carriers NADH or Fd_{red} to proton reduction are endergonic under standard conditions (Figure 6b, supplementary data S2). Decreasing the dissolved H_2 concentration will lower the ΔG' of these H_2 forming reactions, but an extremely low H_2 concentration is required before H_2 formation from NADH becomes exergonic (0.5 μM) (Fd_{red}, 0.2 mM) (Figure 6b). Therefore, at elevated H_2 concentrations this partial reaction raises a thermodynamic barrier. NADH dependent pyruvate reduction to lactate, on the other hand, has a negative Gibbs energy

($\Delta G^{0\prime}$ = −25.0 kJ/reaction) and is independent of the dissolved H_2 concentration. This means that under certain conditions lactate formation is more feasible, despite the more negative $\Delta G'$ for acetate formation compared to lactate formation (Equations 1 and 3), simply because H_2 formation from NADH is energetically unfavorable.

As explained, the thermodynamic barrier associated with H_2 formation from NADH can be lowered by decreasing the dissolved H_2 concentrations, moreover this barrier may also be tackled by an additional input of energy. The latter can be achieved by reverse electron transport catalyzed by an NADH: ferredoxin oxidoreductase, where the transfer of electrons from NADH to Fd_{red} is coupled to a proton or ion gradient [85]. Produced Fd_{red} can subsequently be used for H_2 formation. Alternatively, the energetically more favourable oxidation of Fd_{red} can be used to push the less favourable formation of H_2 from NADH in a bifurcating system (Figure 6b). Such a bifurcating function has been identified for *T. maritima* and was, based on protein sequences, also attributed to the Fe-only hydrogenase (*hyd*) of *C. saccharolyticus* [86]. However, the gene-arrangement of the Fe-only hydrogenase (*hyd*) in *C. saccharolyticus* is identical to that of *Thermoanaerobacter tengcongensis* [33], for which a NADH-dependent hydrogenase activity has been demonstrated [64], thus arguing against a bifurcating systems in *C. saccharolyticus*.

In general, H_2 formation is thermodynamically more favorable at elevated temperatures because (i) $\Delta G^{0\prime}$ values of the involved reactions are lower at increased temperature, and (ii) the RT coefficient in Equation 4 is temperature dependent, thus enhancing the effect of a decreased H_2 concentration (Figure 6a,b). Overall, thermophilic organisms have been shown to be able to produce H_2 at higher yields compared to mesophilic organisms. For thermophiles yields approaching the theoretical limit of 4 H_2 per hexose [83] have been reported, while for mesophiles H_2 yields generally do not exceed 2 H_2 per hexose [9,11,12]. These higher yields reflect the indicated thermodynamic advantage but also the lower diversity of fermentation end products observed for those thermophiles. For a specific organism the diversity of available electron acceptors, and formed end products, depends on the metabolic capabilities of the organism. Whether a specific pathway is operational depends on the regulation of that pathway at the transcription or translation level, but also depends on the kinetic properties and regulation of the enzyme activities of the specific enzymatic steps of that pathway.

6. Factors Limiting H_2 Formation

With an eye to the potential use of complex biomass for H_2 production a multitude of fermentability studies have been performed using *C. saccharolyticus*. An overview of the literature related to fermentability studies on either crop-based feedstock or industrial waste stream derived biomass is given in Table 2. Those investigations mainly focus on the fermentability of various complex substrates, associated H_2 formation and the effect of pretreatment on substrate accessibility and growth, overall demonstrating the bacterium's broad hydrolytic capacity.

Table 2. Literature overview of fermentability studies on *C. saccharolyticus* using crop-based feedstock or industrial waste stream derived biomass. * Extraction methods and mechanical pre-treatments were excluded in this overview, biological pre-treatment indicated pre-treatment by pre-incubation with *Bacillus amyloliquefaciens*; ** Multiple, different pre-treatments used; B, batch cultivation; CB, controlled batch cultivation; $ Minimal and maximal reported H_2 yields are given.

Reference	Substrate	Pre-treatment *	Cultivation method	H_2 yields $/Remarks
[87]	Wheat grains	Enzymatic	B	
	Wheat straw	Acid/Enzymatic	B	
[88]	Barley straw	Acid/Enzymatic **	B	
[29]	Crystalline cellulose	-	B	Proteome data
	Birchwood xylan	-	B	
	Switchgrass	Acid	B	
	Whatman no. 1 filterpaper	-	B	
[89]	Wheat straw	Acid/Enzymatic	B	
	Barley straw	Acid/Enzymatic	B	
	Corn stalk	Acid/Enzymatic	B	
	Corn cob	Acid/Enzymatic	B	
[36]	Poplar	-	B	Microarray data
	Switchgrass	Acid	B	Microarray data
[90]	Beet Molasses	-	CB	0.9–4.2
[91]	Potato steam peels	Enzymatic	CB	1.7–3.4
	Potato steam peels	-	CB	1.1–3.5
[92]	Filter paper	-	B	
	Wheat straw	Biological	B	
	Silphium perfoliatum leaves	Biological	B	
	Maize leaves	Biological	B	
	Sugar cane bagasse	Biological	B	
	Sweet sorghum whole plant	Biological	B	
[93]	Sugar beet	-	CB	3.0
[94]	Sweet sorghum bagasse	Alkaline/Enzymatic **	B/CB	1.3–2.6
[55]	Carrot pulp	Enzymatic	CB	1.3–2.8
	Carrot pulp	-	CB	
[35]	Crystalline cellulose	-	B	Proteome data
	Cellobiose	-	B	Microarray data
[95]	Switchgrass	-	B	
	Poplar	-	B	
[53]	Xylan	-	B	Microarray data
	Xyloglucan	-	B	Microarray data
	Xyloglucan-oligosaccharides	-	B	Microarray data
[96]	Barley straw	Acid/Enzymatic	B	
	Corn stalk	Acid/Enzymatic	B	
	Barley grain	Enzymatic	B	
	Corn grain	Enzymatic	B	
	Sugar beet	-	B	
[97]	Sweet sorghum plant	-	B	
	Sweet sorghum juice	-	B	

Table 2. *Cont.*

Reference	Substrate	Pre-treatment *	Cultivation method	H_2 yields $^{\$}$/Remarks
	Dry sugarcane bagasse	-	B	
	Wheat straw	-	B	
	Maize leaves	-	B	
	Maize leaves	Biological	B	
	Silphium trifoliatum leaves	-	B	
[54]	*Miscanthus giganteus*	Alkaline/Enzymatic	B/CB	2.4–3.4
[52]	Agarose	-	B	With different support matrixes
	Alginic acid	-	B	
	Pine wood shavings	-	B	
[98]	Jerusalem artichoke	-	B	Co-fermentation with natural biogas-producing consortia
	Fresh waste water sludge	-	B	
	Pig manure slurry	-	B	
[78]	Paper sludge	Acid/Enzymatic	CB	
[99]	Paper sludge	Acid/Enzymatic	B	

On the other hand, growth experiments on pure sugar substrates can be used to investigate the specific response associated with a certain substrate or to examine specific pathways involved in the metabolism of a substrate. Table 3 gives an overview of the literature related to growth experiments on pure sugars and pure sugar mixes, including the determined H_2 yields and H_2 productivities. The currently available fermentation data are discussed here to highlight the different factors limiting H_2 formation during the fermentative H_2 production by *C. saccharolyticus*.

Table 3. Literature overview of fermentation studies on *C. saccharolyticus* grown on pure sugar substrates. * Yields in mol H_2/mol hexose; nd, not determined; ** For batch cultivations the maximal productivity is given; nd, not determined; B, batch cultivation; and CB, controlled batch cultivation; Chem, chemostat cultivations.

Reference	Substrate	Substrate load (g/L)	H_2 Yield *	Productivity ** (mmol/(L*h))	Cultivation method	Dilution rate (h^{-1})	Remarks
[100]	Glucose	10	3.0	20.0 mol/(g*h)	CB		
		10	3.4	23.6 mol/(g*h)	CB		No YE in medium
		4	3.5	10.1 mol/(g*h)	Chem	0.05	
		4	3.5	10.4 mol/(g*h)	Chem	0.05	No YE in medium
[75]	Glucose	5	3.5	5.2	Chem	0.05	
		5	2.9	11.0	Chem	0.15	Residual glucose (3 mM)
		5	1.8	2.5	Chem	0.05	no sparging, open gas outlet
		5	wash out	wash out	Chem	0.15	no sparging, open gas outlet
[101]	Glucose	5	nd	nd	B		Extracellular proteome
[91]	Glucose	10	3.4	12.0	CB		
		31	2.8	12.9	CB		Residual glucose
[93]	Sucrose	10	2.9	7.1	CB		
[94]	Glucose/Xylose/Sucrose	10	3.2	10.7	CB		Sugar mix
	(6:2.5:1.5, w/w/w)	20	2.8	9.4	CB		Sugar mix
[55]	Glucose	10	3.2	11.2	CB		
		20	3.4	12.2	CB		
	Fructose	10	2.6	13.2	CB		
		20	2.4	13.4	CB		
	Glucose/Fructose	10	3.0	13.2	CB		Sugar mix
	(7:3, w/w)	20	2.6	12.2	CB		Sugar mix
[35]	Xylose	5	nd	nd	B		Proteome data
	Glucose	5	nd	nd	B		Proteome data
[50]	Glucose	5	nd	nd	B		Extracellular proteome
[79]	Sucrose	4	2.7	23.0	CB		
		4	3.1	11.8	CB		CO_2 sparging
	Glucose	4	3.0	20.0	CB		
		4	2.7	12.0	CB		CO_2 sparging

Table 3. *Cont.*

Reference	Substrate	Substrate load (g/L)	H_2 Yield *	Productivity ** (mmol/(L*h))	Cultivation method	Dilution rate (h^{-1})	Remarks
[53]	Glucose	0.5	nd	nd	B		Microarray data
	Mannose	0.5	nd	nd	B		Microarray data
	Arabinose	0.5	nd	nd	B		Microarray data
	Xylose	0.5	nd	nd	B		Microarray data
	Fructose	0.5	nd	nd	B		Microarray data
	Galactose	0.5	nd	nd	B		Microarray data
	mix (0.5 g/L each)	3	nd	nd	B		Microarray data, Sugar mix
[54]	Glucose/Xylose	10	3.4	12.0	CB		Sugar mix
	(7:3 w/w)	14	3.3	10.1	CB		Sugar mix
		28	2.4	9.7	CB		Sugar mix
[102]	Sucrose	10.3	2.8	22.0	Trickle bed	0.2–0.3	400 L, non-axenic fermentation
[33]	Glucose	4	nd	nd	CB		Microarray data
	Xylose	4	nd	nd	CB		Microarray data
	Rhamnose	4	nd	nd	CB		Microarray data
	Glucose/Xylose (1:1, w/w)	4	nd	nd	CB		Microarray data
[56]	Glucose	4.4	3.3	4.2	Chem	0.05	
		4.4	3.6	8.9	Chem	0.10	
		4.4	2.9	9.5	Chem	0.15	Residual glucose (3.3 mM)
		4.4	2.9	9.1	Chem	0.20	Residual glucose (8.9 mM)
		4.4	3.1	11.0	Chem	0.30	Residual glucose (12.4 mM)
		4.4	3.0	12.4	Chem	0.35	Residual glucose (12.7 mM)
		1.9	4.0	4.0	Chem	0.09	
		1.9	3.3	9.9	Chem	0.30	Residual glucose (0.6 mM)
		4.1	3.5	7.7	Chem	0.09	
		4.1	3.1	11.6	Chem	0.30	Residual glucose (11.9 mM)
[78]	Glucose	10	2.5	10.7	CB		
	Xylose	10	2.7	11.3	CB		
	Glucose/Xylose (11:3, w/w)	8.4	2.4	9.2	CB		Sugar mix
[103]	Sucrose	10	3.3	8.4	CB		

6.1. Comparison between Hydrolysates and Pure Sugar Mixtures

Studies on biomass hydrolysates and mono-saccharide mixtures, mimicking the biomass hydrolysates, showed that, while at low substrate loads fermentation performances were comparable, at higher substrate loads H_2 yields were higher on the mixed mono-saccharides compared to the biomass hydrolysates. The difference in yields was caused by a shift to lactate formation during the growth on high substrate load hydrolysates. Interestingly, for the higher substrate concentrations, the total sugar consumption was higher during growth on hydrolysates compared to growth on sugar mixtures [55,91]. For *C. saccharolyticus*, grown on carrot pulp hydrolysate (20 g/L), a lower cumulative H_2 production was found, compared to growth on a glucose/fructose mixture (20 g/L), while a relatively higher maximal H_2 productivity was observed for the hydrolysate compared to the sugar mixture. During the growth on carrot pulp hydrolysates the relative higher H_2 productivity preceded the switch to lactate formation [55]. These results demonstrate the relation between high H_2 productivity, lactate formation and an overall low H_2 yield. However, these described phenomena were not observed for fermentations on *Miscanthus* hydrolysates. For each tested substrate load (10, 14 and 28 g/L) fermentation performance during growth on the *Miscanthus* hydrolysate was similar to the glucose/xylose sugar mix and only moderate levels of lactate were formed even at high substrate loads [54]. The differences in H_2 production characteristics between the discussed hydrolysates might be related to the difference in sugar composition of the hydrolysates or the differences in pretreatment applied prior to hydrolysis.

In general, biomass derived hydrolysates might contain substances which negatively affect growth or fermentation performance. For example, the dilute-acid pretreatment of lignocellulosic biomass releases undesirable inhibiting compounds like 5-hydroxymethylfurural (HMF), furfural, phenolic compounds and acetate [89] and a growth inhibition of 50% was reported for *C. saccharolyticus* in the presence of 1–2 g/L HMF or furfural [54].

6.2. Incomplete Substrate Conversion

For chemostat cultivation on glucose (4.4 and 5 g/L) it was shown that a dilution rate (D) exceeding 0.1 h^{-1} gave rise to an incomplete substrate conversion but also to a lower H_2 yield. The concomitant decrease in both biomass level and H_2 yield caused the volumetric H_2 productivity (mmol/L*h) to level off at higher dilution rates [56,75]. The observed lower H_2 yield reflects the shift in end product formation, from mainly acetate at a low dilution rate of 0.05 h^{-1} to a mix of acetate and ethanol at a D = 0.15 h^{-1} [75]. Interestingly, no lactate was produced during these chemostat cultivations [56,75]. The observed incomplete substrate conversion indicated that another factor was limiting under those conditions. Indeed, increasing the yeast extract to glucose ratio, from 0.25 to 1 g/g, resulted in the almost complete consumption of glucose and also a doubling of cell density (D = 0.3 h^{-1}), which led to an increase in volumetric H_2 productivity (20 mmol/L*h). The H_2 yield was, however, not affected [56]. This finding indicated that these observed changes in H_2 yields were a function of the growth rate and did not depend on the substrate conversion efficiency.

6.3. End Product Inhibition and Osmotolerance

Incomplete substrate conversions can also be observed in controlled batch fermentations with high initial substrate levels. Growth on 10 g/L of both glucose and fructose resulted in complete substrate consumption, while higher initial substrate levels of glucose (20 and 31 g/L) and fructose (20 g/L) resulted in incomplete conversions [55,91]. Similar observations were done for different sugar mixtures [54,55,94]. These incomplete conversions were attributed to the inhibitory effect of accumulating organic acids like acetate or lactate. Inhibition experiments showed that an acetate concentration of 200 mM and higher prevented acid production by *C. saccharolyticus* grown on glucose (10 g/L) in batch [91], which is in line with the earlier findings of van Niel *et al.*, who observed critical sodium acetate and potassium acetate concentrations of 192 mM and 206 mM, respectively, for batch growth on sucrose [73]. However, similar inhibitory effects were observed for NaCl and KCl, with critical concentration of 216 mM and 250 mM respectively [73], suggesting that increased osmolarity is the cause of inhibition and not the end products per se. This relative low osmotolerance in comparison to marine organisms, like *Thermotoga neapolitana* [91], probably reflects the terrestrial origin of *C. saccharolyticus*. Ljunggren *et al.* designed a kinetic model for the growth of *C. saccharolyticus* incorporating the inhibitory effect of a high osmolarity and determined a critical osmolarity (no growth) in the range of 270 to 290 mM. They also showed that osmolarity is of minor influence on fermentations with low initial glucose levels [74]. This low tolerance to osmotic pressure also prevents the application of CO_2 as a cheaper and more convenient stripping gas than N_2. The use of CO_2 as stripping gas during *C. saccharolyticus* cultivations negatively affects growth rate and hydrogen productivity. CO_2 sparging led to a higher dissolved CO_2 concentration, which required addition of extra base to maintain a constant pH, overall leading to an increase in osmotic pressure [79].

6.4. Medium Requirements

Controlled batch cultivations were used to investigate the influence of NH_4^+ on the performance of *C. saccharolyticus* grown on molasses. These experiments revealed that the omission of NH_4^+ gave rise to a higher H_2 yield and maximal H_2 productivity [90]. Although *C. saccharolyticus* is able to grow on a medium without NH_4^+, containing only YE as a nitrogen source, it becomes very sensitive to changes in P_{H2} ([77], Figure 4b). Chemostat cultivations showed complete glucose (20.7 mM) consumption and high acetate yields (1.87 ± 0.02 mol/mol) under low P_{H2}. However, when the cultivation condition changed to a high P_{H2} a new steady state could be achieved, but substrate consumption was incomplete (55%) and acetate yields decreased (1.68 ± 0.01 mol/mol).

Omission of yeast extract (YE) during growth of *C. saccharolyticus* on molasses did not affect the H_2 yield but led to a lower volumetric H_2 productivity [90]. Similar observations were made for controlled batch fermentations on glucose, where the absence of YE did not affect the H_2 and biomass yields [100]. Contrary to the molasses study the volumetric productivity was not affected [100] but this might be due to the lower substrate load used. *C. saccharolyticus* is able to grow on a defined minimal medium with additional vitamins, but without additional amino acids [100]. Growth in the absence of YE helps to reduce the production costs. However, increased biomass levels and especially growth on high substrate loads might augment medium requirements. So fine-tuning of the medium composition with respect to the specific substrate and substrate load is required.

7. Future Prospects for Improving Biohydrogen Production

7.1. Improving H_2 Yields and H_2 Productivity

C. saccharolyticus has many properties that make it an excellent candidate for biohydrogen production via dark fermentation. However, for biohydrogen production to become economically feasible major improvements should be made with respect to the H_2 productivity [104–106]. Productivity is maximized by improving the substrate consumption rate but also the H_2 yield.

C. saccharolyticus is able to produce H_2 close to the theoretical maximum of 4 H_2 per hexose. In this respect, it is important to realize that the theoretical yield refers to the pure catabolic component of glucose conversion. The glucose used for anabolism should not be incorporated. Generally, this distinction is not considered in literature, and reported experimental data therefore reveal H_2 yields lower than 4. For example, a yield of 3.5 H_2 per consumed glucose has been reported by de Vrije *et al.* (chemostat cultivation with a 23.0 mM glucose load and a dilution rate of 0.1 h^{-1}) [56]. According to their data 16% of the consumed glucose is used only for biomass formation indicating that only ~19.4 mM glucose was available for ATP generation. Given the reported H_2 production this results in a theoretical conversion efficiency of 4.15 mol H_2 per mol glucose, which approximates the theoretical maximum of 4 H_2 per hexose as indicated by Thauer *et al.* [83]. These high yields are only achievable when the organism ferments the substrate solely to acetate. However, non-ideal growth conditions lead to a mixed fermentation profile (acetate, lactate and ethanol), and a consequently lower H_2 yield. From the organism's perspective switching to lactate or ethanol is profitable since it allows the organism to continue to grow under elevated P_{H2} conditions, albeit with a lower growth rate because ATP yields are lower under lactate and ethanol forming conditions. Obviously, the lower H_2 yield under a mixed end-product fermentation, is not desirable from a biotechnological point of view. To maximize the H_2 yield and productivity, the dissolved H_2 concentration should be kept as low as possible, which requires an optimization of the liquid to mass transfer rate [74,76], which is mainly a matter of reactor design. In continuous stirred-tank reactor systems low dissolved H_2 concentrations could be achieved by increasing the sparging rate [107,108] or the stirring speed [109,110]. In addition, reduction of internal reactor pressure [111–113] and enforced bubble formation [113,114] could potentially lead to a lower dissolved H_2 concentration. Addition of zeolite particles, which enhances bubble formation, allowed a reduction of the N_2 stripping rate from 5 L/(h*L) to 1 L/(h*L), without affecting the H_2 productivity and H_2 yield of *C. saccharolyticus*. N_2 stripping could even be completely omitted when an internal reactor pressure of 0.3 bar was used, which sustained a similar fermentation performance compared to cultivation at atmospheric pressure (1 bar) using a N_2 gas stripping rate of 5 L/(h*L) [113].

Most research on *C. saccharolyticus* has been performed in serum bottles and suspended continuous stirred-tank reactor systems, where only relatively low cell biomass levels can be achieved. Higher cell biomass levels would cause an increase in substrate consumption rates leading to an increase in H_2 productivity when H_2 yields are maintained. To realize higher biomass levels a deeper insight into growth limiting medium compounds should be acquired. Additionally, other reactor types, like a trickle bed reactor or a fluidized bed system might allow cell biomass accumulation. *C. saccharolyticus* could be cultivated in a non-axenic 400 L trickle bed reactor [102] out-competing other organisms, with H_2 yields around 2.8 mol H_2/mol hexose and a productivity of 22 mmol H_2/L*h. These results showed that *C. saccharolyticus* can be used in a large scale non-sterile industrial setting.

Combining different organisms in a co-cultivation setup allows the exploitation of the hydrolytic capability of each individual species and could enhance the overall range of useable substrates. Batch co-cultivations of *C. saccharolyticus* with either *C. owensensis*, *C. kristjanssonii* or an enriched compost microflora were performed on a glucose-xylose mixture [115]. The co-cultivation with the enriched compost microflora resulted in a fast, simultaneous consumption of both glucose and xylose with a relatively high specific hydrogen production rate, but with a lower H_2 yield [115]. A stable co-culture consisting of two closely related *Caldicellulosiruptor* species, *C. saccharolyticus* and *C. kristjanssonii*, could be established in a continuous cultivation system [116]. These findings demonstrate the possibility to create co-cultures for H_2 formation and reveal an apparent synergistic effect of the strains, which lead to improved fermentation performances.

Overall H_2 yields from biomass derived substrates can be increased when the dark fermentation is coupled to a second stage like electrohydrogenesis or photofermentation [117,118]. The former system uses a microbial electrolysis cell (MEC), in which electricity is used to convert acetate or other organic acids to hydrogen. In the latter case the main end product of the dark fermentation, acetate, is further converted by an anaerobic non-sulfur purple photosynthetic bacterium, forming a maximum of 4 mol H_2 per mol acetate, giving an overall H_2 yield of 12 mol H_2 per mol glucose. The effluent of *C. saccharolyticus* has been successfully used as a feed for photofermentative growth and H_2 production [90,119,120]. Alternatively, dark fermentation end-products H_2 and acetate could serve as substrates for hydrogenothropic methanogens in a biogas generating system. The addition of *C. saccharolyticus* to natural biogas-producing consortia led to an improvement of biogas production and a stable co-cultivation could be maintained for several months [98].

7.2. Genetic Engineering of Caldicellulosiruptor *Species*

The first steps in the development of a genetic system for *Caldicellulosiruptor* species have been made. Chung *et al.* have shown that methylation with an endogenous unique α-class N4-Cytosine methyltransferase is required for transformation of DNA isolated from *E. coli* into *Caldicellulosiruptor bescii* [121]. Furthermore, an uracil auxotrophic *C. bescii* mutant strain was generated by a spontaneous deletion in the *pyrBCF* locus [121]. This nutritional deficiency was exploited as a selection marker of *C. bescii* transformants [121]. A similar strategy might be applied to develop a genetic system for the other members of this genus.

To improve the H_2 producing capabilities of *C. saccharolyticus* or other *Caldicellulosiruptor* species metabolic engineering strategies could focus on improving the H_2 yields. Additionally it could be aimed at altering intrinsic properties of *Caldicellulosiruptor* species limiting H_2 productivity like the enhancement of their H_2 tolerance or osmotolerance. For example, for industrial applications increased substrate concentrations are favored since it reduces the fresh water demand, thus reducing the overall costs and the environmental impact of the process. However, for *C. saccharolyticus* the maximum substrate load is limited by its sensitivity to osmotic stress [12,73,74].

With respect to the mixed acid fermentation, both lactate and ethanol formations lead to a lowered H_2 yield. Because ethanol formation is not the major reductant sink under redox stress and in general is only produced at low levels, knocking out the alcohol dehydrogenase responsible for ethanol formation will probably not significantly alter the fermentation performance of *C. saccharolyticus*. However, lactate formation can be seen as the main mechanism to alleviate redox stress. Targeting the lactate formation pathway for complete knockout will probably make *C. saccharolyticus* less resilient to fluctuations in dissolved H_2 concentration and is inadvisable.

Alternatively, one could alter glycolysis in such a way that substrate conversion is less energy efficient. So to generate the same amount of ATP, essential for biosynthesis and maintenance, a higher glycolytic flux to acetate is required. This would result in a higher H_2 yield because the glycolytic flux to acetate is increased with respect to the carbon flux to biomass. A less energy efficient glycolysis can be achieved by eliminating some ATP generating steps from the central metabolic pathway. For example, exchanging the NADH-dependent GAPDH in *C. saccharolyticus* with the ferredoxin-dependent GAPOR, would decrease the overall ATP yield of glycolysis. In addition, since the H_2 formation from the generated Fd_{red} is energetically more favourable than H_2 formation from NADH, the organism would become less sensitive to increased H_2 levels [9,12].

H_2 yields on rhamnose and fucose could be increased if the carbon flux from the intermediate lactaldehyde is redirected via methylglyoxal to pyruvate, by the insertion of a lactaldehyde dehydrogenase and a methylglyoxal dehydrogenase. When rhamnose is completely oxidized to acetate, via this pathway, the H_2 yield increases from 1 to 5 H_2 per rhamnose. With respect to alternative substrates, glycerol could serve as a good substrate for H_2 production because of the relative high reduced state of its carbon atoms. A maximum H_2 yield of 3 mol/mol glycerol is achieved if glycerol is completely oxidized to acetate. So far, growth or co-consumption on/of glycerol has, however, not been observed for *C. saccharolyticus* [122].

8. Conclusions

The bacterium *Caldicellulosiruptor saccharolyticus* possesses several features that make it an excellent candidate for biological hydrogen production. With an optimal growth temperature of 70 °C *C. saccharolyticus* is one of the most thermophilic cellulose degrading organisms known to date. The organisms diverse inventory of endo- and exo-glycoside hydrolases allow it to degrade and grow on a variety of cellulose- and hemicellulose-containing biomass substrates. Some of these glycosidases are multi-domain proteins that contain both glycoside hydrolase domains and carbon binding modules, which facilitate the efficient degradation of recalcitrant plant polysaccharides into mono-, di- or oligo-saccharides. The high diversity of transport systems present in the genome confirm the broad substrate preferences of *C. saccharolyticus* and its ability to co-utilization hexoses and pentoses, without any signs of carbon catabolite repression, is a desirable trait for any consolidated bioprocess.

For *C. saccharolyticus* sugar substrates are primarily fermented to acetate, CO_2 and H_2, via the Embden-Meyerhof pathway. Typically, the fermentation of hexose and pentose lead to the generation of the reduced electron carriers NADH and Fd_{red} in a 1:1 ratio. These reduced electron carriers can be reoxidized during two distinct H_2 generating steps, respectively catalyzed by an NADH-dependent cytosolic Fe-only hydrogenase (*hyd*) and the Fd_{red}-dependent membrane-bound [NiFe] hydrogenase (*ech*). Alternatively, rhamnose catabolism is coupled to a different type of redox balancing, where the generated NADH is used for 1,2-propanediol formation and only the Fd_{red} is available for H_2 formation.

Fermentation data reveal that *C. saccharolyticus* is capable of producing H_2 with yields close to the theoretical limit of 4 H_2 per hexose. However, under non-ideal conditions both ethanol and lactate formation act as alternative redox sinks, thus reducing H_2 yields. All possible redox sinks, including the hydrogenases, are upregulated during cultivation under an increased partial hydrogen pressure. The mechanism underlying transcription of the lactate dehydrogenase gene remains elusive, but the transcription of the genes coding for both hydrogenases and the alcohol dehydrogenases, potentially involved in ethanol formation, seem to be under the control of an $NADH/NAD^+$-sensing transcriptional regulator REX.

An increased intracellular $NADH/NAD^+$ ratio, putatively caused by the inhibition of hydrogenase activity at elevated H_2 levels, can hinder glycolysis at the level of glyceraldehyde-3-phosphate dehydrogenase, resulting in the inhibition of growth. Lactate formation serves as an alternative redox sink, alleviating redox stress. Lactate dehydrogenase activity is enhanced by the glycolytic intermediate fructose-1,6-bisphosphate but also modulated by the energy carriers ATP and pyrophosphate. The latter mechanism couples lactate formation to the energy metabolism, where lactate formation is inhibited during exponential growth and inhibition is alleviated during the transition to the stationary phase.

Overall, maintaining low dissolved H_2 levels in the system appeared to be one of the most important factors for optimizing H_2 production. In addition, improvements should be made with respect to the H_2 productivity and osmotolerance of the organism to allow biohydrogen production by *C. saccharolyticus* to become economically feasible.

Acknowledgments

This research was supported by the IPOP program of Wageningen University and a grant from the 6th EU Framework Programme, Priority 6.1: Sustainable Energy Systems, contract 019825 (HYVOLUTION).

References and Notes

1. Lynd, L.R.; Wyman, C.E.; Gerngross, T.U. Biocommodity engineering. *Biotechnol. Prog.* **1999**, *15*, 777–793.

2. Lynd, L.R.; Zyl, W.H.V.; McBride, J.E.; Laser, M. Consolidated bioprocessing of cellulosic biomass: An update. *Curr. Opin. Biotechnol.* **2005**, *16*, 577–583.
3. Agbor, V.B.; Cicek, N.; Sparling, R.; Berlin, A.; Levin, D.B. Biomass pretreatment: Fundamentals toward application. *Biotechnol. Adv.* **2011**, *29*, 675–685.
4. Olson, D.G.; McBride, J.E.; Shaw, A.J.; Lynd, L.R. Recent progress in consolidated bioprocessing. *Curr. Opin. Biotechnol.* **2012**, *23*, 396–405.
5. Hallenbeck, P.C. Fermentative hydrogen production: Principles, progress, and prognosis. *Int. J. Hydrogen Energy* **2009**, *34*, 7379–7389.
6. Bergquist, P.L.; Gibbs, M.D.; Morris, D.D.; Te'o, V.S.; Saul, D.J.; Moran, H.W. Molecular diversity of thermophilic cellulolytic and hemicellulolytic bacteria. *FEMS Microbiol. Ecol.* **1999**, *28*, 99–110.
7. Blumer-Schuette, S.E.; Kataeva, I.; Westpheling, J.; Adams, M.W.W.; Kelly, R.M. Extremely thermophilic microorganisms for biomass conversion: Status and prospects. *Curr. Opin. Biotechnol.* **2008**, *19*, 210–217.
8. VanFossen, A.L.; Lewis, D.L.; Nichols, J.D.; Kelly, R.M. Polysaccharide degradation and synthesis by extremely thermophilic anaerobes. *Incredible Anaerobes Physiol. Genomics Fuels* **2008**, *1125*, 322–337.
9. Kengen, S.W.M.; Goorissen, H.P.; Verhaart, M.R.A.; Stams, A.J.M.; van Niel, E.W.J.; Claassen, P.A.M. Biological hydrogen production by anaerobic microorganisms. In *Biofuels*, Soetaert, W., Verdamme, E.J., Eds.; John Wiley & Sons: Chichester, UK, 2009; pp. 197–221.
10. Verhaart, M.R.A.; Bielen, A.A.M.; van der Oost, J.; Stams, A.J.M.; Kengen, S.W.M. Hydrogen production by hyperthermophilic and extremely thermophilic bacteria and archaea: Mechanisms for reductant disposal. *Environ. Technol.* **2010**, *31*, 993–1003.
11. De Vrije, T.; Claasen, P.A.M. Dark hydrogen fermentations. In *Bio-methane & Bio-hydrogen*, Reith, J.H., Wijffels, R.H., Barten, H., Eds.; Smiet Offset: The Hague, The Netherlands, 2003; pp. 103–123.
12. Willquist, K.; Zeidan, A.A.; van Niel, E.W. Physiological characteristics of the extreme thermophile *Caldicellulosiruptor saccharolyticus*: An efficient hydrogen cell factory. *Microb. Cell Fact.* **2010**, *9*, 89.
13. Rainey, F.A.; Donnison, A.M.; Janssen, P.H.; Saul, D.; Rodrigo, A.; Bergquist, P.L.; Daniel, R.M.; Stackebrandt, E.; Morgan, H.W. Description of *Caldicellulosiruptor saccharolyticus* gen. nov., sp. nov: An obligately anaerobic, extremely thermophilic, cellulolytic bacterium. *FEMS Microbiol. Lett.* **1994**, *120*, 263–266.
14. Sissons, C.H.; Sharrock, K.R.; Daniel, R.M.; Morgan, H.W. Isolation of cellulolytic anaerobic extreme thermophiles from New Zealand thermal sites. *Appl. Environ. Microbiol.* **1987**, *53*, 832–838.
15. Reynolds, P.H.S.; Sissons, C.H.; Daniel, R.M.; Morgan, H.W. Comparison of cellulolytic activities in *Clostridium thermocellum* and three thermophilic, cellulolytic anaerobes. *Appl. Environ. Microbiol.* **1986**, *51*, 12–17.

16. Donnison, A.M.; Brockelsby, C.M.; Morgan, H.W.; Daniel, R.M. The degradation of lignocellulosics by extremely thermophilic microorganisms. *Biotechnol. Bioeng.* **1989**, *33*, 1495–1499.
17. Bredholt, S.; Sonne-Hansen, J.; Nielsen, P.; Mathrani, I.M.; Ahring, B.K. *Caldicellulosiruptor kristjanssonii* sp. nov., a cellulolytic extremely thermophilic, anaerobic bacterium. *Int. J. Syst. Bacteriol.* **1999**, *49*, 991–996.
18. Dwivedi, P.P.; Gibbs, M.D.; Saul, D.J.; Bergquist, P.L. Cloning, sequencing and overexpression in *Escherichia coli* of a xylanase gene, xynA from the thermophilic bacterium Rt8B.4 genus *Caldicellulosiruptor*. *Appl. Microbiol. Biotechnol.* **1996**, *45*, 86–93.
19. Gibbs, M.D.; Reeves, R.A.; Farrington, G.K.; Anderson, P.; Williams, D.P.; Bergquist, P.L. Multidomain and multifunctional glycosyl hydrolases from the extreme thermophile *Caldicellulosiruptor* isolate Tok7B.1. *Curr. Microbiol.* **2000**, *40*, 333–340.
20. Hamilton-Brehm, S.D.; Mosher, J.J.; Vishnivetskaya, T.; Podar, M.; Carroll, S.; Allman, S.; Phelps, T.J.; Keller, M.; Elkins, J.G. *Caldicellulosiruptor obsidiansis* sp. nov., an anaerobic, extremely thermophilic, cellulolytic bacterium isolated from Obsidian pool, Yellowstone national park. *Appl. Environ. Microbiol.* **2010**, *76*, 1014–1020.
21. Huang, C.Y.; Patel, B.K.; Mah, R.A.; Baresi, L. *Caldicellulosiruptor owensensis* sp. nov., an anaerobic, extremely thermophilic, xylanolytic bacterium. *Int. J. Syst. Bacteriol.* **1998**, *48*, 91–97.
22. Miroshnichenko, M.L.; Kublanov, I.V.; Kostrikina, N.A.; Tourova, T.P.; Kolganova, T.V.; Birkeland, N.K.; Bonch-Osmolovskaya, E.A. *Caldicellulosiruptor kronotskyensis* sp. nov. and *Caldicellulosiruptor hydrothermalis* sp. nov., two extremely thermophilic, cellulolytic, anaerobic bacteria from Kamchatka thermal springs. *Int. J. Syst. Evol. Microbiol.* **2008**, *58*, 1492–1496.
23. Mladenovska, Z.; Mathrani, I.M.; Ahring, B.K. Isolation and characterization of *Caldicellulosiruptor lactoaceticus* sp. nov., an extremely thermophilic, cellulolytic, anaerobic bacterium. *Arch. Microbiol.* **1995**, *163*, 223–230.
24. Morris, D.D.; Gibbs, M.D.; Ford, M.; Thomas, J.; Bergquist, P.L. Family 10 and 11 xylanase genes from *Caldicellulosiruptor* sp. strain Rt69B.1. *Extremophiles* **1999**, *3*, 103–111.
25. Nielsen, P.; Mathrani, I.M.; Ahring, B.K. *Thermoanaerobium acetigenum* spec. nov., a new anaerobic, extremely thermophilic, xylanolytic non-spore-forming bacterium isolated from an Icelandic hot spring. *Arch. Microbiol.* **1993**, *159*, 460–464.
26. Onyenwoke, R.U.; Lee, Y.J.; Dabrowski, S.; Ahring, B.K.; Wiegel, J. Reclassification of "*Thermoanaerobium acetigenum*" as *Caldicellulosiruptor acetigenus* comb. nov and emendation of the genus description. *Int. J. Syst. Evol. Microbiol.* **2006**, *56*, 1391–1395.
27. Svetlichnyi, V.A.; Svetlichnaya, T.P.; Chernykh, N.A.; Zavarzin, G.A. *Anaerocellum thermophilum* gen. nov. sp. nov.: An extremely thermophilic cellulolytic eubacterium isolated from hot-springs in the Valley of Geysers. *Microbiology* **1990**, *59*, 598–604.
28. Yang, S.J.; Kataeva, I.; Wiegel, J.; Yin, Y.; Dam, P.; Xu, Y.; Westpheling, J.; Adams, M.W. Reclassification of "*Anaerocellum thermophilum*" as *Caldicellulosiruptor bescii* strain DSM 6725T sp. nov. *Int. J. Syst. Evol. Microbiol.* **2009**, *60*, 2011–2015.

29. Blumer-Schuette, S.E.; Giannone, R.J.; Zurawski, J.V.; Ozdemir, I.; Ma, Q.; Yin, Y.B.; Xu, Y.; Kataeva, I.; Poole, F.L.; Adams, M.W.W.; *et al. Caldicellulosiruptor* core and pangenomes reveal determinants for noncellulosomal thermophilic deconstruction of plant biomass. *J. Bacteriol.* **2012**, *194*, 4015–4028.
30. Blumer-Schuette, S.E.; Ozdemir, I.; Mistry, D.; Lucas, S.; Lapidus, A.; Cheng, J.F.; Goodwin, L.A.; Pitluck, S.; Land, M.L.; Hauser, L.J.; *et al.* Complete genome sequences for the anaerobic, extremely thermophilic plant biomass-degrading bacteria *Caldicellulosiruptor hydrothermalis*, *Caldicellulosiruptor kristjanssonii*, *Caldicellulosiruptor kronotskyensis*, *Caldicellulosiruptor owensensis*, and *Caldicellulosiruptor lactoaceticus*. *J. Bacteriol.* **2011**, *193*, 1483–1484.
31. Elkins, J.G.; Lochner, A.; Hamilton-Brehm, S.D.; Davenport, K.W.; Podar, M.; Brown, S.D.; Land, M.L.; Hauser, L.J.; Klingeman, D.M.; Raman, B.; *et al.* Complete genome sequence of the cellulolytic thermophile *Caldicellulosiruptor obsidiansis* OB47T. *J. Bacteriol.* **2010**, *192*, 6099–6100.
32. Kataeva, I.A.; Yang, S.J.; Dam, P.; Poole, F.L.; Yin, Y.; Zhou, F.F.; Chou, W.C.; Xu, Y.; Goodwin, L.; Sims, D.R.; *et al.* Genome sequence of the anaerobic, thermophilic, and cellulolytic bacterium "*Anaerocellum thermophilum*" DSM 6725. *J. Bacteriol.* **2009**, *191*, 3760–3761.
33. Van de Werken, H.J.G.; Verhaart, M.R.A.; VanFossen, A.L.; Willquist, K.; Lewis, D.L.; Nichols, J.D.; Goorissen, H.P.; Mongodin, E.F.; Nelson, K.E.; van Niel, E.W.J., *et al.* Hydrogenomics of the extremely thermophilic bacterium *Caldicellulosiruptor saccharolyticus*. *Appl. Environ. Microbiol.* **2008**, *74*, 6720–6729.
34. Lamed, R.; Bayer, E.A. The cellulosome of *Clostridium thermocellum*. *Adv. Appl. Microbiol.* **1988**, *33*, 1–46.
35. Blumer-Schuette, S.E.; Lewis, D.L.; Kelly, R.M. Phylogenetic, microbiological, and glycoside hydrolase diversities within the extremely thermophilic, plant biomass-degrading genus *Caldicellulosiruptor*. *Appl. Environ. Microbiol.* **2010**, *76*, 8084–8092.
36. VanFossen, A.L.; Ozdemir, I.; Zelin, S.L.; Kelly, R.M. Glycoside hydrolase inventory drives plant polysaccharide deconstruction by the extremely thermophilic bacterium *Caldicellulosiruptor saccharolyticus*. *Biotechnol. Bioeng.* **2011**, *108*, 1559–1569.
37. Bergquist, P.L.; Love, D.R.; Croft, J.E.; Streiff, M.B.; Daniel, R.M.; Morgan, W.H. Genetics and potential biotechnological applications of thermophilic and extremely thermophilic microorganisms. *Biotechnol. Genet. Eng. Rev.* **1987**, *5*, 199–244.
38. Love, D.R.; Streiff, M.B. Molecular cloning of a beta-glucosidase gene from an extremely thermophilic anaerobe in *Escherichia coli* and *Bacillus subtilis*. *BioTechnology* **1987**, *5*, 384–387.
39. Hudson, R.C.; Schofield, L.R.; Coolbear, T.; Daniel, R.M.; Morgan, H.W. Purification and properties of an aryl beta-xylosidase from a cellulolytic extreme thermophile expressed in *Escherichia coli*. *Biochem. J.* **1991**, *273*, 645–650.

40. Schofield, L.R.; Daniel, R.M. Purification and properties of a beta-1,4-xylanase from a cellulolytic extreme thermophile expressed in *Escherichia coli*. *Int. J. Biochem.* **1993**, *25*, 609–617.
41. Albertson, G.D.; McHale, R.H.; Gibbs, M.D.; Bergquist, P.L. Cloning and sequence of a type I pullulanase from an extremely thermophilic anaerobic bacterium, *Caldicellulosiruptor saccharolyticus*. *Biochimica Et Biophysica Acta-Gene Structure and Expression* **1997**, *1354*, 35–39.
42. Luthi, E.; Jasmat, N.B.; Bergquist, P.L. Xylanase from the extremely thermophilic bacterium "*Caldocellum saccharolyticum*": Overexpression of the gene in *Escherichia coli* and characterization of the gene product. *Appl. Environ. Microbiol.* **1990**, *56*, 2677–2683.
43. Luthi, E.; Bergquist, P.L. A beta-D-xylosidase from the thermophile "*Caldocellum saccharolyticum*" expressed in *Escherichia coli*. *FEMS Microbiol. Lett.* **1990**, *67*, 291–294.
44. Luthi, E.; Jasmat, N.B.; Bergquist, P.L. Overproduction of an acetylxylan esterase from the extreme thermophile "*Caldocellum saccharolyticum*" in *Escherichia coli*. *Appl. Microbiol. Biotechnol.* **1990**, *34*, 214–219.
45. Te'o, V.S.J.; Saul, D.J.; Bergquist, P.L. Cela, another gene coding for a multidomain cellulase from the extreme thermophile "*Caldocellum saccharolyticum*". *Appl. Microbiol. Biotechnol.* **1995**, *43*, 291–296.
46. Park, C.S.; Kim, J.E.; Choi, J.G.; Oh, D.K. Characterization of a recombinant cellobiose 2-epimerase from *Caldicellulosiruptor saccharolyticus* and its application in the production of mannose from glucose. *Appl. Microbiol. Biotechnol.* **2011**, *92*, 1187–1196.
47. Saul, D.J.; Williams, L.C.; Grayling, R.A.; Chamley, L.W.; Love, D.R.; Bergquist, P.L. Celb, a gene coding for a bifunctional cellulase from the extreme thermophile "*Caldocellum saccharolyticum*". *Appl. Environ. Microbiol.* **1990**, *56*, 3117–3124.
48. Morris, D.D.; Reeves, R.A.; Gibbs, M.D.; Saul, D.J.; Bergquist, P.L. Correction of the beta-mannanase domain of the *Celc* pseudogene from *Caldocellulosiruptor saccharolyticus* and activity of the gene product on Kraft pulp. *Appl. Environ. Microbiol.* **1995**, *61*, 2262–2269.
49. Luthi, E.; Jasmat, N.B.; Grayling, R.A.; Love, D.R.; Bergquist, P.L. Cloning, sequence analysis, and expression in *Escherichia coli* of a gene coding for a beta-mannanase from the extremely thermophilic bacterium "*Caldocellum saccharolyticum*". *Appl. Environ. Microbiol.* **1991**, *57*, 694–700.
50. Andrews, G.; Lewis, D.; Notey, J.; Kelly, R.; Muddiman, D. Part I: Characterization of the extracellular proteome of the extreme thermophile *Caldicellulosiruptor saccharolyticus* by GeLC-MS2 (vol 398, pg 377, 2010). *Anal. Bioanal. Chem.* **2010**, *398*, 1837–1837.
51. Ozdemir, I.; Blumer-Schuette, S.E.; Kelly, R.M. S-Layer homology domain proteins Csac_0678 and Csac_2722 are implicated in plant polysaccharide deconstruction by the extremely thermophilic bacterium *Caldicellulosiruptor saccharolyticus*. *Appl. Environ. Microbiol.* **2012**, *78*, 768–777.

52. Ivanova, G.; Rakhely, G.; Kovacs, K.L. Hydrogen production from biopolymers by *Caldicellulosiruptor saccharolyticus* and stabilization of the system by immobilization. *Int. J. Hydrogen Energy* **2008**, *33*, 6953–6961.
53. VanFossen, A.L.; Verhaart, M.R.A.; Kengen, S.M.W.; Kelly, R.M. Carbohydrate utilization patterns for the extremely thermophilic bacterium *Caldicellulosiruptor saccharolyticus* reveal broad growth substrate preferences. *Appl. Environ. Microbiol.* **2009**, *75*, 7718–7724.
54. De Vrije, T.; Bakker, R.R.; Budde, M.A.; Lai, M.H.; Mars, A.E.; Claassen, P.A. Efficient hydrogen production from the lignocellulosic energy crop *Miscanthus* by the extreme thermophilic bacteria *Caldicellulosiruptor saccharolyticus* and *Thermotoga neapolitana. Biotechnol. Biofuels* **2009**, *2*, 12.
55. De Vrije, T.; Budde, M.A.W.; Lips, S.J.; Bakker, R.R.; Mars, A.E.; Claassen, P.A.M. Hydrogen production from carrot pulp by the extreme thermophiles *Caldicellulosiruptor saccharolyticus* and *Thermotoga neapolitana. Int. J. Hydrogen Energy* **2010**, *35*, 13206–13213.
56. De Vrije, T.; Mars, A.E.; Budde, M.A.W.; Lai, M.H.; Dijkema, C.; de Waard, P.; Claassen, P.A.M. Glycolytic pathway and hydrogen yield studies of the extreme thermophile *Caldicellulosiruptor saccharolyticus. Appl. Microbiol. Biotechnol.* **2007**, *74*, 1358–1367.
57. Willquist, K.; van Niel, E.W.J. Lactate formation in *Caldicellulosiruptor saccharolyticus* is regulated by the energy carriers pyrophosphate and ATP. *Metab. Eng.* **2010**, *12*, 282–290.
58. Brown, S.D.; Guss, A.M.; Karpinets, T.V.; Parks, J.M.; Smolin, N.; Yang, S.H.; Land, M.L.; Klingeman, D.M.; Bhandiwad, A.; Rodriguez, M.; *et al.* Mutant alcohol dehydrogenase leads to improved ethanol tolerance in *Clostridium thermocellum. Proc. Natl. Acad. Sci. USA* **2011**, *108*, 13752–13757.
59. Peng, H.; Wu, G.G.; Shao, W.L. The aldehyde/alcohol dehydrogenase (AdhE) in relation to the ethanol formation in *Thermoanaerobacter ethanolicus* JW200. *Anaerobe* **2008**, *14*, 125–127.
60. Bielen, A.A.M.; Verhaart, M.R.A.; VanFossen, A.L.; Blumer-Schuette, S.E.; Stams, A.J.M.; van der Oost, J.; Kelly, R.M.; Kengen, S.M.W. A thermophile under pressure:Transcriptional analysis of the response of *Caldicellulosiruptor saccharolyticus* to different H_2 partial pressures. *Int. J. Hydrogen Energy* **2012**, in press.
61. Ma, K.; Hutchins, A.; Sung, S.J.S.; Adams, M.W.W. Pyruvate ferredoxin oxidoreductase from the hyperthermophilic archaeon, *Pyrococcus furiosus*, functions as a CoA-dependent pyruvate decarboxylase. *Proc. Natl. Acad. Sci. USA* **1997**, *94*, 9608–9613.
62. Kannan, V.; Mutharasan, R. Ethanol fermentation characteristics of *Thermoanaerobacter ethanolicus. Enzyme. Microb. Technol.* **1985**, *7*, 87–89.
63. Desai, S.G.; Guerinot, M.L.; Lynd, L.R. Cloning of L-lactate dehydrogenase and elimination of lactic acid production via gene knockout in *Thermoanaerobacterium saccharolyticum* JW/SL-YS485. *Appl. Microbiol. Biotechnol.* **2004**, *65*, 600–605.
64. Soboh, B.; Linder, D.; Hedderich, R. A multisubunit membrane-bound [NiFe] hydrogenase and an NADH-dependent Fe-only hydrogenase in the fermenting bacterium *Thermoanaerobacter tengcongensis. Microbiology* **2004**, *150*, 2451–2463.

65. Ma, K.S.; Adams, M.W.W. An unusual oxygen-sensitive, iron- and zinc-containing alcohol dehydrogenase from the hyperthermophilic archaeon *Pyrococcus furiosus*. *J. Bacteriol.* **1999**, *181*, 1163–1170.
66. Verhaart, M.R.A. Wageningen University, Wageningen, The Netherlands. Unpublished work, 2010.
67. DeLacey, A.L.; Stadler, C.; Fernandez, V.M.; Hatchikian, E.C.; Fan, H.J.; Li, S.H.; Hall, M.B. IR spectroelectrochemical study of the binding of carbon monoxide to the active site of *Desulfovibrio fructosovorans* Ni-Fe hydrogenase. *J. Biol. Inorg. Chem.* **2002**, *7*, 318–326.
68. Lemon, B.J.; Peters, J.W. Binding of exogenously added carbon monoxide at the active site of the iron-only hydrogenase (CpI) from *Clostridium pasteurianum*. *Biochemistry* **1999**, *38*, 12969–12973.
69. Heinonen, J.K. Biological production of PPi. In *Biological Role of Inorganic Pyrophosphate*; Heinonen, J.K., Ed.; Kluwer Academic Publishers: Boston, MA, USA, Dordrecht, The Netherlands, London, UK, 2001; p. 264.
70. Chen, J.; Brevet, A.; Fromant, M.; Leveque, F.; Schmitter, J.M.; Blanquet, S.; Plateau, P. Pyrophosphatase is essential for growth of *Escherichia coli*. *J. Bacteriol.* **1990**, *172*, 5686–5689.
71. Bielen, A.A.M.; Willquist, K.; Engman, J.; van der Oost, J.; van Niel, E.W.J.; Kengen, S.W.M. Pyrophosphate as a central energy carrier in the hydrogen-producing extremely thermophilic *Caldicellulosiruptor saccharolyticus*. *FEMS Microbiol. Lett.* **2010**, *307*, 48–54.
72. Mertens, E. Pyrophosphate-dependent phosphofructokinase, an anaerobic glycolytic enzyme? *FEBS Lett.* **1991**, *285*, 1–5.
73. Van Niel, E.W.J.; Claassen, P.A.M.; Stams, A.J.M. Substrate and product inhibition of hydrogen production by the extreme thermophile, *Caldicellulosiruptor saccharolyticus*. *Biotechnol. Bioeng.* **2003**, *81*, 255–262.
74. Ljunggren, M.; Willquist, K.; Zacchi, G.; van Niel, E.W.J. A kinetic model for quantitative evaluation of the effect of hydrogen and osmolarity on hydrogen production by *Caldicellulosiruptor saccharolyticus*. *Biotechnol. Biofuels* **2011**, *4*, 31.
75. Willquist, K.; Pawar, S.S.; van Niel, E.W.J. Reassessment of hydrogen tolerance in *Caldicellulosiruptor saccharolyticus*. *Microb. Cell Fact.* **2011**, *10*, 111.
76. Kraemer, J.T.; Bagley, D.M. Supersaturation of dissolved H_2 and CO_2 during fermentative hydrogen production with N_2 sparging. *Biotechnol. Lett.* **2006**, *28*, 1485–1491.
77. Bielen, A.A.M. Wageningen University, Wageningen, The Netherlands. Unpublished work, 2012.
78. Kadar, Z.; de Vrijek, T.; van Noorden, G.E.; Budde, M.A.W.; Szengyel, Z.; Reczey, K.; Claassen, P.A.M. Yields from glucose, xylose, and paper sludge hydrolysate during hydrogen production by the extreme thermophile *Caldicellulosiruptor saccharolyticus*. *Appl. Biochem. Biotechnol.* **2004**, *113–116*, 497–508.

79. Willquist, K.; Claassen, P.A.M.; van Niel, E.W.J. Evaluation of the influence of CO_2 on hydrogen production by *Caldicellulosiruptor saccharolyticus*. *Int. J. Hydrogen Energy* **2009**, *34*, 4718–4726.
80. Amend, J.P.; Plyasunov, A.V. Carbohydrates in thermophile metabolism: Calculation of the standard molal thermodynamic properties of aqueous pentoses and hexoses at elevated temperatures and pressures. *Geochim. Cosmochim. Acta* **2001**, *65*, 3901–3917.
81. Amend, J.P.; Shock, E.L. Energetics of overall metabolic reactions of thermophilic and hyperthermophilic Archaea and Bacteria. *FEMS Microbiol. Rev.* **2001**, *25*, 175–243.
82. Burton, K. Enthalpy change for reduction of nicotinamide-adenine dinucleotide. *Biochem. J.* **1974**, *143*, 365–368.
83. Thauer, R.K.; Jungermann, K.; Decker, K. Energy-conservation in chemotropic anaerobic bacteria. *Bacteriol. Rev.* **1977**, *41*, 100–180.
84. Watt, G.D.; Burns, A. Thermochemical characterization of sodium dithionite, flavin mononucleotide, flavin-adenine dinucleotide and methyl and benzyl viologens as low-potential reductants for biological-systems. *Biochem. J.* **1975**, *152*, 33–37.
85. Biegel, E.; Schmidt, S.; Gonzalez, J.M.; Muller, V. Biochemistry, evolution and physiological function of the Rnf complex, a novel ion-motive electron transport complex in prokaryotes. *Cell. Mol. Life Sci.* **2011**, *68*, 613–634.
86. Schut, G.J.; Adams, M.W.W. The iron-hydrogenase of *Thermotoga maritima* utilizes ferredoxin and NADH synergistically: A new perspective on anaerobic hydrogen production. *J. Bacteriol.* **2009**, *191*, 4451–4457.
87. Panagiotopoulos, I.A.; Bakker, R.R.; de Vrije, T.; Claassen, P.A.M.; Koukios, E.G. Integration of first and second generation biofuels: Fermentative hydrogen production from wheat grain and straw. *Bioresour. Technol.* **2013**, *128*, 345–350.
88. Panagiotopoulos, I.A.; Bakker, R.R.; de Vrije, T.; Claassen, P.A.M.; Koukios, E.G. Dilute-acid pretreatment of barley straw for biological hydrogen production using Caldicellulosiruptor saccharolyticus. *Int. J. Hydrogen Energy* **2012**, *37*, 11727–11734.
89. Panagiotopoulos, I.; Barker, R.; de Vrije, T.; Niel, E.V.; Koukios, E.; Claassen, P. Exploring critical factors for fermentative hydrogen production from various types of lignocellulosic biomass. *J. Jpn. Inst. Energy* **2011**, *90*, 363–368.
90. Özgür, E.; Mars, A.E.; Peksel, B.; Louwerse, A.; Yücel, M.; Gündüz, U.; Claassen, P.A.M.; Eroğlu, I. Biohydrogen production from beet molasses by sequential dark and photofermentation. *Int. J. Hydrogen Energy* **2010**, *35*, 511–517.
91. Mars, A.E.; Veuskens, T.; Budde, M.A.W.; van Doeveren, P.; Lips, S.J.; Bakker, R.R.; de Vrije, T.; Claassen, P.A.M. Biohydrogen production from untreated and hydrolyzed potato steam peels by the extreme thermophiles *Caldicellulosiruptor saccharolyticus* and *Thermotoga neapolitana*. *Int. J. Hydrogen Energy* **2010**, *35*, 7730–7737.
92. Herbel, Z.; Rakhely, G.; Bagi, Z.; Ivanova, G.; Acs, N.; Kovacs, E.; Kovacs, K.L. Exploitation of the extremely thermophilic *Caldicellulosiruptor saccharolyticus* in hydrogen and biogas production from biomasses. *Environ. Technol.* **2010**, *31*, 1017–1024.

93. Panagiotopoulos, J.A.; Bakker, R.R.; de Vrije, T.; Urbaniec, K.; Koukios, E.G.; Claassen, P.A.M. Prospects of utilization of sugar beet carbohydrates for biological hydrogen production in the EU. *J. Cleaner Prod.* **2010**, *18*, S9–S14.
94. Panagiotopoulos, I.A.; Bakker, R.R.; de Vrije, T.; Koukios, E.G.; Claassen, P.A.M. Pretreatment of sweet sorghum bagasse for hydrogen production by *Caldicellulosiruptor saccharolyticus*. *Int. J. Hydrogen Energy* **2010**, *35*, 7738–7747.
95. Yang, S.J.; Kataeva, I.; Hamilton-Brehm, S.D.; Engle, N.L.; Tschaplinski, T.J.; Doeppke, C.; Davis, M.; Westpheling, J.; Adams, M.W.W. Efficient degradation of lignocellulosic plant biomass, without pretreatment, by the thermophilic anaerobe "*Anaerocellum thermophilum*" DSM 6725. *Appl. Environ. Microbiol.* **2009**, *75*, 4762–4769.
96. Panagiotopoulos, I.A.; Bakker, R.R.; Budde, M.A.W.; de Vrije, T.; Claassen, P.A.M.; Koukios, E.G. Fermentative hydrogen production from pretreated biomass: A comparative study. *Bioresour. Technol.* **2009**, *100*, 6331–6338.
97. Ivanova, G.; Rakhely, G.; Kovacs, K.L. Thermophilic biohydrogen production from energy plants by *Caldicellulosiruptor saccharolyticus* and comparison with related studies. *Int. J. Hydrogen Energy* **2009**, *34*, 3659–3670.
98. Bagi, Z.; Acs, N.; Balint, B.; Horvath, L.; Dobo, K.; Perei, K.R.; Rakhely, G.; Kovacs, K.L. Biotechnological intensification of biogas production. *Appl. Microbiol. Biotechnol.* **2007**, *76*, 473–482.
99. Kadar, Z.; de Vrije, T.; Budde, M.A.; Szengyel, Z.; Reczey, K.; Claassen, P.A. Hydrogen production from paper sludge hydrolysate. *Appl. Biochem. Biotechnol.* **2003**, *105–108*, 557–566.
100. Willquist, K.; van Niel, E.W.J. Growth and hydrogen production characteristics of *Caldicellulosiruptor saccharolyticus* on chemically defined minimal media. *Int. J. Hydrogen Energy* **2012**, *37*, 4925–4929.
101. Muddiman, D.; Andrews, G.; Lewis, D.; Notey, J.; Kelly, R. Part II: Defining and quantifying individual and co-cultured intracellular proteomes of two thermophilic microorganisms by GeLC-MS(2) and spectral counting. *Anal. Bioanal. Chem.* **2010**, *398*, 391–404.
102. Van Groenestijn, J.W.; Geelhoed, J.S.; Goorissen, H.P.; Meesters, K.P.; Stams, A.J.; Claassen, P.A. Performance and population analysis of a non-sterile trickle bed reactor inoculated with *Caldicellulosiruptor saccharolyticus*, a thermophilic hydrogen producer. *Biotechnol. Bioeng.* **2009**, *102*, 1361–1367.
103. Van Niel, E.W.J.; Budde, M.A.W.; de Haas, G.G.; van der Wal, F.J.; Claasen, P.A.M.; Stams, A.J.M. Distinctive properties of high hydrogen producing extreme thermophiles, *Caldicellulosiruptor saccharolyticus* and *Thermotoga elfii*. *Int. J. Hydrogen Energy* **2002**, *27*, 1391–1398.
104. Ljunggren, M.; Wallberg, O.; Zacchi, G. Techno-economic comparison of a biological hydrogen process and a 2nd generation ethanol process using barley straw as feedstock. *Bioresour. Technol.* **2011**, *102*, 9524–9531.
105. Ljunggren, M.; Zacchi, G. Techno-economic evaluation of a two-step biological process for hydrogen production. *Biotechnol. Prog.* **2010**, *26*, 496–504.

106. Nath, K.; Das, D. Improvement of fermentative hydrogen production: Various approaches. *Appl. Microbiol. Biotechnol.* **2004**, *65*, 520–529.
107. Kim, D.H.; Han, S.K.; Kim, S.H.; Shin, H.S. Effect of gas sparging on continuous fermentative hydrogen production. *Int. J. Hydrogen Energy* **2006**, *31*, 2158–2169.
108. Kraemer, J.T.; Bagley, D.M. Optimisation and design of nitrogen-sparged fermentative hydrogen production bioreactors. *Int. J. Hydrogen Energy* **2008**, *33*, 6558–6565.
109. Clark, I.C.; Zhang, R.H.H.; Upadhyaya, S.K. The effect of low pressure and mixing on biological hydrogen production via anaerobic fermentation. *Int. J. Hydrogen Energy* **2012**, *37*, 11504–11513.
110. Lamed, R.J.; Lobos, J.H.; Su, T.M. Effects of stirring and hydrogen on fermentation products of *Clostridium thermocellum*. *Appl. Environ. Microbiol.* **1988**, *54*, 1216–1221.
111. Junghare, M.; Subudhi, S.; Lal, B. Improvement of hydrogen production under decreased partial pressure by newly isolated alkaline tolerant anaerobe, *Clostridium butyricum* TM-9A: Optimization of process parameters. *Int. J. Hydrogen Energy* **2012**, *37*, 3160–3168.
112. Mandal, B.; Nath, K.; Das, D. Improvement of biohydrogen production under decreased partial pressure of H_2 by *Enterobacter cloacae*. *Biotechnol. Lett.* **2006**, *28*, 831–835.
113. Sonnleitner, A.; Peintner, C.; Wukovits, W.; Friedl, A.; Schnitzhofer, W. Process investigations of extreme thermophilic fermentations for hydrogen production: Effect of bubble induction and reduced pressure. *Bioresour. Technol.* **2012**, *118*, 170–176.
114. Fritsch, M.; Hartmeier, W.; Chang, J.S. Enhancing hydrogen production of *Clostridium butyricum* using a column reactor with square-structured ceramic fittings. *Int. J. Hydrogen Energy* **2008**, *33*, 6549–6557.
115. Zeidan, A.A.; van Niel, E.W.J. Developing a thermophilic hydrogen-producing co-culture for efficient utilization of mixed sugars. *Int. J. Hydrogen Energy* **2009**, *34*, 4524–4528.
116. Zeidan, A.A.; van Niel, E.W.J. A quantitative analysis of hydrogen production efficiency of the extreme thermophile *Caldicellulosiruptor owensensis* OL(T). *Int. J. Hydrogen Energy* **2010**, *35*, 1128–1137.
117. Claassen, P.A.M.; de Vrije, T. Non-thermal production of pure hydrogen from biomass: HYVOLUTION. *Int. J. Hydrogen Energy* **2006**, *31*, 1416–1423.
118. Liu, H.; Grot, S.; Logan, B.E. Electrochemically assisted microbial production of hydrogen from acetate. *Environ. Sci. Technol.* **2005**, *39*, 4317–4320.
119. Özkan, E.; Uyar, B.; Özgür, E.; Yücel, M.; Eroğlu, I.; Gündüz, U. Photofermentative hydrogen production using dark fermentation effluent of sugar beet thick juice in outdoor conditions. *Int. J. Hydrogen Energy* **2012**, *37*, 2044–2049.
120. Özgür, E.; Afsar, N.; de Vrije, T.; Yücel, M.; Gündüz, U.; Claassen, P.A.M.; Eroğlu, I. Potential use of thermophilic dark fermentation effluents in photofermentative hydrogen production by *Rhodobacter capsulatus*. *J. Cleaner Prod.* **2010**, *18*, S23–S28.

121. Chung, D.; Farkas, J.; Huddleston, J.R.; Olivar, E.; Westpheling, J. Methylation by a unique α-class N4-Cytosine methyltransferase is required for DNA transformation of *Caldicellulosiruptor bescii* DSM6725. *PLoS One* **2012**, *7*, e43844.
122. Bielen, A.A.M. Wageningen University, Wageningen, The Netherlands. Unpublished work, 2009.

MDPI AG
Klybeckstrasse 64
4057 Basel, Switzerland
Tel. +41 61 683 77 34
Fax +41 61 302 89 18
http://www.mdpi.com/

Life Editorial Office
E-mail: life@mdpi.com
http://www.mdpi.com/journal/life

www.ingramcontent.com/pod-product-compliance
Lightning Source LLC
Chambersburg PA
CBHW051925190326
41458CB00026B/6414

* 9 7 8 3 0 3 8 4 2 1 7 7 1 *